Plant Pathogenic Bacteria

Genomics and Molecular Biology

Edited by Robert W. Jackson

School of Biological Sciences
University of Reading
Reading
UK

Caister Academic Press

Copyright © 2009

Caister Academic Press
Norfolk, UK

www.caister.com

British Library Cataloguing-in-Publication Data
A catalogue record for this book is available from the British Library

ISBN: 978-1-904455-37-0

Description or mention of instrumentation, software, or other products in this book does not imply endorsement by the author or publisher. The author and publisher do not assume responsibility for the validity of any products or procedures mentioned or described in this book or for the consequences of their use.

All rights reserved. No part of this publication may be reproduced, stored in a retrieval system, or transmitted, in any form or by any means, electronic, mechanical, photocopying, recording or otherwise, without the prior permission of the publisher. No claim to original U.S. Government works.

Printed and bound in Great Britain

Contents

	List of contributors	v
	Preface	ix
	Acknowledgements	xi
1	Origin and Evolution of Phytopathogenic Bacteria John Stavrinides	1
2	The Impact of Genomic Approaches on our Understanding of Diversity and Taxonomy of Plant Pathogenic Bacteria Boris A. Vinatzer and Carolee T. Bull	37
3	Adaptation to the Plant Apoplast by Plant Pathogenic Bacteria Arantza Rico, Rachel Jones and Gail M. Preston	63
4	The Genomics of *Agrobacterium*: Insights into its Pathogenicity, Biocontrol and Evolution Joao C. Setubal, Derek Wood, Thomas Burr, Stephen K. Farrand, Barry S. Goldman, Brad Goodner, Leon Otten and Steven Slater	91
5	Common Genes and Genomic Breaks: a Detailed Case Study of the *Xylella fastidiosa* Genome Backbone and Evolutionary Insights Alessandro M. Varani, Wanessa C. Lima, Leandro M. Moreira, Mariana C. de Oliveira, Rangel de Souza, Edwin Civerolo, Ana Tereza R. de Vasconcelos and Marie-Anne Van Sluys	113
6	Genome Sequence-based Insights into the Biology of the Sugarcane Pathogen *Leifsonia xyli* subsp. *xyli* Claudia B. Monteiro-Vitorello, Marcelo Marques Zerillo, Marie-Anne Van Sluys and Luis Eduardo Aranha Camargo	135
7	Genomics-driven Advances in *Xanthomonas* biology Damien F. Meyer and Adam J. Bogdanove	147
8	Genomics of the Enterobacterial Plant Pathogens Ian Toth, Leighton Pritchard, Paul Birch and Hui Liu	163
9	*Ralstonia solanacearum* and Bacterial Wilt in the Post-genomics Era Darby Brown	175

10	*Pseudomonas syringae* Genomics Provides Important Insights to Secretion Systems, Effector Genes and the Evolution of Virulence Dawn L. Arnold, Scott A. C. Godfrey and Robert W. Jackson	203
11	MAMPs/PAMPs – Elicitors of Innate Immunity in Plants Gitte Erbs and Mari-Anne Newman	227
12	The Art of Manipulation: Bacterial Type III Effectors and their Plant Targets Jens Boch	241
13	Cyclic Di-GMP Signalling and the Regulation of Virulence in Bacterial Plant Pathogens J. Maxwell Dow, Yvonne Fouhy, Belén Fernandez Garcia and Robert P. Ryan	273
14	Gene traders: Characteristics of Native Plasmids from Plant Pathogenic Bacteria George W. Sundin and Jesús Murillo	295
15	Bioinformatics Aspects of High-throughput Sequencing Technology Dan MacLean and David J. Studholme	311
	Index	327

Contributors

Dawn L. Arnold
Center for Research in Plant Science
University of the West of England
Bristol
UK

Dawn.Arnold@uwe.ac.uk

Paul Birch
Plant Pathology Programme
Scottish Crop Research Institute
Dundee
UK

Paul.Birch@scri.ac.uk

Jens Boch
Institute of Biology
Martin-Luther-University
Halle-Wittenberg
Germany

jens.boch@genetik.uni-halle.de

Adam J. Bogdanove
Department of Plant Pathology
Iowa State University
Ames, IA
USA

ajbog@iastate.edu

Darby Brown
University of Wisconsin-Richland
Department of Biological Sciences
Richland Center, WI
USA

darby.brown@uwc.edu

Carolee T. Bull
United States Department of Agriculture
Agricultural Research Service
Salinas, CA
USA

Carolee.bull@ars.usda.gov

Thomas Burr
College of Agriculture and Life Sciences
Cornell University
Ithaca, NY
USA
and
Department of Plant Pathology
New York State Agricultural Experiment Station
Geneva, NY
USA

tjb1@cornell.edu

Eduardo Aranha Camargo
Escola Superior de Agricultura Luiz de Queiroz
Universidade de São Paulo
Piracicaba, SP
Brazil

leacamar@esalq.usp.br

Edwin Civerolo
United States Department of Agriculture
Agricultural Research Service
Parlier, CA
USA

eciverolo@fresno.ars.usda.gov

J. Maxwell Dow
BIOMERIT Research Centre
Department of Microbiology, BioSciences Institute
National University of Ireland
Cork
Ireland

m.dow@ucc.ie

Gitte Erbs
Faculty of Life Sciences
Department of Plant Biology and Biotechnology
University of Copenhagen
Frederiksberg
Denmark

ger@life.ku.dk

Stephen K. Farrand
Department of Microbiology
University of Illinois at Urbana-Champaign
Urbana, IL
USA

stephenf@life.uiuc.edu

Belén Fernandez Garcia
BIOMERIT Research Centre
Department of Microbiology, BioSciences Institute
National University of Ireland
Cork
Ireland

belenfg@telefonica.net

Scott A.C. Godfrey
Center for Research in Plant Science
University of the West of England
Bristol
UK

Scott.Godfrey@uwe.ac.uk

Barry S. Goldman
Monsanto Company
St. Louis, MO
USA

barry.s.goldman@monsanto.com

Brad Goodner
Department of Biology
Hiram College
Hiram, OH
USA

GoodnerBW@hiram.edu

Robert W. Jackson
School of Biological Sciences
University of Reading
Reading
UK

r.w.jackson@reading.ac.uk

Rachel Jones
Department of Plant Sciences
University of Oxford
Oxford
UK

rachel.jones@plants.ox.ac.uk

Wanessa C. Lima
Genome and Transposable Elements Laboratory (GaTE)
Departamento de Botânica - Instituto de Biociências
Universidade de São Paulo
São Paulo, SP
Brazil

walima@usp.br

Hui Liu
Plant Pathology Programme
Scottish Crop Research Institute
Dundee
UK

Hui.Liu@scri.ac.uk

Dan MacLean
The Sainsbury Laboratory
Norwich
UK

dan.maclean@sainsbury-laboratory.ac.uk

Yvonne McCarthy
BIOMERIT Research Centre
Department of Microbiology, BioSciences Institute
National University of Ireland
Cork
Ireland

y.mccarthy@ucc.ie

Damien F. Meyer
Department of Plant Pathology
Iowa State University
Ames, IA
USA
and

Research Unit on Emerging and Exotic Animal Diseases Control.
French Agricultural Research Centre for International Development (CIRAD)
Petit-Bourg, Guadeloupe
France

damiaen.meyer@cirad.fr

Claudia B. Monteiro-Vitorello
Centro de Ciências Naturais e Humanas
Universidade Federal do ABC
Santo André, SP
Brazil

claudia.vitorello@ufabc.edu.br

Leandro M. Moreira
Genome and Transposable Elements Laboratory (GaTE)
Departamento de Botânica - Instituto de Biociências
Universidade de São Paulo
São Paulo, SP
Brazil

lmmorei@gmail.com

Jesús Murillo
Laboratorio de Patología Vegetal
Departamento de Producción Agraria
Universidad Pública de Navarra
Pamplona
Spain

jesus.murillo@unavarra.es

Mari-Anne Newman
Faculty of Life Sciences
Department of Plant Biology and Biotechnology
University of Copenhagen
Frederiksberg
Denmark

mari@life.ku.dk

Mariana C. de Oliveira
Genome and Transposable Elements Laboratory (GaTE)
Departamento de Botânica - Instituto de Biociências
Universidade de São Paulo
São Paulo, SP
Brazil

mcdolive@ib.usp.br

Leon Otten
Department of Cell Biology
Institut de Biologie Moléculaire des Plantes
Strasbourg
France

leon.otten@ibmp-ulp.u-strasbg.fr

Gail M. Preston
Department of Plant Sciences
University of Oxford
Oxford
UK

gail.preston@plants.ox.ac.uk

Leighton Pritchard
Plant Pathology Programme
Scottish Crop Research Institute
Dundee
UK

Leighton.Pritchard@scri.ac.uk

Arantza Rico
Department of Plant Sciences
University of Oxford
Oxford
UK

arantza.rico@plants.ox.ac.uk

Robert P. Ryan
BIOMERIT Research Centre
Department of Microbiology, BioSciences Institute
National University of Ireland
Cork
Ireland

r.ryan@ucc.ie

Joao C. Setubal
Virginia Bioinformatics Institute and Department of Computer Science
Virginia Polytechnic Institute and State University
Blacksburg, VA
USA

setubal@vbi.vt.edu

Steven Slater
Department of Applied Biological Sciences and The Biodesign Institute
Arizona State University
Tempe, AZ
USA

scslater@asu.edu

Marie-Anne Van Sluys
Genome and Transposable Elements Laboratory
(GaTE)
Departamento de Botânica - Instituto de Biociências
Universidade de São Paulo
São Paulo,SP
Brazil

mavsluys@usp.br

Rangel de Souza
Laboratório de Bioinformática (LABINFO)
Laboratório Nacional de Computação Científica
Petrópolis, RJ
Brazil

rangel@lncc.br

John Stavrinides
University of Arizona
Tucson, AZ
USA

johnstav@email.arizona.edu

David J. Studholme
The Sainsbury Laboratory
Norwich
UK

David.studholme@tsl.ac.uk

George W. Sundin
Department of Plant Pathology and Center for
Microbial Ecology
Michigan State University
East Lansing, MI
USA

sundin@msu.edu

Ian Toth
Plant Pathology Programme
Scottish Crop Research Institute
Dundee
UK

ian.toth@scri.ac.uk

Alessandro M. Varani
Genome and Transposable Elements Laboratory
(GaTE)
Departamento de Botânica - Instituto de Biociências
Universidade de São Paulo
São Paulo,SP
Brazil

amvarani@ib.usp.br

Ana Tereza R. de Vasconcelos
Laboratório de Bioinformática (LABINFO)
Laboratório Nacional de Computação Científica
Petrópolis, RJ
Brazil

atrv@lncc.br

Boris A Vinatzer
Department of Plant Pathology, Physiology and
Weed Science
Virginia Tech
Blacksburg, VA
USA

vinatzer@vt.edu

Derek Wood
Department of Biology
Seattle Pacific University
Seattle, WA
USA
and
Department of Microbiology
University of Washington
Seattle, WA
USA

woodd1@spu.edu

Marcelo Marques Zerillo
Genome and Transposable Elements Laboratory
(GaTE)
Departamento de Botânica - Instituto de Biociências
Universidade de São Paulo
São Paulo,SP
Brazil

mzerillo@usp.br

Preface

Plant disease is a major problem worldwide for agriculture, with bacterial plant pathogens making major contributions. Besides bacterial pathogens that are already established in many areas, one just has to scan through the New Disease Reports in the *Plant Pathology* journal to see on a monthly basis the many instances of pathogens moving to new geographic areas or even the emergence of new pathogen variants. Moreover, bacterial plant pathogens are particularly difficult to control because of the scarcity of chemical control agents for bacteria, with the exception of antibiotics. However, the use of antibiotics is restricted in many countries due to the potential for evolution of antibiotic resistance and their transmission to human pathogens. So there is a critical need to understand the process of bacterial pathogenesis in plants. Fortunately, due to their relative simplicity and tractability, bacterial plant pathogens have been a popular subject for genetic study and significant progress has been made. The advent of genome sequencing has transformed the field providing even more tools and resources for examining the basis of disease. Owing to the nature of the fast-moving field, it is sometimes hard to keep up with all the modern developments – this book aims to address this.

Fifteen chapters are featured in this book from internationally acclaimed experts in the field of bacterial pathogenicity, with a specific focus on molecular biology and genomics. Since many of the pathogens featured in the book can grow rapidly, the capacity for adaptive change is high. Thus, the book starts with two general chapters that examine bacterial evolution, diversity and taxonomy – all of which have been transformed by molecular biology and genomics. The third chapter delves into a crucially understudied area of how pathogens adapt to the plant apoplast environment. The next seven chapters focus on specific plant pathogens: *Agrobacterium*, *Leifsonia*, *Pectobacterium*, *Pseudomonas*, *Ralstonia*, *Xanthomonas*, *Xylella* including chapters written by several of the modern genomics pioneers who published the first plant pathogen genome sequence in 2000. The next four chapters focus on specific, intensively studied areas of research in the plant pathogen field: these include the role of microbe-associated molecular patterns (MAMPs) in triggering innate immunity; how pathogens utilize effectors (virulence factors) to suppress plant defence; how cyclic di-GMP signalling in the pathogens is involved in the regulation of virulence; and how plasmids are microbial gene traders and often responsible in shaping the bacterial genome and spreading virulence factors. The final chapter addresses the bioinformatics associated with high-throughput sequencing technology and highlights the future of genomics in the shape of genome sequencing and resequencing. The logistics and cost of sequencing a genome in the late 1990s to 2000 was in the region of £50 000–100 000 and involved sequencing of thousands of subclones of DNA in separate sequencing reactions and separate runs on gels or capillaries – it is hard to believe that in 7 years this has progressed such that it can now be done in one single reaction for less than £1000. As such, the field of genomics is really starting to open up for the plant pathogen field. From sequencing one isolate per species a few years ago we can now sequence and resequence single strains to look for adaptive changes that occur during the interaction with plants – this, in concert with more traditional genetic approaches, will be a massive technology for the future analyses of these bacteria.

In summary, this book provides a thorough up-to-date review of the field of the genetics and genomics of plant pathogenic bacteria. The book should be a valuable starting point for researchers new to bacterial plant pathogens, for established plant pathogen researchers from the PhD student to the professor to get an up-to-date review of the state of the art of the field, and as a valuable source of information for a broad range of scientists within and outside of the field. I hope that the book conveys the exciting times that lie ahead in the bacterial plant pathogen field, as we now start to obtain an understanding of the strategies and mechanisms used by pathogens for causing disease, how these bacteria evolve to overcome plant resistance and how non-pathogens can evolve to attack plant hosts. From genetics to genomics to mega-genomics, and beyond!

Robert W. Jackson

Acknowledgements

I would like to take this opportunity to thank all the authors for their chapters and for the excellent communication during the editing process. I thank Boris Vinatzer and Dawn Arnold for helpful discussions and Annette Griffin for the initial invitation to edit this book and for her rapid and instructive feedback to my queries. I am grateful to the University of Reading and The Royal Society for financial support.

Other Books of Interest

Lactobacillus Molecular Biology: From Genomics to Probiotics	2009
Mycobacterium: Genomics and Molecular Biology	2009
Real-Time PCR: Current Technology and Applications	2009
Clostridia: Molecular Biology in the Post-genomic Era	2009
Plant Pathogenic Bacteria: Genomics and Molecular Biology	2009
Microbial Production of Biopolymers and Polymer Precursors	2009
Plasmids: Current Research and Future Trends	2008
Vibrio cholerae: Genomics and Molecular Biology	2008
Pathogenic Fungi: Insights in Molecular Biology	2008
Corynebacteria: Genomics and Molecular Biology	2008
Leishmania: After The Genome	2008
Archaea: New Models for Prokaryotic Biology	2008
RNA and the Regulation of Gene Expression	2008
Legionella Molecular Microbiology	2008
Molecular Oral Microbiology	2008
Epigenetics	2008
Animal Viruses: Molecular Biology	2008
Segmented Double-Stranded RNA Viruses	2008
Acinetobacter Molecular Microbiology	2008
Pseudomonas: Genomics and Molecular Biology	2008
Microbial Biodegradation: Genomics and Molecular Biology	2008
The Cyanobacteria: Molecular Biology, Genomics and Evolution	2008
Coronaviruses: Molecular and Cellular Biology	2007
Real-Time PCR in Microbiology: From Diagnosis to Characterisation	2007
Bacteriophage: Genetics and Molecular Biology	2007
Candida: Comparative and Functional Genomics	2007
Bacillus: Cellular and Molecular Biology	2007
AIDS Vaccine Development: Challenges and Opportunities	2007
Alpha Herpesviruses: Molecular and Cellular Biology	2007
Pathogenic *Treponema*: Molecular and Cellular Biology	2007
PCR Troubleshooting: The Essential Guide	2006
Influenza Virology: Current Topics	2006
Microbial Subversion of Immunity: Current Topics	2006
Cytomegaloviruses: Molecular Biology and Immunology	2006
Papillomavirus Research: From Natural History To Vaccines and Beyond	2006
Epstein Barr Virus	2005
HIV Chemotherapy: A Critical Review	2005

Caister Academic Press   www.caister.com

Origin and Evolution of Phytopathogenic Bacteria

John Stavrinides

Abstract

Plant pathogenic bacteria impact innumerable and valuable agricultural crops, causing hundreds of millions of dollars in damage each year. Understanding their evolution is paramount to establishing and effecting practical disease management strategies to reduce or prevent their reproduction and spread. This chapter provides an overview of the evolution of plant pathogens, considering their phylogenetic distribution, and the genetic and evolutionary factors that have contributed to their emergence. The importance of secretion systems as pathogenicity determinants is considered, with examples from a diverse array of characterized plant pathogens. Also addressed are the evolutionary constraints experienced by generalist and specialist pathogens, the ecological factors that determine their relative success, and the coevolutionary arms race that develops between specialist pathogens and their hosts. Finally, several case studies of recurrently emerging pathogens are presented, which not only illustrate the incredible evolutionary potential of phytopathogenic bacteria, but also the major contributions of anthropogenic factors to emerging infectious diseases.

Introduction

Bacteria are responsible for some of the most devastating agricultural diseases worldwide, impacting major agricultural crops, essential fruit trees, and garden ornamentals, and causing massive economic losses annually. Decades of study into the molecular mechanisms underlying plant pathogenesis have been particularly fruitful, with enormous strides in our understanding of the specific genetic factors that contribute to bacterial phytopathogenicity, and the biochemical mechanisms underlying host-encoded defence strategies and pathogen surveillance systems. Research in these areas has permitted significant advances in agricultural practices, including widespread use of genetically modified crops designed to prevent the establishment and spread of numerous bacterial diseases; however, despite our best efforts, phytopathogenic bacteria continue to adapt and thrive, evolving highly sophisticated methods to effectively pinpoint, colonize, and parasitize our vital agricultural crops.

The remarkable evolutionary potential of phytopathogens has prompted extensive research into their evolutionary origins, and how various evolutionary processes have come together to drive the genesis of highly refined and specialized mechanisms for plant pathogenicity. This chapter will address the origin of phytopathogenic bacteria, considering the current phylogeny of both highly diversified and independently evolved plant pathogens, and the molecular and evolutionary mechanisms responsible for their parasitic capabilities. Also reviewed are the origins and evolution of specific bacterial colonization and infection strategies, the evolutionary dynamics of host–pathogen interplay, and the selective pressures that result in the establishment and maintenance of the long-standing host–pathogen coevolutionary arms race. Finally, the emergence and re-emergence of several pathogens is examined, considering the historical and epidemiological factors that have contributed to various disease outbreaks. A fundamental understanding of the evolutionary pressures that drive genetic and genomic changes have an inordinate capacity to yield in depth insight into the emergence and re-emergence of new infectious phytopathogens, and may inevitably lead to the development of practical and effective disease management strategies.

Phylogenetic distribution of phytopathogenic bacteria

An analysis of the distribution of plant pathogens in the grand scheme of bacterial diversity can be particularly informative in understanding the evolution of plant pathogens. The recent advances in molecular phylogenetics have permitted an increased resolution of the evolutionary relationships among bacterial plant pathogens and their non-pathogenic relatives, allowing inferences to be made regarding the possible ancestral state of a given lineage, and when plant pathogenicity may have evolved. Phylogenetic analyses also provide the basis for directed experimental studies aimed at identifying and characterizing the specific genetic factors that facilitate or enable plant pathogenesis (see also chapter by Bull and Vinatzer).

At the level of phylum, plant pathogens appear confined mostly to the Gram-negative Proteobacteria, with the most studied plant pathogens being represented in the α–, β- and γ-subclasses (Fig. 1.1). Within the α-subclass is the tumorigenic *Agrobacterium* (see chapter by Setubal *et al.*), and the citrus greening pathogen *Candidatus* Liberibacter (Garnier *et al.*, 2000), while the β-subclass representatives include *Burkholderia*, which houses several opportunistic plant pathogenic species (Coenye and Vandamme, 2003), and *Ralstonia* (see chapter by Brown), some species of which are known for their aggressive wilting capabilities (Hayward, 1991). The γ-subclass contains the largest complement of plant pathogenic bacteria, including the genera *Erwinia, Pantoea, Pseudomonas, Xanthomonas*, and *Xylella*, all of which have immense agricultural importance (see chapters by Toth *et al.*, Arnold *et al.*, Meyer and Bogdanove, and van Sluys *et al.*). Also in this subclass are several overlooked yet extremely relevant plant pathogens, such as the unculturable phloem-restricted strawberry pathogen, *Candidatus* Phlomobacter (Bove and Garnier, 2002), which has been classified recently as one of the top three containment pathogens in the United Kingdom (DEFRA, 2007).

Although many of the genera described above contain at least one plant pathogenic species, there can be significant variability in the diversity of ecological niches occupied by the strains within each genus. *Agrobacterium*, *Xanthomonas*, and *Xylella*, for example, appear to be exclusively plant pathogens or at least plant-associated, while the others, like *Pseudomonas*, comprise a multitude of ecologically distinct species, including putatively non-pathogenic species (*P. fluorescens, P. putida*) as well as plant pathogens (*P. syringae, P. viridiflava, P. marginalis, P. corrugata*) (Catara, 2007; Goumans and Chatzaki, 1998). The human pathogenic *P. aeruginosa* is of particular interest in this group, having been shown to be a cross-kingdom pathogen capable of infecting both plant and animal hosts (Rahme *et al.*, 1995; Rahme *et al.*, 1997).

Because the members of the Proteobacteria have a disproportionate agricultural impact as a group, they tend to be the main focus of current phytopathological research; however, there are other major groups that are frequently overlooked as significant plant pathogens, even though they are responsible for several diseases of critical staple crops. The phylum Firmicutes is known mostly for its human pathogenic species; yet, it also contains the plant pathogenic phytoplasmas (class Mollicutes) (Fig. 1.1), which are morphologically small, largely unculturable, insect-vectored bacteria that cause a variety of plant yellows diseases of rice, maize, potato, and soybean, as well as numerous other vegetable and fruit crops (Garnier *et al.*, 2001; Lee *et al.*, 2000; Weintraub and Beanland, 2006). There are also numerous plant pathogenic groups within the phylum Actinobacteria (Fig. 1.1), spread across multiple families in the class Actinobacteria, order Actinomycetales. The filament-forming *Streptomyces*, as well as the coryneforme plant pathogens *Rhodococcus* (Goodfellow, 1984), *Rathayibacter* (Davis, 1986; Sasaki *et al.*, 1998), *Clavibacter* (Davis *et al.*, 1984), *Curtobacterium* (Chen *et al.*, 2007; Collins and Jones, 1983), and *Leifsonia* (Evtushenko *et al.*, 2000) (Fig. 1.1; see chapter by Monteiro-Vitorello *et al.*)

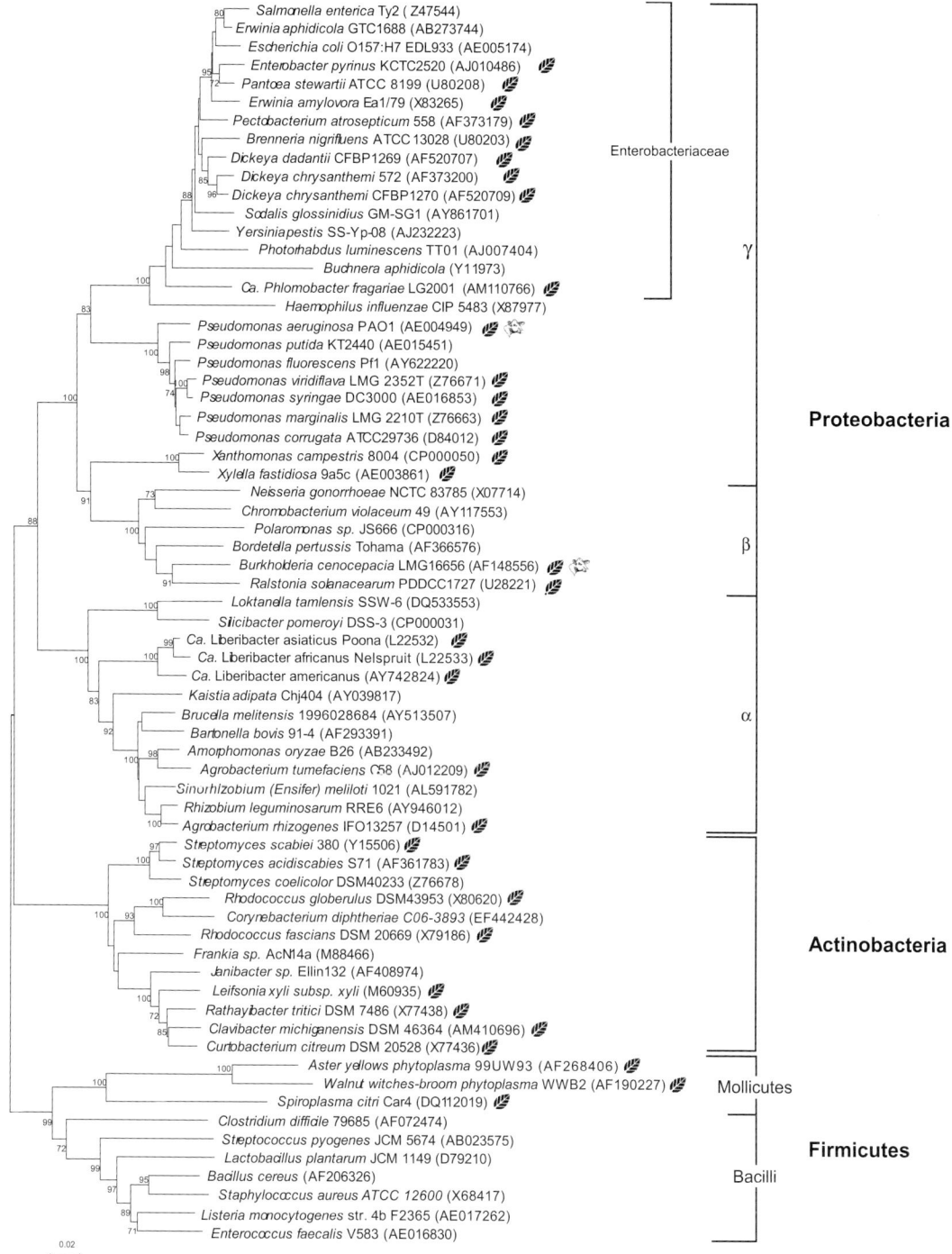

Figure 1.1 Phylogenetic distribution of the major plant pathogens among the Proteobacteria, Actinobacteria, and Firmicutes. Plant pathogens (denoted by a leaf) are interspersed with animal pathogens and free-living bacteria, with the vast majority of described phytopathogenic bacteria being present in the Proteobacteria. Cross-kingdom pathogens *Pseudomonas aeruginosa* and *Burkholderia cenocepacia* are marked with both a leaf and an animal symbol.

cause a variety of diseases on legumes, solanaceous plants, and various monocots, including maize and wheat. Given their phytopathogenic capabilities and phylogenetic position, these organisms may represent independently evolved plant pathogens, and will be discussed later from this perspective. Considering the phylogenetic distribution of extant phytopathogens, it is obvious that the evolution of plant pathogenicity has evolved multiple times, sometimes occurring relatively early in a lineage, such that many related genera exhibit phytopathogenicity, or relatively late, where only one or two genera, or even species gain plant-colonizing abilities.

Pathogenicity and virulence determinants

The terms 'pathogenicity' and 'virulence' have been adopted by many different disciplines, each one retaining a specific meaning and use. Given that the use of these terms is highly subjective, and the fact that they will be utilized extensively throughout this chapter, their use will be defined explicitly here. The term 'pathogenicity' will be used to describe the capacity of a bacterium to cause disease (a bacterium is either pathogenic or non-pathogenic), with 'pathogenicity factors' being those traits that confer disease-causing capabilities (Smith, 1977). In contrast, 'virulence' reflects the degree of pathogenicity, and is often correlated with symptom severity (Watson and Brandly, 1949). By extension, the term 'virulence factor' will refer to any determinant that alters the amount of disease caused by a pathogen.

Horizontal gene transfer and pathogenicity islands

Plant pathogenic bacteria have evolved a diversity of pathogenicity and virulence determinants, including specialized secretion systems, phytotoxins, phytohormones, extracellular proteins, attachment structures, micronutrient uptake systems, as well as specialized structural layers like lipopolysaccharides (LPS) and exopolysaccharides (EPS), all of which may play prominent roles in suppressing the host defence response, preventing pathogen recognition, facilitating nutrient extraction, and promoting pathogen reproduction and spread (Abramovitch and Martin, 2004; Alfano and Collmer, 1996; Arnold et al., 2003; Toth and Birch, 2005). Many such pathogenicity and virulence factors are frequently associated with mobile genetic elements, such as bacteriophage, transposons, integrative-conjugative elements, and plasmids (Canchaya et al., 2003; Dobrindt et al., 2004; Dutta and Pan, 2002; Frost et al., 2005; Hacker et al., 2004; Ochman et al., 2000; Thomas and Nielsen, 2005) (see chapter by Sundin and Murillo). These mobile genetic elements are the main contributors to horizontal (or lateral) gene transfer (HGT), and allow bacteria to acclimate instantly to new, and fluctuating environments (i.e. hosts) (Lawrence and Ochman, 1997; Lawrence and Ochman, 2002; Lerat et al., 2005; Ochman, 2001; Ochman et al., 2000; Osborn and Boltner, 2002). For example, mobile and autonomously replicating plasmids often carry a wealth of virulence factors, such as plant hormone biosynthetic genes (Evidente et al., 1995; Glass and Kosuge, 1988; Watanabe et al., 1998), phytotoxins (Bender et al., 1989), and an assortment of extracellular proteins (Buell et al., 2003; Stavrinides and Guttman, 2004; Sundin et al., 2004; Zhao et al., 2005), which can have an enormous impact on the pathogenic abilities of bacteria (Coplin, 1989; Vivian et al., 2001). Additional information on plasmids and their contributions to pathogenicity is available in the Sundin and Murillo chapter. Likewise, prophage have been implicated in the delivery of virulence-associated genes, consistent with the identification of several plant pathogen virulence factors exhibiting sequence similarity to phage-related genes (Guttman et al., 2002). The defective prophage XacP1 of the citrus blight pathogen *X. axonopodis* has been shown to carry an expansin-like protein that is induced during pathogen colonization of citrus (da Silva et al., 2002), and may promote plant cell wall modifications during pathogenesis (Cosgrove, 1999). Thus, mobile genetic elements may serve as a reservoir of novel virulence factors, sequestered and assimilated by plant pathogenic bacteria from other organisms in the environment through HGT.

Genes associated with pathogenicity and virulence are often found clustered, and encompassing large segments of the genome, which is consistent with inheritance as a single genomic entity. These regions were initially called 'pathogenicity islands' (PAIs), and were defined as niche-specific gene clusters containing numerous virulence genes, typically situated adjacent to tRNA genes, flanked by direct repeats or mobile elements, and having a G+C content that differed from the rest of the host genome (Hacker et al., 1990; Karch et al., 1999). PAIs with these explicit characteristics have been identified in plant pathogens, such as the large toxin biosynthetic clusters that direct the synthesis of coronatine, tabtoxin, and syringomycin and syringopeptin (Arnold et al., 2003). Coordinately regulated gene clusters encoding type III and type IV secretion systems, which are required for pathogenesis in the organisms that carry them, also exhibit the hallmarks of PAIs, reinforcing the notion that PAIs have the capacity to provide bacteria with new phytopathogenic capabilities in 'quantum leaps' (Groisman and Ochman, 1996).

The evolution of secretion systems

All phytopathogenic bacteria presently described utilize secreted proteins to extract nutrients from their hosts and cause disease. There are currently six secretion pathways, type I to type VI, utilized for a variety of export processes by both pathogenic and non-pathogenic bacteria (Saier, 2006; Van Sluys et al., 2002), with the first four types being described as major pathogenicity determinants for plant pathogens (Fig. 1.2; see also chapter by Arnold et al.). The importance of these systems to plant disease is consistent with the numerous studies demonstrating that mutations in key secretion system components render various bacteria non-pathogenic (Andro et al., 1984; Hueck, 1998; Salmond, 1994). Furthermore, there is evidence supporting the evolution of these pathogenicity-associated secretion systems through the recruitment of other ancestral systems, followed by gradual genetic refinement to accommodate virulence factors. The evolution of these secretion systems has therefore been central to the evolution of plant pathogenic bacteria.

Type I secretion

The type I secretion system (T1SS) is ubiquitous among both pathogenic and non-pathogenic bacteria, mediating one-step secretion of various substrates, including proteases, phosphatases, esterases, nucleases, and glucanases (Dinh et al., 1994; Pao et al., 1998). A T1SS consists of a channel or gated pore, formed by the association of an ABC transporter, membrane fusion protein, and an outer membrane protein, which spans both inner and outer membranes (Delepelaire, 2004). The ABC transporter component of the T1SS is responsible for substrate recognition (Delepelaire, 2004; Gerlach and Hensel), consistent with its description as a member of an evolutionarily diverse protein family exhibiting substrate-binding versatility (Fath and Kolter, 1993). Functional and phylogenetic classification reveals multiple evolutionary trajectories leading to ABC transporter diversification (Jack et al., 2001; Pao et al., 1998; Saier, 2000; Saier, 2003; Saier and Paulsen, 2001; Saier et al., 1998). Several groups of ABC transporters contain transmembrane hydrophobic α-helices, each containing as few as two or as many as 30 of these domains (Saier, 2003). The evolution of these proteins has been proposed to have been mediated by intragenic duplication, triplication, and quadruplication events within ancestral transporters containing four, five, and six transmembrane regions, as well as gene fusion, splicing, deletion, and insertion events (Saier, 2003). These intragenic changes have clearly contributed to the diversity of function and specificity of the ABC transporter binding and, thus, substrate specificity of the T1SS secretion complex.

Although not often considered a major contributor to pathogenesis, T1SSs have clearly been adapted for extracellular secretion of virulence factors by several phytopathogens. *E. amylovora* secretes a virulence-associated metalloprotease (PrtA) via a T1SS encoded by *prtD*, *prtE*, and *prtF* (Fig. 1.1), which is necessary for the colonization of apple leaves (Zhang et al., 1999). In *P. syringae*, the pore-forming syringopeptin and syringomycin phytotoxins are secreted by the ABC transporter encoded by SyrD (Quigley et al., 1993), causing extensive plant tissue necrosis during colonization of cherry fruit (Hutchison and Gross, 1997; Scholz-Schroeder et al., 2001). In *X. oryzae* pv. *oryzae*, the *raxA*, *raxB*, and *raxC* genes are responsible for the type I-dependent secretion of the bacterial quorum-sensing molecule AvrXa21 (Burdman et al., 2004; Lee et al., 2006b). AvrXa21 is recognized by the Xa21 receptor-like kinase of rice, which leads

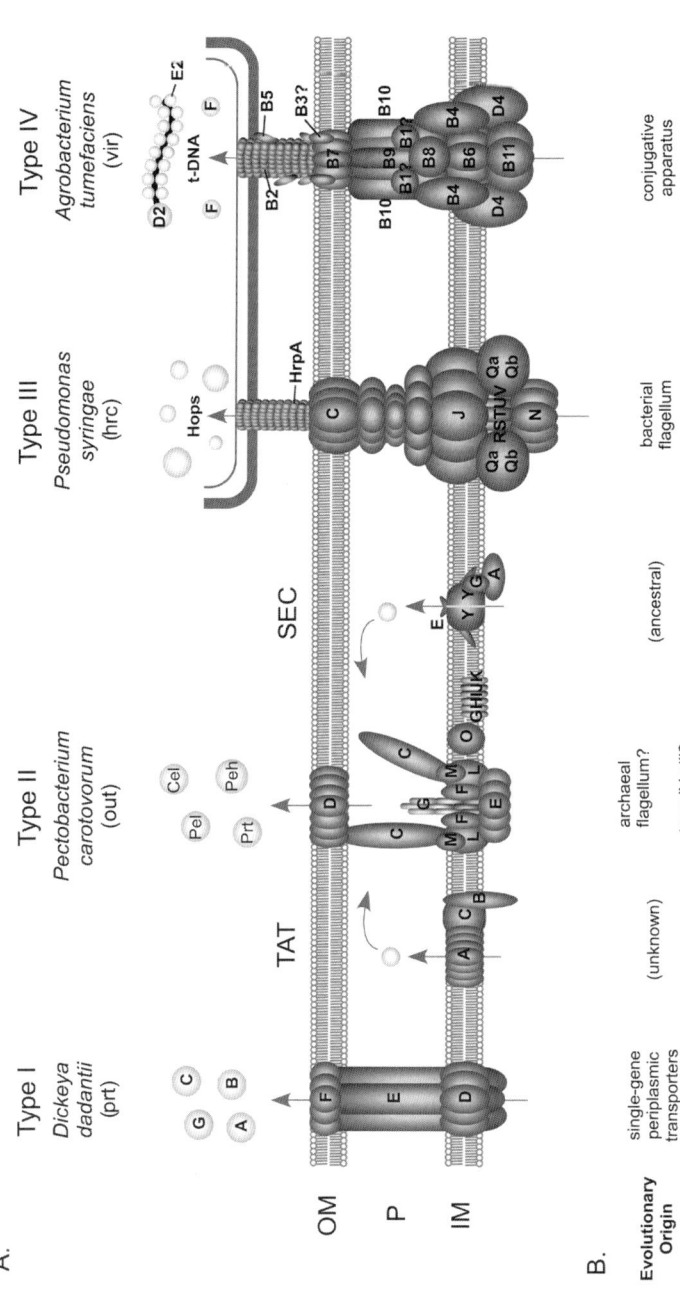

Figure 1.2 (A) Four major secretion systems, type I–IV, utilized by four different phytopathogens for plant colonization. The type I and type II secretion systems export factors directly into the extracellular environment. The structural components of the type I secretion system from *Dickeya dadantii* is composed of PrtD, PrtE, and PrtF, which collectively secrete various enzymes, including PrtA, PrtB, PrtC, and PrtG. The type II secretion system of *Pectobacterium carotovorum* is more structurally complex, and functions in the extracellular export of pectate lyases (Pel), polygalacturonase (Peh), cellulase (Cel), protease (Prt), which are first brought into the periplasmic space by the twin arginine translocation system (TAT) or SEC system. Unlike type I and type II secretion, the type III and type IV secretion systems have evolved to primarily inject proteins (Hops – hrp/hrc outer proteins) or DNA–protein complexes (tDNA) directly into plant cells, using a tube-like conduit or pilus, made of HrpA in the *Pseudomonas syringae* type III secretion system, or VirB2/VirB5 in the *Agrobacterium tumefaciens* type IV secretion system. Placement of each system subunit for secretion systems I–III in the indicated organisms, as well as the TAT and SEC systems, was extrapolated using homologous components from well-characterized systems (Filloux, 2004; Gold et al., 2007; He, 1998; Lee et al., 2006a; Tampakaki et al., 2004), whereas the position of most type IV secretion system components has already been established (McCullen and Binns, 2006). The name of each secretion system is indicated under the bacterial species name, with all major proteins in the structural complex being named according to their gene letter designation. Arrows passing through each secretion system indicate the path of secreted proteins. OM, outer membrane; P, periplasm; IM, inner membrane. (B) The possible evolutionary origin of each secretion system is indicated. The SEC system is distributed among all bacterial (and archaeal) lineages, suggesting it is ancestral.

to partial reduction in *in planta* bacterial growth and significantly smaller lesion size compared with *rax* mutants (Lee et al., 2006b). The maintenance of AvrXa21 in populations of *X. oryzae* is especially puzzling, as there is no discernible difference in growth between wild-type and *rax* mutants on hosts lacking Xa21 (Burdman et al., 2004). Presumably, the maintenance of AvrXa21 is attributable to some role in virulence, perhaps in the induction of specific virulence systems for later stages of plant infection. T1SS are clearly versatile, capable of secreting virulence associated proteases, toxins, and quorum-sensing molecules, illustrating their prominent role in plant colonization for a variety of bacterial pathogens.

It is especially interesting to note that T1SSs, unlike all other secretion systems that will be discussed, also have an established defensive role for phytopathogenic bacteria. Plants produce an array of antimicrobial secondary compounds, including phytoalexins, which are often induced during pathogen colonization, and phytoanticipins, which are preformed secondary metabolites (Vanetten et al., 1994). *X. fastidiosa* mutants deficient in the multidrug efflux system component *tolC* exhibited increased susceptibility to a variety of phytochemicals, including berberine, genistein, and rhein, and were unable to cause disease on grape (Reddy et al., 2007). Thus, the T1SS of *X. fastidiosa* has clearly evolved to bind antimicrobials produced by plants, serving to actively export these toxic compounds out of the bacterial cell. Similar defensive roles for T1SSs have also been demonstrated for *Dickeya dadanti* (formerly *Erwinia chrysanthemi*) (Barabote et al., 2003), *E. amylovora* (Burse et al., 2004), and *A. tumefaciens* (Palumbo et al., 1998). T1SSs clearly serve as both offensive and defensive components of the phytopathogen weapons arsenal.

Type II secretion
The type II secretion system (T2SS) is a two-step secretion pathway that exports proteins from the periplasmic space into the extracellular environment, relying on the prerequisite action of either the general export pathway (GEP)/SEC system (Desvaux et al., 2004) or the twin-arginine translocation (TAT) pathway (Berks et al., 2000) to export the appropriate proteins into the periplasmic space (Stathopoulos et al., 2000).

The GEP/SEC system is distributed among all major evolutionary groups of pathogenic and non-pathogenic bacteria (Cao and Saier, 2003). Proteins targeted to this secretion machinery carry a particular N-terminal signal peptide, which is recognized by the *sec* chaperone and directed across the inner cytoplasmic membrane. En route, the signal peptide is cleaved by a peptidase, and the protein is released into the periplasm where it may undergo folding and/or modification. The mature protein can then be secreted across the outer membrane, via the action of many protein complexes, and subsequently released into the extracellular environment. This same export process can be achieved by the TAT system, although unlike the GEP/SEC system, the TAT system facilitates the transport of folded proteins across the inner membrane (Berks et al., 2000).

The macromolecular structure of the T2SS and the molecular mechanisms of T2SS-dependent transport have been well-documented (Cianciotto, 2005; Filloux, 2004; Johnson et al., 2006; Pugsley, 1993; Sandkvist, 2001a; Sandkvist, 2001b). From this, it has become obvious that the T2SS is homologous to the type IV piliation systems, sharing related constituents and mechanism of function (Pugsley, 1993). Some studies have extended this further, suggesting that several T2SS components are homologous to the constituents of archaeal flagellar systems (Peabody et al., 2003), indicating a possible origin for this secretion system; however, only few proteins retain homology, all of which appear to be members of relatively large protein families. Nonetheless, while the origin of the T2SS remains unknown, it is clear that this secretion system has been harnessed by many plant pathogens to mediate export of various virulence factors (Sandkvist, 2001b). For example, the soft rot pathogens *D. dadanti* and *Pectobacterium carotovorum* (formerly *Erwinia carotovora* subsp. *carotovora*) secrete a repertoire of key virulence-associated pectate lyases, polygalacturonases, pectin methylesterases, and cellulases via a T2SS (Cianciotto, 2005; Johnson et al., 2006; Sandkvist, 2001b) (Fig. 1.2). *D. dadanti* T2SS mutants are significantly attenuated in virulence (Andro et al., 1984; Thurn and Chatterjee, 1985). The T2SS of *Pectobacterium atrosepticum* (formerly *Erwinia carotovora* subsp. *atroseptica*) also appears to be essential for virulence on potato, secreting the virulence-associated Svx protein during pathogenesis (Corbett et al., 2005). Similarly, mutations in the *Xanthomonas oryzae* pv. *oryzae* T2SS results in loss of pathogenicity, with the accumulation of xylanase and cellulase occurring in the periplasmic space (Sun et al., 2005). Xylanase, in particular, has been shown to be a major pathogenicity determinant for *X. oryzae* pv. *oryzae* on rice (Rajeshwari et al., 2005; Ray et al., 2000; Sun et al., 2005), consistent with the polysaccharide composition of rice cell walls which are up to 60% xylan (Takeuchi et al., 1994). It is interesting to note that xylanase secretion could not be detected for *X. campestris* pv. *campestris* 8004, a pathogen of crucifers, although α-amylase, pectinase, and protease were produced in abundance by this pathogen (Sun et al., 2005). This clearly reflects the necessity of the T2SS for pathogenesis and the specialization of the type II secreted repertoire of each pathogen to its respective host. *R. solanacearum* also appears to utilize a T2SS for export of virulence factors (Liu et al., 2005), although unlike the systems described above, it utilizes a TAT system to transport folded proteins into the periplasmic space (Gonzalez et al., 2007). Site directed mutagenesis of several TAT-secreted proteins significantly attenuated virulence and the wilting phenotype, suggesting that this system has become an integral component of the *R. solanacearum* disease strategy (Gonzalez et al., 2007).

Type III secretion
The type III secretion system (T3SS) is an injection apparatus that allows both plant and animal pathogens to deliver virulence proteins, known as type III secreted effectors (T3SEs) directly into the cytoplasm of their eukaryotic hosts. T3SEs function to destabilize host cell processes, thereby compromising defence signalling and allowing for successful pathogen colonization (see chapters by Arnold et al. and Boch). This successful strategy is utilized by numerous plant pathogens, including *X. campestris*, *P. syringae*, *R. solanacearum*, and *E. amylovora* (Hueck, 1998). Upon entering their host, environmental cues activate transcription T3SS (see chapter by Rico et al.) (Jin et al., 2003; Tang et al., 2006), resulting in the assembly of a multi-protein complex that spans both inner and outer membranes, and from which, a syringe-like conduit emerges and penetrates into the plant cell wall (Fig. 1.2). A multitude of co-regulated T3SEs are subsequently exported through this needle, and directly into the host cell where they exert a variety of virulence effects (Grant et al., 2006). Mutations in the T3SS abolish pathogenicity, suggesting an essential role for both the T3SS and T3SEs in pathogenesis (Rahme et al., 1991).

The genomic region encoding the T3SS exhibits several signatures of a prototypical PAI. In *P. syringae*, the T3SS regulon has been described as a modular island composed of a large central gene cluster encoding the structural apparatus of the system, flanked on either end by two smaller gene clusters encoding various secreted virulence proteins (Alfano et al., 2000). This suggests that this PAI may have been acquired by *P. syringae* through HGT in a ready-to-use format, containing both the necessary structural components of the apparatus and the appropriate virulence proteins. Like most prototypical PAIs, it is situated near a tRNA^{LEU} gene; however, it lacks the abnormal G+C content and mobile elements typically found associated with these islands, consistent with a relatively ancient acquisition. Indeed, a phylogenetic analysis of key components of the T3SS of *P. syringae* reveals phylogenies that are consistent with the evolution of the species complex (Sawada et al., 1999), and together with the overall levels of intraspecific variation, is indicative of early acquisition of this PAI, likely before the diversification of *P. syringae*. The acquisition of the T3SS by *P. syringae* may have played an important role in its emergence as a phytopathogen. A similar conclusion has been drawn from an analysis of the *R. solanacearum* T3SS, whereby ancient rather than recent acquisition is favoured largely due to an absence of multiple flanking mobile elements, and an overall G + C content consistent with that of the remainder of the genome (Genin and Boucher, 2004; Salanoubat et al., 2002). In contrast, the T3SS of *P. agglomerans* is located on a 150-kb plasmid, pPATH, and is contained within a 70-kb PAI along with many T3SEs and other virulence factors (Nizan et al.,

1997). The plasmid localization of this PAI, together with the presence of numerous degenerate transposable elements and other PAI-localized genes (Guo et al., 2002), has been suggested to be reflective of relatively recent acquisition and thus, recent evolution of the pathogen (Manulis and Barash, 2003). Thus, acquisition of the T3SS may have enabled bacteria to rapidly expand into new niches and exploit new hosts (Troisfontaines and Cornelis, 2005). It is interesting to note that several animal pathogens carry multiple T3SSs, which have separate, but complementary functions (Coburn et al., 2007); however, there are no reports of multiple T3SS required for pathogenesis in plant pathogens. A secondary T3SS-like system was identified in the bean pathogen, *P. syringae* pv. *phaseolicola* 1448A, which is composed of 12 genes whose closest homologues are those of the T3SS genes found in *Rhizobium*, *Photorhabdus*, *Aeromonas*, and *P. aeruginosa* (Joardar et al., 2005). Because mutations in the primary T3SS of *P. syringae* pv. *phaseolicola* 1448A abolish pathogenicity and the ability of the pathogen to trigger a defence response in the plant, this second secretion system cannot complement loss of the primary system, nor can it secrete known T3SEs (Joardar et al., 2005). While this suggests that it is not a major pathogenicity factor, a role in virulence has been neither excluded nor established.

It is clear that T3SSs play an important role in pathogenesis, and early horizontal acquisition may have enabled bacteria to exploit plants. But from where did the T3SS originate? The evolution of the T3SS itself has been the subject of considerable debate, although there is agreement that many structural, functional, and sequence similarities between the T3SS and the bacterial flagellum reflect an obvious evolutionary relatedness. The flagellum is the primary motility structure for most bacteria, and is broadly distributed among the major bacterial groups (Beatson et al., 2006; Harshey and Toguchi, 1996). The gene cluster encoding the major structural constituents are most frequently chromosomally encoded, and share sequence and syntenic conservation between species. In addition, the genealogical relationships of these proteins are largely congruent with the established bacterial species tree, indicating transmission largely by vertical descent since originating from a common ancestor of all modern bacteria (Rytkonen et al., 2007; Saier, 2004). The broad distribution of flagella across all bacterial clades and the much more limited distribution of T3SSs favours the evolutionary derivation of the T3SS from the flagellar system (Hueck, 1998; Troisfontaines and Cornelis, 2005; Vangijsegem et al., 1995). Some have also proposed that the T3SS and flagellar systems are anciently diverged, based on the accumulation of comparable genetic variation, and phylogenetic support for monophyly of the T3SS and flagellar components (i.e. each has a single ancestor) (Gophna et al., 2003; Nguyen et al., 2000). To explain the current distribution of T3SS in pathogens, this latter hypothesis would require numerous independent losses of the T3SS in many different lineages; a non-parsimonious yet still plausible scenario. Although we may never fully understand the precise evolutionary steps leading to the formation of the T3SS, it is abundantly clear that this system is currently a primary pathogenicity determinant for numerous phylogenetically diverse plant pathogens.

Type IV secretion

The contributions of the type IV secretion system (T4SS) to plant pathogenicity have been characterized extensively, relating mostly to the tumour-inducing (Ti) capabilities of *A. tumefaciens* (Christie, 1997; Christie and Vogel, 2000; Lai and Kado, 2000) (see chapters by Setubal et al. and Sundin and Murillo). The *A. tumefaciens* T4SS is on the Ti plasmid, and responsible for the extracellular export of key virulence-associated DNA–protein complexes across the bacterial membrane. The system itself comprises 11 proteins, VirB1–VirB11, which form a macromolecular secretion apparatus that spans both inner and outer membranes (Fig. 1.2). Much like the T3SS, a tube-like conduit emerges from this membrane-bound complex, connects the bacterial cell cytoplasm to that of the plant host cell, and serves as the delivery route for oncogenic DNA. The structure and function of the T4SS is consistent with its evolutionary derivation from the conjugal transfer apparatus genes of plasmids. This has been suggested to have occurred multiple times based on parsimony-based phylogenetic analysis of the T4SS VirB proteins

(Frank *et al.*, 2005). Consistent with this is the fact that the *A. tumefaciens* T4SS retains its ancestral conjugative abilities, mediating conjugal transfer of plasmid RSF1010 between bacteria, and even delivering RSF1010 into plant cells (Bohne *et al.*, 1998).

Although the prominent role of this system in the pathogenicity of clinically important human pathogens has been demonstrated (Cao and Saier, 2001; Christie, 2001; Seubert *et al.*, 2003), few plant pathogens have been shown to utilize a T4SS for pathogenesis. In addition to *A. tumefaciens*, the functionality and involvement in plant pathogenesis of a plasmid-encoded T4SS has been verified for *Burkholderia cenocepacia*, which facilitates disease in onion by delivering a heat-stable protein into onion tissues that results in plasmolysis of plant protoplasts and the manifestation of tissue watersoaking (Engledow *et al.*, 2004). Although only these two systems have been shown to have a direct role in pathogenesis, bioinformatic analyses have identified T4SS islands or their components in the genomes of *L. xyli, P. syringae, R. solanacearum, X. axonopodis, X. campestris, X. aurantifolii,* and *X. fastidiosa* (Van Sluys *et al.*, 2002) (see chapter by van Sluys *et al.*), suggesting other pathogens may also use a T4SS for virulence. In addition, the relatively broad distribution of this system among phytopathogens may indicate a more prominent role for the T4SS in mediating plant disease than previously thought.

Parallels in secretion system evolution
It is likely that the four virulence-associated secretion systems were co-opted for pathogenicity, through transporter substrate specificity modifications (T1SS), or more complex modifications, exemplified by the T3SS and T4SS, which have evolved quite apparently from ancestral systems. The T2SS may have also originated through the modification of an ancestral system, although the specific identity of the ancestral system remains to be determined. Nonetheless, the adaptation of these systems to the pathogenic lifestyle, in addition to the wealth of virulence factors that they accommodate, has probably been central to the ability of bacteria to expand into new niches and exploit plants as hosts. The contributions of additional secretion systems to pathogenicity will likely follow, given the complexities of phytopathogenic bacteria, the diversity of hosts they attack, the diseases they cause, and the simply astonishing nature of bacterial innovation.

The independent evolution of plant pathogens

Given the diversity of pathogenicity factors and colonization strategies, and prominent role of horizontal transfer in mediating genetic exchange between bacteria, it is not surprising that many pathogens appear to have evolved the ability to colonize plants independently. These uniquely derived phytopathogens can be identified quite readily by considering their evolutionary history, as their plant pathogenic lifestyle is often at odds with the known ecology of their phylogenetically defined bacterial relatives. It is these phytopathogens, in particular, that create numerous and potentially exciting avenues of research for understanding the evolution of plant pathogenicity.

Plant pathogens among the enteric bacteria

The Enterobacteriaceae are considered largely gut-associated bacteria, housing the human pathogenic *E. coli, Shigella, Salmonella* and *Yersinia*, as well as numerous mutualists of insects, including *Buchnera, Sodalis,* and *Wigglesworthia*; however, also within this family was the genus '*Erwinia*', whose species were described collectively as major rot pathogens of various crops including carrot, potato, sugarbeets and pepper, and as major pathogens of humans and fish (Barras *et al.*, 1994; Perombelon, 2002; Starr and Chatterjee, 1972; Toth and Birch, 2005). This group has since been re-evaluated, and is now divided into *Erwinia, Enterobacter, Pectobacterium, Dickeya, Pantoea* and *Brenneria*, all of which contain at least one species exhibiting phytopathogenic properties (Gardan *et al.*, 2003; Hauben *et al.*, 1998; Kado, 2006; Samson *et al.*, 2005; Toth *et al.*, 2006) (Fig. 1.1). The presence of these plant pathogens in a group of seemingly dedicated gut-associating bacteria is particularly intriguing. A recent genomic comparison of *P. atrosepticum* with the enteric animal pathogens reveals acquisition of at least 17 different PAIs by *P. atrosepticum*, including a T3SS, genes for agglutination and adhesion, and polyketide phytotoxin biosynthetic clusters (Toth et al., 2006). Of all coding regions contained within genomic islands, 60% share homology to genes from other plant-associated bacteria, suggesting that HGT from ecologically related bacteria has significantly contributed to their evolution as plant pathogens. Although most organisms that have switched niches often exhibit drastic genomic changes (i.e. genome reduction), the *P. atrosepticum* genome does not exhibit the typical signatures of niche specialization, suggesting that these phytopathogens continue to retain their historical gut associations, perhaps as a transient stage to their newly acquired plant pathogenic lifestyle. It is especially interesting that various *Erwinia*, *Dickeya*, and *Pectobacterium* species retain relatively tight pathogenic and non-pathogenic associations with herbivorous- and plant-associated insects (Capuzzo et al., 2005; de Vries et al., 2001a; de Vries et al., 2001b; Harada et al., 1997). For example, *Erwinia amylovora* is pathogenic to the olive fly and Western flower thrips (Capuzzo et al., 2005; de Vries et al., 2004), but survives at least 5 days in or on the green lacewing and 12 days on aphids. The retention of gut-associating capabilities may enable this pathogen to exploit these and other insects as vectors, and possibly as hosts. The colonization of *Drosophila* by *P. carotovorum* results in a host defence response, characterized by the production of host-induced antibacterials (Basset et al., 2000); however, the bacteria persist as a direct result of a single bacterial gene, *evf*, which enhances survivorship of the bacteria in the gut by preventing insect excretion (Basset et al., 2003). The acquisition and maintenance of genetic determinants that enable the bacteria to form insect associations are likely to facilitate bacterial survival and dispersal. Likewise *D. dadantii* and *Erwinia aphidicola* have been shown to be pathogenic to the pea aphid, colonizing the gut and causing death (Grenier et al., 2006; Harada and Ishikawa, 1997). Thus, these 'enteric' plant pathogens appear to retain their ancestral gut-associations in insects, although the relative importance of these associations to the overall life cycle of these pathogens is not yet clear. Their obvious link to plant-associated insects could suggest that their historical gut associations have become a secondary component of their lifecycle, and the evolution of plant pathogenicity may have followed from frequent insect-mediated deposition on plants. For further information on *Erwinia* pathogenicity, see chapter by Toth et al..

A plant pathogen among mutualists

Agrobacterium appears to be a phylogenetic outlier with respect to its lifestyle, being most closely related to the rhizobial bacteria that form mutualistic associations with plants (Fig. 1.1) (see chapter by Setubal et al.). Other related genera within the Rhizobeaceae include the free-living saprotrophs *Carbophilus* and *Kaistia*, and the nitrogen-fixing (and plant-associating) *Amorphomonas* and *Sinorhizobium* (*Ensifer*) suggesting a complex evolutionary history that may have included the independent evolution of plant pathogenicity (Fig. 1.1). For example, *A. tumefaciens*, and *A. rhizogenes* are crown and root pathogens whose pathogenic properties are attributable to their tumour-inducing (Ti) and root-inducing (Ri) plasmids, respectively (Bevan and Chilton, 1982). Both of these plasmids carry a specialized type IV secretion system that translocates an oncogenic DNA fragment into plant cells to facilitate bacterial pathogenesis (Bevan and Chilton, 1982) (Fig. 1.2). The *Agrobacterium* plasmid replicons themselves share a common evolutionary history with those of *Rhizobium*, reflecting the shared ancestry of their plasmids (Moriguchi et al., 2001; Wong and Golding, 2003). Interestingly, the key host-association factors of *Rhizobium* are also localized on a plasmid, pSymA, including nodulation-promoting factors (Galibert et al., 2001). The presence of shared homologous plasmids among *Rhizobium* and *Agrobacterium* strains suggests that the evolution of plant pathogenicity was not simply due to its acquisition of an entirely foreign Ti or Ri plasmid, but through large-scale rearrangements and incorporation of PAIs into an existing plasmid backbone (Moriguchi et al., 2001). Interestingly, it has been demonstrated that transfer of specific *Rhizobium* megaplasmids to *A. tumefaciens* permits this pathogen to nodulate both bean and clover (Abe et al., 1998; Martinez et al., 1987), although transfer of the Ti plasmid to mutualistic *Sinorhizobium meliloti* does not confer tumorigenicity (Vanveen et al., 1989). This is also consistent with the discovery of some transitionary strains of *A. rhizogenes* (formerly *Rhizobium rhizogenes*), which carry both pSym and pTi/pRi, allowing them to associate mutualistically with bean, and parasitically with other plants (Velazquez *et al.*, 2005). These studies, along with the fact that *Agrobacterium* lacks the typical T3SS that is characteristic of many Gram-negative plant pathogens (Wood *et al.*, 2001), support the independent evolution of *Agrobacterium* plant pathogenicity. For additional information on *Agrobacterium* pathogenicity, see Chapter 7.

Plant pathogenicity among the soil-dwelling actinobacteria

Members of the Actinobacteria are a group of high G + C content, Gram-positive bacteria that are considered largely soil-inhabiting. Only a few groups, and in some cases only a few species within the Actinobacteria are plant pathogenic, reinforcing the independent evolution of plant pathogenicity within this lineage (Fig. 1.1). *Leifsonia xyli* subsp. xyli the causal agent of ratoon stunting disease of sugarcane, was one of the first Gram-positive plant pathogenic bacteria to have its genome sequenced (Davis *et al.*, 1980; Monteiro-Vitorello *et al.*, 2004). *Leifsonia* 'pathogenicity' is marked by rapid bacterial proliferation in the lumen and pits of xylem vessels, ultimately resulting in the restriction of water flow in the plant, and severe stunting of growth (Weaver *et al.*, 1977). A genomic analysis revealed a relatively small suite of virulence genes, and a relatively high proportion of pseudogenes (Monteiro-Vitorello *et al.*, 2004), suggesting a recent switch of ecological niches. Four major PAIs, named LxxGI1- LxxGI4, are present in the genome, three of which harbour putative virulence genes (Monteiro-Vitorello *et al.*, 2004). *Leifsonia* relatives are largely aquatic (Reddy *et al.*, 2003), although some species have been isolated from nematode-induced galls on annual bluegrass (Evtushenko *et al.*, 2000). *Leifsonia xyli*-like bacteria have also been identified as grass endophytes (Mills *et al.*, 2001), possibly indicating evolution and specialization of this genus to grasses. See chapter by Monteiro-Vitorello *et al.* for additional information on *Leifsonia*.

Streptomyces bacteria have begun receiving increasing attention as plant pathogens, despite recognition of their importance in the late 1940s (Hooker, 1949). *Streptomyces scabies* and other related species are important pathogens of various crops, forming scab-like lesions on potato and tap-root plants, including carrot, radish, and beet (Loria *et al.*, 1997; Loria *et al.*, 2006). Large PAIs, likely acquired by HGT, encode various toxins including thaxtomin, which is essential for colonization of plant tissues (Kers *et al.*, 2005; Loria *et al.*, 2006). Likewise, the actinomycete *Clavibacter michiganensis*, a pathogen of tomato, potato, maize, wheat, and alfalfa crops (Gartemann *et al.*, 2003; Jahr *et al.*, 1999), also appears to have acquired several important pathogenicity factors, many of which are localized on two transmissible plasmids, pCM1 and pCM2 (Meletzus *et al.*, 1993). These plant pathogenic bacteria in the Actinomycete lineage, whose lifestyle is largely soil adapted where they frequently encounter plants, have probably evolved independently through the acquisition of appropriate pathogenicity factors from ecologically related microorganisms.

Evolutionary dynamics of host–pathogen interactions

Generalists and specialists

Many of the 'independently evolved' phytopathogens described above appear to have a relatively broad host range, perhaps indicating that a relatively wide host breadth is characteristic of recently evolved plant pathogens. These generalists may be less selectively favoured, experiencing various evolutionary and ecological tradeoffs that may gradually drive them to specialism. Because the evolutionary dynamics of the host–pathogen interaction is dictated by the specificity of association between partners, generalists and specialists are likely to be shaped by differing ecological and environmental conditions, and thus distinct evolutionary pressures. This can have significant consequences on the evolutionary trajectory of their pathogenicity and overall lifestyle.

Generalists

Generalist or multi-host pathogens are capable of infecting and exploiting a variety of different plants. By having a wealth of hosts at their disposal, generalists may be presumed to have an enormous advantage, and are therefore presumed to be more successful. There are, however, several costs associated with this host promiscuity, the first of which is the requirement and deployment of numerous infection, exploitation and transmission strategies, required for contending with the inherent differences in the biology and immunity of different plant hosts (Gandon, 2004). Second, because generalists infect genotypically variable hosts, they may not achieve a single virulence optimum, which is defined as the level of virulence that maximizes aggressiveness and transmission level (Woolhouse et al., 2001). A highly aggressive generalist pathogen may be capable of colonizing all encountered hosts, but will suffer significantly reduced transmissibility if it surpasses the tolerance threshold of its host and induces lethality. Despite these ecological drawbacks, it is particularly interesting that the vast majority of pathogens have been suggested to be generalists (Malpica et al., 2006; Woolhouse et al., 2001). Plant pathogenic species, such as *P. syringae*, *X. campestris*, and *P. agglomerans*, are often further subdivided into pathovars ('pathogenic variety') to indicate host specialization, yet are not considered generalists. They are, however, considered to have a very broad host range when the bacterial species is considered as a single group. In contrast, individual strains of the *Ralstonia solanacearum* (formerly *Pseudomonas solanacearum*, and *Burkholderia solanacearum*) can colonize and infect several plant hosts, often from different plant families (Hayward, 1991). The infection strategy of *R. solanacearum* is notably different from that of most plant pathogens in that colonization is initiated at the plant root instead of the aerial tissues, often at wounds or sites of secondary root emergence (Hayward, 1991). *R. solanacearum* responds chemotactically to specific amino acids and organic acids of certain plant exudates, propelling itself toward its plant host with its flagellum (Yao and Allen, 2006). Flagellar mutants lose their pathogenic potential, as they are unable to locate their plant hosts; however, they retain full pathogenicity if introduced directly into the plant (Tans-Kersten et al., 2001). Following colonization, the bacterium multiplies and spreads through the vascular system, resulting in rapid wilting and host death.

R. solanacearum exhibits substantial ecological diversity and has been subdivided into five races based on its host range, and six biovars based on its ability to utilize specific nutrient sources (Hayward, 1991). The nutritionally versatile race 3 (biovar 2), presumed to have originated in South America is pathogenic to members of the Solanaceae including potato, tomato, pepper, and eggplant, as well as the common garden geranium (Hayward, 1991; Ji et al., 2007; Swanson et al., 2005). Race 2 (biovar 1) evolved in Central America, and causes Moko disease of banana (Buddenhagen, 1960). Because of this geographically wide distribution, broad host range, and extreme versatility, *R. solanacearum* is now a top containment pathogen for the pest management agencies of Canada, the United States, and the United Kingdom. So why is *R. solanacearum* such a successful generalist? A genomic comparison of the tomato strain *R. solanacearum* GMI1000 (race 1/biovar 3) and the potato/tomato/geranium strain UW551 (race 3/biovar 2) does not provide much insight, as these two strains have very few differences in their suites of pathogenicity factors, although both appear to contain all six secretion pathways (Gabriel et al., 2006; Genin and Boucher, 2004; Salanoubat et al., 2002). Of these secretion systems, only two, the T2SS and T3SS, have been directly implicated in pathogenicity and host-association (Boucher et al., 1985; Kang et al., 1994). The GALA7 T3SE, for example, which has been named for the motif that characterizes the entire protein family, has been shown recently to mediate the host-specific interaction of *R. solanacearum* with *Medicago truncatula* (Angot et al., 2006), but this does not preclude the possibility that other secretion mechanisms may contribute to the broad host-associating capabilities of the pathogen.

It is tempting to speculate that the success of this pathogen is not due to any one set of pathogenicity factors, but rather a combination of its unique mode of pathogenesis, lifecycle, and evolutionary origins. Because the induction of host lethality appears to be a major component of its life cycle, its aggressiveness and virulence potential may not have an upper bound. Following host death, the pathogen is rapidly returned to the soil, where remarkable degradative abilities (also characteristic of its kin in the β-Proteobacteria), allow it to survive long term (Genin and Boucher, 2004). Also an enormous benefit is its ability to colonize multiple perennial weeds in the families Solanaceae, Euphorbiaceae, Asteraceae, and Portulacaceae, albeit asymptomatically, enabling its survival until it reaches its next host (Hayward, 1991). Taken together, these attributes may contribute substantially to the success of R. solanacearum as a generalist, the validation of which will require a thorough understanding of its virulence potential, trade-offs between host association and soil-dwelling stages, host range, and life history. The pathogenicity of R. solanacearum is examined further in Chapter 5.

While R. solanacearum exhibits a broad plant host range, the generalist and opportunistic P. aeruginosa has an even broader host range, exhibiting cross-kingdom pathogenicity. P. aeruginosa has been shown to successfully infect humans, nematodes, and fruit flies, in addition to plants (Rahme et al., 1995; Rahme et al., 1997). The plant host range of P. aeruginosa does not appear limited, since it has been shown to colonize lettuce, alfalfa, and even the model organism, Arabidopsis thaliana (Plotnikova et al., 2000; Rahme et al., 1995; Rahme et al., 1997). Colonization strategies of P. aeruginosa appear similar to most specialized plant pathogens, characterized by attachment to the leaf surface and aggregation around stomatal openings, entry and colonization of the intercellular spaces, disruption of plant cell walls and membranes, and local and systemic infection that can result in the rotting of tissues and plant death (Plotnikova et al., 2000). The specific virulence factors that contribute to its colonizing abilities of plants also appear to be essential for animal pathogenicity, since toxA, gacA, plcS, dsbA, and mucD mutants of P. aeruginosa exhibit reduced virulence in both plant and animal models (Prithiviraj et al., 2005; Rahme et al., 1995; Rahme et al., 1997; Yorgey et al., 2001). Thus, the generalist nature of P. aeruginosa could be indicative of a transitionary plant and animal pathogenic state, whereby the acquisition or maintenance of key determinants may confer a broad host range. The biology of P. aeruginosa demonstrates, rather explicitly, that with only a limited subset of virulence factors, pathogens can effectively exploit plants with the possibility of becoming increasingly or even decreasingly specialized. This has significant repercussions for understanding the underlying genetic basis for both host specificity and host switches.

Specialists

Specialist pathogens adapt to exclusively infect one or several closely related plant species, a strategy that may be driven by niche restriction as a result of resource limitation due to patchy distribution of appropriate hosts. The inability of a pathogen to associate with its host may drive the formation of alternative, yet inefficient, associations that permit provisional pathogen survival. Following acquisition of the appropriate genetic material, the bacterium may eventually evolve an increased specificity for that particular host. Over time, the repeated colonization of the host by the pathogen may lead to a steady refinement of the pathogen genetic repertoire and the ability to exploit the host more efficiently.

Evolution has been suggested to favour host specialism, largely due to the fitness costs associated with generalism in any one environment (Woolhouse et al., 2001). The more rapid evolution experienced by specialists in a narrow niche results in a higher probability of fixing beneficial alleles, as well as fewer deleterious alleles drifting to fixation and a lower frequency of deleterious alleles exhibiting mutation-selection balance (Whitlock, 1996). Furthermore, one of the main advantages of host specialization over generalism is the ability of pathogens to achieve optimum virulence while maximizing their transmission potential, a result of coevolutionary processes between pathogen and host. Despite this, specialist pathogens can find themselves at an enormous disadvantage, since their fate is tied directly to one host (Woolhouse et al., 2001).

The ability of a pathogen to interact with particular hosts is often dictated by specific genetic factors, referred to as host specificity determinants, which are often assumed to be virulence-associated proteins that target and exploit specific substrates in a particular host or group of hosts. Host specificity, as discussed here, will not include cultivar-level specificity (the ability of a pathogen to colonize select cultivars of a particular plant species), although many studies examining this have contributed substantially to our understanding of host–pathogen coevolution (Cournoyer et al., 1995; Hunter and Taylor, 2006; Jackson et al., 1999; Jackson et al., 2000; Tsiamis et al., 2000). *P. agglomerans* has been shown to be an excellent system for examining the genetic determinants that govern host-specific interactions. *P. agglomerans* pv. *betae* induces galls on both beet and gypsophila (baby's breath), while *P. agglomerans* pv. *gypsophilae* induces galls on gypsophila but triggers a defence response on beet (Nissan et al., 2006). The homologous T3SEs HsvG (Valinsky et al., 1998) and HsvB (Nissan et al., 2006) are involved in the host specific colonization of gypsophila and beet, respectively, and contain intragenic direct repeats that confer host-specific interactions with each of these hosts. Two direct repeats are required in HsvG to confer gypsophila pathogenicity, whereas only one is required in HsvB for pathogenicity on beet. A switch in host specificity was observed when the repeats of these two proteins were exchanged. Further experimentation revealed that these proteins localize to the plant cell nucleus, and may be responsible for interfering with plant transcription.

P. syringae, without a doubt, has been one of the most intensely studied pathogens with respect to its host-specific interactions with various plant species (see chapter by Arnold et al). Although each strain exhibits a relatively restricted host range, the species as a group can infect an inordinate number of economically relevant crop plants and fruit trees from the families Actinidiaceae, Amaranthaceae, Betulaceae, Brassicaceae, Cucurbitaceae, Fabaceae, Moraceae, Myricaceae, Oleaceae, Poaceae, Rosaceae, Solanaceae, and Theaceae (Sarkar and Guttman, 2004; Wang et al., 2007). Comparative genomic hybridization of an array of 353 virulence-associated genes from *P. syringae*, which included the components of the T3SS, revealed the association of specific T3SEs with the colonization of a particular host (Sarkar et al., 2006). For example, the T3SEs *hopY1*, *hopL1*, *hopT1-2*, *hopX1*, *avrPto1*, *hopR1*, *hopB1*, *hopAK1*, *hopP1*, and *hopO1-3* were significantly associated with isolates pathogenic to tomato, suggesting they may function as host-specific virulence genes. This study also illustrated substantial diversity among the genes encoding the T3SS for a defined set of host-specific strains, implicating the secretion system itself in host-specific adaptation.

The T3SS in other pathosystems has also been implicated as an important determinant of host specificity. Two distinct types of T3SS PAIs, segregating as presence/absence polymorphisms, were identified in populations of the plant pathogen *P. viridiflava* (Araki et al., 2006). A population-level survey revealed that all strains contained a single PAI to the exclusion of the other. One PAI was shown to be associated with enhanced virulence on tobacco and decreased virulence on *Arabidopsis*, while the other confers the opposite phenotypes. These PAIs appear to be maintained by balancing selection, and the host-specific patterns associated with each PAI indicate that this may be due to host distribution. Thus, the T3SS, in addition to its secreted repertoire may be important host-specifying determinants.

Specialism may be both driven and maintained by various genetic and ecological factors related to pathogen dispersal. Many bacterial pathogens rely on specific mechanisms for locating and reaching specific hosts. The tumorigenic *A. tumefaciens* exhibits chemotaxis towards acetosyringone produced by wounded plants, which also functions to induce multiple Ti plasmid-encoded virulence factors (Ashby et al., 1987; Parke et al., 1987; Shaw et al., 1988; Stachel et al., 1985). Some dispersal mechanisms are more complex, and are mediated by other organisms. Herbivorous insects, such as leafhoppers, planthoppers, psyllids, thrips, beetles, and bees often ferry plant pathogens, carrying and delivering the bacteria directly onto and sometimes into their definitive host plants (Garnier et al., 2001; Mitchell, 2004). *Xylella fastidiosa*, the causal agent of Pierce's disease, is vectored primarily by leafhoppers known as sharpshooters, which inject *X. fastidiosa* directly into the xylem vessels of grape vines during feeding (Hopkins, 1989; Purcell and Hopkins, 1996). The bacteria subsequently multiply in these vessel elements, disrupting water flow to the leaves and causing severe wilting. The association of *X. fastidiosa* with its sharpshooter vector is mediated by the *X. fastidiosa* gene *rpfF* (Newman et al., 2004).

X. fastidiosa rpfF mutants are unable to form biofilms inside the sharpshooter, and are rapidly shed. In an interesting twist, *rpfF* mutants are hypervirulent on plants, suggesting that the *rpfF* gene also functions to attenuate virulence or metabolic load on the insect host, further reinforcing the notion that the relatively tight association of *X. fastidiosa* with its insect vector may have established or maintained pathogen–host specificity. Additional information on the pathogenicity of *X. fastidiosa* can be found in the chapter by van Sluys *et al.*

Host immunity and defence

Generalist and specialist pathogens alike present plants with enormous challenges, prompting plants to evolve effective surveillance and defence mechanisms designed to thwart pathogen (or potential pathogen) colonization. Innate immunity, the first line of defence against pathogens, is a combination of non-specific, preformed and induced defences that provide an immediate response against a foreign body (da Cunha *et al.*, 2006; Jones and Dangl, 2006; Nurnberger and Brunner, 2002; Nurnberger *et al.*, 2004). Innate immunity functions to detect the mere presence of potential pathogens, often via the recognition of conserved microbe-associated molecular patterns, or MAMPS (originally referred to as pathogen-associated molecular patterns, PAMPs) (Abramovitch *et al.*, 2006; Chisholm *et al.*, 2006; Nurnberger and Brunner, 2002). MAMPs are generally intrinsic components of the bacterial cell, such as the flg22 epitope of flagellin, harpins, cold shock proteins, lipopolysaccharides, peptidoglycans, and the elongation factor Tu (EF-Tu) (Kunze *et al.*, 2004; Nurnberger *et al.*, 2004). Pathogen recognition receptors on the host cell surface perceive MAMPs, triggering a basal defence response that prevents pathogen colonization (Nurnberger and Brunner, 2002; Nurnberger *et al.*, 2004; Shan *et al.*, 2007). PAMPs/MAMPs are discussed further in the chapter by Erbs and Newman.

The innate immune response, however, is often easily overcome by specialist pathogens (He *et al.*, 2007), largely due to host–pathogen coevolution that provides more opportunity to evolve new strategies to surmount host protective barriers. Resistance (R) proteins in plants have evolved to detect the activity of specific virulence proteins that are injected directly into the plant cell, activating a rapid, localized cell death known as the hypersensitive response (HR) (HammondKosack and Jones, 1997; Heath, 1998; Heath, 2000; Martin *et al.*, 2003; McDowell and Simon, 2006; Shan *et al.*, 2007; Staskawicz *et al.*, 1995; Tang *et al.*, 1996; Truman *et al.*, 2006). The HR is often accompanied by callose deposition, expression and accumulation of pathogenesis-related proteins, and the production of reactive oxygen species, all of which may be produced at the site of bacterial infection (Grant *et al.*, 2006; Heath, 1998; Heath, 2000). Despite this multi-tiered plant defence system, pathogens still continue to evolve strategies for overcoming these barriers, instituting the long-term co-evolutionary arms race with their plant hosts.

The co-evolutionary arms race

The 'arms race' describes the dynamic struggle that occurs between pathogen and host, whereby pathogens are constantly evolving new pathogenicity traits, and hosts evolving corresponding defences (Fig. 1.3; see chapter by Boch) (Dawkins and Krebs, 1979; Stahl and Bishop, 2000). This interaction is most prominent among specialist pathogens, which have a propensity for long-term coevolutionary associations with their plant hosts (Gandon, 2004; Whitlock, 1996; Woolhouse *et al.*, 2001). The selective forces that result from this long-standing interaction are seen most prominently at the molecular level in specific bacterial pathogenicity and virulence factors. These genetic factors may evolve by positive Darwinian selection (the generation and spread of alternate favourable alleles in the population), or diversifying selection (the stable maintenance of multiple alleles in the population), either of which may enable the pathogen to escape detection. Similar signatures of selection can be detected in several components of the plant defence or surveillance system, reflecting the coadaptive responses of the host plant to the pathogen.

Molecular signatures of selection in the type III secretion system

The T3SS is a necessary and indispensable component of the host colonization strategy for many plant pathogens. During pathogenesis, the T3SS pilus is formed by the polymerization of

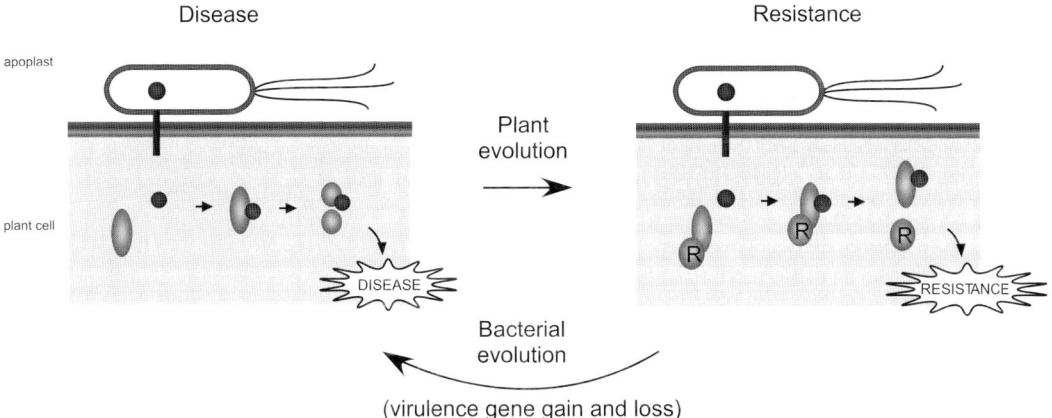

Figure 1.3 Evolution of the host–pathogen interaction and the coevolutionary arms race. Left panel. Phytopathogenic bacteria use a type III secretion system to inject virulence proteins (dark circle) directly into the host cell. These virulence factors target and modify specific host substrates, thereby facilitating disease. Right panel. Over time, plants may evolve resistance proteins (circle labelled 'R'), which function to guard host substrates (oval). Modification of the host target results in R protein-mediated defence induction and plant resistance. In response, pathogens evolve the ability to counter the host defences, permitting successful colonization and the continuation of the cycle.

the protein encoded by *hrpE* in *X. campestris* and *hrpA* in *P. syringae* (Roine et al., 1997; Weber et al., 2005). Portions of this protein, which constitute the outer surface of the pilus, are completely exposed during pathogenesis, and are therefore likely to come into contact with components of the host defence system. A Ka/Ks selection test on the *X. campestris hrpE* gene reveals positively selected sites at the 5′ end of the gene (Roine et al., 1997; Weber and Koebnik, 2006), which correspond to hydrophilic amino acids that may be positioned on the exposed side of the outer pilus, and are therefore in direct contact with the host cytoplasmic milieu. Likewise, an analysis of *hrpA* from 22 strains of *P. syringae* revealed a pattern of genetic variation consistent with diversifying selection (Guttman et al., 2006). In this context, different *hrpA* alleles may be beneficial in different hosts, leading to selection for a diversity of alleles at this locus. The presence of patterns consistent with positive and diversifying selection in the genes responsible for major components of the T3SS are indicative evolutionary changes that are the consequence of the imposition of host-induced selective pressure.

Evolution of type III secreted effectors by HGT and pathoadaptation
The intimate interaction of T3SEs themselves with the intracellular environment of the plant cell subjects them to immense selective pressures

(Arnold et al., 2007; Pitman et al., 2005), requiring rapid evolution to remain ahead of the molecular arms race. Modification of the pathogen T3SE repertoire or diversity is achieved through several different mechanisms, including HGT and pathoadaptation (Fig. 1.4A and B). There are innumerable studies documenting the HGT-mediated acquisition of T3SEs, many providing excellent characterization of the signatures of HGT (Arnold et al., 2001; Kim et al., 1998; Rohmer et al., 2004). Because HGT has been reviewed extensively, pathoadaptive change will be examined here in greater detail. Pathoadaptation refers to the evolution of virulence-associated genes through the introduction of single nucleotide polymorphisms (SNPs), as well as insertions and deletions (indels) (Sokurenko et al., 1998; Sokurenko et al., 1999; Weissman et al., 2003). A significant proportion of studies examining pathogen evolution through pathoadaptive change have been centred around the model plant pathogen *P. syringae*, largely due to its vast repertoire of known and functionally characterized T3SEs (Chang et al., 2005; Collmer et al., 2002; Fouts et al., 2002; Guttman et al., 2002; Lindeberg et al., 2006; Lindeberg et al., 2005; Petnicki-Ocwieja et al., 2002; Schechter et al., 2004). For example, the wild-type allele of the *P. syringae* T3SE HopX T3SE (formerly AvrP-phE) normally induces an R protein-mediated defence response in bean plants; however, seven

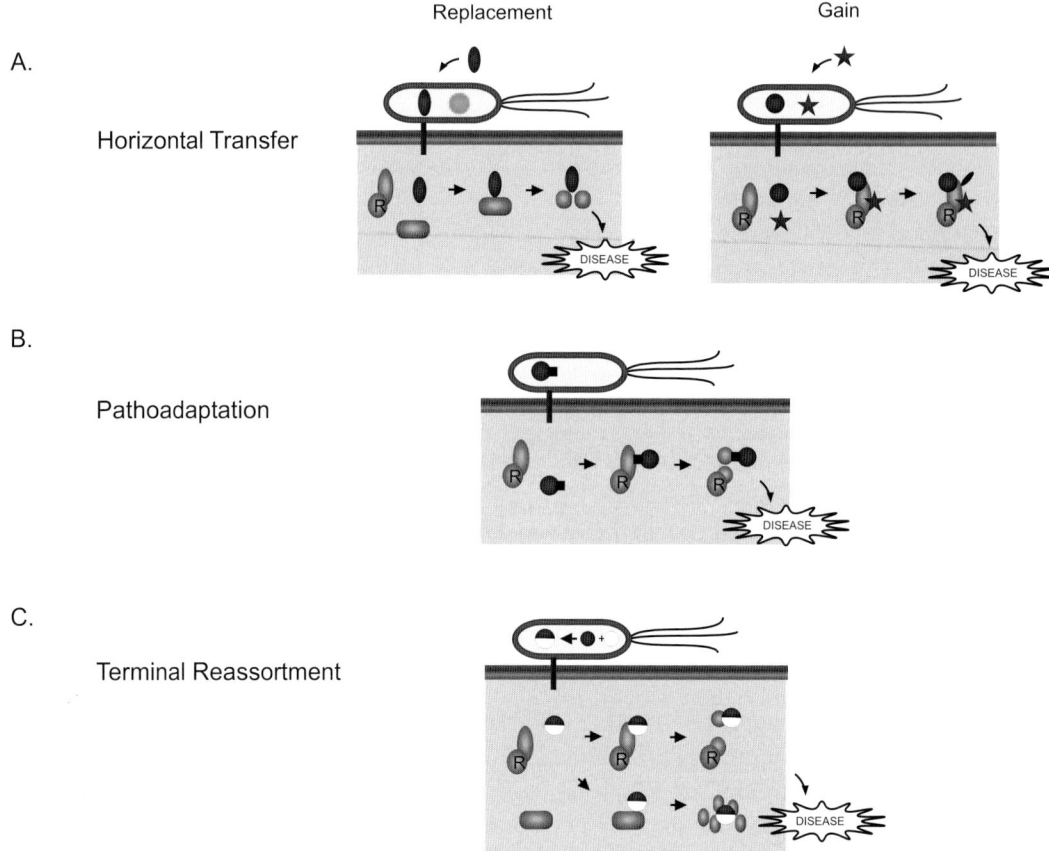

Figure 1.4 Pathogens may respond to host resistance by modifying their suite of type III effectors. (A) Horizontal transfer may facilitate the replacement of the detected effector (faded circle in bacterium) by a different protein (dark oval), which may target an alternate host protein in the same pathway (rectangle). Horizontal transfer may also lead to gain of a new effector (star) that modifies the same substrate, thereby preventing recognition of the original effector. (B) Pathoadaptive changes (indicated as a black square) in the original effector may alter the biochemical specificity of the effector, leading to modification of the host substrate in a manner that does not trigger R protein-mediated defences. (C) Terminal reassortment may drive the exchange of protein domains between two type III effectors (black circle and white circle), forming a chimeric type III effector (half black/half white circle) with an activity that modifies the host substrate, but does not trigger a defence response. Alternatively, the chimeric effector may have a entirely unique biochemical function, targeting a different substrate (rectangle) in the host cell and modifying it to promote disease.

natural allelic variants were identified, including alleles with indels and amino acid substitution, which are no longer recognized by the host (Stevens et al., 1998). An experimental evolution study has also demonstrated the importance of pathoadaptation in response to host-imposed selective pressure. The T3SE HopAR1 (formerly AvrPphB) from *P. syringae* pv. *phaseolicola* 1302A is normally recognized by the *R3* resistance gene, but following *in planta* serial passaging in a host containing the *R3* resistance gene, the *hopAR1* gene is rapidly lost (Pitman et al., 2005). Loss is correlated with the excision of a ~106 kb genomic island, PPHGI-1, which contains 100 predicted open reading frames, including the T3SE *hopAR1*. T3SEs of *P. syringae* are discussed further in the chapter by Boch.

The *R. solanacearum* T3SE *avrA* has been shown to be an important virulence factor whose evolution has been impacted by host-driven pathoadaptive change. The presence of the intact *avrA* gene is correlated with the ability to cause disease on tomato and the elicitation of the HR on tobacco (Robertson et al., 2004). Strains capable of colonizing both hosts were found to carry *avrA* alleles disrupted by one of two miniature transposable elements. Introduction of the intact *avrA* allele into these strains reduced their virulence on both tobacco and tomato, illustrating that pathoadaptive change in *avrA* facilitates host-specific colonization. Interestingly, *avrA* gene inactivation may not be necessarily permanent, and could potentially revert to its functional state under the appropriate conditions. Such mutational reversions have been demonstrated in the *R. solanacearum* pathogenicity factor regulator *phcA*, such that non-pathogenic *phcA* mutants can revert to their pathogenic form *in planta* by pathoadaptive elimination of indels, transposons, and even tandem duplications disrupting the open reading frame (Poussier et al., 2003). The evolutionary plasticity is evident in T3SEs and other virulence genes, with pathoadaptive change leading to inactivation and reactivation of virulence genes that play a significant role in host adaptation.

Pathoadaptation has also been implicated in the evolution of the *Xanthomonas* AvrBs3 T3SE family whose members include AvrXa7, AvrXa10, Avrb6, Hax and Pth, and are characterized by several intragenic tandem repeats (Herbers et al., 1992; Yang et al., 2005; Yang and Gabriel, 1995). The allele of *avrXa7* from the rice pathogen *X. oryzae* pv. *oryzae* induces an HR in rice plants bearing the cognate R gene *Xa7*, but enhances virulence on susceptible plants. Alterations in the direct repeats of *avrXa7* was shown to dramatically influence the virulence phenotype, with some mutant alleles retaining virulence capabilities on a susceptible host while simultaneously losing the ability to be recognized by Xa7 (Yang et al., 2005). Field studies of natural populations of *Xanthomonas axonopodis* pv. *vesicatoria* have revealed similar trends, with the AvrBs2 T3SE that is normally recognized by the R protein Bs2 (Minsavage et al., 1990), evolving through pathoadaptive change in response to the mass introduction of the *Bs2* R gene into cultivated pepper and tomato varieties. Surveyed allelic variants were largely inactivations, caused by transposition events or internal indels (Wichmann et al., 2005).

While the examples above appear to be influenced largely by one evolutionary force, the *P. syringae* HopZ family appears to have been shaped by both pathoadaptive change and HGT. The HopZ family of *P. syringae*, which comprise three homology groups widely distributed within the *P. syringae* species complex (Ma et al., 2006), are members of the YopJ family of cysteine proteases/acetyltransferases found in both plant and animal pathogens (Ciesiolka et al., 1999; Mukherjee et al., 2006; Orth et al., 2000). These include HopZ1 (formerly known as HopPmaD and HopPsyH), HopZ2 (formerly AvrPpiG) and HopZ3 (formerly HopPsyV) (Lindeberg et al., 2005), all of which retain similar biochemical function but exhibit host-specific patterns of recognition. A phylogenetic analysis of this group reveals that the HopZ1 homologue is ancient in *P. syringae*, while both HopZ2 and HopZ3 appear to have been acquired by *P. syringae* via horizontal transfer from other phytopathogenic bacteria. Being evolutionarily old, HopZ1 has evolved at least three functional alleles and several degenerate forms through pathoadaptation. The introduction of the most ancestral allele (*hopZ1a*) into strains harbouring alternate or degenerate alleles results in an R protein-mediated defence response in their respective hosts, suggesting that host-imposed selective processes have led to the diversification of HopZ1. The recognition of specific HopZ1 alleles by the host may have favoured their substitution with functionally equivalent effectors or pathoadaptive change, allowing the pathogen to escape detection. Clearly, the genetic patterns manifested in T3SEs are often prominent molecular signatures of the coevolutionary interactions between the pathogen and its hosts.

Reassortment as an alternative to HGT and pathoadaptation

Although HGT and pathoadaptation remain the prominent methods by which pathogens generate virulence-associated variation, the reassortment of the existing T3SE complement may be an even more efficient means to create new virulence genes (Fig. 1.4C). A recent analysis of all T3SEs from both plant and animal pathogens suggested that one evolutionary process may enable bacteria to rapidly evolve new T3SEs with distinct biochemical function. Terminal reassortment (TR) is a process whereby functional modules are reassorted in a manner analogous to exon shuffling (Stavrinides et al., 2006). This process relies on the modularity of T3SEs, which are characterized by an N-terminal secretion and translocation domain that directs the effector to the T3SS for export, and at least one C-terminal functional domain (Collmer et al., 2002; Guttman and Greenberg, 2001; Lloyd et al., 2001; Mudgett et al., 2000; Petnicki-Ocwieja et al., 2002; Sory et al., 1995; Sory and Cornelis, 1994). In addition, most T3SEs possess a regulatory region immediately upstream that permits coordinate activation of gene expression with the T3SS. In TR, the promoter and 5′ portion of a T3SE may be fused to a different genomic region through the action of mobile elements, indels, and random genomic rearrangements, creating a new fusion protein. This was supported by an analysis of the T3SEs from 23 pathogenic and mutualistic bacterial species, which revealed that on average, 24% of all T3SE families contain chimeric T3SEs. Also identified were apparently truncated T3SEs, usually composed of just an effector N-terminus, which were referred to as an orphaned effector terminus (ORPHET). The number of ORPHETs and chimeric effectors found in a species is significantly correlated with the number of T3SEs carried by that species, and the proportion of chimeras and truncations found among T3SEs (24%) is significantly greater than that seen among non-T3SEs (7%). Much like any stochastic process, TR is presumed to occur relatively frequently, but only fusions that confer a selective advantage to the bacterial host will persist. The stronger impact of TR on T3SEs over other genes is likely due to T3SE modularity, central role in host–pathogen interactions, and their frequent association with mobile elements (Arnold et al., 2001; Kim et al., 1998).

Escalatory tactics in the host–pathogen interaction

There are numerous examples of host–pathogen molecular warfare, and the escalatory tactics employed by each symbiont to maintain ascendancy (Fig. 1.3). The colonization of rice by *X. oryzae* requires several pathogenicity determinants, including cellulase (ClsA), cellobiosidase (CbsA), and lipase/esterase (LipA), all of which are secreted in a type II-dependent manner (Jha et al., 2007). Infection of the rice plant by a T3SS-deficient mutant of *X. oryzae* results in a rapid defence response, characterized largely by callose deposition, cell death, and acquired resistance, and attributed to the activities of the three type II-secreted enzymes. Thus, the T3SS of *X. oryzae* and its effectors can suppress recognition of the essential ClsA, CbsA, and LipA proteins, although the specific T3SEs involved in this process have yet to be identified. Despite this, the cooperative nature of these two distinct secretion systems demonstrates the escalation of colonization tactics instituted by the pathogen. Similarly, colonization of the tomato leaf surface by non-plant pathogens, such as *E. coli*, triggers a MAMP-dependent defence response characterized by rapid closure of stomata (Melotto et al., 2006). A similar response is observed when the tomato pathogen *P. syringae* pv. *tomato* DC3000 colonizes the leaf tissue; however, *P. syringae* counteracts the host response with the phytotoxin coronatine, a plant-hormone mimic (Bender et al., 1999; Weiler et al., 1994) that induces stomatal reopening (Melotto et al., 2006).

One of the most striking examples of the escalatory tactics comes from the interaction between *P. syringae* and *Arabidopsis*. The interaction of the *P. syringae* T3SE AvrRpm1 with the R protein RPM1 of *Arabidopsis* induces an HR (Dangl et al., 1992), preventing pathogens carrying the *avrRpm1* T3SE from successfully colonizing RPM1-carrying plants. In response to this, pathogens have evolved a strategy for silencing R gene-mediated defence signalling through the activity of a different T3SE, AvrRpt2. AvrRpt2 has been shown to completely block the development of the HR that normally follows the interaction between the AvrRpm1 and RPM1 by modifying the host factor RIN4 (Dangl et al., 1992) (Fig. 1.4A). Additional T3SEs have also been shown to suppress, fully or partially, the R gene-mediated defence response (Jamir et al., 2004), demonstrating how the evolution of host surveillance systems has imposed reciprocal selective pressure on the pathogen to evolve new T3SEs capable of suppressing host defences. Plants, in turn, have responded to this pathogen tactic through R protein modification. Signatures of positive selection have been detected in R genes of plants, with evidence for positive selection in key residues responsible for interacting with

T3SEs, as well as a role for balancing selection in maintaining *R* gene allelic diversity (Bergelson *et al.*, 2001; Van der Hoorn *et al.*, 2002). In addition, numerous *R* gene pseudogenes have been identified, implicating a birth-and-death model of *R* gene evolution (Michelmore and Meyers, 1998). Although plants have evolved these detection systems, they are repeatedly trumped by the evolution of new traits that enable highly specific plant pathogens to successfully bypass host defences and initiate plant tissue colonization.

The emergence and re-emergence of infectious bacterial diseases

Given the ability of bacteria to respond rapidly to the selective pressures imposed by potential host plants, it is not surprising that many remain a significant threat to vital agricultural crops. As of January 2007, three bacterial species have been classified as priority pathogens for control and containment by the USDA: *R. solanacearum*, *Candidatus* Liberibacter asiaticus and *X. axonopodis* pv. *citri* (USDA, 2007). The last two citrus pathogens, in particular, exemplify the importance of not only the evolutionary potential of phytopathogens, but also the often careless agricultural practices that contribute to the rapid and irreversible spread of bacterial diseases.

Huanglongbing – *Candidatus* Liberibacter asiaticus

Huanglongbing (HLB), also known as greening disease of citrus, is caused by the genus *Candidatus* Liberibacter (Garnier *et al.*, 2000), a group of uncultured, phloem-restricted bacteria that infect almost all major citrus fruit trees, with sweet oranges, mandarins, and mandarin hybrids being most affected (Bove, 2006). The genus has been given the designation '*Candidatus*' to reflect its description as an uncultured species. The disease is highly catastrophic, since it not only affects tree health, but also gives fruit an acrid taste, making it completely unpalatable and wholly unmarketable.

HLB is considered one of the oldest diseases of citrus, being first described in China in 1919, where it was named for the characteristic dragon-like appearance of infected trees (Huanglongbing means 'yellow dragon disease') (Bove, 2006; Polek, 2007). Historical accounts also place it in the Philippines (1921), South Africa (1928), Thailand (1960s), India (1963), and Indonesia (1965), although under various disease names. In South Africa, HLB was called 'greening disease', since fruit on infected trees did not ripen, while the same disease was referred to in Indonesia as 'vein phloem degeneration', due to abnormal pockets of necrotic phloem that accumulated in mature leaves (Tirtawid *et al.*, 1965). Despite its broad distribution, there are three disparate species recognized today: *Ca.* Liberibacter asiaticus, *Ca.* Liberibacter africanus, and *Ca.* Liberibacter americanus (Fig. 1.5), with the latter being identified in Brazil in 2004.

HLB is vectored by psyllids insects, which inject the bacterium directly into phloem during feeding. In Asia and America, transmission of HLB is mediated by the Asian citrus psyllid, *Diaphorina citri* (Hung *et al.*, 2004), while in Africa, this role is assumed by the African citrus psyllid vector, *Trioza erytreae*. Interestingly, each vector-bacterium pair also exhibit varying degrees of heat tolerance, which significantly influences their epidemiology and geographic distribution. The Asian species of HLB and its insect vector *D. citri* prefer temperatures below 30°C, although they are largely heat tolerant, capable of withstanding temperatures well above 30°C. In contrast, the African HLB bacterium and psyllid are heat intolerant (Hung *et al.*, 2004).

Ca. Liberibacter asiaticus presents the greatest threat to North American citrus. This is largely due to its heat tolerance, broad distribution, and general aggressiveness, along with the fact that its vector, *D. citri*, has already been introduced into Florida, Texas and Hawaii, as well as nearby Brazil and Mexico. In fact, in August 2005, HLB disease was discovered in Florida, only 7 years after the first report of *D. citri* being found in the area, prompting massive efforts to prevent HLB spread to other parts of the United States (Bove, 2006; Polek, 2007). Unfortunately, HLB is now well established in Florida, and is spreading northward from its initial introduction in Miami-Dade County (Fig. 1.6). To date, the state of California, which contributes about 40% to national citrus production, has not reported any incidences of HLB or the *D. citri* (Bove, 2006; Polek, 2007). The emergence and dissemi-

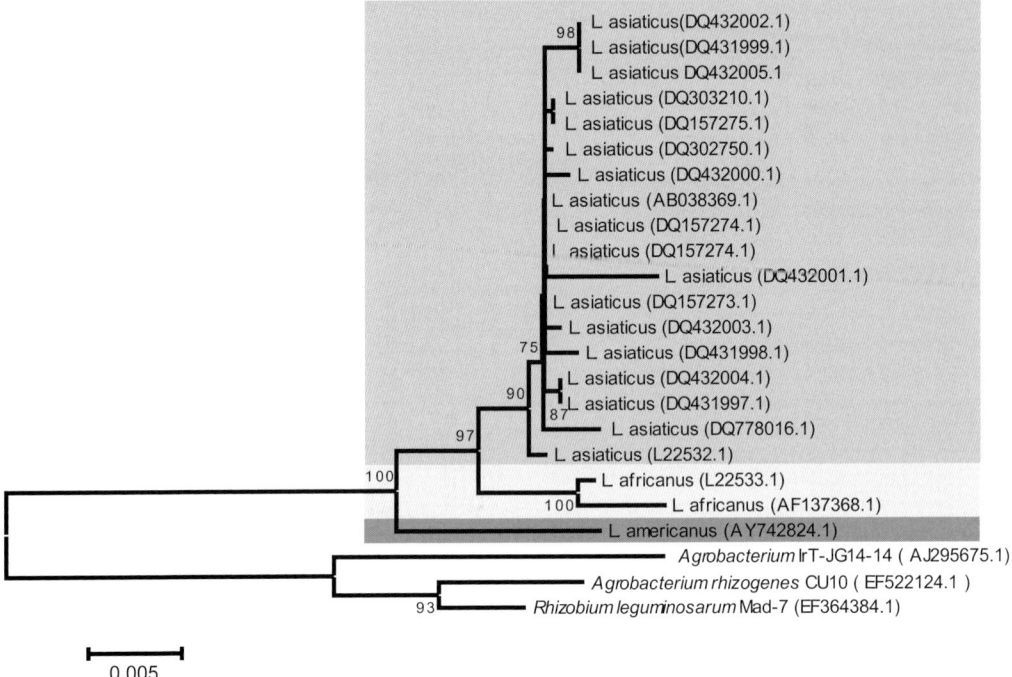

Figure 1.5 Phylogeny of *Candidatus* Liberibacter species inferred through 16S ribosomal DNA. Three major species groups, *Ca.* Liberibacter asiaticus, *Ca.* Liberibacter americanus, and *Ca.* Liberibacter africanus are highlighted as separate lineages, with members of the *Agrobacterium/Rhizobium* group serving as the outgroup.

nation of the HLB pathogen, however, is tightly linked to the fate of its insect vector, such that if *D. citri* becomes established in California, the citrus industry ought to brace itself.

Citrus canker – *Xanthomonas axonopodis* pv. *citri*

X. axonopodis pv. *citri* (*Xac*), the causal agent of citrus canker, is of particular importance to the $9.1 billion dollar citrus industry of Florida (Polek, 2007; Schoulties *et al.*, 1987; Schubert *et al.*, 1996; Schubert and Sun, 1996). *Xac* enters plant tissues through open wounds, gaining access to xylem elements and proliferating, and resulting in the formation of canker lesions and the premature shedding of fruits and leaves. The local and short-distance spread of *Xac* is mediated by wind-driven rain, which can carry a maximum inoculum of 10^8 bacterial cells per droplet (Schubert and Sun, 1996), for a distance of more than 580 m during storms (Anderson *et al.*, 2004). Anthropogenic practices facilitate long-distance transmission, usually through the movement of infected plant tissues and seedlings, or through contaminated farming machinery.

Xac has a relatively complex history in Florida, being introduced in the region around 1910 from contaminated rootstock imported from Japan (Gottwald *et al.*, 2002) (Fig. 1.6). By 1913, *Xac* had spread throughout the Gulf States, and northward to South Carolina. Drastic eradication efforts were undertaken to eliminate the disease, with the last infected tree being removed in 1927. In 1933, citrus canker was declared eradicated from Florida. In 1986, however, citrus canker re-emerged in both residential and commercial groves in Florida, prompting massive removal of both infected and potentially exposed trees. In 1994, citrus canker was declared eradicated, although in 1995 just one year later, citrus canker re-emerged in Miami initially infecting 50 square miles of residential citrus trees. By 2002, the infected area had grown to more than 600 square miles. Since 2004, the citrus industry saw hundreds of millions of dollars in losses due to an increase in storm activities, which facilitated widespread dispersal of *X. axonopodis* (USDA, 2007). In March 2006, the United States Animal and Plant Health Inspection Service declared that eradication of citrus canker was no longer

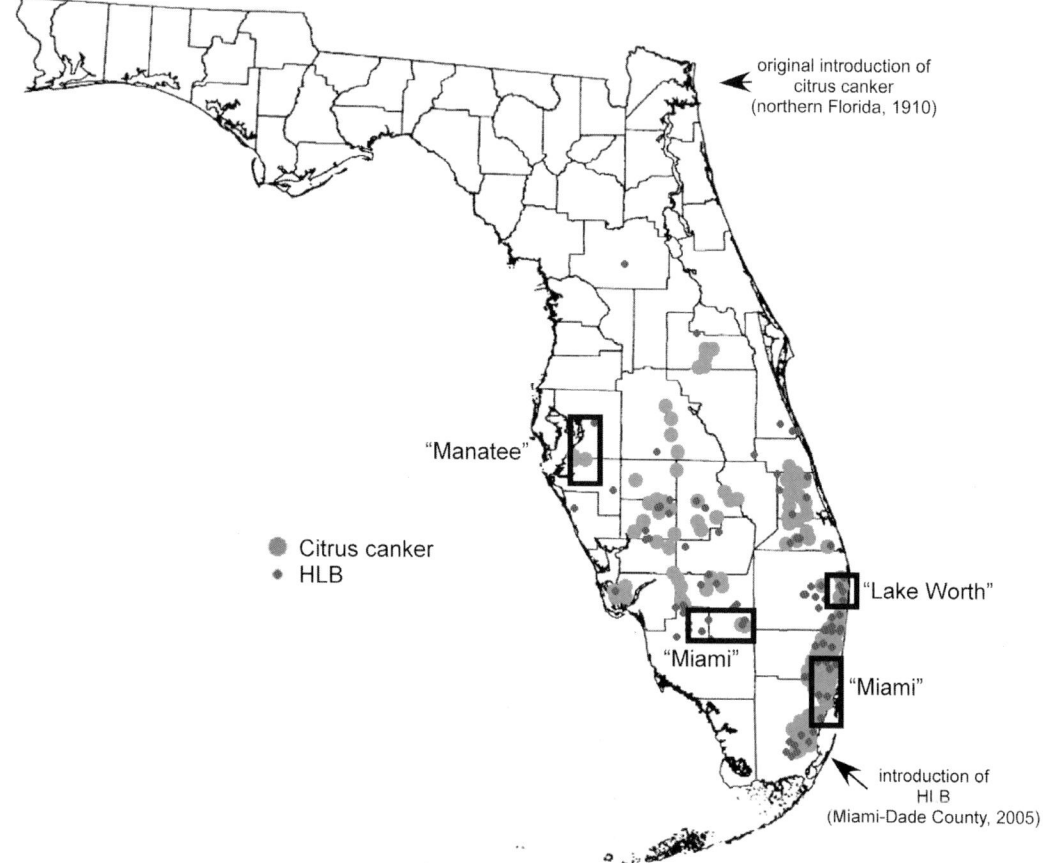

Figure 1.6 Current approximated distribution of bacterial citrus canker disease (*Xanthomonas axonopodis* pv. *citri*) and Huanglongbing (*Ca.* Liberibacter) in Florida. The original introduction of citrus canker occurred in northern Florida in 1910, whereas HLB was introduced in southern Florida in 2006. The 2002 distribution of the three distinct *X. axonopodis* strains, 'Manatee', 'Miami', and 'Lake Worth' is indicated by boxed regions (Gottwald et al., 2002). Current distribution maps are available from the USDA (2007).

possible given the degree to which it had spread, and instead, implemented disease management strategies that included specific procedures for citrus harvesting, handling, and transportation (USDA, 2007).

The repeated re-emergence of Xac prompted in depth analyses of its epidemiology and phylogeography, with the hopes of identifying the source of the new citrus canker outbreaks and preventing them. Were these recurrent appearances of citrus canker due to survival of the original strain in or on an intermediate host, or did they represent a simple case of a separate introduction of a new Xac strain? Molecular typing of the 1995 outbreak strain of Xac revealed a markedly different genotype from that of the 1986 outbreak (Schubert et al., 1996), and collective examination of all the various strains revealed at least three new introductions of Xac since the 1980s (Cubero and Graham, 2002). Some strains, however, like those of the 1997 outbreak, were shown to be identical to the 1986 strain through genetic fingerprinting (Cubero and Graham, 2002; Li et al., 2006; Mavrodieva et al., 2004). Survival of the original strain may have been facilitated by survival on an intermediate host, although it is interesting to note that the virulence factor $pthA$, which is essential for pathogenicity on citrus (Swarup et al., 1991), induces an HR on several other plants typically colonized by *Xanthomonas* species (Swarup et al., 1992); however, several *X. axonopodis* pathovars have been shown to be relatively good epiphytes, capable of enduring on the leaf surfaces of select hosts (Gent et al., 2005). To date, there are at least three distinct genotypically discernable Xac

strains in Florida (Cubero and Graham, 2002) (Fig. 1.6), each having a fairly defined host range. Thus, the epidemiological factors contributing to the repeated re-emergence of *Xac* included both new introductions of the pathogen, and survival of the original strain. Additional information on *Xanthomonas* and its pathogenicity is available in Chapter 2.

Conclusions

Phytopathogenic bacteria exhibit incredible diversity in both their host associations and plant colonizing strategies, attributable to both their unique ancestries and variable suites of pathogenicity factors. Although most phytopathogens are prevalent among the Proteobacteria, many are scattered among numerous phylogenetic lineages, reflecting the independent evolution of plant-colonizing capabilities, which have likely been mediated by the horizontal acquisition of pathogenicity determinants. Although horizontal transfer has long been recognized as the evolutionary force driving phytopathogen evolution, we understand that it serves mainly to disseminate extant pathogenicity determinants. The genesis of various pathogenicity and virulence determinants, however, is not as easily explained by a single process. Secretion systems, in particular, which play a central role in phytopathogenicity, appear to have involved recruitment and refinement of ancestral systems for use as primary weaponry in the exploitation of plant hosts. The T1SS is a simple secretion system, presumably derived from the association of membrane-associated, pore-forming proteins, which has continued to evolve through relatively minor steps to become increasingly optimized for the recognition and export of virulence factors. The evolution of the T2SS is somewhat more ambiguous, since it shares homology to the biogenesis pathway of type IV pili, but is otherwise unrelated to other ancestral systems. In contrast, the T3SS and T4SS are the direct result of modification and refinement of ancestral systems, being used almost exclusively for pathogenesis. Collectively, all four secretion systems and their suite of respective virulence factors contribute substantially to the phytopathogenicity of modern bacteria, and have likely contributed to the emergence and evolution of plant pathogens.

Extant phytopathogens, whether recently or anciently evolved, undergo highly complex and often persistent interactions with their hosts. The repeated association of two organisms and their ability to respond mutually to reciprocal selective pressure creates a dynamic molecular arms race that is manifested in the specific offensive and defensive armament of each partner. The creation and maintenance of novel genetic variation may enable one partner to gain dominance in the relationship, until the other partner evolves a counter-tactic to regain ascendancy. The result of this interaction is often manifested most prominently in the host and pathogen substrates that interact, such as pathogen T3SEs and their cognate host R proteins. The repeated colonization by a plant pathogen leads the host to evolve defence and surveillance systems to detect T3SEs and eliminate the bacterial invader. In response, the bacterium may acquire new T3SEs, or its pre-existing complement may be modified, enabling the bacterium to reinitiate infection. This endless cycle continues, establishing the coevolutionary arms race.

Bacteria remain highly adaptable, capable of responding swiftly to drastic changes in environmental conditions. This is particularly evident in case studies of emerging infectious plant diseases, which are constantly plaguing the agricultural industry. Attempts to eliminate disease by physical removal of infected plants may slow disease spread; however, as in the case of citrus canker, bacterial versatility likely permits the rapid formation of alternate host associations, thus ensuring bacterial survival until environmental conditions favour re-emergence of the pathogen. Unfortunately, bacterial re-emergence is frequently mediated by human practices, which often serve as the most efficient long-distance dispersal mechanisms. Once established, only the implementation of rigorously structured disease management strategies is practical, as attempted 'eradication' of these highly adaptable microorganisms will almost always be ineffective.

With only a small fraction of the current bacterial diversity having been surveyed, we will undoubtedly begin to uncover additional bacterial phytopathogenic species; however, given our relatively limited understanding of bacterial ecology, we may soon discover that an inordinate number of known or characterized bacterial species, currently considered benign saprotrophs or highly host-specific pathogens, also have phytopathogenic potential. The general ecology of such organisms will be particularly valuable, if we are to understand thoroughly the evolution of plant pathogenic bacteria. Finally, we must accept the fact that the evolutionary potential of microorganisms is immense, and despite our repeated and futile attempts to slow or arrest evolution, there is little doubt that phytopathogenic bacteria will repeatedly circumvent our rather simplistic control and containment strategies, continuing to establish and thrive on our most prized crop plants.

Acknowledgements

I would like to thank Drs David Guttman, Mark van Passel and Kerry Oliver for their valuable comments and suggestions, Patrick Degnan for figure critique, and Drs Richard Michelmore, Pamela Ronald, Lori Burrows and Mark Woolhouse for providing insight into various topics. I would also like to thank the editor, Dr. Robert Jackson for providing me with this opportunity to explore most thoroughly, the extraordinary realm of bacterial phytopathogens. I am currently supported by a Natural Sciences and Engineering Research Council of Canada Postdoctoral Fellowship.

References

Abe, M., Kawamura, R., Higashi, S., Mori, S., Shibata, M., and Uchiumi, T. (1998). Transfer of the symbiotic plasmid from *Rhizobium leguminosarum* biovar trifolii to *Agrobacterium tumefaciens*. J. Gen. Appl. Microbiol. 44, 65–74.

Abramovitch, R. B., Anderson, J. C., and Martin, G. B. (2006). Bacterial elicitation and evasion of plant innate immunity. Nat. Rev. Mol. Cell Biol. 7, 601–611.

Abramovitch, R. B., and Martin, G. B. (2004). Strategies used by bacterial pathogens to suppress plant defenses. Curr. Opin. Plant Biol. 7, 356–364.

Alfano, J. R., Charkowski, A. O., Deng, W. L., Badel, J. L., Petnicki-Ocwieja, T., van Dijk, K., and Collmer, A. (2000). The *Pseudomonas syringae* Hrp pathogenicity island has a tripartite mosaic structure composed of a cluster of type III secretion genes bounded by exchangeable effector and conserved effector loci that contribute to parasitic fitness and pathogenicity in plants. Proc. Natl. Acad. Sci. USA 97, 4856–4861.

Alfano, J. R., and Collmer, A. (1996). Bacterial pathogens in plants: Life up against the wall. Plant Cell 8, 1683–1698.

Anderson, P. K., Cunningham, A. A., Patel, N. G., Morales, F. J., Epstein, P. R., and Daszak, P. (2004). Emerging infectious diseases of plants: pathogen pollution, climate change and agrotechnology drivers. Trends Ecol. Evol. 19, 535–544.

Andro, T., Chambost, J. P., Kotoujansky, A., Cattaneo, J., Bertheau, Y., Barras, F., Vangijsegem, F., and Coleno, A. (1984). Mutants of *Erwinia chrysanthemi* defective in secretion of pectinase and cellulase. J. Bacteriol. 160, 1199–1203.

Angot, A., Peeters, N., Lechner, E., Vailleau, F., Baud, C., Gentzbittel, L., Sartorel, E., Genschik, P., Boucher, C., and Genin, S. P. (2006). *Ralstonia solanacearum* requires F-box-like domain-containing type III effectors to promote disease on several host plants. Proc. Natl. Acad. Sci. USA 103, 14620–14625.

Araki, H., Tian, D., Goss, E. M., Jakob, K., Halldorsdottir, S. S., Kreitman, M., and Bergelson, J. (2006). Presence/absence polymorphism for alternative pathogenicity islands in *Pseudomonas viridiflava*, a pathogen of *Arabidopsis*. Proc. Natl. Acad. Sci. USA 103, 5887–5892.

Arnold, D. L., Jackson, R. W., Fillingham, A. J., Goss, S. C., Taylor, J. D., Mansfield, J. W., and Vivian, A. (2001). Highly conserved sequences flank avirulence genes: isolation of novel avirulence genes from *Pseudomonas syringae* pv. pisi. Microbiology-Sgm 147, 1171–1182.

Arnold, D. L., Jackson, R. W., Waterfield, N. R., and Mansfield, J. W. (2007). Evolution of microbial virulence: the benefits of stress. Trends Genet. 23, 293–300.

Arnold, D. L., Pitman, A., and Jackson, R. W. (2003). Pathogenicity and other genomic islands in plant pathogenic bacteria. Mol. Plant Pathol. 4, 407–420.

Ashby, A. M., Watson, M. D., and Shaw, C. H. (1987). A Ti-plasmid determined function is responsible for chemotaxis of *Agrobacterium tumefaciens* towards the plant wound product acetosyringone. FEMS Microbiol. Lett. 41, 189–192.

Barabote, R. D., Johnson, O. L., Zetina, E., San Francisco, S. K., Fralick, J. A., and San Francisco, M. J. D. (2003). *Erwinia chrysanthemi tolC* is involved in resistance to antimicrobial plant chemicals and is essential for phytopathogenesis. J. Bacteriol. 185, 5772–5778.

Barras, F., Vangijsegem, F., and Chatterjee, A. K. (1994). Extracellular enzymes and pathogenesis of soft rot *Erwinia*. Annu. Rev. Phytopathol. 32, 201–234.

Basset, A., Khush, R. S., Braun, A., Gardan, L., Boccard, F., Hoffmann, J. A., and Lemaitre, B. (2000). The phytopathogenic bacteria *Erwinia carotovora* infects *Drosophila* and activates an immune response. Proc. Natl. Acad. Sci. USA 97, 3376–3381.

Basset, A., Tzou, P., Lemaitre, B., and Boccard, F. (2003). A single gene that promotes interaction of a phytopathogenic bacterium with its insect vector, *Drosophila melanogaster*. EMBO Rep. 4, 205–209.

Beatson, S. A., Minamino, T., and Pallen, M. J. (2006). Variation in bacterial flagellins: from sequence to structure. Trends Microbiol. 14, 151–155.

Bender, C. L., Alarcon-Chaidez, F., and Gross, D. C. (1999). *Pseudomonas syringae* phytotoxins: Mode of action, regulation, and biosynthesis by peptide and polyketide synthetases. Microbiol. Mol. Biol. Rev. 63, 266-+.

Bender, C. L., Malvick, D. K., and Mitchell, R. E. (1989). Plasmid-mediated production of the phytotoxin coronatine in *Pseudomonas syringae* pv. *tomato*. J. Bacteriol. 171, 807–812.

Bergelson, J., Kreitman, M., Stahl, E. A., and Tian, D. C. (2001). Evolutionary dynamics of plant R-genes. Science 292, 2281–2285.

Berks, B. C., Sargent, F., and Palmer, T. (2000). The Tat protein export pathway. Mol. Microbiol. 35, 260–274.

Bevan, M. W., and Chilton, M. D. (1982). T-DNA of the *Agrobacterium* Ti-plasmid and Ri-plasmid. Annu. Rev. Genet. 16, 357–384.

Bohne, J., Yim, A., and Binns, A. N. (1998). The Ti plasmid increases the efficiency of *Agrobacterium tumefaciens* as a recipient in *virB*-mediated conjugal transfer of an IncQ plasmid. Proc. Natl. Acad. Sci. USA 95, 7057–7062.

Boucher, C. A., Barberis, P. A., Trigalet, A. P., and Demery, D. A. (1985). Transposon mutagenesis of *Pseudomonas solanacearum* – isolation of Tn5-induced avirulent mutants. J. Gen. Microbiol. 131, 2449–2457.

Bove, J. M. (2006). Huanglongbing: A destructive, newly emerging, century-old disease of citrus. J. Plant Pathol. 88, 7–37.

Bove, J. M., and Garnier, M. (2002). Phloem-and xylem-restricted plant pathogenic bacteria. Plant Science 163, 1083–1098.

Buddenhagen, I. W. (1960). Strains of *Pseudomonas solanacearum* in indigenous hosts in banana plantations of Costa Rica, and their relationship to bacterial wilt of bananas. Phytopathology 50, 660–664.

Buell, C. R., Joardar, V., Lindeberg, M., Selengut, J., Paulsen, I. T., Gwinn, M. L., Dodson, R. J., Deboy, R. T., Durkin, A. S., Kolonay, J. F., et al. (2003). The complete genome sequence of the *Arabidopsis* and tomato pathogen *Pseudomonas syringae* pv. *tomato* DC3000. Proc. Natl. Acad. Sci. USA 100, 10181–10186.

Burdman, S., Shen, Y. W., Lee, S. W., Xue, Q. Z., and Ronald, P. (2004). RaxH/RaxR: A two-component regulatory system in *Xanthomonas oryzae* pv. *oryzae* required for AvrXa21 activity. Mol. Plant–Microbe Interact. 17, 602–612.

Burse, A., Weingart, H., and Ullrich, M. S. (2004). The phytoalexin-inducible multidrug efflux pump AcrAB contributes to virulence in the fire blight pathogen, *Erwinia amylovora*. Mol. Plant–Microbe Interact. 17, 43–54.

Canchaya, C., Proux, C., Fournous, G., Bruttin, A., and Brussow, H. (2003). Prophage genomics. Microbiol. Mol. Biol. Rev. 67, 238-+.

Cao, T. B., and Saier, M. H. (2001). Conjugal type IV macromolecular transfer systems of Gram-negative bacteria: organismal distribution, structural constraints and evolutionary conclusions. Microbiology-Sgm 147, 3201–3214.

Cao, T. B., and Saier, M. H. (2003). The general protein secretory pathway: phylogenetic analyses leading to evolutionary conclusions. Biochimica et Biophysica Acta (BBA) – Biomembranes 1609, 115–125.

Capuzzo, C., Firrao, G., Mazzon, L., Squartini, A., and Girolami, V. (2005). 'Candidatus *Erwinia dacicola*', a coevolved symbiotic bacterium of the olive fly *Bactrocera oleae* (Gmelin). Int. J. Syst. Evol. Microbiol. 55, 1641–1647.

Catara, V. (2007). *Pseudomonas corrugata*: plant pathogen and/or biological resource? Mol. Plant Pathol. 8, 233–244.

Chang, J. H., Urbach, J. M., Law, T. F., Arnold, L. W., Hu, A., Gombar, S., Grant, S. R., Ausubel, F. M., and Dangl, J. L. (2005). A high-throughput, near-saturating screen for type III effector genes from *Pseudomonas syringae*. Proc. Natl. Acad. Sci. USA 102, 2549–2554.

Chen, Y.-F., Yin, Y.-N., Zhang, X.-M., and Guo, J.-H. (2007). *Curtobacterium flaccumfaciens* pv. beticola, a new pathovar of pathogens in sugar beet. Plant Dis. 91, 677–684.

Chisholm, S. T., Coaker, G., Day, B., and Staskawicz, B. J. (2006). Host-microbe interactions: Shaping the evolution of the plant immune response. Cell 124, 803–814.

Christie, P. J. (1997). *Agrobacterium tumefaciens* T-complex transport apparatus: A paradigm for a new family of multifunctional transporters in eubacteria. J. Bacteriol. 179, 3085–3094.

Christie, P. J. (2001). Type IV secretion: intercellular transfer of macromolecules by systems ancestrally related to conjugation machines. Mol. Microbiol. 40, 294–305.

Christie, P. J., and Vogel, J. P. (2000). Bacterial type IV secretion: conjugation systems adapted to deliver effector molecules to host cells. Trends Microbiol. 8, 354–360.

Cianciotto, N. P. (2005). Type II secretion: a protein secretion system for all seasons. Trends Microbiol. 13, 581–588.

Ciesiolka, L. D., Hwin, T., Gearlds, J. D., Minsavage, G. V., Saenz, R., Bravo, M., Handley, V., Conover, S. M., Zhang, H., Caporgno, J., et al. (1999). Regulation of expression of avirulence gene *avrRxv* and identification of a family of host interaction factors by sequence analysis of *avrBsT*. Mol. Plant–Microbe Interact. 12, 35–44.

Coburn, B., Grassl, G. A., and Finlay, B. B. (2007). *Salmonella*, the host and disease: a brief review. Immunol. Cell Biol. 85, 112–118.

Coenye, T., and Vandamme, P. (2003). Diversity and significance of *Burkholderia* species occupying diverse ecological niches. Environ. Microbiol. 5, 719–729.

Collins, M. D., and Jones, D. (1983). Reclassification of *Corynebacterium flaccumfaciens*, *Corynebacterium betae*, *Corynebacterium oortii* and *Corynebacterium poinsettiae* in the Genus *Curtobacterium*, as *Curtobacterium flaccumfaciens* comb nov. J. Gen. Microbiol. 129, 3545–3548.

Collmer, A., Lindeberg, M., Petnicki-Ocwieja, T., Schneider, D. J., and Alfano, J. R. (2002). Genomic mining type III secretion system effectors in *Pseudomonas syringae* yields new picks for all TTSS prospectors. Trends Microbiol. *10*, 462–469.

Coplin, D. L. (1989). Plasmids and their role in the evolution of plant pathogenic bacteria. Annu. Rev. Phytopathol. *27*, 187–212.

Corbett, M., Virtue, S., Bell, K., Birch, P., Burr, T., Hyman, L., Lilley, K., Poock, S., Toth, I., and Salmond, G. (2005). Identification of a new quorum-sensing-control led virulence factor in *Erwinia carotovora* subsp atroseptica secreted via the type II targeting pathway. Mol. Plant–Microbe Interact. *18*, 334–342.

Cosgrove, D. J. (1999). Enzymes and other agents that enhance cell wall extensibility. Annu. Rev. Plant Physiol. Plant Mol. Biol. *50*, 391–417.

Cournoyer, B., Sharp, J. D., Astuto, A., Gibbon, M. J., Taylor, J. D., and Vivian, A. (1995). Molecular characterization of the *Pseudomonas syringae* pv. *pisi* plasmid-borne avirulence gene *avrPpiB* which matches the R3 resistance locus in pea. Mol. Plant–Microbe Interact. *8*, 700–708.

Cubero, J., and Graham, J. H. (2002). Genetic relationship among worldwide strains of *Xanthomonas* causing canker in citrus species and design of new primers for their identification by PCR. Appl. Environ. Microbiol. *68*, 1257–1264.

da Cunha, L., McFall, A. J., and Mackey, D. (2006). Innate immunity in plants: a continuum of layered defenses. Microbes Infect. *8*, 1372–1381.

da Silva, A. C. R., Ferro, J. A., Reinach, F. C., Farah, C. S., Furlan, L. R., Quaggio, R. B., Monteiro-Vitorello, C. B., Van Sluys, M. A., Almeida, N. F., Alves, L. M. C., et al. (2002). Comparison of the genomes of two *Xanthomonas* pathogens with differing host specificities. Nature *417*, 459–463.

Dangl, J. L., Ritter, C., Gibbon, M. J., Mur, L. A. J., Wood, J. R., Goss, S., Mansfield, J., Taylor, J. D., and Vivian, A. (1992). Functional homologs of the *Arabidopsis* Rpm1 disease resistance gene in bean and pea. Plant Cell *4*, 1359–1369.

Davis, M. J. (1986). Taxonomy of plant pathogenic coryneform bacteria. Annu. Rev. Phytopathol. *24*, 115–140.

Davis, M. J., Gillaspie, A. G., Harris, R. W., and Lawson, R. H. (1980). Ratoon stunting disease of sugarcane – isolation of the causal bacterium. Science *210*, 1365–1367.

Davis, M. J., Gillaspie, A. G., Vidaver, A. K., and Harris, R. W. (1984). *Clavibacter* – a new genus containing some phytopathogenic coryneform bacteria, including *Clavibacter xyli* subsp xyli sp nov, subsp nov and *Clavibacter xyli* subsp cynodontis subsp nov, pathogens that cause ratoon stunting disease of sugarcane and Bermudagrass Stunting Disease. Int. J. Syst. Bacteriol. *34*, 107–117.

Dawkins, R., and Krebs, J. R. (1979). Arms Races between and within species. Proc. R. Soc. Lond. B Biol. Sci. *205*, 489–511.

de Vries, E. J., Breeuwer, J. A. J., Jacobs, G., and Mollema, C. (2001a). The association of western flower thrips, *Frankliniella occidentalis*, with a near *Erwinia* species gut bacteria: Transient or permanent? J. Invertebr. Pathol. *77*, 120–128.

de Vries, E. J., Jacobs, G., and Breeuwer, J. A. J. (2001b). Growth and transmission of gut bacteria in the western flower thrips, *Frankliniella occidentalis*. J. Invertebr. Pathol. *77*, 129–137.

de Vries, E. J., Jacobs, G., Sabelis, M. W., Menken, S. B. J., and Breeuwer, J. A. J. (2004). Diet-dependent effects of gut bacteria on their insect host: the symbiosis of *Erwinia* sp and western flower thrips. Proc. R. Soc. Lond. B Biol. Sci. *271*, 2171–2178.

Department of Environment, Food, and Rural Affairs. (2007) Pests and disease identification information. http://www.defra.gov.uk/ (accessed 17August, 2007).

Delepelaire, P. (2004). Type I secretion in gram-negative bacteria. Biochim. Biophys. Acta Mol. Cell Res. *1694*, 149–161.

Desvaux, M., Parham, N. J., Scott-Tucker, A., and Henderson, I. R. (2004). The general secretory pathway: a general misnomer? Trends Microbiol. *12*, 306–309.

Dinh, T., Paulsen, I. T., and Saier, M. H. (1994). A family of extracytoplasmic proteins that allow transport of large molecules across the outer membranes of Gram-negative bacteria. J. Bacteriol. *176*, 3825–3831.

Dobrindt, U., Hochhut, B., Hentschel, U., and Hacker, J. (2004). Genomic islands in pathogenic and environmental microorganisms. Nat. Rev. Microbiol. *2*, 414–424.

Dutta, C., and Pan, A. (2002). Horizontal gene transfer and bacterial diversity. J. Biosc. *27*, 27–33.

Engledow, A. S., Medrano, E. G., Mahenthiralingam, E., LiPuma, J. J., and Gonzalez, C. F. (2004). Involvement of a plasmid-encoded type IV secretion system in the plant tissue watersoaking phenotype of *Burkholderia cenocepacia*. J. Bacteriol. *186*, 6015–6024.

Evidente, A., Dimaio, E., Caponero, A., and Iacobellis, N. S. (1995). Plant-growth regulators from ash strains of *Pseudomonas syringae* subsp *savastanoi*. Experientia *51*, 990–993.

Evtushenko, L. I., Dorofeeva, L. V., Subbotin, S. A., Cole, J. R., and Tiedje, J. M. (2000). *Leifsonia poae* gen. nov., sp nov., isolated from nematode galls on *Poa annua*, and reclassification of '*Corynebacterium aquaticum*' Leifson 1962 as *Leifsonia aquatica* (ex Leifson 1962) gen. nov., nom. rev., comb. nov and *Clavibacter xyli* Davis et al. 1984 with two subspecies as *Leifsonia xyli* (Davis et al. 1984) gen. nov., comb. nov. Int. J. Syst. Evol. Microbiol. *50*, 371–380.

Fath, M. J., and Kolter, R. (1993). ABC transporters – bacterial exporters. Microbiological Reviews *57*, 995–1017.

Filloux, A. (2004). The underlying mechanisms of type II protein secretion. Biochim. Biophys. Acta Mol. Cell Res. *1694*, 163–179.

Fouts, D. E., Abramovitch, R. B., Alfano, J. R., Baldo, A. M., Buell, C. R., Cartinhour, S., Chatterjee, A. K.,

D'Ascenzo, M., Gwinn, M. L., Lazarowitz, S. G., et al. (2002). Genomewide identification of *Pseudomonas syringae* pv. tomato DC3000 promoters controlled by the HrpL alternative sigma factor. Proc. Natl. Acad. Sci. USA 99, 2275–2280.

Frank, A. C., Alsmark, C. M., Thollesson, M., and Andersson, S. G. E. (2005). Functional divergence and horizontal transfer of type IV secretion systems. Mol. Biol. Evol. 22, 1325–1336.

Frost, L. S., Leplae, R., Summers, A. O., and Toussaint, A. (2005). Mobile genetic elements: The agents of open source evolution. Nat. Rev. Microbiol. 3, 722–732.

Gabriel, D. W., Allen, C., Schell, M., Denny, T. P., Greenberg, J. T., Duan, Y. P., Flores-Cruz, Z., Huang, Q., Clifford, J. M., Presting, G., et al. (2006). Identification of open reading frames unique to a select agent: *Ralstonia solanacearum* race 3 biovar 2. Mol. Plant–Microbe Interact. 19, 69–79.

Galibert, F., Finan, T. M., Long, S. R., Puhler, A., Abola, P., Ampe, F., Barloy-Hubler, F., Barnett, M. J., Becker, A., Boistard, P., et al. (2001). The composite genome of the legume symbiont *Sinorhizobium meliloti*. Science 293, 668–672.

Gandon, S. (2004). Evolution of multihost parasites. Evolution 58, 455–469.

Gardan, L., Gouy, C., Christen, R., and Samson, R. (2003). Elevation of three subspecies of *Pectobacterium carotovorum* to species level: *Pectobacterium atrosepticum* sp nov., *Pectobacterium betavasculorum* sp nov and *Pectobacterium wasabiae* sp nov. Int. J. Syst. Evol. Microbiol. 53, 381–391.

Garnier, M., Foissac, X., Gaurivaud, P., Laigret, F., Renaudin, J., Saillard, C., and Bove, J. M. (2001). Mycoplasmas, plants, insect vectors: a matrimonial triangle. Comptes Rendus de l'Academie des Sciences – Series III – Sciences de la Vie 324, 923–928.

Garnier, M., Jagoueix-Eveillard, S., Cronje, P. R., Le Roux, H. F., and Bove, J. M. (2000). Genomic characterization of a liberibacter present in an ornamental rutaceous tree, *Calodendrum capense*, in the Western Cape province of South Africa. Proposal of 'Candidatus Liberibacter africanus subsp capensis'. Int. J. Syst. Evol. Microbiol. 50, 2119–2125.

Gartemann, K. H., Kirchner, O., Engemann, J., Grafen, I., Eichenlaub, R., and Burger, A. (2003). *Clavibacter michiganensis* subsp michiganensis: first steps in the understanding of virulence of a Gram-positive phytopathogenic bacterium. J. Biotechnol. 106, 179–191.

Genin, S., and Boucher, C. (2004). Lessons learned from the genome analysis of *Ralstonia solanacearum*. Annu. Rev. Phytopathol. 42, 107–134.

Gent, D. H., Lang, J. M., and Schwartz, H. F. (2005). Epiphytic survival of *Xanthomonas axonopodis* pv. allii and *X axonopodis* pv. phaseoli on leguminous hosts and onion. Plant Dis. 89, 558–564.

Gerlach, R. G., and Hensel, M. (2008) Protein secretion systems and adhesins: The molecular armory of Gram-negative pathogens. Int. J. Med. Microbiol. (in press).

Glass, N. L., and Kosuge, T. (1988). Role of IAA-lysine synthetase in regulation of IAA pool size and virulence of *Pseudomonas syringae* ssp. savastanoi. J. Bacteriol. 170, 2367–2373.

Gold, V. A. M., Duong, F., and Collinson, I. (2007). Structure and function of the bacterial Sec translocon. Mol. Membr. Biol. 24, 387–394.

Gonzalez, E. T., Brown, D. G., Swanson, J. K., and Allen, C. (2007). Using the *Ralstonia solanacearum* Tat secretome to identify bacterial wilt virulence factors. Appl. Environ. Microbiol. 73, 3779–3786.

Goodfellow, M. (1984). Reclassification of *Corynebacterium fascians* (Tilford) Dowson in the Genus *Rhodococcus*, as *Rhodococcus fascians* Comb Nov. Syst. Appl. Microbiol. 5, 225–229.

Gophna, U., Ron, E. Z., and Graur, D. (2003). Bacterial type III secretion systems are ancient and evolved by multiple horizontal-transfer events. Gene 312, 151–163.

Gottwald, T. R., Graham, J. H., and Schubert, T. S. (2002). Citrus canker: The pathogen and its impact. Plant Health Progress ERN: 10.1094/PHP-2002-0812-01-RV.

Goumans, D. E., and Chatzaki, A. K. (1998). Characterization and host range evaluation of *Pseudomonas viridiflava* from melon, blite, tomato, chrysanthemum and eggplant. Eur. J. Plant Pathol. 104, 181–188.

Grant, S. R., Fisher, E. J., Chang, J. H., Mole, B. M., and Dangl, J. L. (2006). Subterfuge and manipulation: Type III effector proteins of phytopathogenic bacteria. Annu. Rev. Microbiol. 60, 425–449.

Grenier, A. M., Duport, G., Pages, S., Condemine, G., and Rahbe, Y. (2006). The phytopathogen *Dickeya dadantii* (*Erwinia chrysanthemi* 3937) is a pathogen of the pea aphid. Appl. Environ. Microbiol. 72, 1956–1965.

Groisman, E. A., and Ochman, H. (1996). Pathogenicity islands: Bacterial evolution in quantum leaps. Cell 87, 791–794.

Guo, M., Manulis, S., Mor, H., and Barash, I. (2002). The presence of diverse IS elements and an *avrPphD* homologue that acts as a virulence factor on the pathogenicity plasmid of *Erwinia herbicola* pv. gypsophilae. Mol. Plant–Microbe Interact. 15, 709–716.

Guttman, D., Gropp, S., Morgan, R., and Wang, P. (2006). Diversifying selection drives the evolution of the type III secretion system pilus of *Pseudomonas syringae*. Mol. Biol. Evol. 23, 2342–2354.

Guttman, D. S., and Greenberg, J. T. (2001). Functional analysis of the type III effectors AvrRpt2 and AvrRpm1 of *Pseudomonas syringae* with the use of a single-copy genomic integration system. Mol. Plant–Microbe Interact. 14, 145–155.

Guttman, D. S., Vinatzer, B. A., Sarkar, S. F., Ranall, M. V., Kettler, G., and Greenberg, J. T. (2002). A functional screen for the type III (Hrp) secretome of the plant pathogen *Pseudomonas syringae*. Science 295, 1722–1726.

Hacker, J., Bender, L., Ott, M., Wingender, J., Lund, B., Marre, R., and Goebel, W. (1990). Deletions of chromosomal regions coding for fimbriae and hemolysins occur in vitro and in vivo in various extraintestinal *Escherichia coli* isolates. Microb. Pathog. 8, 213–225.

Hacker, J., Hochhut, B., Middendorf, B., Schneider, G., Buchrieser, C., Gottschalk, G., and Dobrindt, U. (2004). Pathogenomics of mobile genetic elements of toxigenic bacteria. International Journal of Medical Microbiology 293, 453–461.

HammondKosack, K. E., and Jones, J. D. G. (1997). Plant disease resistance genes. Annu. Rev. Plant Physiol. Plant Mol. Biol. 48, 575–607.

Harada, H., and Ishikawa, H. (1997). Experimental pathogenicity of *Erwinia aphidicola* to pea aphid, *Acyrthosiphon pisum*. J. Gen. Appl. Microbiol. 43, 363–367.

Harada, H., Oyaizu, H., Kosako, Y., and Ishikawa, H. (1997). *Erwinia aphidicola*, a new species isolated from pea aphid, *Acyrthosiphon pisum*. J. Gen. Appl. Microbiol. 43, 349–354.

Harshey, R. M., and Toguchi, A. (1996). Spinning tails: Homologies among bacterial flagellar systems. Trends Microbiol. 4, 226–231.

Hauben, L., Moore, E. R. B., Vauterin, L., Steenackers, M., Mergaert, J., Verdonck, L., and Swings, J. (1998). Phylogenetic position of phytopathogens within the Enterobacteriaceae. Syst. Appl. Microbiol. 21, 384–397.

Hayward, A. C. (1991). Biology and epidemiology of bacterial wilt caused by *Pseudomonas solanacearum*. Annu. Rev. Phytopathol. 29, 65–87.

He, P., Shan, L., and Sheen, J. (2007). Elicitation and suppression of microbe-associated molecular pattern-triggered immunity in plant–microbe interactions. Cell. Microbiol. 9, 1385–1396.

He, S. Y. (1998). Type III protein secretion systems in plant and animal pathogenic bacteria. Annu. Rev. Phytopathol. 36, 363–392.

Heath, M. C. (1998). Apoptosis, programmed cell death and the hypersensitive response. Eur. J. Plant Pathol. 104, 117–124.

Heath, M. C. (2000). Hypersensitive response-related death. Plant Mol. Biol. 44, 321–334.

Herbers, K., Conradsstrauch, J., and Bonas, U. (1992). Race-specificity of plant resistance to bacterial spot disease determined by repetitive motifs in a bacterial avirulence protein. Nature 356, 172–174.

Hooker, W. J. (1949). Parasitic action of *Streptomyces scabies* on roots of seedlings. Phytopathology 39, 442–462.

Hopkins, D. L. (1989). *Xylella fastidiosa* – xylem-limited bacterial pathogen of plants. Annu. Rev. Phytopathol. 27, 271–290.

Hueck, C. J. (1998). Type III protein secretion systems in bacterial pathogens of animals and plants. Microbiol. Mol. Biol. Rev. 62, 379–433.

Hung, T. H., Hung, S. C., Chen, C. N., Hsu, M. H., and Su, H. J. (2004). Detection by PCR of *Candidatus* Liberibacter asiaticus, the bacterium causing citrus Huanglongbing in vector psyllids: application to the study of vector-pathogen relationships. Plant Pathol. 53, 96–102.

Hunter, P. J., and Taylor, J. D. (2006). Patterns of interaction between isolates of three pathovars of *Pseudomonas syringae* and accessions of a range of host and nonhost legume species. Plant Pathology 55, 46–53.

Hutchison, M. L., and Gross, D. C. (1997). Lipopeptide phytotoxins produced by *Pseudomonas syringae* pv syringae: Comparison of the biosurfactant and ion channel-forming activities of syringopeptin and syringomycin. Mol. Plant–Microbe Interact. 10, 347–354.

Jack, D. L., Yang, N. M., and Saier, M. H. (2001). The drug/metabolite transporter superfamily. Eur. J. Biochem. 268, 3620–3639.

Jackson, R. W., Athanassopoulos, E., Tsiamis, G., Mansfield, J. W., Sesma, A., Arnold, D. L., Gibbon, M. J., Murillo, J., Taylor, J. D., and Vivian, A. (1999). Identification of a pathogenicity island, which contains genes for virulence and avirulence, on a large native plasmid in the bean pathogen *Pseudomonas syringae* pathovar phaseolicola. Proc. Natl. Acad. Sci. USA 96, 10875–10880.

Jackson, R. W., Mansfield, J. W., Arnold, D. L., Sesma, A., Paynter, C. D., Murillo, J., Taylor, J. D., and Vivian, A. (2000). Excision from tRNA genes of a large chromosomal region, carrying *avrPphB*, associated with race change in the bean pathogen, *Pseudomonas syringae* pv. phaseolicola. Mol. Microbiol. 38, 186–197.

Jahr, H., Bahro, R., Burger, A., Ahlemeyer, J., and Eichenlaub, R. (1999). Interactions between *Clavibacter michiganensis* and its host plants. Environ. Microbiol. 1, 113–118.

Jamir, Y., Guo, M., Oh, H. S., Petnicki-Ocwieja, T., Chen, S. R., Tang, X. Y., Dickman, M. B., Collmer, A., and Alfano, J. R. (2004). Identification of *Pseudomonas syringae* type III effectors that can suppress programmed cell death in plants and yeast. Plant J. 37, 554–565.

Jha, G., Rajeshwari, R., and Sonti, R. V. (2007). Functional interplay between two *Xanthomonas oryzae* pv. oryzae secretion systems in modulating virulence on rice. Mol. Plant–Microbe Interact. 20, 31–40.

Ji, P. S., Allen, C., Sanchez-Perez, A., Yao, J., Elphinstone, J. G., Jones, J. B., and Momol, A. T. (2007). New diversity of *Ralstonia solanacearum* strains associated with vegetable and ornamental crops in Florida. Plant Dis. 91, 195–203.

Jin, Q. L., Thilmony, R., Zwiesler-Vollick, J., and He, S. Y. (2003). Type III protein secretion in *Pseudomonas syringae*. Microbes Infect. 5, 301–310.

Joardar, V., Lindeberg, M., Jackson, R., Selengut, J., Dodson, R., Brinkac, L., Daugherty, S., Deboy, R., Durkin, A., Giglio, M., *et al.* (2005). Whole-genome sequence analysis of *Pseudomonas syringae* pv. phaseolicola 1448A reveals divergence among pathovars in genes involved in virulence and transposition. J. Bacteriol. 187, 6488–6498.

Johnson, T. L., Abendroth, J., Hol, W. G. J., and Sandkvist, M. (2006). Type II secretion: from structure to function. FEMS Microbiol. Lett. 255, 175–186.

Jones, J. D. G., and Dangl, J. L. (2006). The plant immune system. Nature 444, 323–329.

Kado, C. (2006). *Erwinia* and related genera, In The Prokaryotes (New York: Springer), pp. 443–450.

Kang, Y. W., Huang, J. Z., Mao, G. Z., He, L. Y., and Schell, M. A. (1994). Dramatically reduced virulence of mutants of Pseudomonas solanacearum defective in export of extracellular proteins across the outer membrane. Mol. Plant–Microbe Interact. 7, 370–377.

Karch, H., Schubert, S., Zhang, D., Zhang, W., Schmidt, H., Olschlager, T., and Hacker, J. (1999). A genomic island, termed high-pathogenicity island, is present in certain non-O157 Shiga toxin-producing *Escherichia coli* clonal lineages. Infect. Immun. 67, 5994–6001.

Kers, J. A., Cameron, K. D., Joshi, M. V., Bukhalid, R. A., Morello, J. E., Wach, M. J., Gibson, D. M., and Loria, R. (2005). A large, mobile pathogenicity island confers plant pathogenicity on *Streptomyces* species. Mol. Microbiol. 55, 1025–1033.

Kim, J. F., Charkowski, A. O., Alfano, J. R., Collmer, A., and Beer, S. V. (1998). Sequences related to transposable elements and bacteriophages flank avirulence genes of *Pseudomonas syringae*. Mol. Plant–Microbe Interact. 11, 1247–1252.

Kunze, G., Zipfel, C., Robatzek, S., Niehaus, K., Boller, T., and Felix, G. (2004). The N terminus of bacterial elongation factor Tu elicits innate immunity in Arabidopsis plants. Plant Cell 16, 3496–3507.

Lai, E. M., and Kado, C. I. (2000). The T-pilus of *Agrobacterium tumefaciens*. Trends Microbiol. 8, 361–369.

Lawrence, J. G., and Ochman, H. (1997). Amelioration of bacterial genomes: Rates of change and exchange. J. Mol. Evol. 44, 383–397.

Lawrence, J. G., and Ochman, H. (2002). Reconciling the many faces of lateral gene transfer. Trends Microbiol. 10, 1–4.

Lee, I. M., Davis, R. E., and Gundersen-Rindal, D. E. (2000). Phytoplasma: Phytopathogenic mollicutes. Annu. Rev. Microbiol. 54, 221–255.

Lee, P. A., Tullman-Ercek, D., and Georgiou, G. (2006a). The bacterial twin-arginine translocation pathway. Annu. Rev. Microbiol. 60, 373–395.

Lee, S. W., Han, S. W., Bartley, L. E., and Ronald, P. C. (2006b). Unique characteristics of *Xanthomonas oryzae* pv. oryzae AvrXa21 and implications for plant innate immunity. Proc. Natl. Acad. Sci. USA 103, 18395–18400.

Lerat, E., Daubin, V., Ochman, H., and Moran, N. A. (2005). Evolutionary origins of genomic repertoires in bacteria. PLoS Biology 3, 807–814.

Li, W., Brlansky, R. H., and Hartung, J. S. (2006). Amplification of DNA of *Xanthomonas axonopodis* pv. citri from historic citrus canker herbarium specimens. J. Microbiol. Methods 65, 237–246.

Lindeberg, M., Cartinhour, S., Myers, C. R., Schechter, L. M., Schneider, D. J., and Collmer, A. (2006). Closing the circle on the discovery of genes encoding Hrp regulon members and type III secretion system effectors in the genomes of three model *Pseudomonas syringae* strains. Mol. Plant–Microbe Interact. 19, 1151–1158.

Lindeberg, M., Stavrinides, J., Chang, J. H., Alfano, J. R., Collmer, A., Dangl, J. L., Greenberg, J. T., Mansfield, J. W., and Guttman, D. S. (2005). Proposed guidelines for a unified nomenclature and phylogenetic analysis of type III Hop effector proteins in the plant pathogen *Pseudomonas syringae*. Mol. Plant–Microbe Interact. 18, 275–282.

Liu, H. L., Zhang, S. P., Schell, M. A., and Denny, T. P. (2005). Pyramiding, unmarked deletions in *Ralstonia solanacearum* shows that secreted proteins in addition to plant cell-wall-degrading enzymes contribute to virulence. Mol. Plant–Microbe Interact. 18, 1296–1305.

Lloyd, S. A., Norman, M., Rosqvist, R., and Wolf-Watz, H. (2001). *Yersinia* YopE is targeted for type III secretion by N-terminal, not mRNA, signals. Mol. Microbiol. 39, 520–531.

Loria, R., Bukhalid, R. A., Fry, B. A., and King, R. R. (1997). Plant pathogenicity in the genus *Streptomyces*. Plant Dis. 81, 836–846.

Loria, R., Kers, J., and Joshi, M. (2006). Evolution of plant pathogenicity in *Streptomyces*. Annu. Rev. Phytopathol. 44, 469–487.

Ma, W. B., Dong, F. F. T., Stavrinides, J., and Guttman, D. S. (2006). Type III effector diversification via both pathoadaptation and horizontal transfer in response to a coevolutionary arms race. PLoS Genet. 2, 2131–2142.

McCullen, C. A., and Binns, A. N. (2006). *Agrobacterium tumefaciens* and plant cell interactions and activities required for interkingdom macromolecular transfer. Annu. Rev. Cell Dev. Biol. 22, 101–127.

McDowell, J. M., and Simon, S. A. (2006). Recent insights into R gene evolution. Mol. Plant Pathol. 7, 437–448.

Malpica, J. M., Sacristan, S., Fraile, A., and Garcia-Arenal, F. (2006). Association and host selectivity in multi-host pathogens. PLoS ONE 1, e41.

Manulis, S., and Barash, I. (2003). *Pantoea agglomerans* pvs. gypsophilae and betae, recently evolved pathogens? Mol. Plant Pathol. 4, 307–314.

Martin, G. B., Bogdanove, A. J., and Sessa, G. (2003). Understanding the functions of plant disease resistance proteins. Annual Review Of Plant Biology 54, 23–61.

Martinez, E., Palacios, R., and Sanchez, F. (1987). Nitrogen-fixing nodules induced by *Agrobacterium tumefaciens* harboring *Rhizobium phaseoli* plasmids. J. Bacteriol. 169, 2828–2834.

Mavrodieva, V., Levy, L., and Gabriel, D. W. (2004). Improved sampling methods for real-time polymerase chain reaction diagnosis of citrus canker from field samples. Phytopathology 94, 61–68.

Meletzus, D., Bermpohl, A., Dreier, J., and Eichenlaub, R. (1993). Evidence for plasmid-encoded virulence factors in the phytopathogenic bacterium *Clavibacter michiganensis* subsp michiganensis NCPPB382. J. Bacteriol. 175, 2131–2136.

Melotto, M., Underwood, W., Koczan, J., Nomura, K., and He, S. Y. (2006). Plant stomata function in innate immunity against bacterial invasion. Cell 126, 969–980.

Michelmore, R. W., and Meyers, B. C. (1998). Clusters of resistance genes in plants evolve by divergent selection and a birth-and-death process. Genome Res. 8, 1113–1130.

Mills, L., Leaman, T. M., Taghavi, S. M., Shackel, L., Dominiak, B. C., Taylor, P. W. J., Fegan, M., and Teakle, D. S. (2001). *Leifsonia xyli*-like bacteria are endophytes of grasses in eastern Australia. Australas. Plant Pathol. 30, 145–151.

Minsavage, G. V., Dahlbeck, D., Whalen, M. C., Kearney, B., Bonas, U., Staskawicz, B. J., and Stall, R. E. (1990).

Gene-for-gene relationships specifying disease resistance in *Xanthomonas campestris* pv vesicatoria – pepper interactions. Mol. Plant–Microbe Interact. 3, 41–47.

Mitchell, P. L. (2004). Heteroptera as vectors of plant pathogens. Neotrop. Entomol. 33, 519–545.

Monteiro-Vitorello, C. B., Camargo, L. E. A., Van Sluys, M. A., Kitajima, J. P., Truffi, D., do Amaral, A. M., Harakava, R., de Oliveira, J. C. F., Wood, D., de Oliveira, M. C., et al. (2004). The genome sequence of the gram-positive sugarcane pathogen *Leifsonia xyli* subsp xyli. Mol. Plant–Microbe Interact. 17, 827–836.

Moriguchi, K., Maeda, Y., Satou, M., Hardayani, N. S. N., Kataoka, M., Tanaka, N., and Yoshida, K. (2001). The complete nucleotide sequence of a plant root-inducing (Ri) plasmid indicates its chimeric structure and evolutionary relationship between tumor-inducing (Ti) and symbiotic (Sym) plasmids in Rhizobiaceae. J. Mol. Biol. 307, 771–784.

Mudgett, M. B., Chesnokova, O., Dahlbeck, D., Clark, E. T., Rossier, O., Bonas, U., and Staskawicz, B. J. (2000). Molecular signals required for type III secretion and translocation of the *Xanthomonas campestris* AvrBs2 protein to pepper plants. Proc. Natl. Acad. Sci. USA 97, 13324–13329.

Mukherjee, S., Keitany, G., Li, Y., Wang, Y., Ball, H. L., Goldsmith, E. J., and Orth, K. (2006). *Yersinia* YopJ acetylates and inhibits kinase activation by blocking phosphorylation. Science 312, 1211–1214.

Newman, K. L., Almeida, R. P. P., Purcell, A. H., and Lindow, S. E. (2004). Cell–cell signaling controls *Xylella fastidiosa* interactions with both insects and plants. Proc. Natl. Acad. Sci. USA 101, 1737–1742.

Nguyen, L., Paulsen, I. T., Tchieu, J., Hueck, C. J., and Saier, M. H. (2000). Phylogenetic analyses of the constituents of type III protein secretion systems. J. Mol. Microbiol. Biotechnol. 2, 125–144.

Nissan, G., Manulis-Sasson, S., Weinthal, D., Mor, H., Sessa, G., and Barash, I. (2006). The type III effectors HsvG and HsvB of gall-forming *Pantoea agglomerans* determine host specificity and function as transcriptional activators. Mol. Microbiol. 61, 1118–1131.

Nizan, R., Barash, I., Valinsky, L., Lichter, A., and Manulis, S. (1997). The presence of hrp genes on the pathogenicity-associated plasmid of the tumorigenic bacterium *Erwinia herbicola* pv gypsophilae. Mol. Plant–Microbe Interact. 10, 677–682.

Nurnberger, T., and Brunner, F. (2002). Innate immunity in plants and animals: emerging parallels between the recognition of general elicitors and pathogen-associated molecular patterns. Curr. Opin. Plant Biol. 5, 318–324.

Nurnberger, T., Brunner, F., Kemmerling, B., and Piater, L. (2004). Innate immunity in plants and animals: striking similarities and obvious differences. Immunol. Rev. 198, 249–266.

Ochman, H. (2001). Lateral and oblique gene transfer. Curr. Opin. Genet. Dev. 11, 616–619.

Ochman, H., Lawrence, J. G., and Groisman, E. A. (2000). Lateral gene transfer and the nature of bacterial innovation. Nature 405, 299–304.

Orth, K., Xu, Z. H., Mudgett, M. B., Bao, Z. Q., Palmer, L. E., Bliska, J. B., Mangel, W. F., Staskawicz, B., and Dixon, J. E. (2000). Disruption of signaling by *Yersinia* effector YopJ, a ubiquitin-like protein protease. Science 290, 1594–1597.

Osborn, A. M., and Boltner, D. (2002). When phage, plasmids, and transposons collide: genomic islands, and conjugative- and mobilizable-transposons as a mosaic continuum. Plasmid 48, 202–212.

Palumbo, J. D., Kado, C. I., and Phillips, D. A. (1998). An isoflavonoid-inducible efflux pump in *Agrobacterium tumefaciens* is involved in competitive colonization of roots. J. Bacteriol. 180, 3107–3113.

Pao, S. S., Paulsen, I. T., and Saier, M. H. (1998). Major facilitator superfamily. Microbiol. Mol. Biol. Rev. 62, 1–34.

Parke, D., Ornston, L. N., and Nester, E. W. (1987). Chemotaxis to plant phenolic inducers of virulence genes is constitutively expressed in the absence of the Ti-plasmid in *Agrobacterium tumefaciens*. J. Bacteriol. 169, 5336–5338.

Peabody, C. R., Chung, Y. J., Yen, M. R., Vidal-Ingigliardi, D., Pugsley, A. P., and Saier, M. H. (2003). Type II protein secretion and its relationship to bacterial type IV pili and archaeal flagella. Microbiology-Sgm 149, 3051–3072.

Perombelon, M. C. M. (2002). Potato diseases caused by soft rot erwinias: an overview of pathogenesis. Plant Pathology 51, 1–12.

Petnicki-Ocwieja, T., Schneider, D. J., Tam, V. C., Chancey, S. T., Shan, L., Jamir, Y., Schechter, L. M., Janes, M. D., Buell, C. R., Tang, X. Y., et al. (2002). Genomewide identification of proteins secreted by the Hrp type III protein secretion system of *Pseudomonas syringae* pv. tomato DC3000. Proc. Natl. Acad. Sci. USA 99, 7652–7657.

Pitman, A. R., Jackson, R. W., Mansfield, J. W., Kaitell, V., Thwaites, R., and Arnold, D. L. (2005). Exposure to host resistance mechanisms drives evolution of bacterial virulence in plants. Curr. Biol. 15, 2230–2235.

Plotnikova, J. M., Rahme, L. G., and Ausubel, F. M. (2000). Pathogenesis of the human opportunistic pathogen *Pseudomonas aeruginosa* PA14 in Arabidopsis. Plant Physiology 124, 1766–1774.

Polek, M. (2007). Citrus bacterial canker disease and Huanglongbing (Citrus Greening). University of California Division of Agriculture and Natural Resources *Publication 8218*.

Poussier, S., Thoquet, P., Trigalet-Demery, D., Barthet, S., Meyer, D., Arlat, M., and Trigalet, A. (2003). Host plant-dependent phenotypic reversion of *Ralstonia solanacearum* from non-pathogenic to pathogenic forms via alterations in the *phcA* gene. Mol. Microbiol. 49, 991–1003.

Prithiviraj, B., Weir, T., Bais, H. P., Schweizer, H. P., and Vivanco, J. M. (2005). Plant models for animal pathogenesis. Cell. Microbiol. 7, 315–324.

Pugsley, A. P. (1993). The complete general secretory pathway in Gram-negative bacteria. Microbiol. Rev. 57, 50–108.

Purcell, A. H., and Hopkins, D. L. (1996). Fastidious xylem-limited bacterial plant pathogens. Annu. Rev. Phytopathol. 34, 131–151.

Quigley, N. B., Mo, Y. Y., and Gross, D. C. (1993). *syrD* is required for syringomycin production by *Pseudomonas syringae* pathovar syringae and is related to a family of ATP-binding secretion proteins. Mol. Microbiol. 9, 787–801.

Rahme, L. G., Mindrinos, M. N., and Panopoulos, N. J. (1991). Genetic and transcriptional organization of the Hrp cluster of *Pseudomonas syringae* pv phaseolicola. J. Bacteriol. 173, 575–586.

Rahme, L. G., Stevens, E. J., Wolfort, S. F., Shao, J., Tompkins, R. G., and Ausubel, F. M. (1995). Common virulence factors for bacterial pathogenicity in plants and animals. Science 268, 1899–1902.

Rahme, L. G., Tan, M. W., Le, L., Wong, S. M., Tompkins, R. G., Calderwood, S. B., and Ausubel, F. M. (1997). Use of model plant hosts to identify *Pseudomonas aeruginosa* virulence factors. Proc. Natl. Acad. Sci. USA 94, 13245–13250.

Rajeshwari, R., Jha, G., and Sonti, R. V. (2005). Role of an *in planta*-expressed xylanase of *Xanthomonas oryzae* pv. oryzae in promoting virulence on rice. Mol. Plant–Microbe Interact. 18, 830–837.

Ray, S. K., Rajeshwari, R., and Sonti, R. V. (2000). Mutants of *Xanthomonas oryzae* pv. oryzae deficient in general secretory pathway are virulence deficient and unable to secrete xylanase. Mol. Plant–Microbe Interact. 13, 394–401.

Reddy, G. S. N., Prakash, J. S. S., Srinivas, R., Matsumoto, G. I., and Shivaji, S. (2003). *Leifsonia rubra* sp nov and *Leifsonia aurea* sp nov., psychrophiles from a pond in Antarctica. Int. J. Syst. Evol. Microbiol. 53, 977–984.

Reddy, J. D., Reddy, S. L., Hopkins, D. L., and Gabriel, D. W. (2007). TolC is required for pathogenicity of *Xylella fastidiosa* in *Vitis vinifera* grapevines. Mol. Plant–Microbe Interact. 20, 403–410.

Robertson, A. E., Wechter, W. P., Denny, T. P., Fortnum, B. A., and Kluepfel, D. A. (2004). Relationship between avirulence gene (*avrA*) diversity in *Ralstonia solanacearum* and bacterial Wilt incidence. Mol. Plant–Microbe Interact. 17, 1376–1384.

Rohmer, L., Guttman, D. S., and Dangl, J. L. (2004). Diverse evolutionary mechanisms shape the type III effector virulence factor repertoire in the plant pathogen *Pseudomonas syringae*. Genetics 167, 1341–1360.

Roine, E., Wei, W. S., Yuan, J., NurmiahoLassila, E. L., Kalkkinen, N., Romantschuk, M., and He, S. Y. (1997). Hrp pilus: An hrp-dependent bacterial surface appendage produced by *Pseudomonas syringae* pv tomato DC3000. Proc. Natl. Acad. Sci. USA 94, 3459–3464.

Rytkonen, A., Poh, J., Garmendia, J., Boyle, C., Thompson, A., Liu, M., Freemont, P., Hinton, J. C. D., and Holden, D. W. (2007). SseL, a *Salmonella* deubiquitinase required for macrophage killing and virulence. Proc. Natl. Acad. Sci. USA 104, 3502–3507.

Saier, M. (2006). Protein secretion and membrane insertion systems in Gram-negative bacteria. J. Membr. Biol. 214, 75–90.

Saier, M. H. (2000). A functional-phylogenetic classification system for transmembrane solute transporters. Microbiol. Mol. Biol. Rev. 64, 354-+.

Saier, M. H. (2003). Tracing pathways of transport protein evolution. Mol. Microbiol. 48, 1145–1156.

Saier, M. H. (2004). Evolution of bacterial type III protein secretion systems. Trends Microbiol. 12, 113–115.

Saier, M. H., and Paulsen, I. T. (2001). Phylogeny of multidrug transporters. Semin. Cell Dev. Biol. 12, 205–213.

Saier, M. H., Paulsen, I. T., Sliwinski, M. K., Pao, S. S., Skurray, R. A., and Nikaido, H. (1998). Evolutionary origins of multidrug and drug-specific efflux pumps in bacteria. FASEB J. 12, 265–274.

Salanoubat, M., Genin, S., Artiguenave, F., Gouzy, J., Mangenot, S., Arlat, M., Billault, A., Brottier, P., Camus, J. C., Cattolico, L., *et al.* (2002). Genome sequence of the plant pathogen *Ralstonia solanacearum*. Nature 415, 497–502.

Salmond, G. P. C. (1994). Secretion of extracellular virulence factors by plant-pathogenic bacteria. Annu. Rev. Phytopathol. 32, 181–200.

Samson, R., Legendre, J. B., Christen, R., Fischer-Le Saux, M., Achouak, W., and Gardan, L. (2005). Transfer of *Pectobacterium chrysanthemi* (Burkholder *et al.* 1953) Brenner *et al.* 1973 and *Brenneria paradisiaca* to the genus *Dickeya* gen. nov as *Dickeya chrysanthemi* comb. nov and *Dickeya paradisiaca* comb. nov and delineation of four novel species, *Dickeya dadantii* sp nov., *Dickeya dianthicola* sp nov., *Dickeya dieffenbachiae* sp nov and *Dickeya zeae* sp nov. Int. J. Syst. Evol. Microbiol. 55, 1415–1427.

Sandkvist, M. (2001a). Biology of type II secretion. Mol. Microbiol. 40, 271–283.

Sandkvist, M. (2001b). Type II secretion and pathogenesis. Infect. Immun. 69, 3523–3535.

Sarkar, S. F., Gordon, J. S., Martin, G. B., and Guttman, D. S. (2006). The comparative genomics of host-specific virulence in *Pseudomonas syringae*. Genetics in press.

Sarkar, S. F., and Guttman, D. S. (2004). The evolution of the core genome of *Pseudomonas syringae*, a highly clonal, endemic plant pathogen. Appl. Environ. Microbiol. 70, 1999–2012.

Sasaki, J., Chijimatsu, M., and Suzuki, K. (1998). Taxonomic significance of 2,4-diaminobutyric acid isomers in the cell wall peptidoglycan of actinomycetes and reclassification of *Clavibacter toxicus* as *Rathayibacter toxicus* comb. nov. Int. J. System. Bacteriol. 48, 403–410.

Sawada, H., Suzuki, F., Matsuda, I., and Saitou, N. (1999). Phylogenetic analysis of *Pseudomonas syringae* pathovars suggests the horizontal gene transfer of *argK* and the evolutionary stability of hrp gene cluster. J. Mol. Evol. 49, 627–644.

Schechter, L. M., Roberts, K. A., Jamir, Y., Alfano, J. R., and Collmer, A. (2004). *Pseudomonas syringae* type III secretion system targeting signals and novel effectors studied with a Cya translocation reporter. J. Bacteriol. 186, 543–555.

Scholz-Schroeder, B. K., Hutchison, M. L., Grgurina, I., and Gross, D. C. (2001). The contribution of syringopeptin and syringomycin to virulence of *Pseudomonas syringae* pv. syringae strain B301D on the basis of *sypA* and *syrB1* biosynthesis mutant analysis. Mol. Plant–Microbe Interact. *14*, 336–348.

Schoulties, C. L., Civerolo, E. L., Miller, J. W., Stall, R. E., Krass, C. J., Poe, S. R., and Ducharme, E. P. (1987). Citrus Canker in Florida. Plant Dis. *71*, 388–395.

Schubert, T. S., Miller, J. W., and Gabriel, D. W. (1996). Another outbreak of bacterial canker on citrus in Florida. Plant Dis. *80*, 1208–1208.

Schubert, T. S., and Sun, X. (1996). Bacterial citrus canker. Florida Department of Agriculture and Conservation Services. ERN:377.

Seubert, A., Hiestand, R., de la Cruz, F., and Dehio, C. (2003). A bacterial conjugation machinery recruited for pathogenesis. Mol. Microbiol. *49*, 1253–1266.

Shan, L. B., He, P., and Sheen, J. (2007). Endless hide-and-seek: Dynamic co-evolution in plant-bacterium warfare. Journal of Integrative Plant Biology *49*, 105–111.

Shaw, C. H., Ashby, A. M., Brown, A., Royal, C., Loake, G. J., and Shaw, C. H. (1988). *virA* and *virG* are the Ti-plasmid functions required for chemotaxis of *Agrobacterium tumefaciens* towards acetosyringone. Mol. Microbiol. *2*, 413–417.

Smith, H. (1977). Microbial surfaces in relation to pathogenicity. Microbiol. Mol. Biol. Rev. *41*, 475–500.

Sokurenko, E. V., Chesnokova, V., Dykhuizen, D. E., Ofek, I., Wu, X. R., Krogfelt, K. A., Struve, C., Schembri, M. A., and Hasty, D. L. (1998). Pathogenic adaptation of *Escherichia coli* by natural variation of the FimH adhesin. Proc. Natl. Acad. Sci. USA *95*, 8922–8926.

Sokurenko, E. V., Hasty, D. L., and Dykhuizen, D. E. (1999). Pathoadaptive mutations: gene loss and variation in bacterial pathogens. Trends Microbiol. *7*, 191–195.

Sory, M., Boland, A., Lambermont, I., and Cornelis, G. R. (1995). Identification of the YopE and YopH domains required for secretion and internalization into the cytosol of macrophages, using the *cyaA* gene fusion approach. Proc. Natl. Acad. Sci. USA *92*, 11998–12002.

Sory, M. P., and Cornelis, G. R. (1994). Translocation of a hybrid YopE-adenylate cyclase from *Yersinia enterocolitica* into HeLa cells. Mol. Microbiol. *14*, 583–594.

Stachel, S. E., Messens, E., Vanmontagu, M., and Zambryski, P. (1985). Identification of the signal molecules produced by wounded plant cells that activate t-DNA transfer in *Agrobacterium tumefaciens*. Nature *318*, 624–629.

Stahl, E. A., and Bishop, J. G. (2000). Plant-pathogen arms races at the molecular level. Curr. Opin. Plant Biol. *3*, 299–304.

Starr, M. P., and Chatterjee, A. K. (1972). Genus *Erwinia* – enterobacteria pathogenic to plants and animals. Annu. Rev. Microbiol. *26*, 389–426.

Staskawicz, B. J., Ausubel, F. M., Baker, B. J., Ellis, J. G., and Jones, J. D. G. (1995). Molecular genetics of plant disease resistance. Science *268*, 661–667.

Stathopoulos, C., Hendrixson, D. R., Thanassi, D. G., Hultgren, S. J., St Geme, J. W., and Curtiss, R. (2000). Secretion of virulence determinants by the general secretory pathway in Gram-negative pathogens: an evolving story. Microbes Infect. *2*, 1061–1072.

Stavrinides, J., and Guttman, D. S. (2004). Nucleotide sequence and evolution of the five-plasmid complement of the phytopathogen *Pseudomonas syringae* pv. maculicola ES4326. J. Bacteriol. *186*, 5101–5115.

Stavrinides, J., Ma, W. B., and Guttman, D. S. (2006). Terminal reassortment drives the quantum evolution of type III effectors in bacterial pathogens. PLoS Pathogens *2*, 913–921.

Stevens, C., Bennett, M. A., Athanassopoulos, E., Tsiamis, G., Taylor, J. D., and Mansfield, J. W. (1998). Sequence variations in alleles of the avirulence gene avrPphE.R2 from *Pseudomonas syringae* pv. phaseolicola lead to loss of recognition of the AvrPphE protein within bean cells and a gain in cultivar-specific virulence. Mol. Microbiol. *29*, 165–177.

Sun, Q. H., Hu, J., Huang, G. X., Ge, C., Fang, R. X., and He, C. Z. (2005). Type-II secretion pathway structural gene *xpsE*, xylanase- and cellulase secretion and virulence in *Xanthomonas oryzae* pv. oryzae. Plant Pathology *54*, 15–21.

Sundin, G. W., Mayfield, C. T., Zhao, Y., Gunasekera, T. S., Foster, G. L., and Ullrich, M. S. (2004). Complete nucleotide sequence and analysis of pPSR1 (72,601 bp), a pPT23A-family plasmid from *Pseudomonas syringae* pv. syringae A2. Mol. Genet. Genom. *270*, 462–475.

Swanson, J. K., Yao, J., Tans-Kersten, J., and Allen, C. (2005). Behavior of *Ralstonia solanacearum* race 3 biovar 2 during latent and active infection of geranium. Phytopathology *95*, 136–143.

Swarup, S., Defeyter, R., Brlansky, R. H., and Gabriel, D. W. (1991). A pathogenicity locus from *Xanthomonas citri* enables strains from several pathovars of *Xanthomonas campestris* to elicit cankerlike lesions on citrus. Phytopathology *81*, 802–809.

Swarup, S., Yang, Y. N., Kingsley, M. T., and Gabriel, D. W. (1992). A *Xanthomonas citri* pathogenicity gene, *pthA*, pleiotropically encodes gratuitous avirulence on nonhosts. Mol. Plant–Microbe Interact. *5*, 204–213.

Takeuchi, Y., Tohbaru, M., and Sato, A. (1994). Polysaccharides in primary cell walls of rice cells in suspension culture. Phytochemistry *35*, 361–363.

Tampakaki, A. P., Fadouloglou, V. E., Gazi, A. D., Panopoulos, N. J., and Kokkinidis, M. (2004). Conserved features of type III secretion. Cell. Microbiol. *6*, 805–816.

Tang, X. Y., Frederick, R. D., Zhou, J. M., Halterman, D. A., Jia, Y. L., and Martin, G. B. (1996). Initiation of plant disease resistance by physical interaction of AvrPto and Pto kinase. Science *274*, 2060–2063.

Tang, X. Y., Xiao, Y. M., and Zhou, J. M. (2006). Regulation of the type III secretion system in phytopathogenic bacteria. Mol. Plant–Microbe Interact. *19*, 1159–1166.

Tans-Kersten, J., Huang, H. Y., and Allen, C. (2001). *Ralstonia solanacearum* needs motility for invasive virulence on tomato. J. Bacteriol. *183*, 3597–3605.

Thomas, C. M., and Nielsen, K. M. (2005). Mechanisms of, and barriers to, horizontal gene transfer between bacteria. Nat. Rev. Microbiol. *3*, 711–721.

Thurn, K. K., and Chatterjee, A. K. (1985). Single-site chromosomal Tn5 insertions affect the export of pectolytic and cellulolytic enzymes in *Erwinia chrysanthemi* Ec16. Appl. Environ. Microbiol. 50, 894–898.

Tirtawid, S., Hadiwidj, T., and Lasheen, A. M. (1965). Citrus vein-phloem degeneration virus a possible cause of citrus chlorosis in Java. Proc. Am. Soc. Hort. Sci. 86, 235–243.

Toth, I. K., and Birch, P. R. J. (2005). Rotting softly and stealthily. Curr. Opin. Plant Biol. 8, 424–429.

Toth, I. K., Pritchard, L., and Birch, P. R. J. (2006). Comparative genomics reveals what makes an enterobacterial plant pathogen. Annu. Rev. Phytopathol. 44, 305–336.

Troisfontaines, P., and Cornelis, G. R. (2005). Type III secretion: More systems than you think. Physiology 20, 326–339.

Truman, W., de Zabala, M. T., and Grant, M. (2006). Type III effectors orchestrate a complex interplay between transcriptional networks to modify basal defence responses during pathogenesis and resistance. Plant J. 46, 14–33.

Tsiamis, G., Mansfield, J. W., Hockenhull, R., Jackson, R. W., Sesma, A., Athanassopoulos, E., Bennett, M. A., Stevens, C., Vivian, A., Taylor, J. D., and Murillo, J. (2000). Cultivar-specific avirulence and virulence functions assigned to *avrPphF* in *Pseudomonas syringae* pv. phaseolicola, the cause of bean halo-blight disease. EMBO J. 19, 3204–3214.

United States Department of Agriculture (2007) Plant pest program information. http://www.aphis.usda.gov (accessed 15 August 2007).

Valinsky, L., Manulis, S., Nizan, R., Ezra, D., and Barash, I. (1998). A pathogenicity gene isolated from the pPATH plasmid of *Erwinia herbicola* pv. gypsophilae determines host specificity. Mol. Plant–Microbe Interact. 11, 753–762.

Van der Hoorn, R. A. L., De Wit, P., and Joosten, M. (2002). Balancing selection favors guarding resistance proteins. Trends Plant Sci. 7, 67–71.

Van Sluys, M. A., Monteiro-Vitorello, C. B., Camargo, L. E. A., Menck, C. F. M., da Silva, A. C. R., Ferro, J. A., Oliveira, M. C., Setubal, J. C., Kitajima, J. P., and Simpson, A. J. (2002). Comparative genomic analysis of plant-associated bacteria. Annu. Rev. Phytopathol. 40, 169–189.

Vanetten, H. D., Mansfield, J. W., Bailey, J. A., and Farmer, E. E. (1994). 2 classes of plant antibiotics – phytoalexins versus phytoanticipins. Plant Cell 6, 1191–1192.

Vangijsegem, F., Gough, C., Zischek, C., Niqueux, Arlat, M., Genin, S., Barberis, P., German, S., Castello, P., and Boucher, C. (1995). The Hrp gene locus of *Pseudomonas solanacearum*, which controls the production of a type III secretion system, encodes 8 proteins related to components of the bacterial flagellar biogenesis complex. Mol. Microbiol. 15, 1095–1114.

Vanveen, R. J. M., Dendulkras, H., Schilperoort, R. A., and Hooykaas, P. J. J. (1989). Ti-plasmid containing *Rhizobium meliloti* are non-tumorigenic on plants, despite proper virulence gene induction and T-strand formation. Arch. Microbiol. 153, 85–89.

Velazquez, E., Peix, A., Zurdo-Pineiro, J. L., Palomo, J. L., Mateos, P. F., Rivas, R., Munoz-Adelantodo, E., Toro, N., Garcia-Benavides, P., and Martinez-Molina, E. (2005). The coexistence of symbiosis and pathogenicity-determining genes in *Rhizobium rhizogenes* strains enables them to induce nodules and tumors or hairy roots in plants. Mol. Plant–Microbe Interact. 18, 1325–1332.

Vivian, A., Murillo, J., and Jackson, R. W. (2001). The roles of plasmids in phytopathogenic bacteria: mobile arsenals? Microbiology-Uk 147, 763–780.

Wang, P. W., Morgan, R. L., Scortichini, M., and Guttman, D. S. (2007). Convergent evolution of phytopathogenic pseudomonads onto hazelnut. Microbiology-Sgm 153, 2067–2073.

Watanabe, K., Nagahama, K., and Sato, M. (1998). A conjugative plasmid carrying the *efe* gene for the ethylene-forming enzyme isolated from *Pseudomonas syringae* pv. glycinea. Phytopathology 88, 1205–1209.

Watson, D. W., and Brandly, C. A. (1949). Virulence and pathogenicity. Annu. Rev. Microbiol. 3, 195–220.

Weaver, L., Teakle, D. S., and Hayward, A. C. (1977). Ultrastructural studies on bacterium associated with ratoon stunting disease of sugarcane. Aust. J. Agric. Res. 28, 843–852.

Weber, E., and Koebnik, R. (2006). Positive selection of the Hrp pilin HrpE of the plant pathogen *Xanthomonas*. J. Bacteriol. 188, 1405–1410.

Weber, E., Ojanen-Reuhs, T., Huguet, E., Hause, G., Romantschuk, M., Korhonen, T. K., Bonas, U., and Koebnik, R. (2005). The type III-dependent Hrp pilus is required for productive interaction of *Xanthomonas campestris* pv. vesicatoria with pepper host plants. J. Bacteriol. 187, 2458–2468.

Weiler, E. W., Kutchan, T. M., Gorba, T., Brodschelm, W., Niesel, U., and Bublitz, F. (1994). The *Pseudomonas* phytotoxin coronatine mimics octadecanoid signaling molecules of higher plants. FEBS Lett. 345, 9–13.

Weintraub, P. G., and Beanland, L. (2006). Insect vectors of phytoplasmas. Annu. Rev. Entomol. 51, 91–111.

Weissman, S. J., Moseley, S. L., Dykhuizen, D. E., and Sokurenko, E. V. (2003). Enterobacterial adhesins and the case for studying SNPs in bacteria. Trends Microbiol. 11, 115–117.

Whitlock, M. C. (1996). The red queen beats the jack-of-all-trades: The limitations on the evolution of phenotypic plasticity and niche breadth. Am. Nat. 148, S65-S77.

Wichmann, G., Ritchie, D., Kousik, C. S., and Bergelson, J. (2005). Reduced genetic variation occurs among genes of the highly clonal plant pathogen *Xanthomonas axonopodis* pv. vesicatoria, including the effector gene *avrBs2*. Appl. Environ. Microbiol. 71, 2418–2432.

Wong, K., and Golding, G. B. (2003). A phylogenetic analysis of the pSymB replicon from the *Sinorhizobium meliloti* genome reveals a complex evolutionary history. Can. J. Microbiol. 49, 269–280.

Wood, D. W., Setubal, J. C., Kaul, R., Monks, D. E., Kitajima, J. P., Okura, V. K., Zhou, Y., Chen, L., Wood, G. E., Almeida, N. F., et al. (2001). The genome of the natural genetic engineer *Agrobacterium tumefaciens* C58. Science 294, 2317–2323.

Woolhouse, M. E. J., Taylor, L. H., and Haydon, D. T. (2001). Population biology of multihost pathogens. Science 292, 1109–1112.

Yang, B., Sugio, A., and White, F. F. (2005). Avoidance of host recognition by alterations in the repetitive and C-terminal regions of AvrXa7, a type III effector of *Xanthomonas oryzae* pv. oryzae. Mol. Plant–Microbe Interact. *18*, 142–149.

Yang, Y. N., and Gabriel, D. W. (1995). *Xanthomonas* avirulence/pathogenicity gene family encodes functional-plant nuclear targeting signals. Mol. Plant–Microbe Interact. *8*, 627–631.

Yao, J., and Allen, C. (2006). Chemotaxis is required for virulence and competitive fitness of the bacterial wilt pathogen *Ralstonia solanacearum*. J. Bacteriol. *188*, 3697–3708.

Yorgey, P., Rahme, L. G., Tan, M. W., and Ausubel, F. M. (2001). The roles of *mucD* and alginate in the virulence of *Pseudomonas aeruginosa* in plants, nematodes and mice. Mol. Microbiol. *41*, 1063–1076.

Zhang, Y. X., Bak, D. D., Heid, H., and Geider, K. (1999). Molecular characterization of a protease secreted by *Erwinia amylovora*. J. Mol. Biol. *289*, 1239–1251.

Zhao, Y. F., Ma, Z. H., and Sundin, G. W. (2005). Comparative genomic analysis of the pPT23A plasmid family of *Pseudomonas syringae*. J. Bacteriol. *187*, 2113–2126.

The Impact of Genomic Approaches on our Understanding of Diversity and Taxonomy of Plant Pathogenic Bacteria

Boris A. Vinatzer and Carolee T. Bull

Abstract

Our understanding of the diversity of bacterial plant pathogens has changed dramatically over the past 100 years. Initially, it was thought that each newly described disease was caused by a distinct bacterial species. Later, similarities in the physiology and codification of nomenclature resulted in the consolidation of taxa regardless of demonstrated diversity and host range. We have now entered an era in which genomic approaches can reveal genetic diversity in much finer detail. This has resulted in the development of phylogenetic trees, which identify theoretical evolutionary groupings at ranks below the level of the currently described bacterial species. There are cases in which these groupings coincide with previously defined pathovars or genomospecies while in other cases these groups represent biologically or ecologically relevant groups that still need to be defined taxonomically. In this chapter we give an overview of the history of systematics and of bacterial species concepts, describe current genomics approaches to uncover diversity, and finally discuss the potential consequences of the already uncovered genomic diversity on the taxonomy of plant pathogenic bacteria.

Introduction

The discipline of bacterial systematics examines the diversity and relationships of bacteria. Systematics involves the analyses of phylogeny as well as the evolutionary processes and genetic mechanisms by which microbial populations develop into ecologically relevant and distinct populations (Simpson, 1961; Goodfellow and O'Donnell, 1993; Goodfellow, 2000; Cohan and Perry, 2007). While population biology may look at short-term changes in population structure, systematics concerns itself with the long-term evolutionary processes that result in distinct stable populations (Cohan and Perry, 2007).

Phytobacteriologists are describing the genetic diversity within plant pathogen populations and are working to correlate the revealed genetic diversity to ecological and functional diversity within those populations. Like other bacteriologists, they are struggling to 'identify ecologically meaningful units of biological diversity' (Cohan and Perry, 2007). Whereas the species (Table 2.1) is the fundamental unit for many biological sciences including botany and zoology, in some cases the current operational bacterial species concept delineates taxa at levels above those of biological, genetic and ecological relevance. Within this context many of the fundamental biological units are described as sub-populations for which there are no codified taxa or nomenclature. A primary goal of systematics, through the practice of taxonomy, is 'to provide a framework of communication among microbiologists by furnishing names of microorganisms and descriptions to give meaning to those names' (Gordon, 1978). Adoption of a new species concept that accounts for ecologically relevant units or codification of taxa below the level of the operational species definition are two options that could help bacterial systematics to fulfil this goal in light of modern studies.

Here we look briefly at the history of bacterial systematics as it relates to phytobacteriology. We describe relevant bacterial species concepts; we explain how genomics approaches can be

Table 2.1 Definitions of bacterial taxa

Taxon	Definition[1]
Codified taxa	
Genus	Taxonomic group above the species level
Species	The basic taxonomic unit in bacterial systematics in general, a bacterial species is a group of strains that cluster due to shared features that may include phenotypic, genotypic and ecological characters
Subspecies	A distinct subgroup within a species as determined by phylogenetic and/or phenotypic analyses
Pathovar	A infrasubspecific group that can be differentiated by pathogenicity. The host range or disease caused by the group are distinct from other groups in the same species or subspecies
Non-codified infrasubspecific taxa	
Genomospecies	A group having 70% DNA-DNA similarity values and 5°C or less ΔT_m but that can not be differentiated by phenotype from other groups
Phylotype	A group that is part of the same phylogenetic cluster based on the analysis of three loci
Sequevar	A group of strains with identical or almost identical DNA sequence in three loci
Race[2]	A group that can be differentiated by host range from other groups within a species or subspecies (at the plant species level for *R. solanacearum* races and at the cultivar level for all other bacterial plant pathogens)
Biovar	A group with the same biochemical properties
Sequence type	A group with the same allelic profile based on an MLST analysis

[1]Definitions of genus, species and subspecies may vary with interpretation and opinion of individual researchers due to the concept of scientific neutrality (Lapage *et al.*, 1992).
[2]The most widely accepted definition of race is pathogenic differentiation on separate cultivars within the same plant species.

used to uncover and quantify diversity, and how these approaches have been used already in some significant taxa of plant pathogenic bacteria. We evaluate the potential impact of these findings on the systematics of these taxa and discuss advantages and disadvantages of using these methods to develop classifications.

The importance of bacterial taxonomy for phytobacteriology

Although taxonomy is often equated with systematics it is generally considered to be a subdiscipline of systematics (Simpson, 1961; Sneath and Sokal, 1973). Taxonomy is the practice of classification, nomenclature, and identification (Staley and Krieg, 1984; Young *et al.*, 1992; Goodfellow and O'Donnell, 1993; Trüper and Schleifer, 2006). The importance of these three elements continues to change, depending on the newest methods available and goals of the scientific era. To understand taxonomy, we must evaluate how our concepts of bacterial species impact its practice of taxonomy because species are the fundamental units of taxonomy (Staley and Krieg, 1984; Wayne *et al.*, 1987; Goodfellow and O'Donnell, 1993; Vandamme *et al.*, 1996).

The subjective era of classification

Historically, the need for classifying and understanding relationships came out of the need to identify pathogens (human and agricultural) to help prevent or cure the diseases caused by these organisms (Starr, 1959; Stolp *et al.*, 1965). Initially, limited phenotypic characters of organisms grown in pure culture under various conditions were the only characteristics available for characterization and classification. Classification based on gross morphological characteristics and later on phenotypical characters including nutrient requirements and disease symptoms were used for over 30 years to distinguish among plant pathogenic bacteria (Dowson, 1949; Starr, 1959, 1983). This basic departure point of evaluating differences in physiological traits was useful to identification of bacteria but not necessarily in their classification (Dowson, 1949; Starr, 1959). Species were monothetic groups for which the traits for identification were both sufficient and necessary for identification (Goodfellow and O'Donnell, 1993). Species were defined by traits chosen to demonstrate differences rather than trying to determine which organisms were most similar.

Identification based on distinguishing characters marked an era of subjective classification that lacked rigorous testing of the validity of the groupings against scientific hypotheses. Artificial or special purpose classification systems developed with taxa based on a few convenient shared characters, irrespective of any relationship. Species were based primarily on the goals, or prejudice of the researcher and resulted in species of unequal ranks (Cowan, 1962; Brenner, 1981; Williams et al., 1984; Goodfellow and O'Donnell, 1993).

Phytobacteriologists placed significantly more importance on pathogenicity rather than other characteristics. Thus, plant diseases became equivalent to the pathogen causing them and resulted in what is known as the 'new host–new species cliché' (Starr, 1959, 1983; Stolp et al., 1965; Dye and Lelliott, 1974). Plant pathologists assigned a new specific epithet to bacteria causing diseases on plants from which bacterial plant pathogens had not previously been isolated and/or diseases that looked different than other bacterial diseases on a particular host. Although host range was supposedly the primary criteria used for distinguishing organisms, Starr (1959) describes attempts to define the host range of many pathogens as feeble. In many cases no attempt was made to compare the causal agent to other plant pathogens, let alone related non-phytopathogenic species. Thus, many descriptions of new diseases had inadequate and imprecise descriptions of the causal organisms and resulted in numerous synonyms for nearly identical pathogens (Starr, 1959, 1983; Stolp et al., 1965; Dye and Lelliott, 1974; Young et al., 1992). Although the goal of this system was utilitarian, its usefulness was limited due to the lack of understanding that one organism could cause multiple diseases.

In addition to the lack of clear description for the causal agents of described diseases, a number of phytobacteriologists practiced a type of discipline insularity that kept them from recognizing that bacterial plant pathogens could be related to other bacteria. Starr (1959) thoroughly describes the differences in opinion among bacteriologists who either proposed taxa based primarily on their ability to cause diseases on plants, regardless of their affiliation with other bacteria, and those who realized that plant pathogenic bacteria could not be separated away from other taxa based on their pathogenicity alone. When Starr (1959) eloquently implored phytobacteriologists to leave the vacuum of their discipline and to evaluate the pathogens in the context of other bacteria, he was voicing the need for an objective classification scheme.

The objective era of classification

With the advent of the objective era of classification the practice of classification moved toward the development of 'natural classification' schemes based on overall resemblance rather than on 'artificial classification' schemes, based on a few traits subjectively chosen to emphasize the differences in bacteria (Sneath and Sokal, 1973: Goodfellow and O'Donnell, 1993; Rosselló-Mora and Amann, 2001). The special purpose classification for plant pathogens discussed above is an example of one type of artificial classification. The term natural classification may be confusing because the meaning of this term has changed over time. Initially, natural classification referred to a phenetic classification system but has also been used to refer to phylogenetic classification systems (Goodfellow and O'Donnell, 1993; Rosselló-Mora and Amann, 2001). In general, natural classification refers to schemes that attempt to be objective and to include all available data. The beginning of the objective era of classification was commensurate with the use of statistical analysis to test hypotheses related to the circumscription of taxa (Sneath and Sokal, 1973; Rosselló-Mora and Amann, 2001).

Applying a phenetic classification, organisms are grouped so that the members of a group have the maximum amount of similarity. All available traits, both phenotypic and genotypic can be used to develop phenetic classifications. Many traits are shared by strains within taxa delineated by these methods but strains are not required to have a particular trait in order to belong to a particular taxon. The Adansonian principles used in phenetic classifications require that equal weight be given to each trait (Sneath and Sokal, 1973; Goodfellow and O'Donnell, 1993). This allows the data to be codified, evaluated statistically and allows for hypotheses testing. Thus, modern phenetic classification is not just based on a few convenient phenotypic traits of organisms but is 'based on overall similarity, as determined by equal weighting of all known characters' (Goodfellow and O'Donnell, 1993). Identification of unknowns can be more difficult since diagnostic tests involving single or even a few traits may be misleading. For example, the lack of fluorescence in some isolates of *Pseudomonas syringae* may be confusing since *P. syringae* strains are considered to be fluorescent pseudomonads.

In early phenetic classification schemes few characters were used because the computational needs of large data sets could not be handled. The advent of more powerful computing technology allowed larger data sets to be analysed readily. Numerical taxonomy is a type of phenetic classification using as many characters as possible, giving equal weight to each (Sneath, 1957). The key advantage of this system of classification according to Sneath and Sokal (1973) is that the scientific method can be applied within its framework because it is repeatable and objective.

The large number of traits and the statistical methods used to group organisms in numerical taxonomy has led researchers to draw phylogenetic inferences from the analysed data (Sneath and Sokel, 1973). However, the resulting analyses are only estimates of evolutionary relationships that are based on the likelihood that the groupings, as determined by statistical analysis, represent evolutionarily defined species. It is assumed that through the analysis of numerous character states, clusters based on resemblances will provide information from which phylogeny might be implied since the sum of the character states should be the observable expression of the genotype.

Phylogenetic classification schemes group organisms based on the evolutionary history of the organisms and involve the analysis of cladistic relationships (Goodfellow and O'Donnell, 1993). The establishment of Darwin's theory of evolution as a core principle of biology has driven systematists to classify organisms phylogenetically to address the evolutionary basis of organismal diversification and speciation. However, the practice of determining evolutionary relationships is not straightforward.

Amino acid sequences of conserved molecules or the nucleotide sequences of the genes from which they are derived have been described as a set of molecular evolutionary clocks that hold the promise of being a 'living fossil record' for bacteria (Woese, 1987). The primary premise of using DNA sequences to infer phylogeny, is that the DNA sequences for closely related organisms will be more similar than those of distantly related organisms. However, researchers are often mistaken in thinking that the use of genotypic data in taxonomy leads directly to phylogenetic classifications. Genotypic and/or phenotypic data can be used in either phenetic or phylogenetic classifications. The distinction is in how the data are approached. Phenetic analyses evaluate sequence similarity and each nucleotide along the DNA segment is given equal weight in the classification. In phenetic classification it doesn't matter whether a sequence difference represents a monophyletic group or was gained by lateral transmission or random convergence. Phylogenetic analyses use cladistic approaches (where each branch in a tree is believed to descend from a common ancestor) to evaluate the evolutionary history of the organisms. The underlying assumption of the approach in phylogenetic classification is that the evolution of DNA molecules approximates the rate and significance of ecological, physiological, and metabolic evolution. The constructed trees are evaluated for the likelihood that they represent the evolutionary relationship among the organisms being studied.

Approximations of DNA similarity began to be used in the late 1950s and by 1990 microbial systematics had firmly adopted molecular tools for use in classification. Measurement of DNA base composition was one of the first tools based on DNA sequences to be employed for classification of bacteria (Lee et al., 1956; Belozersky and Spirin, 1958). This tool is based on the observation that DNA from different phenetic groups differed in their % G + C content and members of the same groups had relatively similar % G + C (Lee et al., 1956). Rosselló-Mora and Amann (2001) give an excellent explanation of this method and how it is applied. The % G + C content of major genera of plant pathogenic bacteria were considered to be salient features (Starr, 1983), but did not help to distinguish species. For example, Starr and Mandel (1969) determined that the % G + C content for *Erwinia* spp. fell within 50–58% range that is common for members of the Enterobacteriaceae. However, this method is not able to distinguish between pectolytic *Erwinia* from *Erwinia amylovora* although these groups can be separated by phenotypic classification methods (Lockhart and Koenig, 1965; Starr and Chatterjee, 1972; Brenner et al., 1974; Hauben et al., 1998).

The advent of DNA–DNA hybridization

DNA–DNA hybridization (DDH) is a primary strategy used for circumscribing bacterial species today. DDH estimates DNA similarity by quantifying the formation of hybrids by single strands of DNA from different organisms in pairwise comparisons (Rosselló-Mora and Amann, 2001; Stackebrandt et al., 2002). This strategy was developed in the 1960s and by the early 1980s several straightforward, though time-consuming, methods were developed (De Ley et al., 1970; Grimont et al., 1980; Johnson, 1984). These methods are time-consuming because of the number of pairwise comparisons that need to be made with highly purified DNA. Newer methods promise to be less labour intensive and maintain the same level of discernment (Christensen et al., 2000; Cho and Tiedje, 2001). As early as the 1970s this strategy was being applied to plant pathogens (De Ley et al., 1970; Brenner et al., 1972). By the 1980s it was recognized that this tool held the promise of providing a harmonized criterion for species across all taxa (Wayne, et al., 1987).

The thermal denaturation midpoint (T_m) of hybrids has been shown to decrease from 1% to 2.2 % for each percent mispairing (Bautz and Bautz 1964; Ullman and McCarthy, 1973) indicating that DNA sequence is related to thermal stability. This information is used in conjunction with DDH data to circumscribe bacterial species. Members of the same bacterial species have at least 70% DDH values and ΔT_m of less than 5°C (Wayne et al., 1987; Stackebrandt et al., 2002). Among the advantages of using this as the minimal criterion for a species is that it imposed consistency in the rank of species over the range of bacterial variation and a quantitative definition that could be used without scientific bias. For the last 15 years this definition has been used as the defining standard for bacterial species including plant pathogens. Recently, acceptance of this approach was reconfirmed and 'for the time being, the parameters of DNA-DNA similarity and, whenever determinable, a ΔT_m remain the acknowledged standard for species delineation' (Stackebrandt et al., 2002).

Use of 16S rDNA sequences for phylogenetic inference

The movement toward using DNA sequences to infer phylogenetic relationships was greatly advanced by analysis of DNA sequences coding for small subunit ribosomal RNAs (16S rDNA). Small subunit ribosomal RNAs are known as robust molecular evolutionary clocks because they are universally present in all organisms, have conserved functions among organisms, but have both conserved and variable sequences (Woese, 1987). The development of nearly universal primers from the conserved regions of the molecules have allowed for directly sequencing 16S rDNA from isolated cultures or directly from bacteria present in environmental samples (Lane et al., 1985; Pace et al., 1986). Thus, this molecule and its coding sequence have become 'the undisputed standard for determining phylogenetic relationships' (Young, 1994). As evidence of this conclusion, in November 2007 the Ribosomal Database Project had 16S rDNA sequences for over 58 000 strains and over 97 000 uncultured organisms and sequences continue to be added at a breakneck pace (Cole et al., 2007).

Although analysis of 16S rDNA sequences can result in phylogenies for cultured and uncultured organisms, this technology has not replaced DDH as the standard for delineation of bacterial species. This is in part because the relationship between DDH and 16S rDNA sequence similarity is not linear. In some cases, sequence identity does not correlate to DDH values of 70% (Fox et al., 1992). According to Stackebrandt and Goebel (1994), in general, DDH values and 16S rDNA sequence similarities can be correlated but there is no threshold value which, when met consistently, predicts species discernment. They concluded, however, that organisms sharing less than 97% 16S rDNA similarity are unlikely to belong to the same species.

Although analysis of 16S rDNA sequences is useful for the evolutionary inferences about higher taxa it is less useful for distinguishing among plant pathogens from the same species causing different diseases. In general, the sequence variation among closely related strains is not high enough to provide markers for discrimination. The use of other genes and the use of multiple genes simultaneously for analyses of phylogenetic relationships and pathogen diversity is discussed below.

Polyphasic classification

Some authors contend that phylogenies inferred from genetic data may not be accurate because of differences in evolutionary rates and bacterial phenotypes hold important information that we are not yet able to decipher from the primary structure of DNA (Vandamme et al., 1996: Goodfellow et al., 1997). As with other scientific fields, taxonomic hypotheses are more rigorous when more types of data support these hypotheses. Polyphasic approaches involve successive or simultaneous taxonomic studies of various types of data including genetic and phenotypic data as well as phylogenetic inference for development of classifications. Like numerical classifications, polyphasic classifications don't form keys for identification. No single criterion or small group of traits is sufficient to place an organism in a taxonomic group including results from DDH and Δ Tm analyses. Extensive and labour-intensive polyphasic classifications are among the most comprehensive and stable for bacterial plant pathogens, but may not always be feasible.

Final thoughts on bacterial classification

Due to the influences of the ideas put forth by Linnaeus and Darwin, systematists strive to develop classification systems, which attempt to arrange the naturally occurring bacterial diversity into a hierarchical system indicating the extent of evolutionary relatedness. This is a difficult task because advances in technology continually change our ability to explore phylogenetic relationships and diversity. Technology has essentially driven the move from a subjective era towards an objective era in taxonomy and genomics is poised to fuel further progress in this field. However, technological progress can inadvertently create problems for researchers. According to Young (2000), 'new methods allow the recognition of distinction among populations but they do not necessarily permit easy identification of isolates or their identification to appropriate taxa.' Additionally, new technologies do not just influence classification but, as we describe below, they also influence our concept of bacterial species.

Each classification system has various merits but the ultimate goal of taxonomy is a clear framework that allows researchers to understand relationships among bacteria and communicate effectively. Each needs to be evaluated on its ability to provide that framework. Through a scientifically neutral process the appropriateness of classifications and the methods used to develop them are evaluated through peer review (Tindall, 1999). Acceptance occurs as researchers agree with the methods and conclusions presented and begin using the associated nomenclature in the literature. This places a huge burden on the average researcher who may not have the taxonomic training needed to make these judgements. They may mistakenly believe that the classification published the most recently is the 'correct' classification (Tindall, 1999). However, there is no 'correct' classification per se but instead a classification system and its associated nomenclature becomes the preferred system as it is adopted by scientific consensus and use in the literature.

Nomenclatural codification

Although there is no official classification, there are specific rules for naming organisms proposed in classification systems. Often, the surprisingly heated controversies in taxonomy revolve around the strong opinions about nomenclature since 'Nomenclature is central to all facets of the microbial sciences ... the name of an organism is the key to its literature, an entry to what is known about it' (Goodfellow, 2000).

The nomenclatural rules are codified in the current version of 'The International Code of Nomenclature of Bacteria' (Lapage et al., 1992). These rules and amendments to them provide the nuts and bolts for naming bacteria. 'The Approved Lists of Bacterial Names' (Approved Lists: Skerman et al., 1980) provided a comprehensive list of names of bacteria that conformed fully to code and a new starting point for nomenclatural priority. Bacterial names present on the lists represented taxa for which there were reference cultures and adequate descriptions of the species. Names not on the lists no longer had any standing in bacterial nomenclature. Species published after 1980 are validated by publication in the 'Validation Lists' published by the 'International Committee on Systematics of Prokaryotes' in the 'International Journal of Systematic and Evolutionary Microbiology' or by primary publication in that journal. The 1980 Approved Lists had a huge impact on phytobacteriology in that many plant pathogen species did not make it on to the Approved Lists. All of the information associated with those species would have been lost if some alternative provision was not made. The International Standards for Naming Pathovars of Phytopathogenic Bacteria (Dye et al., 1980) was published in the same year as the Approved Lists. The infrasubspecific category, pathovar, was created to preserve the information associated with various pathogens until the research could be completed to determine which of these former species could be proposed as legitimate, validly described species. These standards were revised slightly in 2001 (Young et al., 2001). To help plant pathologists and other bacteriologists keep up with the rapidly changing nomenclature of bacterial plant pathogens, the 'International Society for Plant Pathology Committee on the Taxonomy of Plant Pathogenic Bacteria' maintains Lists of Names of Plant Pathogenic Bacteria (http://www.isppweb.org/about_tppb.asp) and has published several articles explaining misconceptions about the nomenclature of plant pathogenic bacteria. In these documents, as with the lists, the most recent nomenclature is not by default the most appropriate to use. Individual researchers must analyse the taxonomic literature for the groups of interest in order to determine which classification scheme is the most appropriate and therefore which nomenclature is appropriate.

Lastly, we need to emphasize that for all taxonomic studies or studies of diversity the appropriate type and pathotype strains should be included and identified in the research. This is not because the type strain is typical of the species or group being studied but because it is the name-bearing strain. The type strain or pathotype needs to be included for nomenclatural purposes. If the appropriate type and pathotype strains are not included in studies on bacterial diversity, no proposals for changes to classification and nomenclature can be made.

Species concepts

In contemplating bacterial species concepts, one is first confronted with debates about the existence of bacterial species (Cowan, 1962; Stolp et al., 1965; Ghiselin, 1974; Van Regenmortel, 1997) and whether a species concept should be universal for all living organisms (Hull, 1997; Mayden, 1997; Rosselló-Mora and Amann, 2001; Ward et al., 2008). Bacteria are significantly different from eukaryotes in many ways, but their asexuality and ability to transfer genetic material between distantly related organisms raises questions about the applicability of eukaryotic species concepts to prokaryotes. Moreover a bacterial species concept may need to be flexible enough to allow that in different environments speciation may occur due to different forces (Ward et al., 2008). In situ and culture-based evidence suggests that bacterial diversity can be circumscribed into genetically and phenotypically distinct clusters recognizable as species (Cohan, 2002; 2006; Hanage et al., 2006; Ward et al., 2008). However, there is no consensus among microbiologists about the nature of bacterial species. Although many variations of bacterial species concepts exist, they are variations and combinations of essentially three species concepts: Biological, Evolutionary, and Ecological. A main difficulty for applied plant pathology is developing a species definition based on a given species concept that allows for the delineation of species and identification procedures. Here we describe a few bacterial species concepts some of which have potential to lead to an understanding of bacterial diversity and a coherent and usable taxonomy.

The 'biological species concept' (BSC) put forth by Mayr (1942) is perhaps the most broadly recognized by biologists and non-scientists alike, but its applicability to bacteria is questionable. According to this concept 'A species is a group of interbreeding natural populations that is a reproductively isolated from other such groups' (Mayr and Ashlock, 1991). The ability to exchange genetic material is proposed to be the main cohesive force (Templeton, 1989). It is a stretch to apply the biological species concept to asexually reproducing organisms, (Stackebrandt and Goebel, 1994; Claridge et al., 1997; Embley and Stackebrandt, 1997). However, authors interpret the BSC in various ways in order to make bacteria fit the concept. In the past, some researchers argued that because bacteria are clonal they are essentially reproductively isolated and therefore fit this concept. Alternatively, because they exchange genetic material some consider them to be 'interbreeding' (Dykhuizen and Green, 1991; Claridge et al., 1997). Since the advent of multilocus sequence typing (MLST) (Maiden et al., 1998), exchange of DNA by homologous recombination between bacteria has been found to be widespread in many species (Feil et al., 2001). Homologous recombination often occurs as a consequence of sexual conjugation (Narra and Ochman, 2006). Because the efficiency of homologous recombination decreases with the increase of sequence divergence (Majevski and Cohan 1999), Fraser et al. (2007) propose that 'sexual isolation' is an important factor in bacterial speciation giving support to a bacterial BSC. However, genetic exchange during conjugation is limited to only a portion of the genome. Therefore, Staley (2006) uses the term 'pseudo-biological species concept' when referring to a BSC in bacteria.

Just as evolutionary theory has influenced applied taxonomy, many bacterial species concepts are based on evolutionary principles. The 'evolutionary species concept' defines species as 'the largest aggregate of individual organisms that evolve as a unit' (Wiley, 1978). This population would by definition share a common evolutionary fate through time (Simpson, 1961). In this concept, prokaryotes and other asexual organisms are not excluded because of constraints on reproductive style. According to Simpson (1961) an evolutionary species is a 'lineage evolving separately from others and with its own unitary evolutionary role and tendencies.' This was the only universally acceptable species concept (Mayden, 1997).

The evolutionary species concept has been further refined. For example, the 'phylogenetic species concept' describes a species as the smallest unit that is diagnosable and monophyletic (Eldredge and Cracraft, 1980; Cracraft, 1983; Nixon and Wheeler, 1990; Wheeler and Platnick, 2000). For this concept species are defined as populations of organisms that are diagnosable by a unique combination of character states (Nixon and Wheeler, 1990; Claridge et al., 1997). Phylogenetic bacterial species have most recently been delineated through polyphasic analyses.

The 'ecological species concept' put forth by Van Valen (1976) is a version of the evolutionary species concept, which includes the criterion that different species exploit different ecological niches. Van Valen suggests that a species consists of organisms from the same or closely related lineages that occupy the same niche. This concept credits the niche as the primary force in cohesion of bacterial lineages (Van Valen, 1976). Organisms occupy different niches by either utilizing different resources or using the same resources in different locations or at different times. Changes in single genes can change functional fitness and ecological niche. Van Valen's assumptions included that selection acts primarily on phenotypes and that genes are of minor importance in evolution. The main difficulty in using this species concept for taxonomy is in defining ecological niches. Although host range would be an easy criterion for measuring differences in ecological niches for plant pathogens, use of this concept could again result in special purpose classifications derived by the favourite niche-differentiating factor of the researcher. Without clear quantitative measurements of niche differences it is difficult to develop a species definition from this concept that can be applied equitably across all groups.

The 'genomic-phylogenetic species concept' (GPSC) is similar to the phylogenetic species concept but allows for the use of sequence-derived phylogenies to determine species boundaries. Sequence-derived phylogenies can detect genetic drift or adaptive radiation that arise due to differences in the biogeography of organisms (Staley, 2004; 2006). Like the ecological species concept the GPSC assumes that the niche of an organism is the primary cohesive feature of a species, and that differences in the molecular sequences directly reflect differences in niches. This methodological approach can account for significant genetic diversity that occurs within some species as defined by the operational species concept, thus its application would result in more species defined with greater precision. Additionally, described genomospecies (see Table 2.1) could probably be proposed as species using this concept. The GPSC is being developed concurrently with the ability to rapidly sequence large amounts of DNA (including whole genomes) for minimal cost. As described below, estimates of average nucleotide identity (ANI) using for example multilocus sequence analysis (MLSA) can be used to infer the phylogenies for this species concept. The genetic data used in this approach, like others based on sequence analysis, is additive, archival and portable (Staley, 2006). However, some researchers are uncomfortable with the reliance on sequence data for taxonomic inference because its use is based on the assumption that the phenotype of the organism is irrelevant since all of the information about the organism is coded in the nucleotide sequence of the bacterial genome (Wayne *et al.*, 1987; Stackebrandt *et al.*, 2002; Staley, 2006; Goris *et al.*, 2007). The current obstacle in its application to taxonomy appears to be in determining what the threshold level is for species identity (Goris *et al.*, 2007; Ward *et al.*, 2008).

The 'ecological-evolutionary species concept' like the GPSC is adapted from a mixture of ecological and evolutionary species concepts. It theorizes that species are 'populations with unique gene assemblages and thus play pivotal roles in regulating community function in space and time according to how environmental changes alter their abundances' or 'unique ecologically adapted populations' (Ward *et al.*, 2008). This concept requires both ecological and genetic divergence between populations before two separate species can be circumscribed. The fundamental unit of biology is referred to as an ecotype. This concept is based on evolutionary and ecological principles and could be exceptionally relevant to plant pathology primarily because the concept is being developed based on observations of genetic diversity of microbes in natural settings. The concept is highly flexible and can accommodate different mechanisms that are proposed to introduce genetic variation into populations as well as difference in environmental influences on those mechanisms (Ward *et al.*, 2008). Species from one environment may have a different level of divergence than species in other environments due to differences in speciation mechanisms and the amount of time that has passed since the species diverged. It also recognizes that even small genetic changes, for example the acquisition of pathogenicity islands, can have significant ecological impacts. Practically, species as conceived here can be based on genetic variation analysed *in situ* or from pure cultures. As with other sequence-based concepts, identification could be based on determining which phylogenetically discrete population novel organisms belong to. However, this concept also indicates that some measure of ecotype specificity also needs to be demonstrated. Demonstration of spatiotemporal specificity of distribution or gene expression can be used to measure ecological distinctions. Correspondence between sequence clusters and ecological parameters are used to predict ecotypes. This framework for taxonomy could again result in species of different ranks and a new subjective era if some standardization or benchmarks are not presented with the results of these studies.

The operational species definition

As discussed, DNA–DNA hybridization (DDH) at or above 70% and ΔT_m of less than 5°C serve as an unofficial species definition for prokaryotes although methods for classification are not regulated and remain scientifically neutral (Wayne et al., 1987; Tindall, 1999; Stackebrandt et al., 2002). This definition is based on the hypothesis that hybridization is directly influenced by the genomic sequences of the pairs being analysed and can therefore be a measure of sequence similarity. DDH continues to be used as a surrogate for comparisons of complete genome sequences in the practice of delineating species based on phylogenetic species concepts.

Although this definition has done much to stabilize bacterial taxonomy it lacks several key elements needed in modern bacterial systematics. In particular, the threshold was derived from and can only be applied to cultured organisms and this threshold may not be consistent for all bacteria (Vandamme et al., 1996; Ward et al., 2008). Also, named species based on this definition do not provide a clear picture of bacterial diversity since these groups often consist of smaller diagnosable ancestral groupings occupying disparate niches (Ward, 1998; Gardan et al., 1999; Cohan, 2002; Staley, 2004; Gevers et al., 2005; Knostantions et al., 2006; Ward et al., 2008). Lastly, pairwise comparisons that are relative only to other comparisons made in the same laboratory or within the same experiment will not allow us to accumulate data electronically and are not portable in that other researchers can not use them in additional experiments. Our experience with 16S rDNA databases has demonstrated the portability of electronic databases for taxonomic analysis and this is now an important criterion by which to evaluate technologies used for species delineation. Sequence data only need to be generated once and can be used inexhaustibly for every subsequent experiment.

The thresholds set by this definition were determined empirically but genomic analysis is revealing that these thresholds correspond well to sequence-derived phylogenetic groupings. The definition threshold corresponds well to organisms that have approximately 95% average nucleotide identity (ANI: Konstantinidis and Tiedje, 2005; Goris et al., 2007, explained in detail below). With further testing it is likely that 95% ANI may become the new definition (Konstantinidis et al., 2006) though it may not have relevance to any species concept based on evolutionary and ecological principles.

Conclusions on bacterial species concepts

Bacteriologists continue to require determinative strategies to identify animal and plant pathogens to diagnose disease. Many taxonomists are reluctant to give up phenotypic determinative tests for practical reasons. Many assume we do not understand the blueprints well enough to define the impact on the finished product and thus are reluctant to rely on sequence-based phylogenies for classification and identification (Konstantinidis and Tiedje 2005; Konstantinidis et al., 2006). However, the acceptance of the status *Candidatus*, circumscribed using molecular taxonomy (Murray and Stackebrandt, 1995) may provide the experience needed to allow researchers to become comfortable with sequence-dependant taxonomy. Regardless, species concepts and the taxonomic strategies must practically delineate species and allow organisms to be identified. Species based on both evolutionary and ecological principles appear to do the best job at describing the fundamental units in bacterial diversity. Currently, it is difficult to develop species definitions that can be used for delineation and identification of species from concepts requiring demonstration of niche separation. Identification of 'species-specific diagnostic genetic signatures' (Konstantinidis and Tiedje, 2005) should help us to develop species definitions based on these species concepts for diagnostic purposes (see below).

In practice, bacterial classification currently relies on a mixture of theoretical species concepts (Ward, 1998). Phytobacteriologists are using the operational species concept to develop classifications based on the currently accepted standard, however, much of plant pathogen research consists of evaluating bacterial diversity and its impacts on disease at taxonomic levels far lower than those distinguished by the operational species definition. Therefore, the current taxonomic situation regarding plant pathogenic bacteria is unsatisfactory.

Genomic approaches to estimate or measure diversity

Phylogenetic analysis using multilocus sequence typing (MLST)

Multilocus sequence typing (MLST) was first described in 1998 (Maiden et al., 1998). MLST is a molecular typing method that consists of sequencing 400–600 bp long fragments of at least six housekeeping genes, i.e., genes that code for proteins that perform basic metabolic functions, are present in most bacteria, and are under purifying selection. Arbitrary numbers are assigned to each sequenced allele for each gene fragment. Organisms with identical alleles in each sequenced locus are given the same allele profile designation referred to as sequence type (ST). The two most important advantages of MLST over 16S rDNA sequencing are:

1. The higher variability of housekeeping genes compared to the 16S rDNA sequence and the increased length of the total analysed sequence allow differentiation of strains at taxonomic levels below the operational species level.
2. Sequencing at least six genes reduces the risk that horizontal gene transfer obscures the resulting phylogeny.

Phylogenies based on MLST were found to be very similar to phylogenies based on all genes shared between analysed genomes (Konstantinidis et al., 2006) and somewhat different from phylogenies based on 16S rDNA squences alone. The term MLSA (multilocus sequence analysis) is used when instead of STs the concatenated set of obtained DNA sequences is used for analysis. While MLST is used in epidemiological studies for strains within a known species, MLSA is usually used to determine to which species strains of uncertain identity belong or to better define the phylogenetic relationship between named species within a genus (Gevers et al., 2005).

The advantage of MLST and MLSA over molecular typing methods that are not DNA sequence-based, for example, amplified fragment length polymorphism (AFLP) (Janssen et al., 1996) or DDH, is their portability, i.e., once data are obtained for strains and deposited in a internet-accessible database (for example at www.mlst.net), any lab around the world can compare their own strains to the strains in the database (Maiden, 2006). With non-sequence-based molecular methods or with DDH this is not possible as every lab has to repeat the pairwise comparisons, for type and pathotype strains in every analysis. Moreover, with MLST/MLSA, every lab around the world can add data obtained on their strains to the central database so that the database is continuously expanded.

MLSA was very successful in attributing isolates to well-separated clusters with each cluster corresponding to a separate species. This was possible even in the frequently recombining genus *Neisseria* (Hanage et al., 2005). Therefore, MLSA is a promising approach to reliably determine the phylogenetic relationships between bacteria. The strength and drawback of this method is that clusters can be identified at various levels irrespective of species demarcations. Therefore, the simple fact that a group of strains form a cluster does not by itself support the description of this cluster as a species. Depending on the species concept one applies, further evidence will be needed: for example, genetic distance between strains within the cluster and genetic distance from the next cluster may be required to meet certain thresholds or all strains in the cluster need to have the same phenotypic characteristic that is absent from all strains in the neighbouring clusters.

Estimation or measurement of average nucleotide identity (ANI)

While sequencing of a single fragment, the 16S rDNA, was the standard for estimating phylogenetic relationships between strains in the pre-genomics era, the ANI (Konstantinidis and Tiedje, 2005) between genomes can be considered the standard measurement in the genomics era. ANI is computed for pairs of genomes by either comparing all genes conserved between the two genomes (ANI_g) or by comparing only those genes that are conserved between all sequenced genomes of the entire group of bacteria that is being analysed (ANI_o; Konstantinidis et al., 2006). Obviously, ANI can only be calculated when the entire genome of at least two strains has been sequenced. However, once two or more strains of a species have been sequenced, ANI can be estimated for additional strains using a small number of genes after it has been verified that these genes in the sequenced strains have an ANI close to the overall ANI (Konstantinidis et al., 2006). Below we discuss ANI values in regard to *P. syringae*, *Xanthomonas*, and *Ralstonia solanacearum*. For these discussions the ANI values are always estimates since ANI has only been calculated for a minority of genomes of these species.

MLSA is one good approach to estimate ANI. Alternatively, ANI can be estimated based on careful selection of three genes that give a phylogenetic tree most similar to a tree built on all shared genes between genomes. This approach can give an even better estimate of ANI and a phylogeny that is closer to a phylogeny based on all genes in a genome than an MLST analysis using eight randomly selected housekeeping genes (total sequenced length 9500 nt) because genes subject to frequent recombination are avoided (Konstantinidis et al., 2006). For the purpose of typing bacteria to develop classification schemes, estimating ANI using three genes or performing an MLST analysis based on six or eight genes are very similar approaches and the advantages of one over the other are subtle. The choice of which approach to use will mainly depend on the additional specific objectives of the research. For example, if a recombination analysis is the goal (see below), MLST will be chosen over an analysis of three genes. Regardless, to be able to compare values of ANI obtained in independent studies and to use ANI when describing new species, a strict standardization on how to estimate or calculate ANI will need to be established.

Comparing ANI and DDH values, it was found that an ANI of 94–95% corresponds well to a DDH value of 70% (Konstantinidis and Tiedje, 2005; Goris et al., 2007). However, this cannot be generalized. The limitation of ANI is that it is calculated based on genes that are shared between the two genomes that are compared. How many genes are shared between genomes compared to the total number of genes in each of the two genomes does not influence ANI. Therefore, ANI is not a sufficient measurement of relatedness. For example, since the two *Xanthomonas campestris* pv *campestris* strains sequenced so far (da Silva et al., 2002; Qian et al., 2005) are almost 100% identical in the DNA sequence of their housekeeping genes, they have an estimated ANI of almost 100%. However, 12 % of their genomes do not align and the two strains have different host ranges (Qian et al., 2005). Therefore, to unravel diversity and to identify differences between strains, estimation of ANI needs to be integrated with approaches that determine the presence or absence of strain-specific genes. This can be done by calculating or estimating the percentage of conserved DNA (Goris et al., 2007).

Calculation or estimation of '% of conserved DNA'

The percentage of conserved DNA expresses the ratio between the length of the DNA that can be aligned between two genomes and the total genome length (Goris et al., 2007). The percentage of conserved DNA is a value that can only be obtained from whole genome sequences or at least high-quality draft genome sequences. It can be calculated by using the BLAST algorithm as described by Goris and co-workers (Goris et al., 2007) or it could be calculated using pairwise or multiple genome alignment programs like MUMmer (Kurtz et al., 2004) or MAUVE (Darling et al., 2004). As with ANI, this value can also be estimated applying other approaches, first of all comparative genome hybridization (CGH) using microarrays, see, for example, the recent study by Guidot and colleagues of *Ralstonia solanacearum* (Guidot et al., 2007). In CGH, short overlapping oligonucleotides or DNA fragments corresponding to all genes or to the entire genome (genes + intergenic regions) of one or more strains of a species are spotted on a chip and hybridized with genomic DNA of another bacterium used as 'probe'. It is very difficult to interpret results from CGH experiments correctly. Presence/absence of genes is determined by the hybridization strength between the DNA molecule on the chip and the DNA in the probe. Known probes need to be used to determine the threshold signal strength to be considered a positive result. Choosing a correct threshold, becomes the more challenging for genomes with lower ANI since the hybridization strength depends on DNA identity. When the identity between a gene on the chip and a gene in the probe becomes too low (around 80%) it becomes difficult (or even impossible) to distinguish between a positive and a negative signal. Guidot et al. (2007) performed CGH for *R. solanacearum* with oligonucleotides corresponding to the completely sequenced *R. solanacearum* GMI1000 isolate while Sarkar et al. (2006) performed CGH for *P. syringae* with a custom gene array containing type III effector genes and other virulence genes of the completely sequenced *PtoDC3000* isolate. While the *R. solanacearum* array performed well when using as hybridization probe genomic DNA of the isolate IPO1609 for which a draft genome sequence had been obtained (over 95% of genes known to be present in this strain from its genome sequence were detected and only 5.9% false positives were found), the *P. syringae* array did not perform as well. When genomic DNA of the sequenced *P. syringae* isolate *Psy*B728a was used as a probe, only 16 out of 19 (84%) *PtoDC3000* effector genes with known orthologues in *Psy*B728a were detected and six (27%) false-positives were recorded, thus revealing the challenges of CGH.

It needs to be stressed that, as in the case of ANI, estimation or even calculation of the '% of conserved DNA' requires standardization in order for this measurement to become a useful approach to compare bacterial diversity between strains. The value obtained for the '% of conserved DNA' will depend on the algorithm and the parameters chosen when comparing whole genome sequences or the hybridization parameters and the statistical analysis when using CGH. Only after establishing precise standards for these approaches can '% of conserved DNA' become a useful measure of diversity that can be used towards the description of bacterial species.

Identification of 'species-specific diagnostic genetic signatures'

While '% of conserved DNA' is a very useful measurement of diversity when calculated between two individual strains, this value is hugely influenced by the selection of the two strains that are compared and cannot be taken as an estimate of the differences between two species in general. Only by comparing multiple strains from two species (or two groups of bacteria that could potentially be considered different species) does it become possible to identify the '% of conserved DNA' within each species and between species. It must be stressed here that inclusion of the appropriate type and pathotype strains is absolutely necessary if the resulting work is to be used to classify bacteria for taxonomic purposes. By analyzing multiple strains, it can also be determined if there are genes present in all analysed strains of one species that are absent in all analysed strains of the second species. If such conserved genetic differences, or 'species-specific diagnostic genetic signatures', as Konstantinidis and Tiedje (2005) define it, are found between two groups of bacteria, then these two groups probably occupy different ecological niches and can be defined as separate species when applying some evolutionary and ecological species concepts. Genetic signatures are most easily found at the whole genome level using programs such as OrthoMCL (Li et al., 2003) that compare all genes of several genomes against each other using the BLAST algorithm and then cluster genes into orthologous groups. Of course, standard BLAST parameters need to be used and conventions need to be established to determine the minimum number of strains that need to be analysed and the minimum number of species-specific diagnostic genetic signatures that need to exist to definitely conclude that two groups of bacteria should belong to two different ecological species. We will explore below how this approach could be applied to plant pathogens.

Estimation of recombination rates

In addition to its use in typing bacteria, MLSA is a tremendously powerful population genetics tool. MLSA can be used to estimate the contribution of recombination to the evolution of analysed strains (Maiden, 2006; Pérez-Losada et al., 2006). Applying a 'pseudo-biological' species concept, as described above (Staley, 2006), one could argue that bacterial species consist of bacteria that frequently recombine with each other and that they are 'sexually isolated' from the next most closely related species because of the absence or almost absence of recombination. Using MLSA, recombination rates between bacteria can be inferred from the sequence of the analysed housekeeping genes. The presence of recombination between strains of the same cluster and the absence of recombination between two different clusters could then be used as indication that the two clusters constitute two different species, using the 'pseudo-biological' species definition (Fraser et al., 2007).

Genomics approaches and plant pathogens

The most important change genomic approaches offer taxonomy is that they may abolish the need for easily testable phenotypic differences between species. As Staley (2006) points out, the existence of differences between species when applying biochemical tests was important when biochemical tests were the easiest way to identify a bacterium in the pre-genomics era, but today's technology makes it in most cases easier to sequence DNA fragments than to run biochemical tests. Since the absence of distinguishing phenotypic differences have prevented some genomospecies of *P. syringae* (Gardan et al., 1999) and other plant pathogens from being proposed as named species, applying a genomics-based taxonomy to plant pathogenic bacteria has the potential to harmonize the current classification with what is known about genetic diversity and relationships among bacterial plant pathogens. Using genomic approaches it is feasible to describe bacterial plant pathogen species by applying: (1) a genomic-phylogenetic species concept by estimating or measuring ANI and determining the existence of phylogenetic clusters corresponding to putative species using MLSA, (2) an ecological species concept by identifying the % of conserved genome and the presence of species-specific diagnostic genetic signatures that correspond to putative species occupying different ecological niches (e.g. host ranges), and (3) a pseudo-biological species concept by using MLSA data to determine the rate of recombination between strains of a putative species and between putative species.

We will now describe the diversity so far uncovered in the species *Pseudomonas syringae* and *Ralstonia solanacearum* and in the genus *Xanthomonas* using genomics approaches and discuss how these results could be used to argue for or against the description of new species within each of these plant pathogen groups (see also chapters by Arnold et al., Brown, and Meyer and Bogdanove).

The *Pseudomonas* group

Before the advent of DDH and 16S rDNA sequencing, plant pathogenic bacteria of the genus *Pseudomonas* included a diverse group of bacterial species from the beta- and gamma-proteobacteria divided in different species simply based on the plant of isolation. In a *tour de force*, Anzai and colleagues (Anzai et al., 2000) sequenced 16S rDNA in many strains that were up to then considered to be members of the genus *Pseudomonas* (including plant and human pathogens and environmental bacteria) and, as a result, reduced the number of *Pseudomonas* species by 26. One of the organisms transferred from *Pseudomonas* was *P. syzygii*, the pathogenic agent of Sumatra disease of cloves. Prior to this work, the major change for the plant pathogens in the genus *Pseudomonas* was the proposed transfer of the species *Pseudomonas solanacearum* from *Pseudomonas* and into the newly established genus *Ralstonia* (Yabuuchi et al., 1995), which now includes *R. syzygii* (Vaneechoutte et al., 2004).

As discussed before, many pathogens that were distinguishable by differences in pathogenicity were consolidated into a small number of species with one of them being *P. syringae* upon publication of the Approved List of Bacterial Names (Skerman et al., 1980). Dye et al. (1980) proposed the pathovar designation to preserve the continuity and access to the literature related to these pathogens. For example, the previously named species *Pseudomonas tomato* became *Pseudomonas syringae* pv. *tomato*. They predicted that upon further study, some of the former species distinguishable by pathovar epithets would be proposed as authentic species. Seminal work by Gardan et al. (1992; 1999) established that several genetically distinct groups within *P. syringae* could be proposed as species based on the operational species definition. Based on DDH and polyphasic studies, they first proposed elevating *P. syringae* subsp. *savastanoi* to the species *P. savastanoi* (Gardan et al., 1992) and later defined nine genomospecies within the *P. syringae* group (Gardan et al., 1999). Because no distinguishing phenotypic traits were identified for seven of the genomospecies (including genomospecies 2, which included *P. savastanoi*) they followed the current taxonomic recommendation (Wayne et al., 1987) and proposed only two named species (*P. cannabina* and *P. tremae*).

Each of the genomospecies comprises strains from several pathovars. Moreover, this work allocated to different genomospecies, strains with the same pathovar designation indicating heterogeneity among strains with similar hosts or overlapping host ranges. For most pathovars evaluated, only the pathotype strain was analysed. Analysis of additional strains within each pathovar will allow evaluation of heterogeneity within additional pathovars. This work, and others showing overlapping host ranges among members of different genomospecies, suggests that adaptation to a given plant species may have arisen independently in different genomospecies, as was later inferred by Sarkar and Guttman (2004) from MLSA data. For example, while some organisms identified as *P. syringae* pv. *maculicola* strains (in part because they were isolated from Brassicaceae) belong to genomospecies 3 others belong to *P. cannabina* (Bull *et al.*, unpublished data).

Sequencing of the three *P. syringae* strains *P. syringae* pv. *tomato* DC3000 (Buell *et al.*, 2003), *P. syringae* pv. *syringae* B728a (Feil *et al.*, 2005), and *P. syringae* pv. *phaseolicola* 1448A (Joardar *et al.*, 2005), which probably represent genomospecies 3, 1 and 2, respectively (Gardan *et al.*, 1999), and an extensive MLST analysis (Sarkar and Guttman, 2004; Hwang *et al.*, 2005) are revolutionizing our understanding of the diversity among genomospecies within *P. syringae*. Estimated ANI (on average 93.5%) and % of conserved DNA (on average 78.5%) calculated by Vinatzer (unpublished data) for all three sequenced *P. syringae* strains and DDH values of 38% between *Pto*DC3000 and *Psy*B728a obtained by Goris and colleagues (Goris *et al.*, 2007) support the findings by Gardan *et al.* (1999) that these pathovars belong to three separate genomospecies. Additionally, the three genomospecies form three distinct clusters based on MLSA (Sarkar and Guttman, 2004). The data from these and other studies indicate that pathogenicity and the mechanisms by which these groups interact with plants are not predictive of their phylogenetic affinities. Available data indicate that there is no specific infection strategy used by all strains of one genomospecies compared to all strains of another genomospecies. Additionally, in regard to the production of toxins, there is no strict correlation between production of specific toxins and phylogeny (Hwang *et al.*, 2005). Neither is there a strict correlation between the presence/absence of virulence effector genes and the strains belonging to different genomospecies (Sarkar *et al.*, 2006). However, there is promise that genomic approaches may help find conserved phenotypic differences between genomospecies: sequencing several genomes of each genomospecies and pathovars within those genomospecies and/or performing CGH experiments may lead to the identification of conserved gene differences predictive of conserved phenotypic differences. Using the terminology introduced by Konstantinidis and Tiedje (2005) the identification of 'diagnostics genetic signatures' for each genomospecies would justify their proposal as named species based on ecological-evolutionary and genomic-phylogenetic species concepts. Importantly, these sequences would also provide a list of candidate genes that allow strains of the different genomospecies to occupy different ecological niches greatly enhancing our knowledge of the biology of the *P. syringae* group. Thus, these analyses may further delineate the genomospecies into more refined and meaningful biological units that can be correlated to host range or other niche specific phenotypes. Genomic signatures could also provide genetic fingerprints that could be easily developed into diagnostic tools to reduce our dependence on phenotypic characters for identification.

Finally, using analyses appropriate for distinguishing among species applying a 'pseudo-biological' species concept provides additional support for the hypothesis that the *P. syringae* genomospecies are authentic species. Sarkar and Guttman (2004) suggested that *P. syringae sensu lato* is a clonal species, a conclusion that is somewhat misleading since based on genomic similarities, *P. syringae sensu lato* is not a homogeneous bacterial species but represents a temporary artificial grouping of quite distantly related genomospecies (Gardan *et al.*, 1999; Goris *et al.*, 2007). Moreover, the pathotype strains were not evaluated (except for *P. syringae* pv. *tomato*, i.e. DC3000) making it difficult to know what the analysed strains represent in relation to the *P. syringae* genomospecies. Since homologous recombination between bacteria decreases with increasing DNA sequence diversity (Majewski and Cohan, 1999), recombination between *P. syringae* strains of different genomospecies can be assumed to be less likely than recombination within a genomospecies, but only few strains of the same genomospecies were analysed by Sarkar and Guttman (2004). Therefore, it cannot be excluded that closely related bacteria within individual genomospecies frequently recombine with each other. In fact, results obtained by the authors (Vinatzer, unpublished data) indicate high recombination rates between strains of genomospecies 3. Thus, the use of a 'pseudo-biological species concept' might provide a framework that could explain the biological and genetic diversity within *P. syringae*.

From a plant pathologist point of view, adoption of named species corresponding to the genomospecies or further delineated species could be very useful if species-specific diagnostic signatures were available. Due to the history of phytobacterial taxonomy, there is no reason to suspect that the organisms currently associated with named pathovars are monophyletic. In fact, there are several pathovars that are composed of strains from two or even three different genomospecies making these pathovars polyphyletic, for example *P. syringae* pv. *maculicola* described above. The use of MLST could help to rapidly assign isolates from various pathovars to their appropriate genomospecies and potentially to the appropriate pathovar or other significant biological/ecological unit within the genomospecies. This is an important first step in developing monophyletic taxa that are predictive of the genotypic identity and the host range of the strains identified by that combination. This could be accomplished by MLST if type and pathotype strains used by Gardan *et al.*, (1999) were included in the developed *P. syringae* MLST scheme (Sarkar and Guttman, 2004; Hwang *et al.*, 2005).

The *Xanthomonas* group

The classification and corresponding nomenclature for many economically important *Xanthomonas* species and pathovars is complicated. This is in part because of the lack of coordination between the Bacteriological Code (Lapage *et al.*, 1992) and Standards (Dye *et al.*, 1980). Additionally, over 100 named species from the genus *Xanthomonas* were united into the species *Xanthomonas campestris* as a result of the Approved Lists (Skerman *et al.*, 1980) because they formed a 'remarkably uniform group as shown by cultural, physiological and biochemical characters' (Dye, 1962). Similar to *P. syringae*, pathovar epithets for these previously named species were derived from former species epithets (Dye *et al.*, 1980), which often reflected the host from which it was isolated regardless of the complete host range of the organisms or host specificity.

The first genomic evaluations were made by Murata and Starr (1973) using DDH. Over 20 years later Vauterin *et al.* (1995) gave the classification of *Xanthomonas* a more solid foundation by performing over 700 DDHs. The genus *Xanthomonas* was divided into 20 species based on DDH results and distinguishing phenotypic characters. Because of the poor criteria initially used to assign organisms to *Xanthomonas* species prior to 1980 and later to pathovars of *X. campestris*, it was not very surprising that different individuals of some pathovars were shown to belong to different DDH groupings. For example, the strains of *X. campestris* pv. *vesicatoria* analysed fell into two separate DDH groups. The sheer numbers of DDH analysis done by this group was remarkable, but the taxonomic position of several pathovars remained unresolved. These strains were by necessity retained in *X. campestris* though they did not fit the emended description of this species given by Vauterin *et al.* (1995).

One major proposal from this work was the transfer of 37 pathovars from *X. campestris* to *X. axonopodis*. Because of the huge number of pairwise comparisons needed just to appropriate type and pathotype strains, only a few additional strains could be evaluated (Vauterin *et al.*, 1995). Thus, once this framework was developed, further analysis of additional strains resulted in proposed changes leading to the description of several new species, for example *X. euvesicatoria* (Jones *et al.*, 2004) and *X. citri* (Gabriel *et al.*, 1989; Schaad *et al.*, 2005).

Two strains, *Xanthomonas axonopodis* pv. *citri* 306 and *Xanthomonas campestris* pv. *campestris* ATCC33913 (the type strain of the species *X. campestris* as well as the pathotype of the pathovar *Xanthomonas campestris* pv. *campestris*), were among the first plant pathogenic bacteria to be completely sequenced (da Silva *et al.*, 2002). Since then, six additional *Xanthomonas* strains have been sequenced completely: *X. campestris* pv. *campestris* 8004 (Qian *et al.*, 2005), *X. campestris* pv. *vesicatoria* (Thieme *et al.*, 2005), *Xanthomonas oryzae* pv. *oryzae* KACC10331 (Lee *et al.*, 2005), *Xanthomonas oryzae* pv. *oryzae* MAFF 311018 (unpublished), *X. campestris* pv. *armoriaceae* 756C (unpublished), and *X. oryzae* pv. *oryzicola* BLS256 (unpublished). Very informative for the determination of intra-species diversity is the comparison of the two *Xanthomonas campestris* pv. *campestris* genomes ATCC33913 and 8004. These two strains have an estimated ANI of 100% (based on the sequence of the *gyrB*, *gltA*, and *gapA* genes) indicating a very recent common ancestor. Nonetheless, only 88% of their genomes align (using the multiple genome alignment program MAUVE; Darling *et al.*, 2004) and the two strains have clearly distinguishable host ranges (Qian *et al.*, 2005). This strengthens earlier data from DDH–16S rDNA comparisons (Hauben *et al.*, 1997), which showed that *Xanthomonas* species with DDH values under 70% could have identical 16S rDNA sequences, and suggests that some Xanthomonads seem to be undergoing a rapid genomic diversification because of frequent gene acquisitions and losses. However, this is probably not universal since there are data to suggest that other Xanthomonads may not undergo rapid diversification. Comparing the genomes of *X. euvesicatoria* 85-10 and *X. citri* 306, these genomes have an estimated ANI of only 97.5% but the % of conserved DNA is 87%. Since this value is almost as high as the 88% of conserved genome obtained for the two *X. campestris* pv. *camesptris* genomes that have an estimated ANI of 100%, this comparison reveals that *X. euvesicatoria* 85-10 and *X. citri* 306 may have diversified much more slowly than the two sequenced *Xanthomonas campestris* pv. *campestris* strains.

An MLST study of *Xanthomonas axonopodis* strains in preparation by Vinatzer *et al.* (unpublished) confirms in large part the DDH data within the *Xanthomonas axonopodis* group and supports the hypothesis that this species is heterogeneous containing members that could be proposed as additional species. However, MLST does not provide as much detail as genome sequencing or CGH and provides less delineation of the diversity within *Xanthomonas* species. It can be expected that the species that were found to be identical based on 16S rDNA sequencing but with less than 70% DDH (Hauben *et al.*, 1997), will probably also be almost identical based on MLST analyses. Therefore, additional analysis of strain-specific genes will be important before concluding that two *Xanthomonas* strains that have almost identical MLST STs really should be considered the same species.

No sequence-based recombination analysis for the genus *Xanthomonas* has been published so far and, therefore, it is not yet possible to say if applying the pseudo-biological species concept would support the current *Xanthomonas* species or suggest the existence of additional species. It is important to note that some strains of *Xanthomonas* were found to frequently exchange DNA in controlled plant infections (Basim *et al.*, 1999) suggesting that recombination may be frequent between some strains in this plant pathogen group possibly indicating the applicability of the pseudo-biological species concept.

The *Ralstonia* species complex

Bacterial wilts caused by *Ralstonia solanacearum* are among the most economically important bacterial diseases of plants world-wide partially because of the sheer number of different plant families that are hosts (Hayward, 1991). This species, like many others, was transferred first from the genus *Pseudomonas* to *Burkholderia* (Yabuuchi *et al.*, 1992) and then to *Ralstonia* (Yabuuchi *et al.*, 1995) based on phenotypic characterization, lipid analysis, 16S rDNA sequencing, and rRNA–DNA hybridization. Although *Pseudomonas solanacearum*, was included in the Approved Lists of Names (Skerman *et al.*, 1980), and the species was known to be heterogeneous (Buddenhagen *et al.*, 1962; Palleroni and Doudoroff, 1971; He *et al.*, 1983), no pathovars were originally designated for this pathogen (Dye *et al.*, 1980). This may be one reason that the term pathovar was never adopted for use by the community studying *R. solanacearum*. However, the designation of race in *R. solanacearum* is essentially the same as the pathovar designation in other plant pathogenic bacteria (Alvarez, 2005). The five races described for this pathogen can be distinguished based on their host ranges at the plant species level and not based on differential reactions on cultivars of a single species as races are defined for pathogenic variants of other bacterial species, such as *P. syringae* (Alvarez, 2005; Buddenhagen, et al., 1962; He et al., 1983). In addition to races, phenotypic diversity among *R. solanacearum* strains has been used to classify organisms into six biovars based on biochemical properties (Hayward, 1964; Hayward, 1994). This makes the taxonomy for this group more difficult since there are no internationally recognized type strains for the individual races or biovars or codification of the associated nomenclature. As we will see, researchers are having some difficulty aligning these two schemes for categorizing the genetic diversity of *R. solanacearum* strains.

Ralstonia solanacearum is described as a species complex (Gillings and Fahy, 1994; Fegan and Prior, 2005), because as early as 1971 it was shown that DDH values for many pairs of *P. solanacearum* strains was less than 70% DDH (Palleroni and Doudoroff, 1971). Two additional plant pathogenic species are often discussed with the *R. solanacearum* species complex because of similarities in their 16S rDNA sequences. *Ralstonia syzygii* is a xylem-inhabiting insect-transmitted pathogen of clove in Indonesia that causes a disease known as Sumatra disease and was first decribed as a unique species *Pseudomonas syzygii* in 1990 (Roberts et al., 1990) and later transferred to the genus *Ralstonia* (Vaneechoutte et al., 2004). Also, the agent causing Blood Disease of Banana (BDB) is part of the *R. solanacearum* complex (Fegan and Prior, 2005). This pathogen was originally named *P. celebensis* but this name did not appear in the Approved Lists (Skerman et al. 1980). The BDB pathogen is in the same 16S rDNA sequence group as *R. solancearum* race 2 and *R. syzygii* and is therefore considered part of the *R. solancearum* complex.

Recently, the phylogenetic relationships among strains in this group have been studied. Based on the sequence of three gene fragments (the 16S-23S rDNA intergenic spacer region, the *hrpB* gene, and the endoglucanasae gene) and MLSA, four phylogenetic groups (phylotypes) were identified within the *R. solanacearum* complex (Castillo and Greenberg, 2006; Fegan and Prior, 2005). Interestingly, each phylotype corresponds to a different geographic location (America, Africa including Indian Ocean, Asia, and Indonesia including some strains from Japan and Australia) suggesting that *R. solanacearum* strains were geographically separated at some point during evolution and have been evolving independently from each other since then. This finding could be useful in the application and evaluation of the genomic-phylogenetic species concept. Geographic specificity was not found among *P. syringae* strains (Sarkar and Guttman, 2004) and has not been evaluated in *Xanthomonas*. Within each *R. solanacearum* phylotype, Fegan and Prior (2005) further distinguish sequevars, which comprise very closely related strains.

From our estimates of ANI based on sequences from fragments of the three genes, *gyrB*, *gapA*, and *adk*, sequenced by Castillo and Greenberg (2006), the ANI between *R. solanacearum* strains from different phylotypes is approximately 96%, which is higher than the 94% value found to generally correspond to 70% DDH threshold (Konstantinidis et al., 2006). Therefore, based on our estimate of ANI, the overall diversity among the organisms in the *R. solanacearum* species complex is within the limits of a species applying the operational species concept. It is even lower than the diversity in *P. syringae* (which includes organisms with an estimated ANI value as low as 93.5%).

So far, only one *R. solanacearum* genome sequence has been completed and published (Salanoubat et al., 2002), while three additional genomes are at the draft status, one of which is publicly available (Gabriel et al., 2006). Additionally, a very informative study on the diversity among *R. solanacearum* and *R. syzygii* applied CGH technology using a microarray carrying oligonucleotides corresponding to genes of the completely sequenced *R. solanacearum* GMI1000 isolate (Guidot et al., 2007). Analysing 17 *R. solanacearum* strains and one *R. syzygii* isolate, it was found that *R. solanacearum* strains contain between 68% and 98% of the *R. solanacearum* GMI1000 genes depending on the isolate. Also, the *R. solanacearum* chromosome and megaplasmid are co-inherited and differences in gene content are either due to gene acquisition or gene loss. The contribution of horizontal gene transfer between the 17 analysed strains seems relatively small. The data confirmed that strains identified as *R. syzygii* belong to phylogroup IV and was missing only 185 genes compared to the core genome of all other analysed strains. This is surprising since *R. syzygii* appears to have a specialized lifestyle as an insect-transmitted xylem-inhabiting bacterium and such a specialized life style has been hypothesized in other cases to lead to gene loss (Moreira *et al.*, 2005). This may indicate that *R. syzygii* has only very recently adapted its current lifestyle. This may be reflected in the close relationship between *R. syzygii* and other members of phylotype IV within *R. solanacearum*. From these data it does not appear that *R. syzygii* has significantly diverged. It does not appear to meet the criterion of a separate species according to a genomic-phylogenetic species concept. However, only genome sequencing of additional *R. syzygii* strains including the type strain will reveal if there are diagnostics genetic signatures that differentiate *R. syzygii* strains from other *R. solanacearum* strains of phylotype IV, which would support its existence as its own species. The same reasoning can be applied to BDB, which is also a member of phylotype IV.

As in the case of *P. syringae*, no specific host–pathogen interactions can be consistently attributed to members of individual phylotypes of *R. solanacearum*. Moreover, strains from different phylotypes cause disease on the same plants while closely related strains from the same phylotype can cause disease on different plant species. Only genome sequencing of several strains from each of the four phylotypes will reveal if there are diagnostics genetic signatures that will distinguish phylotypes. Only if such genetic signatures were found, could strains in the four phylotypes be assumed to occupy different ecological niches and a division into four different species could be supported according to the ecological-evolutionary and genomic-phylogenetic species concepts. One potential may be the 22-kb region found in GMI1000 that is absent from UW551 and encodes enzymes that distinguish biovars 3 and 4 from biovars 1 and 2 (Gabriel *et al.*, 2006).

Regarding the pseudo-biological species concept, recombination has been found to play a minor role in the evolution of *R. solanacearum* based on MLSA of a selection of *R. solanacearum* strains (Castillo and Greenberg, 2006). On the other hand, it has been found that *R. solanacearum* is naturally competent and frequently acquires DNA during plant infection (Bertolla *et al.*, 1999). This suggests that recombination could potentially be frequent between some strains of *R. solanacearum*. Only a more extensive MLSA analysis of many *R. solanacearum* strains with different degrees of relatedness from many different hosts will clarify if subgroups of *R. solanacearum* frequently recombine and could be considered separate species based on a pseudo-biological species concept.

Conclusions

Although sequence-based phylogenies have increased our understanding of the genetic diversity in named plant pathogen species, these have not been used to propose novel classifications. This is partially because so few strains have been analysed and in general type and pathotype strains have not been included in these studies. It would be useful to our discipline if the recently acquired knowledge of genetic diversity and the taxonomy of plant pathogens corresponded. In order to align these two disciplines molecular biologists will need to develop a firm understanding of the Bacteriological Code (Lapage *et al.*, 1992) and The Standards (Dye *et al.*, 1980; Young *et al.*, 2001) or work closely with taxonomists to select appropriate strains for comparisons and to integrate these data into a coherent taxonomy that builds upon previous taxonomic findings. We cannot predict what the outcome will be, but it is hard to imagine what technology will develop that could provide a more predictable taxonomy than one based on whole genome sequences and gene expression.

Plant pathologists are at the threshold of applying a genomic-phylogenetic species concept to plant pathogenic bacterial species. This would support the description of several new species and move toward a classification and nomenclature that matches our understanding of the genetic diversity of these groups. However, species concepts are likely to change as we evaluate more strains. It is likely that the genomic-phylogenetic species concept will begin to be integrated with an evolutionary – ecological species concept and concurrent changes in classification and nomenclature will need to keep up. For example, we will soon have enough genome sequences or CGH data (including data from appropriate type strains) to determine if there are any genetic signatures diagnostic for the genomospecies identified within *P. syringae* or the phylotypes identified within the *R. solanacearum* species complex. Finding diagnostic genetic signatures for phylogenetic groups will not only be important in circumscribing species from genetically related groups, but may also suggest that these groups employ different infection strategies or that other differences regarding their biology and ecology exist. Finding these differences could in turn be instrumental in developing new control strategies. Additional genomic research with bacterial plant pathogens (in particular because of the amount that is known about their ecology) has the potential to test the various species concepts and to evaluate their usefulness for adopting an effective taxonomy.

In conclusion, the advent of genomics has made us appreciate the heterogeneity within plant pathogenic bacterial species and pathovars. Dozens or even hundreds of new plant pathogen genome sequences will become available over the next few years because of recent dramatic reductions in genome sequencing costs (Margulies *et al.*, 2005; Bentley, 2006). These data will allow us to clarify the taxonomic positions of many organisms previously assigned haphazardly to plant pathogenic bacterial species. These sequence analyses will permit the application of the evolutionary-ecological and/or the genomic-phylogenetic species concepts, resulting in the identification of ecologically meaningful clades. This will certainly help us overcome the inconsistent results obtained with different isolates of the same species. Thus, we will evaluate factors that are universal or variable within individual species and pathovars, and develop a strong scientific basis for strategically deploying management tools.

References

Alvarez, A.M. (2005). Diversity and diagnosis of *Ralstonia solanacearum*. In Bacterial Wilt, the Disease and the *Ralstonia solanacearum* Species Complex, C. Allen, P. Prior, and A.C. Hayward, eds. (St. Paul, MN: APS Press), pp. 437–447.

Anzai, Y., Kim, H., Park, J.-Y., Wakabayashi, H., and Oyaizu, H. (2000). Phylogenetic affiliation of the pseudomonads based on 16S rRNA sequence. Int. J. Syst. Evol. Microbiol. 50, 1563–1589.

Basim, H., Stall, R.E., Minsavage, G.V., and Jones, J.B. (1999). Chromosomal gene transfer by conjugation in the plant pathogen *Xanthomonas axonopodis* pv. *vesicatoria*. Phytopathology 89, 1044–1049.

Bautz, E.K.F., and Bautz, F.A. (1964). The influence of noncomplementary bases on the stability of ordered polynucleotides. Proc. Natl. Acad. Sci. U.S.A. 52, 1476–1481.

Belozersky, A.N., and Spirin, A.S. (1958). A correlation between the compositions of deoxyribonucleic and ribonucleic acids. Nature 182, 111–112.

Bentley, D.R. (2006). Whole-genome resequencing. Curr. Opin. Genet. Dev. 16, 545–552.

Bertolla, F., Frostegård, Å., Brito, B., Nesme, X., and Simonet, P. (1999). During infection of its host, the plant pathogen *Ralstonia solanacearum* naturally develops a state of competence and exchanges genetic material. Mol. Plant Microbe Interact. 12, 467–472.

Brenner, D.J. (1981). Introduction to the family Enterobacteriaceae. In The Prokaryotes: A Handbook on Habitats, Isolation, and Identification of Bacteria, Vol. II, M.P. Starr, H. Stolp, H.G. Trüper, A. Balows, and H.G. Schlegel, eds (New York: Springer-Verlag), pp. 1105–1127.

Brenner, D.J., Fanning, G.R. and Steigerwalt, A.G. (1972). Deoxyribonucleic acid relatedness among species of *Erwinia* and between other enterobacteria. J. Bacteriol. 110, 12–17.

Brenner, D.J., Fanning, G.R. and Steigerwalt, A.G. (1974). Deoxyribonucleic acid relatedness among Erwiniae and other *Enterobacteriaceae*: The gall, wilt, and dry-necrosis organisms (Genus *Erwinia* Winslow et al., sensu stricto). Int. J. System. Bacteriol. 24, 197–204.

Buddenhagen, I., Sequeira, L., and Kelman, A. (1962). Designation of races in *Pseudomonas solanacearum*. Phytopathology 52, 726.

Buell, C.R., Joardar, V., Lindeberg, M., Selengut, J., Paulsen, I.T., Gwinn, M.L., Dodson, R.J., Deboy, R.T., Durkin, A.S., Kolonay, J.F., *et al.* (2003). The complete genome sequence of the *Arabidopsis* and tomato pathogen *Pseudomonas syringae* pv. *tomato* DC3000. Proc. Natl. Acad. Sci. U.S.A. 100, 10181–10186.

Castillo, J.A., and Greenberg, J.T. (2006). Evolutionary dynamics of *Ralstonia solanacearum*. Appl. Environ. Microbiol. 73, 1225–1238.

Cho, J.-C., and Tiedje, J.M. (2001). Bacterial species determination from DNA-DNA hybridization by using genome fragments and DNA microarrays. Appl. Environ. Microbiol. 67, 3677–3682.

Christensen, H., Angen, Ø., Mutters, R., Olsen, J.E., and Bisgaard, M. (2000). DNA-DNA hybridization determined in micro-wells using covalent attachment of DNA. Int. J. Syst. Evol. Microbiol. 50, 1095–1102.

Claridge, M.F., Dawah, H.A., Wilson, M.R. (1997). Practical approaches to species concepts for living organisms. In Species: The Units of Biodiversity, M.F. Claridge, H.A. Dawah, and M.R. Wilson eds. (London: Chapman & Hall), pp. 1–15.

Cohan, F.M. (2002). What are bacterial species? Annu. Rev. Microbiol. 56, 457–487.

Cohan, F.M. (2006). Towards a conceptual and operational union of bacterial systematics, ecology, and evolution. Phil. Trans. R. Soc. B 361, 1985–1996.

Cohan, F.M., and Perry, E.B. (2007). A systematics for discovering the fundamental units of bacterial diversity. Curr. Biol. 17, R373-R386.

Cole, J.R., Chai, B., Farris, R.J., Wang, Q., Kulam-Syed-Mohideen, A.S., McGarrell, D.M., Bandela, A.M., Cardenas, E., Garrity, G.M., and Tiedje, J.M. (2007). The ribosomal database project (RDP-II): Introducing myRDP space and quality controlled public data. Nucleic Acids Res. 35, D169-D172.

Cowan, S.T. (1962). The microbial species – a macromyth? Symp. Soc. Gen. Microbiol. 12, 433–455.

Cracraft, J. (1983). Species concepts and speciation analysis. Curr. Ornithol. 1, 159–187.

da Silva, A.C.R., Ferro, J.A., Reinach, F.C., Farah, C.S., Furlan, L.R., Quaggio, R.B., Monteiro-Vitorello, C.B., Van Sluys, M.A., Almeida, N.F., Alves, L M.C., et al. (2002). Comparison of the genomes of two Xanthomonas pathogens with differing host specificities. Nature 417, 459–463.

Darling, A.C.E., Mau, B., Blattner, F.R., and Perna, N.T. (2004). Mauve: multiple alignment of conserved genomic sequence with rearrangements. Genome Res. 14, 1394–1403.

De Ley, J., Cattoir, H., and Reynaerts, A. (1970). The quantitative measurement of DNA hybridization from renaturation rates. Eur. J. Biochem. 12, 133–142.

Dowson, W.J. (1949). Manual of Bacterial Plant Diseases. (London: Adam and Charles Black).

Dye, D.W. (1962). The inadequacy of the usual determinative tests for the identification of Xanthomonas spp. New Zealand J. Sci. 5, 393–416.

Dye, D.W., Bradbury, J.F., Goto, M., Hayward, A.C., Lelliott, R.A., and Schroth, M.N. (1980). International standards for naming pathovars of phytopathogenic bacteria and a list of pathovar names and pathotype strains. Rev. Plant Pathol. 59, 153–168.

Dye, D.W., and Lelliott, R.A. (1974). Genus II. Xanthomonas. In Bergey's Manual of Determinative Bacteriology, 8th ed., R.E. Buchanan, N.E. Gibbons, eds. (Baltimore: Williams & Wilkins), pp. 243–249.

Dykhuizen, D.E., and Green, L. (1991) Recombination in Escherichia coli and the definition of biological species. J. Bacteriol. 173, 7257–7268.

Eldredge, N., and Cracraft, J. (1980). Phylogenetic Analysis and the Evolutionary Process: Method and Theory in Comparative Biology (New York: Columbia University Press).

Embley, T.M., and Stackebrandt, E. (1997). Species in practice: Exploring uncultured prokaryote diversity in natural samples. In Species: The Units of Biodiversity, M.F. Claridge, H.A. Dawah, and M.R. Wilson, eds. (London: Chapman & Hall), pp. 61–81.

Fegan, M., and Prior, P. (2005). How complex is the 'Ralstonia solanacearum species complex'? In Bacterial Wilt, the Disease and the Ralstonia solanacearum Species Complex, C. Allen, P. Prior, and A.C. Hayward, eds. (St. Paul: APS Press), pp. 449–461.

Feil, H., Feil, W. S., Chain, P., Larimer, F., DiBartolo, G., Copeland, A., Lykidis, A., Trong, S., Nolan, M., Goltsman, E., et al. (2005). Comparison of the complete genome sequences of Pseudomonas syringae pv. syringae B728a and pv. tomato DC3000. Proc. Natl. Acad. Sci. U.S.A. 102, 11064–11069.

Feil, E.J., Holmes, E.C., Bessen, D.E., Chan, M.S., Day, N.P., Enright, M.C. Goldstein, R., Hood, D.W., Kalia, A., Moore, C.E., Zhou, J., and Spratt, B.G. (2001). Recombination within natural populations of pathogenic bacteria: short-term empirical estimates and long-term phylogenetic consequences. Proc. Natl. Acad. Sci. USA 98, 182–187.

Fox, G.E., Wisotzkey, J.D. and Jurtshuk, P., Jr. (1992). How close is close: 16S rRNA sequence identity may not be sufficient to guarantee species identity. Int. J. Syst. Bacteriol. 42, 166–170.

Fraser, C., Hanage, W.P., and Spratt, B.G. (2007). Recombination and the nature of bacterial speciation. Science 315, 476–480.

Gabriel, D.W., Allen, C., Schell, M., Denny, T.P., Greenberg, J.T., Duan, Y.P., Flores-Cruz, Z., Huang, Q., Clifford, J.M., Presting, G., et al. (2006). Identification of open reading frames unique to a select agent: Ralstonia solanacearum race 3 biovar 2. Mol. Plant Microbe Interact. 19, 69–79.

Gabriel, D.W., Kingsley, M.T., Hunter, J.E., Gottwald, T. (1989). Reinstatement of Xanthomonas citri (ex Hasse) and X. phaseoli (ex Smith) to species and reclassification of all X. campestris pv. citri strains. Int. J. Syst. Bacteriol. 39, 14–22.

Gardan, L., Bollet, C., Abu Ghorrah, M., Grimont, F., and Grimont, P.A.D. (1992). DNA relatedness among the pathovar strains of Pseudomonas syringae subsp. savastanoi Janse (1982) and proposal of Pseudomonas savastanoi sp. nov. Int. J. Syst. Bacteriol. 42, 606–612.

Gardan, L., Shafik, H., Belouin, S., Broch, R., Grimont, F., and Grimont, P.A.D. (1999). DNA relatedness among the pathovars of Pseudomonas syringae and description of Pseudomonas tremae sp. nov. and Pseudomonas cannabina sp. nov. (ex Sutic and Dowson 1959). Int. J. Syst. Bacteriol. 49, 469–478.

Gevers, D., Cohan, F.M., Lawrence, J.G., Spratt, B.G., Coenye, T., Feil, E.J., Stackebrandt, E., Van de Peer, Y., Vandamme, P., Thompson, F.L., and Swings, J. (2005). Re-evaluating prokaryotic species. Nat. Rev. 3, 733–739.

Ghiselin, M.T. (1974). A radical solution to the species problem. Syst. Zool. 23, 536–544.

Gillings, M.R., and Fahy, P. (1994). Genomic fingerprinting: Towards a unified view of the *Pseudomonas solanacearum* species complex, In Bacterial Wilt: The Disease and Its Causative Agent, *Pseudomonas solanacearum*, A.C. Hayward, and G L. Hartman, eds. (Wallingford, UK: CAB International), pp. 95–112.

Goodfellow, M. (2000). Microbial systematics: Background and uses. In Applied Microbial Systematics, F.G. Priest, M. Goodfellow, eds. (Dordrecht, The Netherlands: Kluwer Academic Publishers), pp. 1–18.

Goodfellow, M., Manfio, G.P. and Chun, J. (1997). Towards a practical species concept for cultivable bacteria. In Species: The Units of Biodiversity, M.F. Claridge, H.A. Dawah and M.R. Wilson, eds. (London: Chapman & Hall), pp. 25–59.

Goodfellow, M. and O'Donnell, A. G. (1993). Roots of bacterial systematics. In Handbook of New Bacterial Systematics, M. Goodfellow, and A.G. O'Donnell, eds. (London: Academic Press Ltd.), pp. 3–54.

Gordon, R.E. (1978). A species definition. Int. J. Syst. Bact. 28, 605–607.

Goris, J., Konstantinidis, K.T., Klappenbach, J.A., Coenye, T., Vandamme, P., and Tiedje, J.M. (2007). DNA-DNA hybridization values and their relationship to whole-genome sequence similarities. Int. J. Syst. Evol. Microbiol. 57, 81–91.

Grimont, P.A.D., Popoff, M.Y., Grimont, F., Coynault, C., and Lemelin, M. (1980). Reproducibility and correlation study of three deoxyribonuceic acid hybridization procedures. Curr. Microbiol. 4, 325–330.

Guidot, A., Prior, P., Schoenfeld, J., Carrére, S., Genin, S., and Boucher, C. (2007). Genomic structure and phylogeny of the plant pathogen *Ralstonia solanacearum* inferred from gene distribution analysis. J. Bacteriol. 189, 377–387.

Hanage, W.P., Fraser, C., and Spratt, B.G. (2005). Fuzzy species among recombinogenic bacteria. BMC Biol. 3, 6.

Hanage, W.P., Fraser, C., and Spratt, B.G. (2006). Sequences, sequence clusters and bacterial species. Phil. Trans. R. Soc. B 361, 1917–1927.

Hauben, L., Vauterin, L., Swings, J., and Moore, E.R.B. (1997). Comparison of 16S ribosomal DNA sequences of all *Xanthomonas* species. Int. J. Syst. Bacteriol. 47, 328–335.

Hauben, L., Moore, E.R.B., Vauterin, L., Steenackers, M., Mergaert, J., Verdonck, L., and Swings, J. (1998). Phylogenetic position of phytopathogens within the *Enterobacteriaceae*. Syst. Appl. Microbiol. 21, 384–397.

Hayward, A.C. (1964). Characteristics of *Pseudomonas solanacearum*. J. Appl. Bacteriol. 27, 265–277.

Hayward, A.C. (1991) Biology and epidemiology of a bacterial wilt caused by *Pseudomonas solanacearum*. Annu. Rev. Phytopathol. 29, 65–87.

Hayward, A.C. (1994). Systematics and phylogeny of *Pseudomonas solanacearum* and related bacteria. In Bacterial Wilt: The Disease and Its Causative Agent, *Pseudomonas solanacearum*, A.C. Hayward and G.L. Gartman, eds, (Wallingford, UK: CAB International), pp. 123–135.

He, L.Y., Sequeira, L., and Kelman, A. (1983). Characteristics of strains of *Pseudomonas solanacearum* from China. Plant Dis. 67, 1357–1361.

Hull, D.L. (1997). The ideal species concept – and why we can't get it. In Species: The Units of Biodiversity, M.F. Claridge, H.A. Dawah, and M.R. Wilson eds. (London: Chapman & Hall), pp. 357–380.

Hwang, M.S.H., Morgan, R.L., Sarkar, S.F., Wang, P.W., and Guttman, D.S. (2005). Phylogenetic characterization of virulence and resistance phenotypes of *Pseudomonas syringae*. Appl. Environ. Microbiol. 71, 5182–5191.

Janssen, P., Coopman, R., Huys, G., Swings, J., Bleeker, M., Vos, P., Zabeau, M., and Kersters, K. (1996). Evaluation of the DNA fingerprinting method AFLP as an new tool in bacterial taxonomy. Microbiology 142, 1881–1893.

Joardar, V., Lindeberg, M., Jackson, R.W., Selengut, J., Dodson, R., Brinkac, L.M., Daugherty, S C., Deboy, R., Durkin, A.S., Giglio, M.G., et al. (2005). Whole-genome sequence analysis of *Pseudomonas syringae* pv. *phaseolicola* 1448A reveals divergence among pathovars in genes involved in virulence and transposition. J. Bacteriol. 187, 6488–6498.

Johnson, J.L. (1984). Nucleic acids in bacterial classification. In Bergey's Manual of Systematic Bacteriology, Vol. 1, N.R. Kreig, and J.G. Holt eds. (Baltimore: Williams & Wilkins), pp 8–11.

Jones, J.B., Lacy, G.H., Bouzar, H., Stall, R.E., and Schaad, N.W. (2004). Reclassification of the xanthomonads associated with bacterial spot disease of tomato and pepper. Syst. Appl. Microbiol. 27, 755–762.

Konstantinidis, K.T., Ramette, A., and Tiedje, J.M. (2006). Toward a more robust assessment of intraspecies diversity, using fewer genetic markers. Appl. Environ. Microbiol. 72, 7286–7293.

Konstantinidis, K.T., and Tiedje, J.M. (2005). Genomic insights that advance the species definition for prokaryotes. Proc. Natl. Acad. Sci. USA 102, 2567–2572.

Kurtz, S., Phillippy, A., Delcher, A.L., Smoot, M., Shumway, M., Antonescu, C., and Salzberg, S.L. (2004). Versatile and open software for comparing large genomes. Genome Biol. 5, R12.

Lane, D.J., Pace, B., Olsen, G.J., Stahl, D.A., Sogin, M.L. and Pace, N.R. (1985). Rapid determination of 16S ribosomal RNA sequences for phylogenetic analyses. Proc. Natl. Acad. Sci. USA 82, 6955–6959.

Lapage, S.P., Sneath, P.H.A., Lessel, E.F., Skerman, V.B.D., Seeliger, H.P.R., and Clark, W.A., (1992). International Code of Nomenclature of Bacteria (1990 Revision) Bacteriological Code (Washington, D.C.: American Society for Microbiology).

Lee, B.-M., Park, Y.-J., Park, D.-S., Kang, H.-W., Kim, J.-G., Song, E.-S., Park, I.-C., Yoon, U.-H., Hahn, J.-H., Koo, B.-S., et al. (2005). The genome sequence of *Xanthomonas oryzae* pathovar *oryzae* KACC10331, the bacterial blight pathogen of rice. Nucleic Acids Res. 33, 577–586.

Lee, K.Y., Wahl, R., and Barbu, E. (1956). Contenu en bases puriques et pyrimidiques des acide désoxyribonucléiques des bactéries. Ann. Inst. Pasteur. 91, 212–237.

Li, L., Stoeckert, C.J., Jr., and Roos, D.S. (2003). OrthoMCL: Identification of ortholog groups for eukaryotic genomes. Genome Res. 13, 2178–2189.

Lockhart, W.R., and Koenig, K., (1965). Use of secondary data in numerical taxonomy of the genus Erwinia. J. Bacteriol. 90, 1638–1644.

Maiden, M.C. (2006). Multilocus sequence typing of bacteria. Annu. Rev. Microbiol. 60, 561–588.

Maiden, M.C., Bygraves, J.A., Feil, E., Morelli, G., Russell, J.E., Urwin, R., Zhang, Q., Zhou, J., Zurth, K., Caugant, D.A., et al. (1998). Multilocus sequence typing: a portable approach to the identification of clones within populations of pathogenic microorganisms. Proc. Natl. Acad. Sci. U.S.A. 95, 3140–3145.

Majewski, J., and Cohan, F.M. (1999). DNA sequence similarity requirements for interspecific recombination in Bacillus. Genetics 153, 1525–1533.

Margulies, M., Egholm, M., Altman, W.E., Attiya, S., Bader, J.S., Bemben, L.A., Berka, J., Braverman, M.S., Chen, Y.-J., Chen, Z., et al. (2005). Genome sequencing in microfabricated high-density picolitre reactors. Nature 437, 376–380.

Mayden, R.L. (1997). A hierarchy of species concepts: The denouement in the saga of the species problem. In Species: The Units of Biodiversity, M.F. Claridge, H.A. Dawah, and M.R. Wilson, eds. (London: Chapman & Hall), pp. 381–324.

Mayr, E. (1942). Systematics and the Origin of Species (New York: Columbia Univ. Press).

Mayr, E., and Ashlock, P.D. (1991). Principles of Systematic Zoology, 2nd edn (New York: McGraw-Hill).

Moreira, L.M., De Souza, R.F., Digiampietri, L.A., Da Silva, A.C.R., and Setubal, J.C. (2005). Comparative analyses of Xanthomonas and Xylella complete genomes. Omics 9, 43–76.

Murata, N., and Starr, M.P., (1973). A concept of the genus Xanthomonas and its species in the light of segmental homology of deoxyribonucleic acids. Phytopath. Z. 77, 285–323.

Murray, R.G.E., and Stackebrandt, E. (1995). Taxonomic note: Implementation of the provisional status Candidatus for incompletely described procaryotes. Int. J. Syst. Bacteriol. 45, 186–187.

Narra, H.P. and Ochman, H. (2006). Of what use is sex to bacteria? Curr. Biol. 16, R705–R710.

Nixon, K.C., and Wheeler, Q D. (1990). An amplification of the phylogenetic species concept. Cladistics 6, 211–223.

Pace, N.R., Stahl, D.A., Lane, D.J., Olsen, G.J. (1986). The analysis of natural microbial populations by ribosomal RNA sequences. Adv. Microbiol. Ecol. 9, 1–55.

Palleroni, N.J., and Doudoroff, M. (1971). Phenotypic characterization and deoxyribonucleic acid homologies of Pseudomonas solanacearum. J. Bacteriol. 107, 690–696.

Pérez-Losada, M., Browne, E.B., Madsen, A., Wirth, T., Viscidi, R.P., and Crandall, K.A. (2006). Population genetics of microbial pathogens estimated from multilocus sequence typing (MLST) data. Infect. Genet. Evol. 6, 97–112.

Qian, W., Jia, Y., Ren, S.-X., He, Y.-Q., Feng, J.-X., Lu, L.-F., Sun, Q., Ying, G., Tang, D.-J., Tang, H., et al. (2005). Comparative and functional genomic analyses of the pathogenicity of phytopathogen Xanthomonas campestris pv. campestris. Genome Res. 15, 757–767.

Roberts, S.J., Eden-Green, S.J., Jones, P., and Ambler, D.J. (1990). Pseudomonas syzygii sp. nov., the cause of Sumatra disease of cloves. Syst. Appl. Microbiol. 13, 34–43.

Rosselló-Mora, R., and Amann, R. (2001). The species concept for prokaryotes. FEMS Microbiol. Rev. 25, 39–67.

Salanoubat, M., Genin, S., Artiguenave, F., Gouzy, J., Mangenot, S., Arlat, M., Billault, A., Brottier, P., Camus, J.C., Cattolico, L., et al. (2002). Genome sequence of the plant pathogen Ralstonia solanacearum. Nature 415, 497–502.

Sarkar, S.F., Gordon, J.S., Martin, G.B., and Guttman, D. S. (2006). Comparative genomics of host-specific virulence in Pseudomonas syringae. Genetics 174, 1041–1056.

Sarkar, S.F., and Guttman, D.S. (2004). Evolution of the core genome of Pseudomonas syringae, a highly clonal, endemic plant pathogen. Appl. Environ. Microbiol. 70, 1999–2012.

Schaad, N.W., Postnikova, E., Lacy, G.H., Sechler, A., Agarkova, I., Stromberg, P.E., Stromberg, V.K., and Vidaver, A.K. (2005). Reclassification of Xanthomonas campestris pv. citri (ex Hasse 1915) Dye 1978 forms A, B/C/D, and E as X. smithii subsp. citri (ex Hasse) sp. nov. nom. rev. comb. nov., X. fuscans subsp. aurantifolii (ex Gabriel 1989) sp. nov. nom. rev. comb. nov., and X. alfalfae subsp. citrumelo (ex Riker and Jones) Gabriel et al., 1989 sp. nov. nom. rev. comb. nov.; X. campestris pv malvacearum (ex smith 1901) Dye 1978 as X. smithii subsp. smithii nov. comb. nov. nom. nov.; X. campestris pv. alfalfae (ex Riker and Jones, 1935) Dye 1978 as X. alfalfae subsp. alfalfae (ex Riker et al., 1935) sp. nov. nom. rev.; and 'var. fuscans' of X. campestris pv. phaseoli (ex Smith, 1987) Dye 1978 as X. fuscans subsp. fuscans sp. nov. Syst. Appl. Microbiol. 28, 494–518.

Simpson, G.G. (1961). Principles of Animal Taxonomy (New York: Columbia University Press).

Skerman, V.B.D., McGowan, V., and Sneath, P.H.A. (1980). Approved lists of bacterial names. Int. J. Syst. Bacteriol. 30:225–420.

Sneath, P.H. (1957). Some thoughts on bacterial classification. J. Gen. Microbiol. 17, 184–200

Sneath, P.H.A., and Sokal, R.R. (1973). Numerical Taxonomy (San Francisco: W.H. Freeman and Co.).

Stackebrandt, E., Frederiksen, W., Garrity, G.M., Grimont, P.A.D., Kampfer, P., Maiden, M.C. J., Nesme, X., Rosselló-Mora, R., Swings, J., Trüper, H.G., Vauterin, L., Ward, A.C., and Whitman, W.B. (2002). Report of the ad hoc committee for the re-evaluation of the species definition in bacteriology. Int. J. Syst. Evol. Microbiol. 52, 1043–1047.

Stackebrandt, E., and Goebel, B.M. (1994). Taxonomic Note: A place for DNA-DNA reassociation and

16S rRNA sequence analysis in the present species definition in bacteriology. Int. J. Syst. Bacteriol. *44*, 846–849.

Staley, J.T. (2004). Speciation and bacterial phylospecies. In Microbial Diversity and Bioprospecting, A.T. Bull, ed. (Washington, DC: ASM Press), pp. 40–48.

Staley, J.T. (2006). The bacterial species dilemma and the genomic-phylogenetic species concept. Phil. Trans. R. Soc. B *361*, 1899–1909.

Staley, J.T., and Krieg, N.R. (1984). Classification of prokaryotic organisms: an overview. In Bergey's Manual of Systematic Bacteriology, Vol. 1, N.R. Krieg and J.G. Hold, eds. (Baltimore: Williams & Wilkins Co.), pp. 1–4.

Starr, M.P., (1959). Bacteria as plant pathogens. Annu. Rev. Microbiol. *13*, 211–238.

Starr, M.P. (1983). Prokaryotes as plant pathogens. In The Prokaryotes: A Handbook on Habitats, Isolation, and Identification of Bacteria, M.P. Starr, H. Stolp, H.G. Trüper, A. Balows, and H.G. Schlegel, eds. (New York: Springer-Verlag), pp 123–134.

Starr, M.P., and Chatterjee, A.K. (1972). The genus *Erwinia*: Enterobacteria pathogenic to plants and animals. Annu. Rev. Microbiol. *26*, 389–426.

Starr, M.P., and Mandel, M., (1969). DNA base composition and taxonomy of phytopathogenic and other enterobacteria. J. Gen. Microbiol. *56*, 113–123.

Stolp, H., Starr, M.P., and Baignet, N.L. (1965). Problems in speciation of phytopathogenic pseudomonads and xanthomonads. Annu. Rev. Phytopathol. *3*, 231–264.

Templeton, A.R. (1989). The meaning of species and speciation: A genetic perspective. In Speciation and Its Consequences, D. Otte and J. A. Endler, eds. (Sunderland, MA: Sinauer), pp. 3–27.

Thieme, F., Koebnik, R., Bekel, T., Berger, C., Boch, J., Büttner, D., Caldana, C., Gaigalat, L., Goesmann, A., Kay, S., et al. (2005). Insights into genome plasticity and pathogenicity of the plant pathogenic bacterium *Xanthomonas campestris* pv. *vesicatoria* revealed by the complete genome sequence. J. Bacteriol. *187*, 7254–7266.

Tindall, B.J. (1999). Misunderstanding the bacteriological code. Int. J. Syst. Bacteriol. *49*, 1313–1316.

Trüper, H.G., and Schleifer, K.-H. (2006). Prokaryote characterization and identification. In The Prokaryotes, Vol. 1: Symbiotic Associations, Biotechnology, Applied Microbiology. M. Dworkin, S. Falkow, E. Rosenberg, K.-H. Schleifer, and E. Stackebrandt, eds. (New York: Springer), pp. 58–79.

Ullman, J.S., and McCarthy, B.J. (1973). The relationship between mismatched base pairs and the thermal stability of DNA duplexes. II. Effects of deamination of cytomine. Biochim. Biophys. Acta *294*, 416–424.

Van Regenmortel, M.H.V. (1997). Viral species. In Species: The Units of Biodiversity, M.F. Claridge, H.A. Dawah, and M.R. Wilson, eds.. (London: Chapman and Hall.), pp. 17–24.

Vandamme, P. Pot, B., Gillis, M., De Vos, P., Kersters, K., and Swings, J. (1996). Polyphasic taxonomy, a consensus approach to bacterial systematics. Microbiol. Rev. *60*, 407–438.

Vaneechoutte, M., Kämpfer, P., De Baere, T., Falsen, E., and Verschraegen, G. (2004). *Wautersia* gen. nov., a novel genus accommodating the phylogenetic lineage including *Ralstonia eutropha* and related species, and proposal of *Ralstonia* [*Pseudomonas*] *syzygii* (Roberts et al. 1990) comb. nov. Int. J. Syst. Evol. Microbiol. *54*, 317–327.

Van Valen, L. (1976). Ecological species, multispecies, and oaks. Taxon *25*, 233–239.

Vauterin, L., Hoste, B., Kersters, K., and Swings, J. (1995). Reclassification of *Xanthomonas*. Int. J. Syst. Bacteriol. *45*, 472–489.

Ward, D.M. (1998). A natural species concept for prokaryotes. Curr. Opin. Microbiol. *1*, 271–277.

Ward, D.M., Cohan, F.M., Bhaya, D., Heidelberg, J.F., Kühl, M., and Grossman, A. (2008). Genomics, environmental genomics and the issue of microbial species. Heredity *100*, 207–219.

Wayne, L.G., Brenner, D.J., Colwell, R.R., Grimont, P.A.D., Kandler, O., Krichevsky, L., Moore, L.H., Moore, W.E.C., Murray, R.G.E., Stackebrandt, E., Starr, M.P., and Trüper, H.G. (1987). Report of the ad hoc committee on reconciliation of approaches to bacterial systematics. Int. J. Syst. Bacteriol. *37*, 463–464.

Wheeler, Q.D., and Platnick, N.I. (2000). The phylogenetic species concept. In Species Concepts and Phylogenetic Theory: A Debate, Q.D. Wheeler and R. Meier, eds. (New York: Columbia University Press), pp. 55–69.

Wiley, E.O. (1978). The evolutionary species concept reconsidered. Syst. Zool. *27*, 17–26.

Williams, S.T., Goodfellow, M., and Vickers, J.C. (1984). New microbes from old habitats? In The microbe. Part II. Prokaryotes and Eukaryotes, D.P. Kelly and N.G. Carr, eds. (Cambridge: Cambridge University Press), pp. 219–256.

Woese, C.R. (1987). Bacterial evolution. Microbiol. Rev. *51*, 221–271.

Yabuuchi, E., Kosako, Y., Oyaizu, H., Yano, I., Hotta, H., Hashimoto, Y., Ezaki, T., and Arakawa, M. (1992). Proposal of *Burkholderia* gen. nov. and transfer of seven species of the genus *Pseudomonas* homology group II to the new genus, with the type species *Burkholderia cepacia* (Palleroni and Holmes 1981) comb. nov. Microbiol. Immunol. *36*, 1251–1275.

Yabuuchi, E., Kosako, Y., Yano, I., Hotta, H., and Nishiuchi, Y. (1995). Transfer of two *Burkholderia* and an *Alcaligenes* species to *Ralstonia* gen. Nov.: Proposal of *Ralstonia pickettii* (Ralston, Palleroni and Doudoroff 1973) comb. Nov., *Ralstonia solanacearum* (Smith 1896) comb. Nov. and *Ralstonia eutropha* (Davis 1969) comb. Nov. Microbiol Immunol *39*, 897–904.

Young, J.M. (2000). Recent developments in systematics and their implications for plant pathogenic bacteria. In Applied Microbial Systematics, F.G. Priest and M. Goodfellow, eds. (Dordrecht, The Netherlands: Kluwer Academic Publishers), pp. 135–163.

Young, J.M., Bull, C.T. De Boer, S.H., Firrao, G., Gardan, L., Saddler, G.E., Stead, D.E., and Takikawa, Y. (2001). Committee on the Taxonomy of Plant

Pathogenic Bacteria International Standards for Naming Pathovars of Phytopathogenic Bacteria. http://www.isppweb.org/about_tppb_naming.asp

Young, J.M., Takikawa, Y., Gardan, L., and Stead, D.E. (1992). Changing concepts in the taxonomy of plant pathogenic bacteria. Annu. Rev. Phytopathol. *30*, 67–105.

Young, J.P.W. (1994). All those new names: An overview of the molecular phylogeny of plant-associated bacteria. In Advances in Molecular Genetics of Plant–Microbe Interactions, Vol. 3., M.J. Daniels *et al.*, eds. (The Netherlands: Kluwer Academic Publishers), pp. 73–80.

Adaptation to the Plant Apoplast by Plant Pathogenic Bacteria

Arantza Rico, Rachel Jones and Gail M. Preston

Abstract

Many plant pathogenic bacteria spend most of their parasitic life in the apoplast, which is the intercellular space of the plants. The apoplast is a nutrient-limited environment that is guarded by plant defences, so plant pathogenic bacteria have evolved several strategies to successfully colonize this niche, which include the type III secretion system and its effectors, toxins and cell wall degrading enzymes, among others. Genomic and nutritional assays suggest that some apoplast-colonizing pathogens show nutritional specialization to the plant host and it is possible that some of the keys to apoplast colonization reside in bacterial adaptation to, and modulation of the nutritional and physiological characteristics of the plant apoplast. In this chapter, we offer a review of bacterial strategies for colonization of the plant apoplast, and discuss the evolutionary processes that may have affected the evolution of these strategies. We also discuss evidence for the hypothesis that successful pathogens modulate plant metabolism for their benefit. Finally we propose strategies and avenues for research that will promote further understanding of the complex picture of apoplast physiology during disease development.

Introduction

The plant apoplast is the intercellular space that surrounds plant cells and the niche where many necrotrophic and biotrophic bacterial pathogens establish their parasitic lifestyle. Plant surfaces are colonized by both pathogenic and non-pathogenic bacteria, but plant pathogens have the unique ability to multiply to high densities in the apoplast of susceptible plants, which they access through natural openings such as stomata and hydathodes, or through wounds (Beattie and Lindow, 1995; Hirano and Upper, 2000; Lu et al., 2001). Compared with the plant surface, the apoplast provides more protection from environmental stresses such as UV radiation or water deficit (Beattie and Lindow, 1995; Hirano and Upper, 2000). However, endophytic bacteria encounter different challenges upon entering the plant, which range from environmental factors such as osmotic stress, pH stress and nutrient limitation (Van den Ackerveken et al., 1994; Outlaw and De Vlieghere-He, 2001; Solomon and Oliver, 2001; Felle, 2006), to antimicrobial compounds and other plant defence responses (Menezes and Jared, 2002; Bittel and Robatzek, 2007; Day and Graham, 2007). To counteract this, bacteria have developed several strategies, including the type III secretion system and type III-secreted effectors, toxins, cell wall-degrading enzymes and extracellular polysaccharides, which collectively suppress or overcome host defences and enable bacteria to modulate the cellular functions and metabolism of host cells (Alfano and Collmer, 1996; Kimura et al., 2001; Alfano and Collmer, 2004; Nomura et al., 2005; Abramovitch et al., 2006; Chisholm et al., 2006; Grant et al., 2006; Jones and Dangl, 2006). The molecular basis of these mechanisms have been the subject of intense research in the last 20 years, and dramatic advances in understanding the multilayered process of the plant–pathogen interaction have been achieved, thanks in part to the availability of full genome sequences for both plant-associated bacteria and the model plant host *Arabidopsis*.

However, there are still many unanswered questions with regard to the nutritional and physiological factors that account for the specific adaptation of plant pathogens to the plant apoplast, and it is likely that many factors other than secreted pathogenicity and virulence factors contribute to the successful establishment of a plant parasitic lifestyle (Boch *et al.*, 2002; Shinohara *et al.*, 2005; Tamir-Ariel *et al.*, 2007). In this chapter we describe the environmental and nutritional characteristics of the apoplast and plant cell wall and discuss how plant pathogens may have adapted for growth in this environment, and how the apoplast may be modulated during pathogenesis for the benefit of parasitic bacteria. Finally, we discuss how researchers can take advantage of advances in high-throughput analytical techniques developed in the post-genomic era, including transcriptomics, proteomics and metabolomics, to develop new approaches to tackle questions regarding the specialization and evolution of bacterial pathogens for growth in plant tissues and the nutritional and physiological changes that occur during plant–pathogen interactions.

Bacteria that colonize the apoplast

Table 3.1 depicts the characteristics of representative plant pathogenic bacteria that preferentially infect different parts of the plant (from roots to stem bases and aerial parts), and which use diverse pathogenicity mechanisms to infect plant tissues, but which have all been shown to colonize the plant apoplast or xylem vessels. All are Gram-negative and belong to the class Proteobacteria and are assigned to different subclasses and families. All the species listed in Table 3.1, except *Xylella fastidiosa*, contain a pathogenicity island that encodes a type III secretion system (TTSS). Traditionally, plant pathogenic bacteria have been classified as 'brute force' or 'stealth' pathogens, based on the pathogenic features and strategies they deploy to colonize and invade their host. The 'brute force' phenotype is typically associated with the ability of bacteria to secrete extracellular proteins that degrade components of the plant cell wall, and is exemplified by species such as *Pectobacterium atrosepticum* (formerly *Erwinia carotovora* subsp. *atroseptica*; see also chapter by Toth *et al.*). Cell wall degradation leads to tissue maceration and bacteria behave as necrotrophs, deriving nutrients from killed plant cells. Conversely, bacteria whose main pathogenic features are TTSS-secreted effectors and extracellular toxins are generally defined as 'stealth' pathogens, because their primary mechanism of parasitism is not to kill and destroy the cell, but to modify plant physiology and plant defence responses to their benefit. These bacteria can multiply to high densities in plant tissues before causing host cell death and thus are considered hemibiotrophs. The best-studied hemibiotrophs belong to the genus *Pseudomonas* and *Xanthomonas* (for detailed reviews of *Pseudomonas syringae* and *Xanthomonas campestris* pathogenicity mechanisms see chapters by Arnold *et al.*, Meyer and Bogdanove; Preston, 2000; Buttner *et al.*, 2003; Alfano and Collmer, 2004; Espinosa and Alfano, 2004; Nomura *et al.*, 2005; Grant *et al.*, 2006; Gurlebeck *et al.*, 2006; Hann and Rathjen, 2007) in which many questions regarding the molecular, biochemical and physiological basis of plant–pathogen interactions have been addressed.

The TTSS is essential for pathogenesis in hemibiotrophic pathogens, such as *P. syringae*, but is dispensable in necrotrophs such as *Dickeya dadantii* (formerly *Erwinia chrysanthemi*) and *Pectobacterium* spp., although it does contribute to infectivity and pathogenesis at low inoculum densities (Bauer *et al.*, 1994; Rantakari *et al.*, 2001; Toth and Birch, 2005). When TTSS genes are expressed a syringe-like pilus is assembled that crosses the plant cell wall, through which effectors are translocated into the plant cytoplasm (Li *et al.*, 2002; see also chapters by Arnold *et al.* and Boch). TTSS-secreted effectors interact with various intracellular targets, leading to parasitism and disease in susceptible plants or to a programmed cell death response known as the hypersensitive response (HR) in some resistant plants (Alfano and Collmer, 2004). It is now well established that a primary role of these effectors is to suppress host defence recognition of microbial associated molecular patterns (MAMPS) (Alfano and Collmer, 2004; Navarro *et al.*, 2004; see also chapter by Erbs and Newman) and in

Table 3.1 Characteristics of representative apoplast and xylem-colonizing plant pathogenic bacteria

Pathogen	Taxonomy	Host[a]	Disease and/or symptoms	Lifestyle[b]	Genome sequence[c]
Pseudomonas syringae pvs.	γ-Proteobacteria	Wide; pathovars are host specific; tomato, *Arabidopsis*, legumes	Leaf spots, blights and cankers	Facultative: leaf surfaces, seed, plant debris, water. Pathogenic: intercellular spaces of leaves, stems, bark and fruit.	CP000075, CP000058, AE016853
Xanthomonas campestris pvs.	γ-Proteobacteria	Pathovars are host specific; pepper, tomato, brassicas	Leaf spots and blights; black rot	Facultative: seed, plant debris, leaf surfaces. Pathogenic: intercellular spaces and vascular tissue of leaves, stems, roots and fruit.	AE008922, CP000050, AM039952
Pectobacterium atrosepticum	γ-Proteobacteria	Potato	Blackleg; stem necrosis, tuber rots	Facultative: rhizosphere, plant debris, water. Pathogenic: intercellular spaces and vascular tissue of stems, roots and tubers.	BX980851
Ralstonia solanacearum	β-Proteobacteria	Wide; tomato, potato, banana	Wilts, chlorosis, necrosis	Facultative: soil, plant debris, water. Pathogenic: intracellular spaces and vascular tissue of stems roots and tubers.	AL646052
Xylella fastidiosa	γ-Proteobacteria	Wide; citrus, grapevine	Chlorosis	Obligate: insect vector. Pathogenic: xylem.	AE003849, AE009442
Pseudomonas savastanoi pv. *savastanoi*	γ-Proteobacteria	Olive and oleander	Olive knots, galls	Facultative: leaf surfaces, bark. Pathogenic: intercellular spaces of stems and branches.	No genome sequence available
Xanthomonas axonopodis pv. *citri*	γ-Proteobacteria	Citrus	Cankers	Weakly facultative: lesions, plant debris. Pathogenic: intracellular spaces of leaves and fruit.	AE008923

[a] Representative host plants are listed.
[b] Facultative: colonizes a variety of niches in addition to entering into a pathogenic lifestyle (Morris et al. 2007); obligate: only colonizes plant host and insect vector; niches that are epidemiologically important for subsequent infection are indicated.
[c] Genbank ID for genome sequence.

susceptible plants some of these effectors are able to suppress effector-triggered immunity by the plant resistance surveillance system (for detailed reviews see: Alfano and Collmer, 2004; Nomura et al., 2005; Abramovitch et al., 2006; Chisholm et al., 2006; Jones and Dangl, 2006). The coordinated secretion of effectors and toxins may also help bacteria to obtain nutrients from the cells, although the nature of the effectors and cellular targets responsible for changes in nutrient availability are unclear (He et al., 2004).

In the following sections, we discuss current evidence for modulation of apoplast physiology by plant pathogenic bacteria, and review the features of the plant apoplast that modulate the bacterial behaviour inside the plant. Although both biotrophic and necrotrophic pathogens can colonize the plant apoplast, current evidence suggests that foliar hemibiotrophs such as *P. syringae* spend proportionately more time in the apoplastic compartment of plant tissues during disease development, when compared to soil-borne necrotrophs such as *P. atrosepticum* and *Ralstonia solanacearum* (see chapter by Brown), and may be expected to show a greater degree of specialization for growth in this environment. We therefore focus our discussion on well-studied foliar hemibiotrophs such as *P. syringae*, drawing examples from other pathogens where relevant.

Survival on plant surfaces

Before considering specific features of the plant apoplast, and traits associated with adaptation to growth in the apoplast, it is important to consider how apoplast-colonizing bacteria enter into the plant apoplast, and the physiological status of bacteria prior to apoplast colonization. In contrast to fungal pathogens that can deploy appressoria to disrupt the plant cuticle and invade epidermal cells (Mendgen and Hahn, 2002), apoplast-colonizing bacterial pathogens must rely on natural openings and wounds to enter into the intercellular space and must be able to survive in environments such as plant surfaces, the plant rhizosphere, water, or in association with insect vectors prior to entry into plant tissues (Beattie and Lindow, 1995; Hirano and Upper, 2000; Genin and Boucher, 2004; Redak et al., 2004; Grenier et al., 2006; Morris et al., 2007). The plant surface is also an important habitat with respect to dissemination of plant pathogenic bacteria, which have been observed to colonize plant surfaces in large numbers following successful endophytic colonization (Beattie and Lindow, 1995; Brunings and Gabriel, 2003). Every bacterium listed in Table 3.1 displays some ability to grow in non-apoplastic environments, and the genotypic and phenotypic traits of each of these bacteria have been shaped by growth not only in the apoplast, but also in these diverse ecological niches.

In order to survive on leaves, bacteria must be able to cope with a range of environmental stresses that are subject to intense diurnal fluctuations (Hirano and Upper, 2000). These stresses range from ultraviolet (UV) radiation, exposure to desiccation, osmotic stress and temperature changes. Mechanisms to avoid UV damage include the production of siderophores and pigments (Sundin and Murillo, 1999; Hirano and Upper, 2000; Poplawsky et al., 2000; Lindow and Brandl, 2003). Leaves are not even surfaces and the distribution and concentration of nutrients on the leaves is highly variable (Mercier and Lindow, 2000; Leveau and Lindow, 2001; Krimm et al., 2005). Bacteria seem to localize and aggregate in sites that are conducive for their survival, such as the base of trichomes, substomatal cavities or depressions and cracks in veins and cuticles (Leveau and Lindow, 2001; Monier and Lindow, 2004; Krimm et al., 2005; Fig. 3.1). The base of trichomes has been suggested to be a site where water is retained under dessication stress and it has been speculated that these sites provide more nutrient availability to support bacterial growth (Monier and Lindow, 2004; Krimm et al., 2005). Interestingly, some bacteria are able to change the permeability of the cuticle, which in turn may increase the availability of water and nutrients in the phyllosphere (Krimm et al., 2005).

Another strategy for survival on the plant surfaces is aggregation, which protects bacteria from osmotic stress and plant antimicrobial compounds. Aggregation also helps bacteria to adhere to the plant surface and may help bacteria alter epidermal cell physiology to obtain more nutrients (Monier and Lindow, 2003; Boureau et al., 2004; Monier and Lindow, 2004; Jacques et al., 2005; Fig. 3.1). The reversible transition from

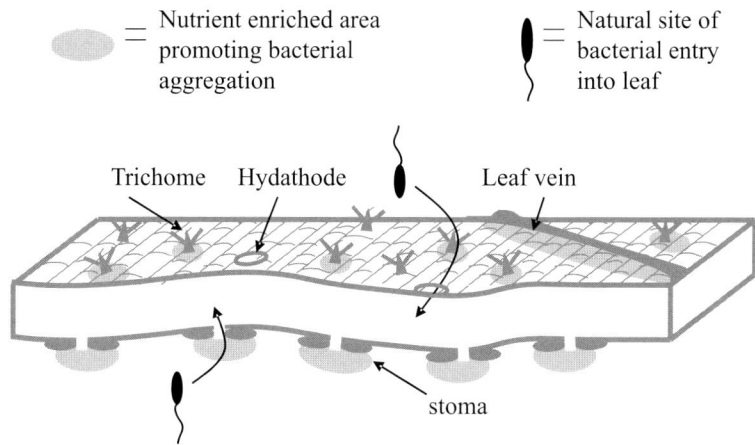

Figure 3.1 Sites of increased nutrient concentration and bacterial aggregation on the leaf surface.

planktonic to biofilm-associated lifestyle is likely to be pivotal in the different colonization phases of bacterial plant pathogens (Dow et al., 2003; Von Bodman et al., 2003; Jacques et al., 2005), as shown for saprophytic fluorescent *Pseudomonas*, which can resist stressful conditions in biofilms, but can explore new colonizable sites as planktonic bacteria (Boureau et al., 2004).

Some features shown to be important for epiphytic growth have advantages for both epiphytic and intercellular lifestyles. For example, genomic and experimental analyses of plant pathogenic and non-pathogenic *Pseudomonas* have shown that both produce lipopeptide biosurfactants, which have been linked to swarming motility, surface colonization and lysis of eukaryotic cells (Berti et al., 2007; de Bruijn et al., 2007; Tran et al., 2007). Similarly, it was shown that mutants of *P. syringae* with reduced epiphytic fitness and desiccation tolerance performed worse than wild-type bacteria on wet leaves and when infiltrated into leaves (Beattie and Lindow, 1994). Production of extracellular polysaccharides (EPS) not only protects bacteria from desiccation and other stresses, but also contributes to the intercellular growth of *X. fastidiosa*, *X. campestris*, *R. solanacearum*, *Erwinia amylovora*, *Pantoea stewartii* subsp. *stewartii* and *P. syringae* (Osman et al., 1986; Fett and Dunn, 1989; Bellemann and Geider, 1992; da Silva et al., 2001; Von Bodman et al., 2003).

Although some traits are important for bacterial growth on plant surfaces and in intercellular spaces, other features seem to give specific advantages in one niche or the other, or at different stages of the infection process. For example, in rhizosphere-colonizing bacteria such as *R. solanacearum* and *Agrobacterium tumefaciens*, chemotaxis and motility are required for the initial stages of colonization (Brencic and Winans, 2005; Yao and Allen, 2006; see also chapters by Brown, and Setubal et al.), but seem to be less important for the later stages of infection, where density-dependent activities such as production of pectolytic enzymes, antibiotics and other virulence factors are required (Von Bodman et al., 2003; Quinones et al., 2005). Similarly, motility promotes virulence in the foliar pathogen *E. amylovora* when bacteria are sprayed onto leaves, but flagellar biogenesis genes are negatively regulated by the alternate sigma factor HrpL in TTSS-inducing conditions (Cesbron et al., 2006).

Another feature whose regulation is complex and which has a variable role in pathogenesis between species is the production of high-affinity iron-scavenging siderophores. It is known that nutrients and phenolic compounds present in leaf exudates induce siderophore production because they sequester iron (Karamanoli and Lindow, 2006). Siderophore production enables bacteria to access iron and increase their epiphytic fitness (Bultreys and Gheisen, 2000; Karamanoli and Lindow, 2006). However, recent studies have provided conflicting evidence regarding the importance of siderophores in virulence. Production of chrysobactin and achromobactin is required for systemic virulence by the necrotrophic pathogen *D. dadantii* (Franza et al. 2005), but it is unclear whether the siderophores yersiniabactin and pyoverdin produced by *P. syringae* are required for virulence (Jones *et al.* 2007).

Entry into the apoplast

A high inoculum density on the plant surface favours invasion of intercellular spaces (Weller and Saettler, 1980; Melotto *et al.*, 2006), and therefore large epiphytic populations contribute to disease outbreaks (Beattie and Lindow, 1995; Hirano and Upper, 2000). Epiphytic population densities increase at high humidity and following rainfall, which not only provides moisture, but also contributes to cracks in leaf surfaces that increase nutrient availability and provide points of entry. A few bacterial pathogens, including strains of *P. syringae*, actively cause damage to plant leaves by producing ice nucleating proteins that raise the temperature at which frost damage occurs (Gurian-Sherman and Lindow, 1993).

Fig. 3.2 shows a schematic diagram depicting a cross-section of a leaf and some of the entry sites that are known to be used by foliar pathogens. Bacteria have been shown to be concentrated in the base of open stomata (Melotto *et al.*, 2006), which can contain sucrose concentrations of up to 150 mM when high transpiration and active photosynthesis occur, due to the movement of sucrose through the transpiration stream (Outlaw and De Vlieghere-He, 2001; Fig. 3.2 inset). High humidity promotes stomatal opening, favouring the entry of bacteria. However stomata are not passive sites of entry and several factors modulate closing and opening of stomata including bacterial and fungal elicitors (Lee *et al.*, 1999; Melotto *et al.*, 2006; Underwood *et al.*, 2007). Melotto and collaborators (2006) found that abscisic acid (ABA)-mediated induction of stomatal closure in response to detection of MAMPs was an integral part of the salicylic acid-dependent defence response. ABA regulates stomatal closure upon desiccation stress in the roots and is connected with stomatal responses to humidity (Xie *et al.*, 2006). Coronatine, a well-described toxin in *P. syringae* that is required for full virulence in the plant (Bender *et al.*, 1999), was able to suppress MAMP-induced stomatal closure. Factors other than coronatine may also

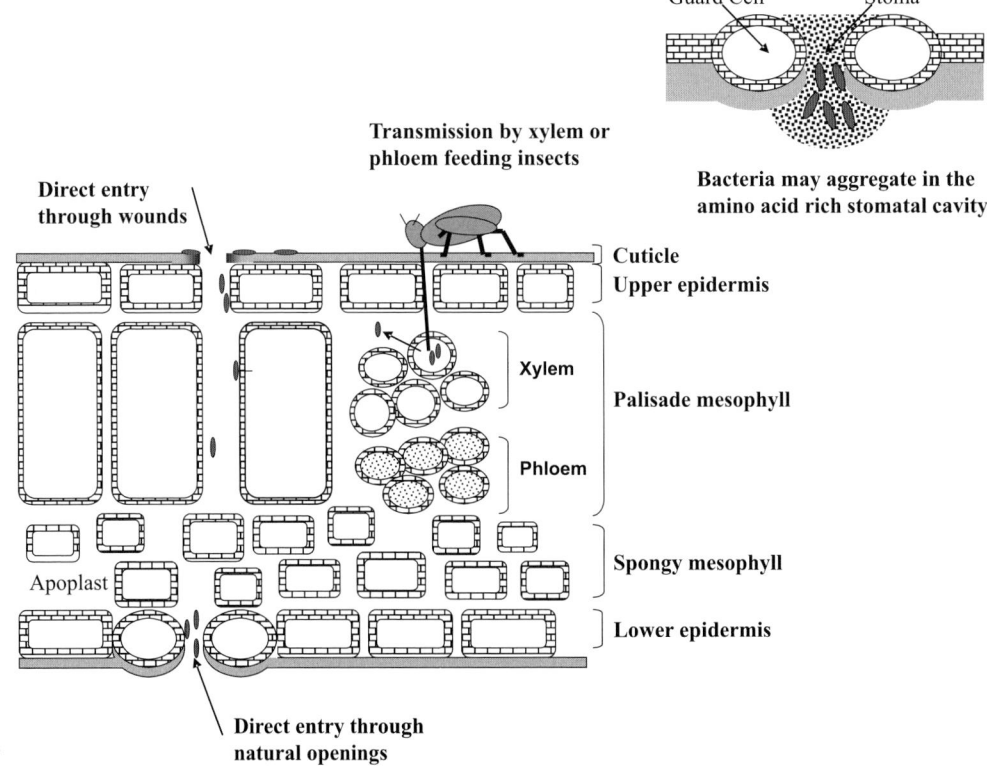

Figure 3.2 Bacterial entry into the apoplast from aerial parts of the plant.

be able to overcome stomatal defence. For example, the tobacco pathogen *P. syringae* pv. *tabaci*, which does not produce coronatine, was also able to re-open stomata (Melotto *et al.*, 2006).

Differences in entry sites have been linked to distinct adaptive features and strategies in different plant pathogenic bacteria. Bacteria that commonly enter through natural entry sites where active basal defences are present, such as stomata and hydathodes (Hugouvieux *et al.*, 1998; Melotto *et al.*, 2006), may benefit from mechanisms, such as coronatine-mediated stomatal opening, which counteract these primary defences. The existence of both hydathode and stomata-specific entry strategies is supported by the observation that some pathogens preferentially enter plant leaves through different routes. For example *X. campestris* pv. *campestris* preferentially enters leaves via hydathodes, while *X. campestris* pv. *armoraciae* enters through stomata (Hugouvieux *et al.*, 1998). Hydathodes, like stomata, are sites where nutrients and water may be released from the intercellular spaces, most notably in the form of guttation fluid, which may be exuded under conditions of high humidity and taken up as the atmosphere and soil becomes drier, thereby drawing in hydathode-colonizing bacteria. The composition of guttation fluid is distinctly different from apoplastic fluid, indicating that some nutrients are retained within the plant leaf, while others are exuded (Pilot *et al.*, 2004). The hydathode entry mechanism of some bacteria, such as *X. campestris* pv. *campestris*, seems to require extracellular factors, such as EPS and lipopolysaccharides (LPS), but is independent of the TTSS (Hugouvieux *et al.*, 1998). One feature that may promote invasion through specific entry sites is attachment. Both *X. campestris* pv. *hyacinthi* and *P. syringae* pv. *phaseolicola* have been shown to preferentially attach to stomata (Romantschuk, 1992). Conversely, bacteria that enter through wounds or active growth sites such as lateral root structures need chemotactic and nutrient utilization abilities that allow them to move towards plant exudates and to competitively colonize infection sites (Romantschuk, 1992; Genin and Boucher, 2004; Brencic and Winans, 2005; Yao and Allen, 2006; Zolobowska and Van Gijsegem, 2006).

Characteristic features of the plant apoplast

Bacteria entering the plant apoplast encounter an environment that is distinctly different from their previous environment, whether that environment is a wound, a natural opening or an insect vector.

The apoplast is a dynamic environment in which many metabolic and transport processes take place, as well as changes in redox and osmotic conditions, pH, nutrient availability, and hormone signalling in response to biotic interactions and environmental conditions such as light, water and mineral nutrient availability. Physiologists have long known that the apoplast is acidic, which helps to maintain the flux of nutrients through proton pumps (Grignon and Sentenac, 1991; Sattelmacher, 2001). This is particularly relevant in the case of amino acids and sugars that are transported from the apoplast to the cytoplast by H^+ symport (Felle, 2006).

The major structural component of the apoplast is the plant cell wall and in the context of plant–pathogen interactions, it constitutes the primary physical barrier to bacterial access to cellular nutrients. One of the primary defences upon recognition of MAMPs is the production of reactive oxygen species by extracellular peroxidases and the strengthening of the cell wall through cross-linking and deposition of papillae or callose (Nurnberger and Brunner, 2002). Successful pathogens are able to suppress this defence mechanism and several TTSS effectors have been identified as being involved in this process (Hauck *et al.*, 2003; DebRoy *et al.*, 2004; de Torres *et al.*, 2006; Truman *et al.*, 2006; see also chapter by Boch).

Additional defence mechanisms expressed in the plant apoplast include extracellular enzymes that are produced during early basal resistance (EBR) upon detection of bacterial elicitors (Ott *et al.*, 2006). These authors isolated two soluble extracellular chitinases that could be involved in degrading the peptidoglycan from the bacterial cell wall, which could also help to release further elicitors from the invading pathogen, hence inducing a stronger defence (Ott *et al.*, 2006). The plant cell wall is also rich in secondary metabolites with known antimicrobial activities against herbivores, fungal and bacterial pathogens. Their biochemical nature varies between plant species and the best described are saponins, cyanogenic glycosides, glucosinolates and cinnamic acids (Osbourn, 1996; Fontaniella *et al.*, 2007). They can have a direct antimicrobial effect or contribute to the activation of plant defences (Kim and Sano, 2008).

Sugars can move from photosynthetic sources to sink organs through the apoplast in some species and this takes place through differences in osmotic potential (Schaarschmidt *et al.*, 2007). Extracellular invertases, which are involved in breaking down sucrose molecules into glucose and fructose, have an essential role in this process, but at the same time they could also contribute to provide carbohydrates to the invading pathogens (Walters and McRoberts, 2006). Sugars can also be transported upwards and accumulate in stomatal cavities through the transpiration stream (Outlaw and De Vlieghere-He, 2001).

Nitrogen is taken up as nitrate or ammonium in the root and must be reduced to ammonia before amino acid biosynthesis (Lea and Azevedo, 2006). The reduction of nitrate can occur in several parts of the plant and the apoplast of the leaf may contain different types of nitrogen compounds, including newly synthesized amino acids that are transported to other sites (Lea *et al.*, 2007). Depending on the plant species, the plant organ, and the physiological state of the plant, different amounts and types of nitrogen could be available to the pathogen (Walters and Bingham, 2007). The fact that hemibiotrophic pathogens do not destroy cells to obtain nutrients has prompted the speculation that apoplast-colonizing pathogens might face starvation, particularly with respect to nitrogen. However, several studies indicate that at least for some pathogens, this is not the case. For example, Farrar (1995) showed that the total leaf nitrogen in barley leaves exceeded the nutritional requirements of the brown rust pathogen (*Puccinia hordei* Otth.). Similarly, the concentration of amino acids and other nitrogen compounds in tomato leaf apoplast extracts from healthy leaves has been shown to be high enough to support the requirements of bacterial and fungal pathogens (Solomon *et al.*, 2003; Rico and Preston, 2008).

Adaptation for growth in the apoplastic environment

The plant apoplast provides a nutritionally distinct habitat compared to other habitats colonized by bacteria, so it is likely that plant pathogens that spend a large proportion of their existence in an apoplastic niche will show distinct adaptations to the nutrients, stresses and challenges present in this environment, as well as possessing adaptations that allow them to colonize alternate niches such as plant surfaces and vectors, or to colonize diverse host plants. These adaptations may take four main forms:

i Acquisition of genes that encode beneficial traits such as the ability to modulate plant metabolism, uptake of apoplastic nutrients, tolerance to apoplastic stresses, or mechanisms to promote dissemination from one susceptible host plant to a new host plant. Novel genes may be acquired through mutation, recombination or lateral gene transfer.
ii Amplification of the copy number or transcription of genes that encode beneficial traits. Traits that are required for growth in the apoplast, but not in other habitats may be transcriptionally or post-translationally up-regulated in response to apoplastic signals.
iii Loss of genes that confer little or no benefit in an apoplastic environment.
iv Loss of genes that encode traits which are detrimental in an apoplastic environment, e.g. transporters that promote uptake of plant-derived antimicrobial chemicals.

Some recent studies have even suggested that the process of genome evolution in bacterial plant pathogens may be accelerated by stresses present in the plant environment (Arnold *et al.*, 2007). In the following sections we consider how selective pressures present in the plant apoplast may have affected the evolution of bacterial plant pathogens, and in particular the model pathogen *P. syringae*, and how genomic and experimental data can be linked to niche-specific adaptation to an apoplastic environment.

Nutritional adaptation to life in the apoplast

Taxonomic studies of fluorescent pseudomonads have long been grounded in the knowledge that plant pathogenic species such as *P. syringae* show distinct phenotypic characteristics from non-plant pathogenic rhizosphere colonizing bacteria such as *P. fluorescens* (see chapters by Stavrinides, and Vinatzer and Bull). Early studies established that the oxidase-negative plant pathogenic pseudomonads, which include all *P. syringae* pathovars, were less nutritionally versatile than oxidase-positive saprophytic pseudomonads, particularly with respect to organic acids and certain amino acids (Stanier *et al.*, 1966; Misaghi and Grogan, 1969; Sands *et al.*, 1970). These phenotypic differences between pseudomonads suggested a nutritional adaptation to different lifestyles, which in the case of *P. syringae*, could be linked to growth on the specific nutrients present in the plant apoplast and leaf exudates, rather than in the complex array of nutrients available to rhizosphere bacteria. Further speculation about nutrient-specific adaptation to specific hosts was generated by the observation that specific groups within *P. syringae*, such as *P. syringae* pv. *phaseolicola*, could be separated form other *P. syringae* pathovars because they gave negative results in phenotypic tests for which other *P. syringae* strains gave positive results (Misaghi and Grogan, 1969; Sands *et al.*, 1970; Young and Triggs, 1994). Thus, one could suggest that the narrow host range of *P. syringae* pv. *phaseolicola*, which infects only bean, soybean and kudzu, could be correlated with restricted nutritional versatility.

As noted in the previous section, adaptation to specific host plants could arise through both loss of function and gain of function changes in bacteria, and be driven by both positive and negative selection for use of specific nutrients. Hildebrand (1972) hypothesized that the almost unique ability of *P. syringae* pv. *pisi* to utilize homoserine as a carbon source, in contrast to other *P. syringae* pathovars and saprophytic pseudomonads, could be correlated with its ability to cause disease in pea, a plant with high levels of homoserine. Hildebrand (1972) also speculated that some amino acids could be selectively toxic to plant pathogens and restrict their ability to infect plants. Amino acid-dependent inhibition of pathogen growth could arise in several different ways. The presence of excessive amounts of a single amino acid could suppress the biosynthesis or uptake of other amino acids (De Felice *et al.*, 1977; Jackson *et al.*, 1993; Braun *et al.*, 2008), or could repress induction of pathogenicity and virulence factors needed for plant colonization. The hypothesis that colonization ability could be linked to the ability to use specific plant nutrients was further explored by Coplin *et al.* (1974), who showed that differences in amino acid composition in tobacco and tomato xylem fluid could explain differences in susceptibility to methionine and leucine auxotrophs of *R. solanacearum* K60. These results agreed with what was described as the 'nutrition-inhibition' hypothesis of pathogenicity, which stated that a nutritional deficiency could account for the avirulence of a pathogen, and that amino acid utilization pathways in bacteria can be inhibited by excess amounts of one or more amino acids (Sands and Zucker, 1976). However, the role of differential nutritional abilities in host adaptation has been largely unexplored since the publication of these early works, and is only now beginning to be readdressed, as researchers begin to exploit the availability of genome sequence data and new tools for studying plant-bacteria interactions.

Rico and Preston (2008) showed that the tomato pathogen *P. syringae* pv. *tomato* DC3000, the tobacco pathogen *P. syringae* pv. *tabaci* 11528 and the non-pathogen *Pseudomonas fluorescens* SBW25 could grow in apoplast extracts from healthy tobacco and tomato (Rico and Preston, 2008), which argues against a significant contribution of apoplast composition in defining the ability of *Pseudomonas* to colonize these host plants (Hildebrand, 1972). They then proceeded to determine the nutrient utilization abilities of each strain, and of other model genome-sequenced *Pseudomonas*, and compared the nutrient utilization profiles of these strains to the composition of the tomato apoplast. The comparison showed that *P. syringae* pv. *tomato* DC3000 is able to use a wide variety of metabolites that are abundant in the apoplast, but also showed that most of the amino acids, organic acids and sugars present in this environment could be used by a wide range of non-host and non-pathogenic *Pseudomonas*, again supporting the idea that nutritional compatibility does not restrict colonization. Nevertheless, it is interesting to note that the most abundant amino acid in tomato apoplast extracts and in tomato fruits is γ-amino-butyric acid (GABA), and that the genomes of *P. syringae* pv. *tomato* DC3000 and *P. syringae* pv. *syringae* B728a contain three copies of each of the two genes required for GABA assimilation [*gabT* (γ-aminobutyrate transferase) and *gabD* (succinic semialdehyde dehydrogenase)], compared with the single copies present in other *Pseudomonas*. This suggests that these bacteria have acquired additional genes to facilitate assimilation of this abundant carbon and nitrogen source (Buell *et al*., 2003).

Another recent example of how genome sequence data can be used to help explain strain and pathovar-specific differences in nutrient utilization comes from studies of the genome of the bean pathogen *P. syringae* pv. *phaseolicola* 1448A, which, as noted above, displays distinctive differences in nutrient utilization compared to other strains of *P. syringae* (Young and Triggs, 1994; Rico and Preston, 2008). Most of the genes encoding core metabolic pathways in *P. syringae* pv. *tomato* DC3000, *P. syringae* pv. *syringae* B728a and *P. syringae* pv. *phaseolicola* 1448A are highly conserved in all three strains. Joardar *et al*. (2005) noted only one difference between *P. syringae* pv. *tomato* DC3000 and *P. syringae* pv. *phaseolicola* 1448A with respect to 154 biological processes; an authentic point mutation in the gene encoding N-formylglutamate amidohydrolase (PSPPH4868), which would account for the inability of *P. syringae* pv. *phaseolicola* strains to use histidine (Young and Triggs, 1994). However, the genome and plasmids of *P. syringae* pv. *phaseolicola* 1448A contain 184 pseudogenes, compared with nine in *P. syringae* pv. *tomato* DC3000 and 47 in *P. syringae* pv. *syringae* B728a. Some of these pseudogenes correspond to mobile elements and virulence factors, but many encode regulators, membrane transporters and enzymes that are predicted to be involved in nutrient utilization. Rico and Preston (2008) showed that lack of utilization of D-sorbitol, D-galacturonic acid and L-histidine, which are used by *P. syringae* pv. *tomato* DC3000 and *P. syringae* pv. *syringae* B728a, could be correlated with frameshift mutations in genes predicted to be involved in the metabolism of these substrates. Interestingly, two of the pseudogenes present in the genome of *P. syringae* pv. *phaseolicola* 1448A correspond to *gabT* (PSPPH3457) and *gabD* (PSPPH5038), which, as noted above, are over-represented in the three *P. syringae* genomes compared to other *Pseudomonas*. This could suggest that the increase in *gabT/D* genes occurred in the common ancestor of all three genome-sequenced *P. syringae* strains, followed by gene loss in *P. syringae* pv. *phaseolicola*.

Rico and Preston (2008) also identified instances where lack of nutrient assimilation in *P. syringae* compared with other *Pseudomonas* or in *P. syringae* pv. *phaseolicola* 1448A compared with other *P. Syringae* could be linked to the absence of specific genes, such as genes predicted to be involved in D-mannitol and L-leucine utilization (Rico and Preston, 2008). Genomic evidence of recent gene loss, in the form of pseudogenes and absence of genes present in closely related organisms, is frequently speculated to be due to recent changes in environment, such as a switch from one host to another, or increased dependence on one niche (Ochman and Davalos, 2006; Raskin *et al*., 2006; Pallen and Wren, 2007). Changes in host range or niche typically originate in a small population of bacteria, and therefore result in a bottleneck effect during which both neutral, beneficial and mildly deleterious mutations can be rapidly fixed (Maurelli, 2007). It is tempting to speculate that differences between *P. syringae* pv. *phaseolicola* and studies of gene expression in apoplast colonizing bacteria are linked to recent changes in the host range or epidemiology of these bacteria, such as a host shift to bean, or increased host availability through agricultural practices, and it will be interesting to determine the extent to which nutritional functions retained in *P. syringae* pv. *phaseolicola* mirror nutrient availability in host tissues.

At the opposite pole of nutritional versatility is the soil pathogen *R. solanacearum*, which can survive in soil and water, contains a wide variety of nutrient utilization pathways and membrane transporters, and is able to infect a wide variety of hosts (Salanoubat *et al*., 2002; Genin and Boucher, 2004; see also chapter by Brown).

Recently, it was demonstrated that a folate biosynthetic mutation rendered R. solanacearum strain OE1-1 unable to multiply in the intercellular space and to subsequently produce systemic infections, indicating that folate levels in the apoplast are not high enough to support the growth of this pathogen (Shinohara et al., 2005). Interestingly, the mutant could produce disease symptoms when inoculated straight into the xylem, which suggests that the bacterium uses different nutritional strategies and adaptations in different stages of its parasitic life.

Apoplast-induced nutrient utilization pathways

Although in vitro studies of nutrient utilization can provide significant insight into the phenotypic attributes of plant pathogenic bacteria, particularly when combined with genome sequence data, they have limitations with regard to understanding adaptation for growth in the plant apoplast. Firstly, nutrient utilization assays carried out in vitro generally require bacteria to use a single metabolite as their sole source of carbon or nitrogen; and secondly, they ignore the effect the environment can have on activation or inhibition of nutrient assimilation pathways. It is likely that some nutrients that can be assimilated in the apoplast are toxic, or unable to support bacterial growth when tested in isolation. Alternatively, bacteria may express the ability to use a particular nutrient in vitro or in alternate niches such as leaf surfaces, but not when present in the apoplast. Two approaches have been used to overcome these limitations: metabolic profiling of apoplast-colonizing bacteria, and studies of gene expression in apoplast-colonizing bacteria.

Metabolic profiling of apoplast-colonizing bacteria

Many different technologies can be used to assess the metabolic activities of bacteria in vitro and in complex environments such as the plant apoplast, including isotope-labelling, NMR, GC-MS and HPLC, and spectroscopic techniques such as Raman microspectroscopy and Fourier transform infrared (FTIR) spectroscopy (Scharff et al., 2003; Stehfest et al., 2005; Huang et al., 2007; Mashego et al., 2007; Neufeld et al., 2007). However, few of these technologies have been applied to plant pathogens and to the specific question of bacterial adaptations for apoplastic growth, with the exception of targeted studies of specific proteins, toxins or polysaccharides. Recently, Osiro and collaborators (2004) used ^{13}C-NMR to study the metabolism of plant pathogenic bacteria and demonstrated the presence of an intact pathway for fatty acid biosynthesis in the xylem-colonizing pathogen X. fastidiosa, thereby over-turning genome-based predictions that this pathway was absent in Xylella. Schmelz and collaborators (2003), demonstrated that GC-MS could be used to profile both plant- and bacteria-derived organic volatile compounds produced by plants infected with P. syringae.

Rico and Preston (2008) recently used a modified phenoarray technique to examine nutrient utilization by P. syringae pv. tomato DC3000 during growth in apoplast extracts. Bacteria were incubated in apoplast extracts or synthetic media, and tetracycline was added prior to inoculation of bacteria into a Biolog GN2 MicroPlate™ (Biolog), which assesses the ability of bacteria to metabolize specific substrates as carbon sources as indicated by reduction of the indicator tetrazolium violet. Antibiotic treatment inhibited de novo protein synthesis in the Biolog plate, so results obtained reflected apoplast-induced metabolic activities, in contrast to synthetic media-induced activities or non-induced activities. Inhibitor treated bacteria used a range of carbon sources, many of which correspond to metabolites known to be present in the plant apoplast, including GABA, glutamine, aspartate, fructose and citrate. Some of the pathways involved in using these substrates were constitutively active in a wide range of media, but others, including pathways for fructose and citrate utilization were shown to be specifically up-regulated during growth in apoplast extracts. Interestingly, most of the pathways that were constitutively expressed in apoplast extracts and synthetic media were pathways present in all Pseudomonas, while many of the apoplast-induced pathways, such as trehalose utilization, showed a variable distribution, or were primarily found in P. syringae, suggesting that P. syringae expresses a core set of constitutively expressed pathways, and a niche-specific set of inducible pathways. This study also showed that *P. syringae* contains numerous utilization pathways that can be used *in vitro*, but which were not active during growth in apoplast extracts from healthy tomato plants. These un-induced pathways may correspond to nutrients used in other niches, such as guttation fluid and leaf surfaces, or to metabolites produced during *Pseudomonas*–plant interactions as a result of the activation of plant defences and changes to plant physiology.

Apoplast-induced gene expression

A second approach that has been more widely used to study bacterial adaptation to the plant apoplast is the use of functional genomic, transcriptomic and proteomic techniques to study bacterial gene expression during plant colonization, or during growth in plant extracts. Boch and collaborators (2002) used *in vivo* expression technology (IVET), to identify a set of genes that were up-regulated during plant colonization by *P. syringae* pv. *tomato* DC3000. This study pre-dates the publication of the *P. syringae* genome, but a list of up-regulated genes can be downloaded as an Artemis feature file from the *Pseudomonas*–plant interaction website (*http://pseudomonas-syringae.org/pst_gen_analy.htm*). Plant-induced genes identified using this method can be divided into four main categories: (i) pathogenicity and virulence factors, such as the type III secretion system, coronatine, alginate and a non-ribosomal peptide synthetase similar to syringomycin synthetase; (ii) nutrient assimilation genes, including genes predicted to be involved in amino acid, organic acid and fatty acid assimilation; (iii) stress tolerance genes, such as catalase, which may be used to detoxify reactive oxygen species; and (iv) biosynthetic genes, including genes for isoleucine, valine, threonine, lysine, serine and thiamine biosynthesis. Several of the biosynthetic pathways shown to be up-regulated in *P. syringae* pv. *tomato* DC3000 during plant colonization correspond to amino acids shown to be present in relatively low levels in the tomato apoplast (Rico and Preston, 2008). IVET has also been used to identify plant-induced genes in other plant pathogens, including *X. campestris* pv. *campestris* (Osbourn *et al.*, 1987), *Erwinia amylovora* (Zhao *et al.*, 2005) and *R. solanacearum* (Brown and Allen, 2004). *R. solanacearum* genes induced during tomato xylem colonization included a homologue of the GABA assimilation gene *gabD*, as well as genes predicted to be involved in virulence, metabolism, nutrient uptake and detoxification of reactive oxygen species.

Both transcriptomic and IVET approaches have been used to identify genes up-regulated in *D. dadantii* 3937 during plant colonization, giving similar results to those identified by Boch and collaborators (2002), and additionally identifying genes predicted to be involved in purine and phenylalanine synthesis, xenobiotic degradation, iron uptake, and transcriptional regulation (Okinaka *et al.*, 2002; Yang *et al.*, 2004). Bacterial samples used for transcriptomic analyses were extracted from the intercellular fluid of African violet leaves, and should therefore reflect bacterial gene expression in the leaf apoplast. In both studies, selected plant-induced genes were further characterized by examining the *in planta* performance of mutants knocked out in the corresponding gene, confirming that several of these up-regulated genes do play a significant role in plant colonization.

A modified version of IVET known as RIVET (recombinase-based IVET), has been used to study plant-induced genes in *X. campestris* pv. *vesicatoria* (Tamir-Ariel *et al.*, 2007). Although both *X. campestris* pv. *vesicatoria* and *P. syringae*. pv. *tomato* DC3000 share a common host, tomato, there is relatively little overlap between the results obtained in this screen and those obtained by Boch and collaborators (2002), indicating either that there are distinct differences in the plant-induced genes expressed by these two pathogens, or, perhaps more likely, that both screens are far from being saturated, and only represent a subset of the genes up-regulated during plant colonization in these two pathogens. For example, one gene shown to be up-regulated in *X. campestris* pv. *vesicatoria* was a citrate transporter (CitH), which was not identified in the IVET screen performed by Boch and collaborators (2002), even though this transporter is conserved in *P. syringae*. pv. *tomato* DC3000 and citrate utilization has been shown to be up-regulated in *P. syringae* pv. *tomato* DC3000 during growth in apoplast extracts (Rico and Preston, 2008).

Regulation of pathogenicity and virulence factors in response to apoplastic signals

Studies of apoplast induced genes clearly demonstrate that differential regulation of gene expression is an important feature of bacterial adaptation for growth in the apoplast. However, relatively little is known regarding the environmental factors and molecular mechanisms that underlie these changes in gene expression. Most studies of the molecular mechanisms underlying apoplast-induced gene expression have focused on the environmental factors and regulatory mechanisms that promote expression of pathogenicity and virulence genes, such as the TTSS, toxins, EPS and exoenzymes. Results obtained in these studies provide further insight into the environment experienced by bacteria during plant colonization.

The type III secretion system

Environmental regulation of the TTSS has been studied in detail, although there are still unanswered questions as to how environmental signals are perceived. The TTSS is expressed rapidly when bacteria are infiltrated into the plant (Rahme et al., 1992). *In vitro*, factors contributing to TTSS expression are a low pH (~5.5), low osmolarity and predominance of sugars such as sucrose, fructose and mannitol over hexose sugars such as glucose, organic acids such as succinate, and some amino acids (Huynh et al., 1989; Rahme et al., 1992; Xiao et al., 1992; Brencic and Winans, 2005). Rich media such as KB repress the expression of the TTSS (Rahme et al., 1992; Xiao et al., 1992). Plant-derived signals may also contribute to the expression of TTSS, since, in some cases, *in planta* expression of the TTSS tends to be higher than in minimal media (Rahme et al., 1992). Moreover, microscopic and immunogold labelling observations of the type III pilus found that a higher number of pili were produced by bacteria that were in close contact with the cell wall (Hu et al., 2001) and a non-diffusible cell wall compound was found to be responsible for TTSS induction in *R. solanacearum* (Aldon et al., 2000; Brencic and Winans, 2005).

Rico and Preston (2008) found that tomato apoplast extracts induced the expression of the TTSS of *P. syringae* pv. *tomato in vitro* to a lesser extent than artificial *hrp*-inducing media (HIM) (Huynh et al. 1989), but more than in the rich medium LB. This is similar to results obtained by Xiao et al. (1994), who found that TTSS expression was induced by tobacco sap, but to a lesser extent than in HIM. Interestingly, tomato apoplast extracts supported fairly rapid growth of *P. syringae* pv. *tomato* in contrast to poor growth observed in HIM, while still supporting significant levels of TTSS expression. This suggests that nutrient limitation is not a prerequisite for TTSS induction, and that other features of the apoplastic environment such as the presence of specific metabolites, or the relative abundance of different classes of metabolites also affect TTSS induction. It would make biological sense for bacteria to express the TTSS only when in the plant apoplast, and not simply in response to starvation, because TTSS expression is metabolically expensive. Apoplast-specific metabolites, or metabolite ratios could provide that signal. Another factor affecting TTSS expression could be the host response to infection, since it has been shown that the timing and intensity of TTSS expression is different in host and non-host plants (Brencic and Winans, 2005).

Toxin production

The production of phytotoxins, a characteristic feature of *P. syringae* strains, is also environmentally regulated. For example, production of the phytotoxin coronatine (COR) has been shown to affected by pH, temperature, osmolarity, carbon sources, nutrient levels, amino acids, and the presence of complex carbon and nitrogen sources (Palmer and Bender, 1993; Palmer et al., 1997). However, the effect of environmental factors on toxin production can vary significantly between strains. COR production by the soybean pathogen *P. syringae* pv. *glycinea* PG4180 has a peak of production at 18°C, while at 28°C, the optimal growth temperature for *P. syringae* pv. *glycinea*, production is minimal (Budde et al., 1998). In contrast, temperature has little effect on coronatine production by *P. syringae* pv. *tomato*, in which coronatine has been shown to be induced by tomato leaf extracts, and in particular by organic acids and intermediates of the shikimic acid pathways, such as shikimate and quinate (Li *et al.*, 1998). This pathway is the route for phenylalanine and other phenolic compounds, including the defence molecule salicylic acid (Brencic and Winans, 2005; Uppalapati *et al.*, 2007). COR has been proposed to act in all three stages of infection: invasion, establishment and persistence. COR mutants of *P. syringae* pv. *tomato* were shown to be significantly impaired in the ability to colonize plants following dip or spray inoculation, as compared to direct infiltration (Mittal and Davis, 1995; Penaloza-Vazquez *et al.*, 2000), which may be linked to the ability of COR to suppress ABA-mediated stomatal closure, and thereby promote invasion (Melotto *et al.*, 2006). COR is structurally similar to jasmonic acid (JA) and its activity *in planta* has been linked to the activation of JA pathways, which, in turn, suppress or delay salicylic acid (SA)-mediated defence pathways (Laurie-Berry *et al.*, 2006). Coronatine may also have an SA-independent role in later stages of infection, as COR mutants are impaired for persistence in both wild-type tomato plants and in plants silenced for isochorismate synthase, an enzyme involved in SA biosynthesis (Uppalapati *et al.*, 2007).

Other phytotoxins produced by strains of *P. syringae* are phaseolotoxin, syringomycin and tabtoxin. Syringomycin production is induced by plant metabolites such as phenolic glycosides and sugars (Mo *et al.*, 1995). Syringomycin is a pore-forming toxin, which induces ion fluxes in cell membranes creating a disruption of the H^+/K^+ efflux, which in turn leads to alkalinization of the apoplast and permeability of the membrane to sugars and other organic compounds, which favour bacterial growth (Hutchison *et al.*, 1995). Phaseolotoxin production is induced only at 18–20°C (Nuske and Fritsche, 1989) and is responsible for the chlorotic symptoms observed in halo blight of beans caused by *P. syringae* pv. *phaseolico*la. Phaseolotoxin inhibits the enzyme ornithine carbamoyl transferase (ORT), which converts ornithine to citrulline in the arginine biosynthetic pathway. The accumulation of ornithine causes a phenotypic requirement for arginine, which leads to chlorosis. Phaseolotoxin is not essential for bacterial multiplication during infection (Tamura *et al.*, 2002), and strains lacking the phaseolotoxin gene cluster and responsible for epidemics in beans have been isolated and characterized (Oguiza *et al.*, 2003; Rico *et al.*, 2003). It is probable that phaseolotoxin, and many other phytotoxins, contribute to virulence by causing local and systemic changes in plant metabolism (Bender *et al.*, 1999). For example, tabtoxin produced by *P. syringae* pv. *tabaci* is cleaved in the plant to form tabtoxinine β-lactam, which inhibits the enzyme glutamine synthetase and leads to abnormal accumulation of ammonia in the plants and the consequent visible chlorosis symptoms. In addition, most of the toxins produced by *P. syringae* exhibit broad-spectrum toxicity to a wide range of eukaryotes, which has led to speculation that some of these toxins act against other endophytic and epiphytic microorganisms that attempt to colonize niches previously colonized by *P. syringae* (Voelksch and Weingart, 1998).

Exoenzymes and EPS

The timing and context in which cell wall degrading enzymes are expressed is essential for successful pathogenesis by necrotrophic pathogens such as *Pectobacterium* and *Dickeya*, as premature expression of pectinolytic enzymes can elicit host cell defences before bacterial numbers are high enough to overcome them. The pectinolytic enzymes (Pels) produced by *P. atrosepticum* are cell-density regulated, which ensures that they are only expressed when bacteria reach high population densities (Brencic and Winans, 2005; Toth and Birch, 2005). The structural and biochemical properties of the plant apoplast may have a significant effect on quorum sensing mechanisms, which depend on the localized accumulation of quorum sensing molecules (Waters and Bassler, 2005). The stability of quorum-sensing molecules can be significantly altered by pH, with the lactone ring of acyl homoserine lactones being stable at acidic and neutral pH, and unstable at alkaline pH, while some quorum sensing molecules, such as AI-2 undergo spontaneous reversible rearrangements that are affected by the chemical nature of the environment (Horswill *et al.*, 2007). The hydrophobicity of the environment may affect AHL diffusion, and bacterial- or plant-derived polysaccharides may act as AHL-sequestering matrices (Horswill *et al.*, 2007).

Quorum-sensing mechanisms typically form part of a signal transduction network that integrates diverse information about bacterial physiology and the environment (see chapter by Dow et al.). In *D. dadantii*, growth phase-dependent, environment-dependent and density-related mechanisms act in a coordinated manner to regulate the timing and expression of pectinolytic enzymes. Lautier and collaborators (2007) have recently described the role of the nucleotide associated protein Fis in repressing Pels expression at the beginning of exponential growth. Pels expression is also regulated by KdgR, a repressor of pectinolytic genes, the activity of which is modulated by the presence of pectic compounds (Nasser *et al.*, 1992). This has led to a model of Pels expression in which nutritional starvation in the initial stages of colonization activates Fis expression, which in turn represses Pels expression. The low availability of cell wall-derived degradation products means that KdgR represses Pels expression as well. Fis is also involved in the activation of the type II secretion system that secretes the pectinolytic enzymes, and after a certain time, bacterial multiplication and the basal production of Pels contributes to the development of a more favourable environment for Pels expression, and ultimately allows the bacteria to deploy a strong attack in the plant.

Quorum sensing mechanisms also regulate pathogenicity and virulence factors in other plant pathogenic bacteria, such as the wilt pathogen *R. solanacearum*. *R. solanacearum* enters plant tissues through openings in root active-growth sites and wounds, where it grows in the intercellular space. It has been speculated that once a sufficient cell density has been reached, the quorum-sensing regulated production of EPS and exoenzymes facilitates entry into the xylem vessels (Genin and Boucher, 2004). *R. solanacearum*, unusually, produces a potentially volatile quorum sensing molecule, 3-OH-palmitic acid methyl ester (3-OH-PAME), which gives the potential for airborne as well as liquid-borne signalling in soil and in plant tissues (Flavier *et al.*, 1997). Quorum sensing mechanisms also regulate EPS production in other pathogens such as *Pantoea stewartii* (Koutsoudis *et al.*, 2006), and the onion pathogen *Pantoea ananatis* (Morohoshi *et al.*, 2007).

Most plant pathogens use two or three of the pathogenicity and virulence factors discussed above to cause infection, with each factor acting in a hierarchy that creates the environmental conditions that allows the next factor to be expressed and used (Mole *et al.*, 2007; see also chapter by Dow *et al.*). Thus, the TTSS of necrotrophs such as *P. carotovorum* and *D. dadantii* is thought to suppress plant defences to enable bacteria to multiply *in planta*, and ultimately, use type I and type II secretion systems to secrete exoenzymes that degrade cell walls, thereby killing plant cells. Similarly, *X. axonopodis* pv. *citri*, the causal agent of citrus canker disease, uses a plant-induced TTSS to inject the TTSS secreted effector PthA into plant cells, which elicits water soaking and hyperplasia in host cells. Cell swelling draws water from the xylem through capillary action, which brings about further cell disruption, which in turn provides additional nutrients for growth, and aids in dispersal when bacteria egress to the plant epidermis. Finally, bacterial growth may in turn activate QS-regulated mechanisms, such as the production of xantham gum (Pettersson *et al.*, 1996; Brunings and Gabriel, 2003).

Modulation of the plant apoplast by plant pathogenic bacteria

It is logical to presume that the mere presence of bacterial pathogens in the apoplast will provoke changes in apoplast metabolism even before the deployment of the TTSS, due to the plant perception of bacterial MAMPs and the utilization by bacteria of the available nutrients. Fig. 3.3 provides a schematic overview of the different strategies deployed by bacterial plant pathogens to survive in the apoplast and modulate plant metabolism, and the responses elicited by bacteria in plant tissues. Extensive research has been carried out into the production and role of antimicrobial and signalling metabolites in pathogenesis and plant defence responses, and into cell wall modifications that occur during infection, a comprehensive review of which is beyond the scope of this chapter. For further information on these the reader is directed to several excellent reviews: ROS (Apel and Hirt, 2004; Mur *et al.*, 2006; Tsanko *et al.*, 2006; Moller *et al.*, 2007); calcium (Lecourieux *et al.*,

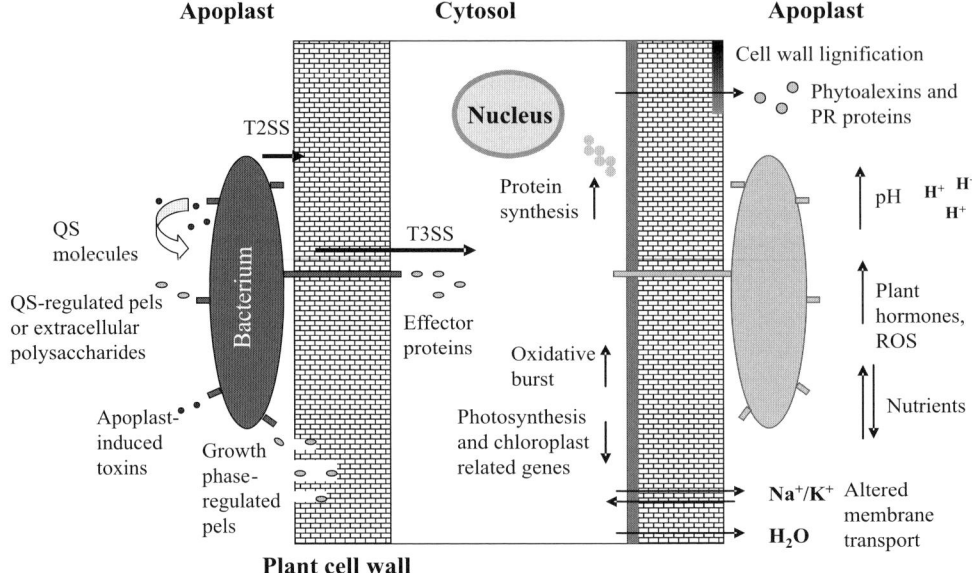

Figure 3.3 Bacterial mechanisms of infection (left) and biochemical and physiological plant responses (right).

2006; Ma and Berkowitz, 2007); hormones (Conrath et al., 2006; Loake and Grant, 2007; Robert-Seilaniantz et al., 2007); phytoalexins (Dixon, 2001; Maor and Shirasu, 2005), and cell wall modification (Hardham et al., 2007; Huckelhoven, 2007; Minic, 2008). Instead we focus on the limited number of studies that have examined the nutritional and physiological characteristics of the plant apoplast during disease and plant defence (Joosten et al., 1990; Solomon and Oliver, 2001; Solomon and Oliver, 2002; Solomon et al., 2003; Abu-Nada et al., 2007; Prabhavathi and Rajam, 2007; Tavernier et al., 2007; Walters and Bingham, 2007).

pH

Under well-watered conditions and in the absence of other abiotic stresses, the pH of the apoplastic compartment of healthy plants is mildly acidic. This acidic pH maintains the proton and ionic gradient between the cell and the extracellular space and is essential for cellular function (Felle, 2006). An acidic apoplast is associated with active photosynthesis, open stomata and rapidly growing leaves (Wilkinson et al., 2007). One of the first observations made regarding the physiological features of the HR elicited by *P. syringae* in non-host plants was the observation that it involved alkalinization of the apoplast, which seems to be due to disruption of the activity of the K^+/H^+ pumps in the plant cell wall–apoplast interface. Apoplast alkalinization has also been observed during compatible interactions and this increase in pH was postulated to be favourable for bacterial growth (Atkinson and Baker, 1987a; Atkinson and Baker, 1987b; He et al., 1994). This pH change may also be correlated with other physiological changes occurring in the plant that could benefit the plant pathogen. For example, the action of syringomycin, which produces pores in artificial membrane bilayers by disrupting the H^+ channels, would not only increase pH, but would also lead to an increased release of sugars into the plant apoplast, which can then be utilized by bacteria (Hutchison et al., 1995).

Nitrogen metabolism and mobilization

pH changes may also occur as a result of changes in nitrogen (N) metabolism and N mobilization (Olea et al., 2004), such as ammonium accumulation and export of amino acids to the apoplast during senescence. Tabtoxin, produced by the tobacco pathogen *P. syringae* pv. *tabaci* irreversibly inhibits glutamine synthetase (GS), which catalyses the first reaction in the main path of ammonia assimilation to glutamine (Lea et al., 1990) and leads to senescence-like chlorosis (Gross,

1991; Bender et al., 1999). In the course of infection by *P. syringae* pv. *tomato* in tomato plants, chloroplast degeneration occurs, which leads to the degradation of the chloroplastic GS isoform (GS2) and glutamate synthase (Fd-GOGAT) (Perez-Garcia et al., 1995), and could contribute to an increase in ammonium levels released in photorespiration. A cytoplasmic GS isoform (GS1) was activated in parallel with degradation of GS2 (Perez-Garcia et al., 1995), followed by accumulation of glutamine, which suggests that GS1 may have been activated by ammonium accumulating as a result of chloroplast inhibition and pathogenesis. In a later study, the authors showed that *P. syringae*-infected tomato leaves contained high levels of asparagine and increased asparagine synthetase (AS) activity. AS catalyses the conversion of glutamine to asparagine, which suggests that asparagine is exported from plant cells in response to infection by *P. syringae* (Olea et al., 2004) in a similar way as when leaf senescence or N deficiency occur (Pageau et al., 2006).

It is not clear whether pathogen-induced N mobilization provides greater benefits for the plant or the pathogen, as N mobilization may provide nutrients to pathogens, but also allows plants to export nutrients to uninfected tissues, prior to necrosis and abscission of infected leaves. Both asparagine and glutamine can be rapidly used by many bacterial pathogens and pathways for using these amino acids have been shown to be expressed in bacteria incubated in tomato apoplast extracts (Rico and Preston, 2008). Similarly, GABA, which is the predominant amino acid in the tomato apoplast (Solomon and Oliver, 2001; Rico and Preston, 2008) has been shown to increase during infection by the hemibiotroph *Cladosporium fulvum* (Solomon and Oliver, 2001), and GABA utilization mechanisms are expressed in *P. syringae* and *R. solanacearum* in apoplast extracts and *in planta*, respectively (Brown and Allen, 2004; Rico and Preston, 2008). However, Tavernier et al. (2007) demonstrated that the increase of GS1 gene expression in bean plants infected with the biotrophic fungus *Colletotrichum lindemuthianum* was parallel to that of two defence genes, and the consequent increase in glutamine occurred in plants infected with non-pathogenic mutants of the fungus as well as in response to pathogenic isolates, which suggests an active role for N-remobilization in plant defence mechanisms.

Changes in enzymes involved in N metabolism have also been observed in sweet orange trees infected with the xylem-colonizing pathogen *X. fastidiosa* (Purcino et al., 2007). The authors examined xylem and leaf extracts and found that the content of nitrate was higher in diseased leaves, but the activity of nitrate reductase was the same in healthy and diseased leaves (Purcino et al., 2007). On the other hand GS activity was higher in diseased leaves. The authors suggested that senescence-linked proteolysis events, and protease activity from *Xylella*, may be responsible for an increase of ammonium and hence an increase in GS activity. The authors also observed changes in the amino acid and polyamine composition of the xylem sap, in particular an increase in arginine and putrescine. The increase in these two compounds could be explained as responses to the pathogen, or as a response to the water stress imposed by xylem occlusion caused by the pathogen, which would promote the biosynthesis of arginine as a precursor for putrescine biosynthesis. Putrescine is known to accumulate in water-stressed plants and has been associated with control of stomatal aperture and protection against oxidative damage (Liu et al., 2000). Putrescine has also been shown to increase in sugarcane plants infected with *X. albilineans*, in both a soluble and a phenolic acid-conjugated form. The accompanying decrease in polyamines and free phenolic acids was correlated with plant susceptibility to bacterial infection (Fontaniella et al., 2007). Polyamines are commonly conjugated to cinnamic acids, such as p-coumaric, ferulic and caffeic acids. The hydrolysis of these conjugated forms upon infection may release phenolic acids, which have a role in plant defence (Nicholson and Hammerschmidt, 1992).

The effect of soil nitrogen on apoplast physiology

It has been long known that mineral nutrition can have a profound effect on disease development (Walters and Bingham, 2007). N is a limiting nutrient in the soil and conventional agricultural systems rely on application of N fertilizer to support plant growth and development. Interestingly, both deficiency and excess of N can influence disease susceptibility and the effect of N on susceptibility seems to be pathogen-specific (Hoffland et al., 2000). It is known that under limiting N conditions, plants synthesize more secondary metabolites that act as defence compounds. This may be related to the C/N balance, since phenolic compounds and cell wall-based depositions are rich in carbon. In this sense, Hoffland et al. (1999) found that with high N availability, the concentration of defence-related phenols such as α-tomatin in tomato plants was lower than in limiting nitrogen conditions. However, when these authors examined the susceptibility of plants to *P. syringae* pv. *tomato* and to two fungal pathogens, the foliar fungus *Oidium lycopersicum* and the soil wilt agent *Fusarium oxysporum* f. sp. *lycopersici*, a correlation between increased N concentration in leaves and susceptibility was found only for the foliar pathogens. N fertilization had no effect on resistance to *Fusarium*. The same authors found that plants with decreased N availability showed increased susceptibility to the necrotroph *Botrytis cinerea* (Hoffland et al., 1999). Therefore variation in the C/N ratio has opposite effects on different pathogens. One explanation for the increased susceptibility of high C/N plants to *B. cinerea* could be the availability of soluble carbohydrates, which may be a source of energy for the pathogen and provide substrates for fungal oxidases (Hoffland et al., 1999). An alternative explanation comes from studies of constitutive and induced resistance in *Arabidopsis* under different N regimes, which showed that N-containing defence molecules, such as chitinase and peroxidase, were lower under limiting N conditions, mainly due to lower soluble protein content, suggesting that in these conditions trade-offs between proteins of the primary and secondary metabolism could occur (Dietrich et al., 2004). It is clear that the effect of N fertilization on plant susceptibility will depend on the pathosystem studied.

N fertilization influences a range of physiological processes and we can draw parallels between plant responses to abiotic factors and those occurring during the infection process. One example is the effect of N fertilization and desiccation stress on apoplastic pH and ABA-mediated responses. It is known that soil desiccation activates ABA-mediated stomata closure to prevent water loss and this can be enhanced when N is low in the soil, due to an increase in nitrate reductase activity, which increases the levels of malate in the xylem (Wilkinson and Davies, 1997; Patonnier et al., 1999). However, an excess of N can lead to similar effects. Wilkinson et al. (2007) examined the effect of excess soil N on leaf growth and transpiration rate and found that an excess of N reduced leaf growth and induced stomatal closure in a pH and ABA-dependent manner. An alkaline pH leads to an accumulation of ABA in the guard cell because the lack of protons impairs the mobilization of ABA from the apoplast to the cell cytoplasm.

The accumulation of ABA is known to increase susceptibility to fungal and bacterial pathogens and ABA-deficient mutants show increased resistance to pathogens (Audenaert et al., 2002). ABA levels may increase as a result of apoplast alkalinization, changes in N metabolism, or as a direct result of the action of T3SS-secreted effectors. De Torres-Zabala et al. (2007) showed that the type III-secreted effector AvrPtoB of *P. syringae* pv. *tomato* induced *de novo* synthesis of ABA, which suppressed defence responses.

Interestingly, ABA can also induce callose accumulation and increase basal resistance against some necrotrophic fungi (Ton and Mauch-Mani, 2004; Felle, 2006), providing further evidence that the physiological responses to pathogenic attack are specific for each pathogen and for the MAMPs and effectors produced by each microbe (de Torres-Zabala et al., 2007). It also seems likely that the plant physiological responses, and the effectors used to counter them, vary over the course of infection. For example, antagonism between ABA and jasmonic acid (JA) signalling pathways means that *P. syringae* pv. *tomato* can suppress ABA-mediated signalling by producing the JA mimicking toxin coronatine during infection, which may help to promote stomatal opening and suppress basal defences in the early stages of infection (Melotto et al., 2006).

Carbon metabolism

Sucrose, the main product of photosynthesis, is produced in green leaves and can be stored, metabolized or transported to the phloem. Apoplastic invertases break down sucrose into glucose and fructose and in some species carbohydrate transport to the phloem occurs through the apoplast (Riotsch and Gonzalez, 2004; Walters and McRoberts, 2006). These cell wall bound-invertases may also have key roles in pathogenesis and disease resistance. The defence response that is activated upon pathogen attack requires the activation of many genes and production of secondary metabolites. Therefore, invertases may play an important role in supplying carbohydrates as precursors for synthesis of anti-microbial metabolites (Riotsch and Gonzalez, 2004; Walters and McRoberts, 2006). The accumulation of hexoses in the apoplast may also act as an extracellular indicator of pathogen infection and induce the defence system (Roitsch and Gonzalez, 2004). Overexpresssion of genes coding for extracellular invertases leads to accumulation of pathogenesis-related (PR) proteins (Schaarschmidt et al., 2007). However, the availability of hexoses could also provide an energy source that aids pathogen growth and multiplication (Fotopoulas et al. 2003).

There is limited information on changes in carbon metabolites during infection. Some studies have suggested that fungal pathogens can convert carbohydrates present in the apoplast into sugars that are unavailable for the plant, such as mannitol, trehalose and glycogen (Lewis and Harley, 1965). Joosten and collaborators (1990) observed a depletion of sucrose in tomato plants infected by the fungus *C. fulvum*, followed by a transient increase in glucose and fructose and further accumulation of mannitol. Mannitol could be an energy source for the pathogen or have a role in osmoregulation and protection from oxidative stress.

Although there is no direct evidence of similar processes in the tomato-*P. syringae* pathosystem, results published by Rico and Preston (2008) suggest that *P. syringae* pv. *tomato* could elicit changes in the carbohydrate pool of tomato plants, and may, like fungi, divert carbohydrates into pools of exopolysaccharide or osmoprotectant molecules. *P. syringae* pv. *tomato* is able to use sucrose, glucose and fructose as carbon sources, along with trehalose, which cannot be metabolized by plants. Trehalose has a role in osmoprotection during desiccation stress in some plants although it rarely occurs in higher plants (Avonce et al., 2004), and it may play a similar role in *P. syringae* pv. tomato, which also contains trehalose synthase genes. It would be interesting to assess if bacterial infection promotes the diversion of hexoses into exopolysaccharides or metabolites, which, such as trehalose, can be beneficial to the pathogen. Interestingly, *Arabidopsis* seedlings overexpressing trehalose 6-P synthase (*AtTPS1*) did not accumulate trehalose at high levels, but showed desiccation tolerance and glucose and ABA-insensitive phenotypes, allowing the seedlings to grow and develop normally in high glucose or ABA concentrations. This led the authors to postulate a signalling role for this molecule in growth and development (Avonce *et al.*, 2004).

Conclusions and future avenues for research

In order to gain a comprehensive understanding of the adaptive features and evolutionary mechanisms that explain the specialization of plant pathogenic bacteria to life in their hosts it is important to take a holistic approach in which comparative genomics and gene expression analyses of plant pathogenic bacteria are combined with knowledge of the molecular, metabolic and physiological responses of the hosts they colonize. In this chapter, we have tried to assemble some of the pieces that form the complex jigsaw of plant–pathogen interactions in the apoplast, and in this final section we highlight key conclusions and prospects for future research.

The availability of genome sequence data for a wide range of bacterial pathogens has already begun to revolutionize our ability to dissect pathogen metabolism and study gene expression during infection, and will continue to do so as more sequences become available through rapid genome sequencing, and as researchers combine genome sequence data with metabolomic and transcriptomic analyses of bacterial plant metabolism. Further exploration of genomic and experimental data can help to understand key questions such as: Do pathogenic bacteria preferentially assimilate nutrients available in the apoplast? Do they use toxins and type III secreted effectors to modulate nutrient availability in the apoplast? Do they use apoplast-specific metabolites to adapt to pH and osmotic stresses?

What is the role of house-keeping genes and global regulators in such adaptation? Which evolutionary processes best explain the adaptation shown by plant pathogens for growth in the apoplast?

Many of these questions can be addressed using model interactions such as the interaction of *P. syringae* pv. *tomato* DC3000 with the model plant *Arabidopsis thaliana*. However, it is important to remember that the degree to which bacteria are specialized for growth in the apoplast of a specific host plant, or even host cultivar, will vary according to the history of that bacterial lineage, and the extent to which growth has been confined to the apoplast, or to a specific host. Although *P. syringae* pv. *tomato* DC3000 is able to colonize the *Arabidopsis* apoplast and cause disease symptoms in a laboratory environment, some of the features of this organism may be specific to interactions with tomato, the host from which it was originally isolated. Comparative analyses of bacterial interactions with diverse host plants are needed to establish whether traits observed in one interaction are characteristic of all interactions. Pathogens such as *P. syringae* pv. *tomato* DC3000, which infect multiple hosts may provide a useful entry point for understanding how bacteria become specialized for growth on specific host species. We can test experimentally how changes in pathogen epidemiology, such as the ability to move rapidly from host to host, without a prolonged epiphytic phase, or a shift from one host to a new host, affect the evolution of plant pathogen genomes by monitoring bacterial performance over many generations.

Studies of plant physiology in response to diverse stresses clearly show that there is significant overlap between physiological responses to abiotic stresses such as drought and nutrient availability and responses to biotic stresses such as bacterial plant pathogens. A model case is mineral nutrition, and in particular N metabolism. Both excess and deficiency can modulate plant susceptibility. However, in this context, very little is known about plant defence gene expression and the production of cell-wall derived defence molecules in either laboratory or field conditions. Such knowledge may have a wide range of practical applications, from improving disease resistance while maximizing plant yield, or for the identification of biomarkers of disease susceptibility and disease onset, that could even be used for asymptomatic plants.

Finally, there remains a surprising lack of knowledge of the way in which apoplast metabolites and the apoplastic environment affect gene expression, enzyme activity and protein stability in plant pathogenic bacteria. It is still difficult to obtain reliable transcriptomic data from bacteria growing in plant tissues, and such data may overlook key differences in the physiology of individual bacteria. There is considerable scope for the development of biosensor technology to explore the environmental conditions experienced by bacteria during infection, and an urgent need for metabolite profiling technologies to identify key changes in apoplast metabolites and in cell wall composition and structure during infection. Ultimately, the only way to understand how bacteria adapt and respond to the apoplastic environment, is to develop a bacterial 'eye' view.

References

Abramovitch, R.B., Anderson, J.C., and Martin, G.B. (2006). Bacterial elicitation and evasion of plant innate immunity. Nat. Rev. Mol. Cell Biol. 7, 601–611.

Abu-Nada, Y., Kushalappa, A., Marshall, W., Al-Mughrabi, K., and Murphy, A. (2007). Temporal dynamics of pathogenesis-related metabolites and their plausible pathways of induction in potato leaves following inoculation with *Phytophthora infestans*. Eur. J. Plant Pathol. 118, 375–391.

Aldon, D., Brito, B., Boucher, C., and Genin, S. (2000). A bacterial sensor of plant cell contact controls the transcriptional induction of *Ralstonia solanacearum* pathogenicity genes. EMBO J. 19, 2304–2314.

Alfano, J.R., and Collmer, A. (1996). Bacterial pathogens in plants: life up against the wall. Plant Cell 8, 1683–1689.

Alfano, J.R., and Collmer, A. (2004). Type III secretion system effector proteins: double agents in bacterial disease and plant defense. Ann. Rev. Phytopathol. 42, 385–414.

Apel, K., and Hirt, H. (2004). Reactive oxygen species: metabolism, oxidative stress, and signal transduction. Annu. Rev. Plant. Biol. 55, 373–399.

Arnold, D.L., Jackson, R.W., Waterfield, N.R., and Mansfield, J.W. (2007). Evolution of microbial virulence: the benefits of stress. Trends Genet. 23, 293–300.

Atkinson, M.M., and Baker, C.J. (1987a). Alteration of plasmalemma sucrose transport in *Phaseolus vulgaris* by *Pseudomonas syringae* pv. *syringae* and its association with K^+/H^+ exchange. Phytopathology 77, 1573–1578.

Atkinson, M.M., and Baker, C.J. (1987b). Association of host plasma membrane K^+/H^+ exchange with multiplication of *Pseudomonas syringae* pv. *syringae* in *Phaseolus vulgaris*. Phytopathology 77, 1273–1279.

Audenaert, K., De Meyer, G.B., and Hofte, M.M. (2002). Abscisic acid determines basal susceptibility of tomato to *Botrytis cinerea* and suppresses salicylic acid-dependent signaling mechanisms. Plant Physiol. 128, 491–501.

Avonce, N., Leyman, B., Mascorro-Gallardo, J.O., Van Dijck, P., Thevelein, J.M., and Iturriaga, G. (2004). The *Arabidopsis* trehalose-6-P synthase *AtTPS1* gene is a regulator of glucose, abscisic acid, and stress signaling. Plant Physiol. 136, 3649–3659.

Bauer, D.W., Bogdanove, A.J., Beer, S.V., and Collmer, A. (1994). *Erwinia chrysanthemi hrp* genes and their involvement in soft rot pathogenesis and elicitation of the hypersensitive response. Mol. Plant-Microbe Interact. 7, 573–581.

Beattie, G.A., and Lindow, S.E. (1995). The secret life of foliar bacterial pathogens on leaves. Ann. Rev. Phytopathol. 33, 145–172.

Beattie, G.A., and Lindow, S.E. (1994). Comparison of the behavior of epiphytic fitness mutants of *Pseudomonas syringae* under controlled and field conditions. Appl. Environ. Microbiol. 60, 3799–3808.

Bellemann, P., and Geider, K. (1992). Localization of transposon insertions in pathogenicity mutants of *Erwinia amylovora* and their biochemical characterization. J. Gen. Microbiol. 138, 931–940.

Bender, C.L., Alarcón-Chaidez, F., and Gross, D.C. (1999). *Pseudomonas syringae* phytotoxins: mode of action, regulation, and biosynthesis by peptide and polyketide synthetases. Microbiol. Mol. Biol. Rev. 63, 266–292.

Berti, A.D., Greve, N.J., Christensen, Q.H., and Thomas, M.G. (2007). Identification of a biosynthetic gene cluster and the six associated lipopeptides involved in swarming motility of *Pseudomonas syringae* pv. *tomato* DC3000. J. Bacteriol. 189, 6312–6323.

Bittel, P., and Robatzek, S. (2007). Microbe-associated molecular patterns (MAMPs) probe plant immunity. Curr. Opin. Plant Biol. 10, 335–341.

Boch, J., Joardar, V., Gao, L., Robertson, T.L., Lim, M., and Kunkel, B.N. (2002). Identification of *Pseudomonas syringae* pv. tomato genes induced during infection of *Arabidopsis thaliana*. Mol. Microbiol. 44, 73–88.

Boureau, T., Jacques, M.A., Berruyer, R., Dessaux, Y., Dominguez, H., and Morris, C.E. (2004). Comparison of the phenotypes and genotypes of biofilm and solitary epiphytic bacterial populations on broad-leaved endive. Microb. Ecol. 47, 95.

Braun, P.R., Al-Younes, H., Gussmann, J., Klein, J., Schneider, E., and Meyer, T.F. (2008). Competitive inhibition of amino acid uptake suppresses chlamydial growth: Involvement of the chlamydial amino acid transporter BrnQ. J. Bacteriol. 190, 1822–1830.

Brencic, A., and Winans, S.C. (2005). Detection of and response to signals involved in host-microbe interactions by plant-associated bacteria. Microbiol. Mol. Biol. Rev. 69, 155–194.

Brown, D.G., and Allen, C. (2004). *Ralstonia solanacearum* genes induced during growth in tomato: an inside view of bacterial wilt. Mol. Microbiol. 53, 1641–1660.

Brunings, A.M., and Gabriel, D.W. (2003). *Xanthomonas citri*: breaking the surface. Mol. Plant Pathol. 4, 141–157.

Budde, I.P., Rohde, B.H., Bender, C.L., and Ullrich, M.S. (1998). Growth phase and temperature influence promoter activity, transcript abundance, and protein stability during biosynthesis of the *Pseudomonas syringae* phytotoxin coronatine. J. Bacteriol. 180, 1360–1367.

Buell, C.R., Joardar, V., Lindeberg, M., Selengut, J., Paulsen, I.T., Gwinn, M.L., Dodson, R.J., Deboy, R.T., Durkin, A.S., Kolonay, J.F. et al. (2003). The complete genome sequence of the *Arabidopsis* and tomato pathogen *Pseudomonas syringae* pv. tomato DC3000. Proc. Natl. Acad. Sci. USA 100, 10181–10186.

Bultreys, A., and Gheysen, I. (2000). Production and comparison of peptide siderophores from strains of distantly related pathovars of *Pseudomonas syringae* and *Pseudomonas viridiflava* LMG 2352. Appl. Environ. Microbiol. 66, 325–331.

Buttner, D., Noel, L., Thieme, F., and Bonas, U. (2003). Genomic approaches in *Xanthomonas campestris* pv. vesicatoria allow fishing for virulence genes. J. Biotechnol. 106, 203–214.

Cesbron, S., Paulin, J.P., Tharaud, M., Barny, M.A., and Brisset, M.N. (2006). The alternative sigma factor HrpL negatively modulates the flagellar system in the phytopathogenic bacterium *Erwinia amylovora* under *hrp*-inducing conditions. FEMS Microbiol. Lett. 257, 221–227.

Chisholm, S.T., Coaker, G., Day, B., and Staskawicz, B.J. (2006). Host-microbe interactions: shaping the evolution of the plant immune response. Cell 124, 803–814.

Conrath, U., Beckers, G.J., Flors, V., Garcia-Agustin, P., Jakab, G., Mauch, F., Newman, M.A., Pieterse, C.M., Poinssot, B., Pozo, M.J. et al. (2006). Priming: getting ready for battle. Mol. Plant-Microbe Interact. 19, 1062–1071.

Coplin, D.L., Sequeira, L., and Hanson, R.S. (1974). *Pseudomonas solanacearum*: virulence of biochemical mutants. Can. J. Microbiol. 20, 519–529.

da Silva, F.R., Vettore, A.L., Kemper, E.L., Leite, A., and Arruda, P. (2001). Fastidian gum: the *Xylella fastidiosa* exopolysaccharide possibly involved in bacterial pathogenicity. FEMS Microbiol. Lett. 203, 165–171.

Day, B., and Graham, T. (2007). The plant host pathogen interface: cell wall and membrane dynamics of pathogen-induced responses. Ann. N. Y. Acad. Sci. 1113, 123–134.

de Bruijn, I., de Kock, M.J., Yang, M., de Waard, P., van Beek, T.A., and Raaijmakers, J.M. (2007). Genome-based discovery, structure prediction and functional analysis of cyclic lipopeptide antibiotics in *Pseudomonas* species. Mol. Microbiol. 63, 417–428.

De Felice, M., Squires, C., Levinthal, M., Guardiola, J., Lamberti, A., and Iaccarino, M. (1977). Growth inhibition of *Escherichia coli* K-12 by L-valine: a consequence of a regulatory pattern. Mol. Gen. Genet. 156, 1–7.

de Torres, M., Mansfield, J.W., Grabov, N., Brown, I.R., Ammouneh, H., Tsiamis, G., Forsyth, A., Robatzek, S., Grant, M., and Boch, J. (2006). *Pseudomonas syringae* effector AvrPtoB suppresses basal defence in *Arabidopsis*. Plant J. 47, 368–382.

de Torres-Zabala, M., Truman, W., Bennett, M.H., Lafforgue, G., Mansfield, J.W., Rodriguez Egea, P., Bogre, L., and Grant, M. (2007). *Pseudomonas syringae* pv. *tomato* hijacks the *Arabidopsis* abscisic acid signalling pathway to cause disease. EMBO J. 26, 1434–1443.

DebRoy, S., Thilmony, R., Kwack, Y., Nomura, K., and He, S.Y. (2004). A family of conserved bacterial effectors inhibits salicylic acid-mediated basal immunity and promotes disease necrosis in plants. Proc. Natl. Acd. Sci. USA 101, 9927–9932.

Dietrich, R., Ploss, K., and Heil, M. (2004). Constitutive and induced resistance to pathogens in *Arabidopsis thaliana* depends on nitrogen supply. Plant, Cell Environ. 27, 896–906.

Dixon, R.A. (2001). Natural products and plant disease resistance. Nature 411, 843–847.

Dow, J.M., Crossman, L., Findlay, K., He, Y., Feng, J., and Tang, J. (2003). Biofilm dispersal in *Xanthomonas campestris* is controlled by cell–cell signaling and is required for full virulence to plants. Proc. Natl. Acad. Sci. USA 100, 10995–11000.

Espinosa, A., and Alfano, J.R. (2004). Disabling surveillance: bacterial type III secretion system effectors that suppress innate immunity. Cell Microbiol 6, 1027–1040.

Farrar, J.F. (1995). Just another sink? Sources of assimilate for foliar pathogens. In Aspects of Applied Biology. Physiological Responses of plants to pathogens, D. R. Walters *et al.* eds., The Association of Applied Biologists, UK, pp. 81–89.

Felle, H.H. (2006). Apoplastic pH during low-oxygen stress in barley. Ann. Bot. 98, 1085–1093.

Fett, W.F., and Dunn, M.F. (1989). Exopolysaccharides produced by phytopathogenic *Pseudomonas syringae* pathovars in infected leaves of susceptible hosts. Plant Physiol. 89, 5–9.

Flavier, A.B., Clough, S.J., Schell, M.A., and Denny, T.P. (1997). Identification of 3-hydroxypalmitic acid methyl ester as a novel autoregulator controlling virulence in *Ralstonia solanacearum*. Mol. Microbiol. 26, 251–259.

Fontaniella, B., Vicente, C., de Armas, R., and Legaz, M. (2007). Effect of leaf scald (*Xanthomonas albilineans*) on polyamine and phenolic acid metabolism of two sugarcane cultivars. Eur. J. Plant Pathol. 119, 401–409.

Fotopoulos, V., Gilbert, M.J., Pittman, J.K., Marvier, A.C., Buchanan, A.J., Sauer, N., Hall, J.L., and Williams, L.E. (2003). The monosacharide transporter gene, At STP4, and the cell wall invertase, At-β-fruct1, are induced in Arabidopsis during infection with the fungal biotroph *Erysiphe cichoracearum*. Plant Physiol. 132, 821–829.

Franza, T., Mahe, B., and Expert, D. (2005). *Erwinia chrysanthemi* requires a second iron transport route dependent of the siderophore achromobactin for extracellular growth and plant infection. Mol. Microbiol. 55, 261–275.

Genin, S., and Boucher, C. (2004). Lessons learned from the genome analysis of *Ralstonia solanacearum*. Ann. Rev. Phytopathol. 42, 107–134.

Grant, S.R., Fisher, E.J., Chang, J.H., Mole, B.M., and Dangl, J.L. (2006). Subterfuge and manipulation: type III effector proteins of phytopathogenic bacteria. Annu. Rev. Microbiol. 60, 425–449.

Grenier, A.M., Duport, G., Pages, S., Condemine, G., and Rahbe, Y. (2006). The phytopathogen *Dickeya dadantii* (*Erwinia chrysanthemi* 3937) is a pathogen of the pea aphid. Appl. Environ. Microbiol. 72, 1956–1965.

Grignon, C., and Sentenac, H. (1991). pH and ionic conditions in the apoplast. Annu. Rev. Plant Physiol. Plant Mol. Biol. 42, 103–128.

Gross, D.C. (1991). Molecular and genetic analysis of toxin production by pathovars of *Pseudomonas syringae*. Annu. Rev. Phytopathol. 29, 247–278.

Gurian-Sherman, D., and Lindow, S.E. (1993). Bacterial ice nucleation: significance and molecular basis. FASEB J. 7, 1338–1343.

Gurlebeck, D., Thieme, F., and Bonas, U. (2006). Type III effector proteins from the plant pathogen *Xanthomonas* and their role in the interaction with the host plant. J. Plant Physiol. 163, 233–255.

Hann, D.R., and Rathjen, J.P. (2007). Early events in the pathogenicity of *Pseudomonas syringae* on *Nicotiana benthamiana*. Plant J. 49, 607–618.

Hardham, A.R., Jones, D.A., and Takemoto, D. (2007). Cytoskeleton and cell wall function in penetration resistance. Curr. Opin. Plant Biol. 10, 342–348.

Hauck, P., Thilmony, R., and He, S.Y. (2003). A *Pseudomonas syringae* type III effector suppresses cell wall-based extracellular defense in susceptible *Arabidopsis* plants. Proc. Natl. Acd. Sci. USA 100, 8577–8582.

He, P., Chintamanani, S., Chen, Z., Zhu, L., Kunkel, B.N., Alfano, J.R., Tang, X., and Zhou, J.M. (2004). Activation of a COI1-dependent pathway in *Arabidopsis* by *Pseudomonas syringae* type III effectors and coronatine. Plant J. 37, 589–602.

He, S.Y., Bauer, D.W., Collmer, A., and Beer, S.V. (1994). Hypersensitive response elicited by *Erwinia amylovora* harpin requires active plant metabolism. Mol. Plant-Microbe Int. 7, 289–292.

Hildebrand, D.C. (1972). Tolerance of homoserine by *Pseudomonas pisi* and implications of homoserine in plant resistance. Phytopathol. 63, 301–302.

Hirano, S.S., and Upper, C.D. (2000). Bacteria in the leaf ecosystem with emphasis on *Pseudomonas syringae*-a pathogen, ice nucleus, and epiphyte. Microbiol. Mol. Biol. Rev. 64, 624–653.

Hoffland, E., Jeger, M., and van Beusichem, M. (2000). Effect of nitrogen supply rate on disease resistance in tomato depends on the pathogen. Plant Soil 218, 239–247.

Hoffland, E., van Beusichem, M., and Jeger, M. (1999). Nitrogen availability and susceptibility of tomato leaves to *Botrytis cinerea*. Plant Soil 210, 263–272.

Horswill, A.R., Stoodley, P., Stewart, P.S., and Parsek, M.R. (2007). The effect of the chemical, biological, and physical environment on quorum sensing in structured microbial communities. Anal. Bioanal Chem. 387, 371–380.

Hu, W., Yuan, J., Jin, Q.L., Hart, P., and He, S.Y. (2001). Immunogold labeling of Hrp pili of *Pseudomonas syringae* pv. tomato assembled in minimal medium and in planta. Mol. Plant–Microbe Interact. 14, 234–241.

Huang, W.E., Bailey, M.J., Thompson, I.P., Whiteley, A.S., and Spiers, A.J. (2007). Single-cell Raman spectral profiles of *Pseudomonas fluorescens* SBW25 reflects in vitro and in planta metabolic history. Microb. Ecol. 53, 414–425.

Huckelhoven, R. (2007). Cell wall-associated mechanisms of disease resistance and susceptibility. Annu. Rev. Phytopathol. 45, 101–127.

Hugouvieux, V., Barber, C.E., and Daniels, M.J. (1998). Entry of *Xanthomonas campestris* pv. *campestris* into hydathodes of *Arabidopsis thaliana* leaves: A system for studying early infection events in bacterial pathogenesis. Mol. Plant–Microbe Interact. 11, 537–543.

Hutchison, M.L., Tester, M.A., and Gross, D.C. (1995). Role of biosurfactant and ion channel-forming activities of syringomycin in transmembrane ion flux: a model for the mechanism of action in the plant–pathogen interaction. Mol. Plant-Microbe Interact. 8, 610–620.

Huynh, T.V., Dahlbeck, D., and Staskawicz, B.J. (1989). Bacterial blight of soybean: regulation of a pathogen gene determining host cultivar specificity. Science 245, 1374–1377.

Jackson, J.H., Herring, P.A., Patterson, E.B., and Blatt, J.M. (1993). A mechanism for valine-resistant growth of *Escherichia coli* K-12 supported by the valine-sensitive acetohydroxy acid synthase IV activity from ilvJ662. Biochimie 75, 759–765.

Jacques, M.A., Josi, K., Darrasse, A., and Samson, R. (2005). *Xanthomonas axonopodis* pv. phaseoli var. fuscans is aggregated in stable biofilm population sizes in the phyllosphere of field-grown beans. Appl. Environ. Microbiol. 71, 2008–2015.

Joardar, V., Lindeberg, M., Jackson, R.W., Selengut, J., Dodson, R., Brinkac, L.M., Daugherty, S.C., Deboy, R., Durkin, A.S., Giglio, M.G. et al. (2005). Whole-genome sequence analysis of *Pseudomonas syringae* pv. phaseolicola 1448A reveals divergence among pathovars in genes involved in virulence and transposition. J. Bacteriol. 187, 6488–6498.

Jones, A.M., Lindow, S.E., and Wildermuth, M.C. (2007). Salicylic acid, yersiniabactin, and pyoverdin production by the model phytopathogen *Pseudomonas syringae* pv. tomato DC3000: synthesis, regulation, and impact on tomato and *Arabidopsis* host plants. J. Bacteriol. 189, 6773–6786.

Jones, J.D., and Dangl, J.L. (2006). The plant immune system. Nature 444, 323–329.

Joosten, M.H.A.J., Hendrickx, L.J.M., and De Wit, P.J.G.M. (1990). Carbohydrate composition of apoplastic fluids isolated from tomato leaves inoculated with virulent or avirulent races of *Cladosporium fulvum* (syn. Fulvia fulva). Eur. J. Plant Pathol. 96, 103–112.

Karamanoli, K., and Lindow, S.E. (2006). Disruption of N-Acyl homoserine lactone-mediated cell signaling and iron acquisition in epiphytic bacteria by leaf surface compounds. Appl. Environ. Microbiol. 72, 7678–7686.

Kim, Y.S., and Sano, H. (2008). Pathogen resistance of transgenic tobacco plants producing caffeine. Phytochemistry 69, 882–888

Kimura, M., Anzai, H., and Yamaguchi, I. (2001). Microbial toxins in plant–pathogen interactions: biosynthesis, resistance mechanisms, and significance. J. Gen. Appl. Microbiol. 47, 149–160.

Koutsoudis, M.D., Tsaltas, D., Minogue, T.D., and von Bodman, S.B. (2006). Quorum-sensing regulation governs bacterial adhesion, biofilm development, and host colonization in *Pantoea stewartii* subspecies *stewartii*. Proc. Natl. Acad. Sci. USA 103, 5983–5988.

Krimm, U., Abanda-Nkpwatt, D., Schwab, W., and Schreiber, L. (2005). Epiphytic microorganisms on strawberry plants (*Fragaria ananassa* cv. Elsanta): identification of bacterial isolates and analysis of their interaction with leaf surfaces. FEMS Microbiol. Ecol. 53, 483–492.

Laurie-Berry, N., Joardar, V., Street, I.H., and Kunkel, B.N. (2006). The *Arabidopsis thaliana* JASMONATE INSENSITIVE 1 gene is required for suppression of salicylic acid-dependent defenses during infection by *Pseudomonas syringae*. Mol. Plant-Microbe Interact. 19, 789–800.

Lautier, T., Blot, N., Muskhelishvili, G., and Nasser, W. (2007). Integration of two essential virulence modulating signals at the *Erwinia chrysanthemi* pel gene promoters: a role for Fis in the growth-phase regulation. Mol. Microbiol. 66, 1491–1505.

Lea, P.J., Robinson, S.A., and Stewart, G.R. (1990). The enzymology and metabolism of glutamine, glutamate, and asparagine. In Intermediary nitrogen metabolism, B. J. Miflin, and P. J. Lea eds., (New York: Academic Press) pp. 121–159.

Lea, P.J., and Azevedo, R.A. (2006). Nitrogen use efficiency. 1. Uptake of nitrogen from the soil. Ann. Appl. Biol. 149, 243–247.

Lea, P.J., Sodek, L., Parry, M.A.J., Shewry, P.R., and Halford, N.G. (2007). Asparagine in plants. Ann. Appl. Biol. 150, 1–26.

Lecourieux, D., Ranjeva, R., and Pugin, A. (2006). Calcium in plant defence-signalling pathways. New Phytol. 171, 249–269.

Lee, S., Choi, H., Suh, S., Doo, I., Oh, K., Jeong-Choi, E., Schroeder-Taylor, A., Low, P., and Lee, Y. (1999). Oligogalacturonic acid and chitosan reduce stomatal aperture by inducing the evolution of reactive oxygen species from guard cells of tomato and *Commelina communis*. Plant Physiol. 121, 147–152.

Leveau, J.H., and Lindow, S.E. (2001). Appetite of an epiphyte: quantitative monitoring of bacterial sugar consumption in the phyllosphere. Proc. Natl. Acad. Sci. USA 98, 3446–3453.

Lewis, D.M., and Harley, J.L. (1965). Carbohydrate physiology of mycorrhizal roots of beech. III. Movement of sugars between host and fungus. New Phytol. 64, 256–269.

Li, X.Z., Starratt, A.N., and Cuppels, D.A. (1998). Identification of tomato leaf factors that activate toxin gene expression in *Pseudomonas syringae* pv. tomato DC3000. Phytopathol. 88, 1094–1100.

Li, C., Brown, I., Mansfield, J., Stevens, C., Boureau, T., Romantschuk, M., and Taira, S. (2002). The Hrp pilus of *Pseudomonas syringae* elongates from its tip and acts as a conduit for translocation of the effector protein HrpZ. EMBO J. 21, 1909–1915.

Lindow, S.E., and Brandl, M.T. (2003). Microbiology of the phyllosphere. Appl. Env. Microbiol. 69, 1875–1883.

Liu, K., Fu, H., Bei, Q., and Luan, S. (2000). Inward potassium channel in guard cells as a target for polyamine regulation of stomatal movements. Plant Physiol. 124, 1315–1326.

Loake, G., and Grant, M. (2007). Salicylic acid in plant defence—the players and protagonists. Curr. Opin. Plant Biol. 10, 466–472.

Lu, M., Tang, X., and Zhou, J.M. (2001). *Arabidopsis NHO1* is required for general resistance against *Pseudomonas* bacteria. Plant Cell 13, 437–447.

Ma, W., and Berkowitz, G.A. (2007). The grateful dead: calcium and cell death in plant innate immunity. Cell. Microbiol. 9, 2571–2585.

Maor, R., and Shirasu, K. (2005). The arms race continues: battle strategies between plants and fungal pathogens. Curr. Opin. Microbiol. 8, 399–404.

Mashego, M.R., Rumbold, K., De Mey, M., Vandamme, E., Soetaert, W., and Heijnen, J.J. (2007). Microbial metabolomics: past, present and future methodologies. Biotechnol. Lett. 29, 1–16.

Maurelli, A.T. (2007). Black holes, antivirulence genes, and gene inactivation in the evolution of bacterial pathogens. FEMS Microbiol. Lett. 267, 1–8.

Melotto, M., Underwood, W., Koczan, J., Nomura, K., and He, S.Y. (2006). Plant stomata function in innate immunity against bacterial invasion. Cell 126, 969–980.

Mendgen, K., and Hahn, M. (2002). Plant infection and the establishment of fungal biotrophy. Trends Plant Sci. 7, 352–356.

Menezes, H., and Jared, C. (2002). Immunity in plants and animals: common ends through different means using similar tools. Comp. Biochem. Physiol. C. Toxicol. Pharmacol. 132, 1–7.

Mercier, J., and Lindow, S.E. (2000). Role of leaf surface sugars in colonization of plants by bacterial epiphytes. Appl. Environ. Microbiol. 66, 369–374.

Minic, Z. (2008). Physiological roles of plant glycoside hydrolases. Planta 227, 723–740.

Misaghi, I., and Grogan, R.G. (1969). Nutritional and biochemical comparisons of plant-pathogenic and saprophytic fluorescent pseudomonads. Phytopathology 59, 1436–1450.

Mittal, S., and Davis, K.R. (1995). Role of the phytotoxin coronatine in the infection of *Arabidopsis thaliana* by *Pseudomonas syringae* pv. tomato. Mol. Plant–Microbe Interact. 8, 165–171.

Mo, Y.Y., Geibel, M., Bonsall, R.F., and Gross, D.C. (1995). Analysis of sweet cherry (*Prunus avium* L.) leaves for plant signal molecules that activate the *syrB* gene required for synthesis of the phytotoxin, syringomycin, by *Pseudomonas syringae* pv syringae. Plant Physiol. 107, 603–612.

Mole, B.M., Baltrus, D.A., Dangl, J.L., and Grant, S.R. (2007). Global virulence regulation networks in phytopathogenic bacteria. Trends Microbiol. 15, 363–371.

Moller, I.M., Jensen, P.E., and Hansson, A. (2007). Oxidative modifications to cellular components in plants. Annu. Rev. Plant. Biol. 58, 459–481.

Monier, J.M., and Lindow, S.E. (2004). Frequency, size, and localization of bacterial aggregates on bean leaf surfaces. Appl. Environ. Microbiol. 70, 346–355.

Monier, J.M., and Lindow, S.E. (2003). Differential survival of solitary and aggregated bacterial cells promotes aggregate formation on leaf surfaces. Proc. Natl. Acad. Sci. USA 100, 15977–15982.

Morohoshi, T., Nakamura, Y., Yamazaki, G., Ishida, A., Kato, N., and Ikeda, T. (2007). The plant pathogen *Pantoea ananatis* produces N-acylhomoserine lactone and causes center rot disease of onion by quorum sensing. J. Bacteriol. 189, 8333–8338.

Morris, C.E., Kinkel, L.L., Xiao, K., Prior, P., and Sands, D.C. (2007). Surprising niche for the plant pathogen *Pseudomonas syringae*. Infect. Genet. Evol. 7, 84–92.

Mur, L.A., Kenton, P., Atzorn, R., Miersch, O., and Wasternack, C. (2006). The outcomes of concentration-specific interactions between salicylate and jasmonate signaling include synergy, antagonism, and oxidative stress leading to cell death. Plant Physiol. 140, 249–262.

Nasser, W., Reverchon, S., and Robert-Baudouy, J. (1992). Purification and functional characterization of the KdgR protein, a major repressor of pectinolysis genes of *Erwinia chrysanthemi*. Mol. Microbiol. 6, 257–265.

Navarro, L., Zipfel, C., Rowland, O., Keller, I., Robatzek, S., Boller, T., and Jones, J.D. (2004). The transcriptional innate immune response to flg22. Interplay and overlap with *Avr* gene-dependent defense responses and bacterial pathogenesis. Plant Physiol. 135, 1113–1128.

Neufeld, J.D., Dumont, M.G., Vohra, J., and Murrell, J.C. (2007). Methodological considerations for the use of stable isotope probing in microbial ecology. Microb. Ecol. 53, 435–442.

Nicholson, R.L., and Hammerschmidt, R. (1992). Phenolic compounds and their role in disease resistance. Annu. Rev. Phytopathol. 30, 369–389.

Nomura, K., Melotto, M., and He, S. (2005). Suppression of host defense in compatible plant-*Pseudomonas syringae* interactions. Curr. Opin. Plant Biol. 8, 361–368.

Nurnberger, T., and Brunner, F. (2002). Innate immunity in plants and animals: emerging parallels between the recognition of general elicitors and pathogen-associated molecular patterns. Curr. Opin. Plant Biol. 5, 318–24.

Nuske, J., and Fritsche, W. (1989). Phaseolotoxin production by *Pseudomonas syringae* pv. phaseolicola: the influence of temperature. J. Basic Microbiol. 29, 441–447.

Ochman, H., and Davalos, L.M. (2006). The nature and dynamics of bacterial genomes. Science 311, 1730–1733.

Oguiza, J.A., Rico, A., Rivas, L.A., Sutra, L., Vivian, A., and Murillo, J. (2003). *Pseudomonas syringae* pv. *phaseolicola* can be separated into genetic lineages distinguished by the possession of the phaseolotoxin biosynthetic cluster. Microbiology 150, 473–482.

Okinaka, Y., Yang, C.H., Perna, N.T., and Keen, N.T. (2002). Microarray profiling of *Erwinia chrysanthemi* 3937 genes that are regulated during plant infection. Mol. Plant–Microbe Interact. 15, 619–629.

Olea, F., Perez-Garcia, A., Canton, F.R., Rivera, M.E., Canas, R., Avila, C., Cazorla, F.M., Canovas, F.M., and de Vicente, A. (2004). Up-regulation and localization of asparagine synthetase in tomato leaves infected by the bacterial pathogen *Pseudomonas syringae*. Plant Cell Physiol. 45, 770–780.

Osbourn, A.E. (1996). Preformed antimicrobial compounds and plant defense against fungal attack. Plant Cell 8, 1821–1831.

Osbourn, A.E., Barber, C.E., and Daniels, M.J. (1987). Identification of plant-induced genes of the bacterial pathogen *Xanthomonas campestris* pathovar campestris using a promoter-probe plasmid. EMBO J. 6, 23–28.

Osiro, D., Muniz, J.R.C., Coleta Filho, H.D., de Sousa, A.A., Machado, M.A., Garratt, R.C., and Colnago, L.A. (2004). Fatty acid synthesis in *Xylella fastidiosa*: correlations between genome studies, ^{13}C NMR data, and molecular models. Biochem. Biophys. Res. Comm. 323, 987–995.

Osman, S.F., Fett, W.F., and Fishman, M.L. (1986). Exopolysaccharides of the phytopathogen *Pseudomonas syringae* pv. *glycinea*. J. Bacteriol. 166, 66–71.

Ott, P.G., Varga, G.J., Szatmari, A., Bozso, Z., Klement, E., Medzihradszky, K.F., Besenyei, E., Czelleng, A., and Klement, Z. (2006). Novel extracellular chitinases rapidly and specifically induced by general bacterial elicitors and suppressed by virulent bacteria as a marker of early basal resistance in tobacco. Mol. Plant-Microbe Interact. 19, 161–172.

Outlaw, W.H., Jr, and De Vlieghere-He, X. (2001). Transpiration rate. An important factor controlling the sucrose content of the guard cell apoplast of broad bean. Plant Physiol. 126, 1716–1724.

Pageau, K., Reisdorf-Cren, M., Morot-Gaudry, J., and Masclaux-Daubresse, C. (2006). The two senescence-related markers, GS1 (cytosolic glutamine synthetase) and GDH (glutamate dehydrogenase), involved in nitrogen mobilization, are differentially regulated during pathogen attack and by stress hormones and reactive oxygen species in *Nicotiana tabacum* L. leaves. J. Exp. Bot. 57, 547–557.

Pallen, M.J., and Wren, B.W. (2007). Bacterial pathogenomics. Nature 449, 835–842.

Palmer, D.A., and Bender, C.L. (1993). Effects of environmental and nutritional factors on production of the polyketide phytotoxin coronatine by *Pseudomonas syringae* pv. *glycinea*. Appl. Environ. Microbiol. 59, 1619–1626.

Palmer, D.A., Bender, C.L., and Sharma, S.B. (1997). Use of Tn5-*gusA5* to investigate environmental and nutritional effects on gene expression in the coronatine biosynthetic gene cluster of *Pseudomonas syringae* pv. *glycinea*. Can. J. Microbiol. 43, 517–525.

Patonnier, M., Peltier, J., and Marigo, G. (1999). Drought-induced increase in xylem malate and mannitol concentrations and closure of *Fraxinus excelsior* L. stomata. J. Exp. Bot. 50, 1223–1231.

Penaloza-Vazquez, A., Preston, G.M., Collmer, A., and Bender, C.L. (2000). Regulatory interactions between the Hrp type III protein secretion system and coronatine biosynthesis in *Pseudomonas syringae* pv. *tomato* DC3000. Microbiol. 146, 2447–2456.

Perez-Garcia, A., Canovas, F.M., Gallardo, F., Hirel, B., and de Vicente, A. (1995). Differential expression of glutamine synthetase isoforms in tomato detached leaflets infected with *Pseudomonas syringae* pv. tomato. Mol. Plant–Microbe Interact. 8, 96–103.

Pettersson, J., Nordfelth, R., Dubinina, E., Bergman, T., Gustafsson, M., Magnusson, K.E., and Wolf-Watz, H. (1996). Modulation of virulence factor expression by pathogen target cell contact. Science 273, 1231–1233.

Pilot, G., Stransky, H., Bushey, D.F., Pratelli, R., Ludewig, U., Wingate, V.P., and Frommer, W.B. (2004). Overexpression of GLUTAMINE DUMPER1 leads to hypersecretion of glutamine from hydathodes of *Arabidopsis* leaves. Plant Cell 16, 1827–1840.

Poplawsky, A.R., Urban, S.C., and Chun, W. (2000). Biological role of xanthomonadin pigments in *Xanthomonas campestris* pv. campestris. Appl. Environ. Microbiol. 66, 5123–5127.

Prabhavathi, V., and Rajam, M.V. (2007). Mannitol-accumulating transgenic eggplants exhibit enhanced resistance to fungal wilts. Plant Sci. 173, 50–54.

Preston, G.M. (2000). *Pseudomonas syringae* pv. tomato: the right pathogen, of the right plant, at the right time. Mol. Plant Pathol. 1, 263–275.

Purcino, R.P., Medina, C.L., Martins, D., Winck, F.V., Machado, E.C., Novello, J.C., Machado, M.A., and Mazzafera, P. (2007). *Xylella fastidiosa* disturbs nitrogen metabolism and causes a stress response in sweet orange *Citrus sinensis* cv. Pera. J. Exp. Bot. 58, 2733–2744.

Quinones, B., Dulla, G., and Lindow, S.E. (2005). Quorum sensing regulates exopolysaccharide production, motility, and virulence in *Pseudomonas syringae*. Mol. Plant-Microbe Interact. 18, 682–693.

Rahme, L.G., Mindrinos, M.N., and Panopoulos, N.J. (1992). Plant and environmental sensory signals control the expression of *hrp* genes in *Pseudomonas syringae* pv. phaseolicola. J. Bacteriol 174, 3499–3507.

Rantakari, A., Virtaharju, O., Vahamiko, S., Taira, S., Palva, E.T., Saarilahti, H.T., and Romantschuk, M. (2001). Type III secretion contributes to the pathogenesis of the soft-rot pathogen *Erwinia carotovora*: partial characterization of the *hrp* gene cluster. Mol. Plant Microbe Interact. 14, 962–968.

Raskin, D.M., Seshadri, R., Pukatzki, S.U., and Mekalanos, J.J. (2006). Bacterial genomics and pathogen evolution. Cell 124, 703–714.

Redak, R.A., Purcell, A.H., Lopes, J.R.S., Blua, M.J., Mizell III, R.F., and Andersen, P.C. (2004). The biology of xylem fluid-feeding insect vectors of *Xylella fastidiosa* and their relation to disease epidemiology. Ann. Rev. Entomol. *49*, 243–270.

Rico, A., and Preston, G.M. (2008). *Pseudomonas syringae* pv. tomato DC3000 uses constitutive and apoplast-induced nutrient assimilation pathways to catabolise nutrients that are abundant in the tomato apoplast. Mol. Plant–Microbe Interact. *21*, 269–282.

Rico, A., López, R., Asensio, C., Aizpún, M., Asensio-S.-Manzanera, C., and Murillo, J. (2003). Nontoxigenic strains of *P. syringae* pv. *phaseolicola* are a main cause of halo blight of beans in Spain and escape current detection methods. Phytopathology *93*, 1553–1559.

Robert-Seilaniantz, A., Navarro, L., Bari, R., and Jones, J.D. (2007). Pathological hormone imbalances. Curr. Opin. Plant Biol. *10*, 372–379.

Roitsch, T., and Gonzalez, M.C. (2004). Function and regulation of plant invertases: sweet sensations. Trends Plant Sci. *9*, 606–613.

Romantschuk, M. (1992). Attachment of plant pathogenic bacteria to plant surfaces. Annu. Rev. Phytopathol. *30*, 225–243.

Salanoubat, M., Genin, S., Artiguenave, F., Gouzy, J., Mangenot, S., Arlat, M., Billault, A., Brottier, P., Camus, J.C., Cattolico, L. *et al.* (2002). Genome sequence of the plant pathogen *Ralstonia solanacearum*. Nature *415*, 497–502.

Sands, D.C., Schroth, M.N., and Hildebrand, D.C. (1970). Taxonomy of phytopathogenic pseudomonads. J. Bacteriol *101*, 9–23.

Sands, D.C., and Zucker, M. (1976). Amino acid inhibition of pseudomonads and its reversal by biosynthetically related amino acids. Physiol. Plant Pathol. *9*, 127–133.

Sattelmacher, B. (2001). The apoplast and its significance for plant mineral nutrition. New Phytol. *149*, 167–192.

Schaarschmidt, S., Kopka, J., Ludwig-Muller, J., and Hause, B. (2007). Regulation of arbuscular mycorrhization by apoplastic invertases: enhanced invertase activity in the leaf apoplast affects the symbiotic interaction. Plant J. *51*, 390–405.

Scharff, A.M., Egsgaard, H., Hansen, P.E., and Rosendahl, L. (2003). Exploring symbiotic nitrogen fixation and assimilation in pea root nodules by in vivo ^{15}N nuclear magnetic resonance spectroscopy and liquid chromatography-mass spectrometry. Plant Physiol. *131*, 367–378.

Schmelz, E.A., Engelberth, J., Alborn, H.T., O'Donnell, P., Sammons, M., Toshima, H., and III, T.J.H. (2003). Simultaneous analysis of phytohormones, phytotoxins, and volatile organic compounds in plants. Proc. Natl. Acad. Sci. USA *100*, 10552–10557.

Shinohara, R., Kanda, A., Ohnishi, K., Kiba, A., and Hikichi, Y. (2005). Contribution of folate biosynthesis to *Ralstonia solanacearum* proliferation in intercellular spaces. Appl. Environ. Microbiol. *71*, 417–422.

Solomon, P.S., and Oliver, R.P. (2001). The nitrogen content of the tomato leaf apoplast increases during infection by *Cladosporium fulvum*. Planta *213*, 241–249.

Solomon, P.S., and Oliver, R.P. (2002). Evidence that gamma-aminobutyric acid is a major nitrogen source during *Cladosporium fulvum* infection of tomato. Planta *214*, 414–420.

Solomon, P.S., Tan, K., and Oliver, R.P. (2003). The nutrient supply of pathogenic fungi; a fertile field for study. Mol. Plant Pathol. *4*, 203–210.

Stanier, R.Y., Palleroni, N.J., and and Doudoroff, M. (1966). The aerobic Pseudomonads: a taxonomic study. J. Gen. Microbiol. *43*, 159–271.

Stehfest, K., Toepel, J., and Wilhelm, C. (2005). The application of micro-FTIR spectroscopy to analyze nutrient stress-related changes in biomass composition of phytoplankton algae. Plant Physiol. Biochem. *43*, 717–726.

Sundin, G.W., and Murillo, J. (1999). Functional analysis of the *Pseudomonas syringae rulAB* determinant in tolerance to ultraviolet B (290–320 nm) radiation and distribution of *rulAB* among *P. syringae* pathovars. Env. Microbiol. *1*, 75–88.

Tamir-Ariel, D., Navon, N., and Burdman, S. (2007). Identification of genes in *Xanthomonas campestris* pv. vesicatoria induced during its interaction with tomato. J. Bacteriol. *189*, 6359–6371.

Tamura, K., Imamura, M., Yoneyama, K., Kohno, Y., Takikawa, Y., Yamaguchi, I., and Takakashi, H. (2002). Role of phaseolotoxin production of *Pseudomonas syringae* pv. *actinidiae* in the formation of halo lesions of kiwifruit canker disease. Physiol. Mol. Plant Pathol. *60*, 207–214.

Tavernier, V., Cadiou, S., Pageau, K., Lauge, R., Reisdorf-Cren, M., Langin, T., and Masclaux-Daubresse, C. (2007). The plant nitrogen mobilization promoted by *Colletotrichum lindemuthianum* in *Phaseolus* leaves depends on fungus pathogenicity. J. Exp. Bot. *58*, 3351–3360.

Ton, J., and Mauch-Mani, B. (2004). Beta-amino butyric acid-induced resistance against necrotrophic pathogens is based on ABA-dependent priming for callose. Plant J. *38*, 119–130.

Toth, I.K., and Birch, P.R. (2005). Rotting softly and stealthily. Curr. Opin. Plant Biol. *8*, 424–429.

Tran, H., Ficke, A., Asiimwe, T., Hofte, M., and Raaijmakers, J.M. (2007). Role of the cyclic lipopeptide massetolide A in biological control of *Phytophthora infestans* and in colonization of tomato plants by *Pseudomonas fluorescens*. New Phytol. *175*, 731–742.

Truman, W., de Zabala, M.T., and Grant, M. (2006). Type III effectors orchestrate a complex interplay between transcriptional networks to modify basal defence responses during pathogenesis and resistance. Plant J. *46*, 14–33.

Tsanko, S.G., Van Breusegem, F., Stone, J.M., Denev, I., and Laloi, C. (2006). Reactive oxygen species as signals that modulate plant stress responses and programmed cell death. BioEssays *28*, 1091–1101.

Underwood, W., Melotto, M., and He, S.Y. (2007). Role of plant stomata in bacterial invasion. Cell. Microbiol. *9*, 1621–1629.

Uppalapati, S.R., Ishiga, Y., Wangdi, T., Kunkel, B.N., Anand, A., Mysore, K.S., and Bender, C.L. (2007). The phytotoxin coronatine contributes to pathogen fitness and is required for suppression of salicylic acid accumulation in tomato inoculated with *Pseudomonas* syringae pv. tomato DC3000. Mol. Plant–Microbe Interact. *20*, 955–965.

Van den Ackerveken, G.F., Dunn, R.M., Cozijnsen, A.J., Vossen, J.P., Van den Broek, H.W., and De Wit, P.J. (1994). Nitrogen limitation induces expression of the avirulence gene *avr9* in the tomato pathogen *Cladosporium fulvum*. Mol. Gen. Genet. *243*, 277–285.

Voelksch, B., and Weingart, H. (1998). Toxin production by pathovars of *Pseudomonas syringae* and their antagonistic activities against epiphytic microorganisms. J. Basic Microbiol. *38*, 135–145.

Von Bodman, S.B., Bauer, W.D., and Coplin, D.L. (2003). Quorum sensing in plant-pathogenic bacteria. Annu. Rev. Phytopathol. *41*, 455–482.

Walters, D.R., and Bingham, I.J. (2007). Influence of nutrition on disease development caused by fungal pathogens: implications for plant disease control. Ann. Appl. Biol. *151*, 307–324.

Walters, D.R., and McRoberts, N. (2006) Plants and biotrophs: a pivotal role for cytokinins? Trends Plant Sci. *11*, 581–586.

Waters, C.M., and Bassler, B.L. (2005). Quorum sensing: cell-to-cell communication in bacteria. Annu. Rev. Cell Dev. Biol. *21*, 319–346.

Weller, D.M., and Saettler, A.W. (1980). Colonization and distribution of *Xanthomonas phaseoli* and *Xanthomonas phaseoli* var. *fuscans* in field-grown navy beans. Phytopathology *70*, 500–506.

Wilkinson, S., Bacon, M.A., and Davies, W.J. (2007). Nitrate signalling to stomata and growing leaves: interactions with soil drying, ABA, and xylem sap pH in maize. J. Exp. Bot. *58*, 1705–1716.

Wilkinson, S., and Davies, W.J. (1997). Xylem sap pH increase: A drought signal received at the apoplastic face of the guard cell that involves the suppression of saturable abscisic acid uptake by the epidermal symplast. Plant Physiol. *113*, 559–573.

Xiao, Y., and Hutcheson, S.W. (1994). A single promoter sequence recognised by a newly identified alternative sigma factor directs expression of pathogenicity and host range determinants in *Pseudomonas syringae*. J. Bacteriol. *176*, 3089–3091.

Xiao, Y., Lu, Y., Heu, S., and Hutcheson, S.W. (1992). Organization and environmental regulation of the *Pseudomonas syringae* pv. syringae 61 *hrp* cluster. J. Bacteriol. *174*, 1734–1741.

Xie, X., Wang, Y., Williamson, L., Holroyd, G., Tagliavia, C., Murchie, E., Theobald, J., Knight, M., Davies, W., Leyser, H.M.O., and Hetherington, A. (2006). The identification of genes involved in the stomatal response to reduced atmospheric relative humidity. Current Biol. *16*, 882–887.

Yang, S., Perna, N.T., Cooksey, D.A., Okinaka, Y., Lindow, S.E., Ibekwe, A.M., Keen, N.T., and Yang, C.H. (2004). Genome-wide identification of plant-up-regulated genes of *Erwinia chrysanthemi* 3937 using a GFP-based IVET leaf array. Mol. Plant-Microbe Interact. *17*, 999–1008.

Yao, J., and Allen, C. (2006). Chemotaxis is required for virulence and competitive fitness of the bacterial wilt pathogen *Ralstonia solanacearum*. J. Bacteriol. *188*, 3697–3708.

Young, J.M., and Triggs, C.M. (1994). Evaluation of determinative tests for pathovars of *Pseudomonas syringae* van Hall 1902. J. Appl. Bacteriol. *77*, 195–207.

Zhao, Y., Blumer, S.E., and Sundin, G.W. (2005). Identification of *Erwinia amylovora* genes induced during infection of immature pear tissue. J. Bacteriol. *187*, 8088–8103.

Zolobowska, L., and Van Gijsegem, F. (2006). Induction of lateral root structure formation on petunia roots: a novel effect of GMI1000 *Ralstonia solanacearum* infection impaired in Hrp mutants. Mol. Plant–Microbe Interact. *19*, 597–606.

The Genomics of *Agrobacterium*: Insights into its Pathogenicity, Biocontrol and Evolution

4

Joao C. Setubal, Derek Wood, Thomas Burr, Stephen K. Farrand, Barry S. Goldman, Brad Goodner, Leon Otten and Steven Slater

Abstract

The genus *Agrobacterium* belongs to the Rhizobiaceae family of α-Proteobacteria and includes widely found plant pathogens that cause the crown gall and hairy root diseases. *A. tumefaciens* is also well-known as a natural biotechnological tool thanks to its ability to transfer part of its DNA into plant cells. In this chapter we review current genomic knowledge about three species in this genus: *A. tumefaciens* C58, *A. vitis* S4, and *A. radiobacter* K84, focusing on the genomic basis for virulence determinants, and symbiotic interactions. These organisms have a relatively complex genome architecture, with each genome containing several replicons, and this provides an interesting view into bacterial evolution. In the second half of the chapter we describe this architecture as well as the syntenic relationships between these three species and closely related *Rhizobium* genomes. An analysis of the secondary large replicons and the smaller plasmids concludes the chapter.

Introduction

The year 2007 marked the centennial of the discovery of the plant pathogen *Agrobacterium* and the 30th anniversary of the molecular proof of its unique virulence mechanism: the interkingdom transfer of a piece of DNA from the bacterium into a plant cell. While the past three decades have seen substantial formative research on single genes and small sets of genes (Nester et al., 2004), the recent determination of genome sequences for *Agrobacterium* and its sister genera have provided a global context within which to view the biology of this unique organism and its relatives. In this chapter, we honour the past 100 years of *Agrobacterium* research by bringing together information from the complete genome sequences of three different *Agrobacterium* strains and those of several other members of the Rhizobiales. These genomes serve as the foundation for comparative genomic approaches that have allowed us to generate new hypotheses related to the evolution of pathogenicity, inter-strain parasitism, and genome evolution in this fascinating group of plant-associated microbes.

Organisms within the family Rhizobiaceae of the α-Proteobacteria are disseminated in soils worldwide and are primarily plant-associated species that include both plant pathogens (*Agrobacterium*) and nitrogen-fixing symbionts (*Rhizobium*, *Sinorhizobium*, *Allorhizobium*). Among these, select *Agrobacterium* strains are responsible for the economically damaging diseases crown gall and hairy root that affect more than 40 commercially important dicotyledonous plants worldwide (De Cleene, 1979; Burr et al., 1998; Escobar and Dandekar, 2003). These diseases result in growth reduction, decreased yields and, under some circumstances, plant death as well as additional economic impacts for potential customers who reject plants carrying unsightly galls or unusual root masses. Tumorigenic and non-tumorigenic strains of *Agrobacterium* co-exist in soils and plant material; the biovar and prominence of tumorigenic forms vary depending on the plant, soil, environment and cropping history (Bouzar and Moore, 1987; Bouzar et al., 1993; Burr et al., 1995; Krimi et al., 2002). *Agrobacterium* spp. also colonize the vascular system of many plant species without inducing visible disease (Cubero *et al.*, 2006). Thus, contaminated plants may be disseminated unknowingly with subsequent crown gall symptoms developing in response to conditions such as freezing or nematode injury.

The process by which *Agrobacterium* causes disease has been studied extensively from both the pathogen and the host perspectives and involves complex inter and intra-organismal chemical signalling (Gelvin, 2000; Tzfira and Citovsky, 2006). Chromosomally determined characteristics including motility, chemotaxis and attachment to plant cells at wound sites all have been shown to play roles in early phases of tumorigenesis. These activities culminate with transfer of a segment of DNA (designated the transferred-DNA or T-DNA) from the tumour-inducing plasmid (Ti plasmid) to the host where it is integrated into the plant genome. The products of genes carried on this T-DNA lead to auxin and cytokinin imbalances, which results in tumour formation or the production of adventitious roots, depending on the genes carried by the T-DNA. The virulence plasmid in isolates that cause hairy root disease is called Ri (for 'root-inducing') plasmid. Tumours and hairy roots produce and secrete specific amino acid and sugar derivatives, called opines, that serve as selective nutrients for the invading bacteria and promote conjugative transfer of their Ti/Ri plasmids (Dessaux *et al.*, 1998). The inducing Ti plasmids encode uptake systems specific to the opines they produce. It is these systems that serve as the basis for biological control by *Agrobacterium radiobacter* K84 and its derivatives (see below). The ability of *Agrobacterium* to transfer DNA to plants and other eukaryotes has been widely exploited as a tool for research and biotechnology (Tzfira and Citovsky, 2006). Although much remains to be discovered, recent research has provided significant insights into host mechanisms involved in the ability of *Agrobacterium* to successfully transform plants and other organisms (Gelvin, 2003; Citovsky *et al.*, 2007).

Historically, *Agrobacterium* has been grouped into five species, based on the disease phenotype associated with the resident Ti or Ri plasmid: *A. tumefaciens* causes crown gall on a wide variety of dicotyledonous plants including stone fruit and nut trees; *A. rubi* causes crown gall on raspberry; *A. vitis* is limited to crown gall on grape (*Vitis* spp.); *A. rhizogenes* causes hairy root disease, and *A. radiobacter* strains are avirulent. A subsequent classification scheme grouped *Agrobacterium* into three biovars based on physiological and biochemical properties. However, the species and biovar classification schemes do not coincide well, in large part because the virulence plasmids are readily transmissible among and between the different biovars. This transferability of phenotypes is consistent with whole genome and molecular markers comparisons, which indicate that *Agrobacterium* strains are derived from diverse chromosomal lineages (see below).

The genome of *A. tumefaciens* C58, a representative of biovar 1, was sequenced and analysed by two separate teams in 2001 (Goodner *et al.*, 2001; Wood *et al.*, 2001). Recently, the genomes of *A. vitis* S4 (a representative of biovar 3) and *A. radiobacter* K84 (a representative of biovar 2 and a biocontrol agent) have also been sequenced and analysed (Slater *et al.* unpublished; Farrand *et al.* unpublished; Burr *et al.* unpublished). In addition, the genomes of four other members of the Rhizobiaceae, *Sinorhizobium meliloti* (Galibert *et al.*, 2001), *Sinorhizobium medicae* WSM419 (GenBank NC_009636), *Rhizobium etli* (Gonzalez *et al.*, 2006) and *Rhizobium leguminosarum* (Young *et al.*, 2006), as well as the genomes of several members belonging to the Rhizobiales, have been published and/or made publicly available. Non-rhizobiaceae rhizobiales that have been sequenced include (keeping in mind that in many of these genera more than one species or strain has been sequenced): *Azorhizobium caulinodans* (GenBank NC_009937), *Bartonella henselae* (GenBank NC_005956), *Bradyrhizobium japonicum* (GenBank NC_004463), *Brucella suis* (NC_004310, NC_004311), *Mesorhizobium loti* (NC_002678), *Nitrobacter hamburgensis* (NC_007964), *Parvibaculum lavamentivorans* (NC_009719), *Ochrobactrum anthropi* (NC_009667, NC_009668), and *Rhodopseudomonas palustris* (NC_007925). These genomes have provided a wealth of data and have acted as the starting point for many functional and comparative studies. In this review we discuss insights from these data related to pathogenicity and its biotechnology offshoots. In addition, we discuss the evolution of the complex bacterial genomes of the *Agrobacterium* species.

Pathogenicity

Virulence determinants in *A. tumefaciens* C58

From the point of view of pathogenicity strain C58 is the best studied of all Agrobacteria. In this section we briefly note key virulence determinants discovered over the past 25 years with a primary focus on new findings published after the 2001 sequencing of the C58 genome (Goodner *et al.*, 2001; Wood *et al.*, 2001). A more detailed overview of this information can be found elsewhere (Slater *et al.*, 2008).

The C58 genome contains four replicons: a circular and a linear chromosome as well as two plasmids, pTiC58 and pAtC58. Genes involved in plant transformation and tumorigenesis can be found on all four replicons. The Ti plasmid harbours the T-DNA and the *vir* (virulence) genes that mediate T-DNA transfer to the plant host. Induction of the *vir* genes occurs through the VirA/G two-component regulatory system in response to phenolic and monosaccharide compounds produce by wounded plants (Gelvin, 2000). These compounds interact with the VirA protein kinase to initiate the induction of the *vir* regulon whose products are required for T-DNA processing and transfer.

The circular chromosome includes the *chvAB* genes required for synthesis and transport of the extracellular β-1,2-glucan involved in binding the bacterial cell to plant cells, the *ros*, *chvD*, *chvE*, *chvH* and *chvG/I* genes that help to regulate the Ti plasmid *vir* genes required for T-DNA transfer, plus *acvB* (discussed below). The linear chromosome encodes the cellulose synthesis (*cel*) genes and the *exoC* (*pgm*) gene required for synthesis of the extracellular β-1,2-glucan and succinoglycan polysaccharides.

Several attachment-related genes mediate the early interaction of *Agrobacterium* with its plant host (Matthysse and Kijne, 1998). The *att* (attachment) genes have been reported to be involved in early attachment of the bacterium to plant cells, and these genes were thought to be located on the chromosome (Binns and Thomashow, 1988). However, the C58 genome sequence showed the *att* genes to be located on pAtC58, along with a second, partial *att* locus. Since pAtC58 was earlier shown to be dispensable for virulence (Rosenberg and Huguet, 1984), a re-evaluation of the role of the *att* genes was required. A more recent study determined that neither pAtC58 nor two specific *att* genes (*attR* and *attD*) are required for T-DNA transfer to plants (Nair *et al.*, 2003). This study also showed that pAtC58 has a positive effect on *vir* gene induction and T-DNA transfer, but did not define the precise manner through which this positive effect is initiated.

A number of genes encoding proteins similar to known plant virulence factors were identified in the original genome papers (Goodner *et al.*, 2001; Wood *et al.*, 2001). These proteins include xylanase, ligninase (*ligE*), and pectinase (*kdgF*), and regulators of cellulase and pectinase production (*pecS/M*). These enzymes may be used to degrade the plant cell wall prior to T-DNA transfer. Additionally, several orthologues of genes required for virulence on animal hosts were identified. These include the highly conserved *mviN* gene implicated in *Salmonella* virulence, the putative adhesin *icmF* associated with macrophage killing in *Rickettsia*, genes encoding two members of the widely conserved HtrA family of serine proteases implicated in response to oxidative stress in *Salmonella* and *Yersinia*, plus an orthologue of the *bacA* locus of *Brucella* and *S. meliloti*. Invasion-related genes include orthologues of the *ialA/ialB* genes of *Bartonella henselae*, and five haemolysin-like proteins with associated type I secretion systems. The plasmid pAtC58 encodes two autotransporting virulence factor family members. This class of proteins can cross the plasma membrane via the signal peptide-dependent pathway, self-insert into the outer membrane, and present a large extracellular domain that is associated with modification of cell adhesion or host cell functions.

Another potential virulence locus includes the genes Atu4334, Atu4337, Atu4340, Atu4341, and Atu4343, in the linear chromosome. It is possible that this locus in *A. tumefaciens* encodes proteins that are similar to members of the IcMF-associated homologous protein (IAHP) group (Das and Choudhuri, 2003). Orthologues in *Pseudomonas aeruginosa* and *Vibrio cholerae* encode protein exporters implicated in pathogenesis of mammalian hosts (Mongkolsuk *et al.*, 1998; Pukatzki *et al.*, 2006). C58 also encodes as many as six different iron uptake systems but produces only one siderophore (Rondon *et al.*, 2004).

Responses to plant defences in *A. tumefaciens* C58

As discussed above, T-DNA transfer from *Agrobacterium* to the plant host is regulated by the VirA/G two-component system, which responds to certain plant phenolics and cell wall monosaccharides. Thus, inhibition of this system would block T-DNA transfer and therefore *Agrobacterium*-associated disease. Recent data demonstrate that salicylic acid (SA), a key regulator of plant defence against pathogens, can block *vir* gene induction by inhibition of VirA (Yuan *et al.*, 2007). The SA apparently interacts directly with the kinase domain of VirA and inhibits signal transfer to VirG. In addition, plants that overproduce SA are more resistant to *Agrobacterium* infection than are wild type plants, and plants that underproduce SA are more susceptible to infection.

γ-Aminobutyric acid (GABA) is a well-known signalling molecule in bacteria, plants and animals that is involved in cell–cell interactions in development and response to stress (see also chapter by Rico *et al.*). A recent study showed that GABA produced by the plant in wounded tissue can affect virulence of *A. tumefaciens* (Chevrot *et al.*, 2006). Evidently, GABA is taken up by the bacteria and converted to γ-butyrolactone (GBL), which then induces expression of the *blcABC* (formerly *attKLM*) operon encoded by pAtC58. This operon confers degradation of GBL to succinate, allowing the lactone to be utilized by the bacteria as a source of carbon. Strain C58 shows a moderate decrease in tumorigenicity on transgenic tobacco plants that overproduce GABA (Chevrot *et al.*, 2006). This attenuation is dependent upon *blcC*. Presumably, degrading the GABA lowers the concentration of this signal molecule, thereby sensitizing the plant to infection by the bacterium. The mechanism by which such signal transduction regulates virulence and susceptibility remains to be determined.

Hydrogen peroxide (H_2O_2) production is a major plant defence that can kill invading organisms. Several studies have investigated the response of *Agrobacterium* to oxidative agents. Ceci *et al.* (2003) solved the crystal structure of the *A. tumefaciens* Dps protein, which protects DNA from oxidative damage. The *E. coli* version of this protein is required for viability in stationary phase, and its protective qualities have been attributed to DNA-binding activity and oxidation of bound iron (Almiron *et al.*, 1992; Zhao *et al.*, 2002). However, Ceci *et al.* (2003) showed that the *Agrobacterium* protein protects against oxidation without binding DNA.

Additional work has focused on regulation of the oxidative response in *A. tumefaciens*, and the role of catalase in H_2O_2 resistance (Ceci *et al.*, 2003; Eiamphungporn *et al.*, 2003; Nakjarung *et al.*, 2003; Prapagdee *et al.*, 2004; Chuchue *et al.*, 2006). The regulatory genes *oxyR* and *oxyS* were disrupted, as well as the *katA* and *catE* genes that respectively encode a bifunctional catalase-peroxidase and monofunctional catalase. Both the *katA* and *catE* genes are induced by superoxide via the OxyR protein. The KatA protein is primarily responsible for resistance to H_2O_2, and the CatE protein serves a supplementary function. A mutation in *katA* results in an avirulent strain (Xu *et al.*, 2001). Recently, Saenkham *et al* (2007) have identified and characterized three isozymes of superoxide dismutase (SOD) coded for by the genome of C58. All are Fe-containing enzymes, and two are cytoplasmic while one is located in the periplasm. Mutants lacking *sodBI* but not *sodBII* or *sodBIII* are attenuated, while the triple mutant is completely avirulent.

The *Xanthomonas campestris ohr* gene is an organic hydroperoxidase resistance protein (Mongkolsuk *et al.*, 1998). A recent study (Chuchue *et al.*, 2006) determined that the *A. tumefaciens ohr* orthologue performs the same function, and that it is regulated by the adjacent *ohrR* gene. They also disrupted five more genes predicted to have a similar activity, demonstrating that *ohr* is the main but not the only mechanism of resistance to organic hydroperoxides of *A. tumefaciens*. The study did not assign specific activities to any of the five additional genes, so all remain candidates for resistance to hydroperoxides.

Virulence determinants in *A. vitis* S4

The genome of *Agrobacterium vitis* S4 contains two chromosomes and five plasmids (see the Genome Evolution section below for more details). The 259-kb Ti plasmid of S4 (pTiS4) is atypical in a number of respects. As described below, the plasmid has two complete replication systems and encodes four discrete T-regions. Moreover, while the plasmid contains a complete Vir system, the organization of the operons differs from that seen in other Ti and Ri plasmids. Most notably, the *virE* operon is located between the *virA* and the *virB* operons. The entire Vir regulon, with the exception of *virE*, is most closely related to the corresponding genes of pTiAB2/73 from a limited host-range pathogen and pRi1724, a mikimopine-type Ri plasmid. In contrast, the *virE* operon of pTiS4 is most closely related to that of pTiC58. Interestingly, the operon is flanked by degenerate IS elements, perhaps accounting for its anomalous phylogeny and location. The conjugative transfer system of pTiS4 shows a similar chimerism; the type IV secretion system is most closely related to that of the octopine-type Ti plasmids while the two conjugative DNA metabolism loci are most closely related to orthologues from the nopaline-type Ti plasmids pTiC58 and pTiSAKURA.

Three T-regions had previously been identified on pTiS4 and the analyses of the genome have revealed a fourth (from 136795 bp to 140056 bp), called T4, that is situated 60 kb from the T1-T3 cluster and separated from it by the *vir* region (L. Otten, unpublished data). The sequence of this T-region is highly similar (90%) to that of the *A. tumefaciens* Lippia strain AB2/73 T-DNA (Otten and Schmidt, 1998). These data and the conservation of the *vir* region of pTiS4 with that of AB2/73 suggest that pTiS4 arose from a fusion event between two progenitor plasmids, one similar to pTiAB2/73. The T-region of AB2/73 and T4 of pTiS4 contain only two genes: *lso* (for Lippia strain oncogene, Avi8194) and *lsn* (Lippia strain nopaline synthase-like gene, Avi8195). The *lso* gene of AB2/73 belongs to the *plast* family and is sufficient to induce small tumours on *N. rustica* leaf fragments. It is not known whether the S4 *lsn* gene or AB2/73 *lsn* encode opine synthesis, although tumours induced by AB2/73 and S4 do not contain nopaline (Unger et al., 1985). The *lsn* gene could encode synthesis of ridéopine (Chilton et al., 2001), a second opine found in S4 tumours.

How the combined activities of the four S4 mini T-DNAs lead to tumours is unknown. However, individual transfer of the four different types of oncogenes (*iaa*, *ipt*, *6b* and *lso*) may allow selection for efficient tumour induction by changes in individual T-DNA transfer efficiency (in addition to changes in oncogene promoter activity or oncoprotein activity). Whether mini-T-DNAs represent a primitive condition or result from fragmentation of a large ancestor T-DNA remains to be determined.

A. vitis is unusual among *Agrobacterium* spp. in that it also induces a host-specific necrosis on grape and a hypersensitive response (HR) on non-host plants such as tobacco (Herlache and Triplett, 2002). Induction of HR and necrosis is regulated by a complex quorum-sensing system (Zheng et al., 2003; Hao et al., 2005). The S4 genome includes at least ten genes belonging to the LuxR family of transcriptional regulators. Thus far, four of them, Avi4374 (*aviR*), Avi0918 (*avhR*), Avi1890 (*avsR*) and Avi3273 (*avxR*), all residing on chromosome I, are known to be essential for necrosis and HR. Three S4 genes belong to the LuxI family of autoinducer synthases; Avi1889 (*avsI*) on chromosome I resides immediately downstream of *avsR* and is also essential for the HR and necrosis. The *avsR*-*avsI* gene pair is homologous to the *sinR*–*sinI* gene pair in *Sinorhizobium meliloti* that regulates production of long-chain AHLs and exopolysaccharide II synthesis (Marketon et al., 2002). *avsI* is regulated by *avsR* and, like *sinI*, is essential for production of long-chain autoinducer molecules (Hao and Burr, 2006; Wang et al., 2007). Two genes immediately downstream of *avsI*, Avi1888 and Avi1887 (of unknown function) are also essential for HR and necrosis (Hao and Burr, 2006). Homologues of the LuxR family of quorum-sensing genes, except for *avsR*-*avsI*, exist on the C58 circular chromosome, but their functions have not been determined.

The LysR-type transcriptional regulators represent a large class of transcription factors that are widely distributed in bacterial species where they are known to play roles in regulation of diverse functions, including interactions with plants (Schell, 1993). In *A. vitis* S4, a *lysR* homologue, Avi1106 on chromosome I, is essential for expression of HR and necrosis (T. Burr, personal communication). Avi1106 is part of a gene cluster that includes downstream ORFs, Avi1107, Avi1110 and Avi1112 encoding hypothetical proteins.

Other symbiotic interactions

Biological control properties of *A. radiobacter* K84

Agrobacterium radiobacter K84 contains one chromosome, a second large (2.6Mb) replicon (whose status as chromosome or plasmid is discussed below), and three plasmids: pAgK84, pAtK84b, and pAtK84c (see the Genome Evolution section below for more details on this organization).

Biological control of crown gall disease by strain K84 is linked to three traits: production of agrocins 84 and 434, and iron sequestration. Synthesis of agrocin 84, an adenine nucleoside antibiotic, is conferred by pAgK84, the sequence of which has been described recently (Kim *et al.*, 2006). Little is known about the synthesis of agrocin 434, a cytidine analogue; however derivatives of K84 lacking pAtK84c do not produce this antibiotic (McClure *et al.*, 1998). Based on the genome annotation, a set of five contiguous genes on pAtK84c (Arad12092, 12094, 12096, 12097 and 12098) could be involved in synthesis and secretion of agrocin 434 (Farrand *et al.*, unpublished data).

Mutants of K84 that do not synthesize the hydroxamate siderophore ALS84 also fail to efficiently control crown gall disease (Penyalver *et al.*, 2001). The sequence of a gene fragment identified from one such mutant, which is similar to gene Arad8230, is similar to a non-ribosomal peptide synthetase responsible for production of the rhizobial iron siderophore vicibactin (Carter *et al.*, 2002). Arad8230 lies within a cluster of genes on the 2.65 Mb replicon of K84 that is similar in organization to the vicibactin locus on pRL12 of *Rhizobium leguminosarum*.

Potential impact of the type III secretion system found in most biovar 2 strains

Genome sequencing of the three *Agrobacterium* biovars (as well as the still unpublished sequence of the biovar 2 pathogen *A. rhizogenes* A4) has revealed the presence of a well-conserved type III secretion system encoded by the secondary large replicon of *A. radiobacter* K84 and *A. rhizogenes* A4 (Farrand *et al.*, unpublished data), but not in *A. tumefaciens* C58 or *A. vitis* S4. Type III secretion systems are found in a wide array of plant and animal pathogens where they mediate host interactions (Mota and Cornelis, 2005; Grant *et al.*, 2006). The core of this system consists of 10 proteins that span the membrane and, in association with a pilus, mediate the translocation of 'effector proteins' into the eukaryotic host and/or the extracellular space through the formation of an injectosome complex (Cornelis, 2006; Galan and Wolf-Watz, 2006). Although initially associated with animal and plant pathogenesis, type III systems have been found in biological control strains and symbiotic nitrogen-fixing organisms, including *R. leguminosarum* and *R. etli*, thus broadening their role in host-microbe interactions. The type III system of *Rhizobium* sp. NGR234, a nitrogen-fixing symbiont with a broad host range that includes over 112 leguminous plants and the non-legume *Paraspnia andersonii*, stimulates or represses nodulation in a host-dependent manner (Pueppke and Broughton, 1999; Marie *et al.*, 2001; Skorpil *et al.*, 2005). In contrast, the type III system in fluorescent pseudomonads used in biological control, while apparently not involved in root colonization, may play a role in targeting plant pathogens (Grant *et al.*, 2006).

Southern blot analysis of 99 *Agrobacterium* strains representing all three biovars with a probe directed against *hrcQ* of K84 (Arad8765) indicates that the type III system is present in most, but not all, biovar 2 strains (Slater *et al.* unpublished). Phylogenetic analyses of the complete type III systems of K84 and the related pathogen *A. rhizogenes* A4 reveal that these systems contain several unique genes and may form a new class of this macromolecular transporter family. No hybridization signal was detected in any of the biovar 1 or biovar 3 strains tested, although *A. vitis* strains are known to produce HR on non-host plants, a phenotype typically associated with type III secretion in plant pathogens (see above). The biovar 2 strains used in the screen are diverse and include pathogens and biological control strains isolated from a variety of plants worldwide. This conservation suggests a functional role for these systems, a possibility that is currently under investigation.

Genome evolution

Overall genome structure

The three sequenced *Agrobacterium* species have distinct genome structures (Table 4.1). The sequenced biovar 1 representative, *A. tumefaciens* C58, contains four replicons: one circular and one linear chromosome and two plasmids, pTiC58 and pAtC58 (Goodner et al., 2001; Wood et al., 2001). The biovar 2 representative *A. radiobacter* K84 contains five replicons: a single circular chromosome (4 Mb), a large circular replicon (2.65 Mb) and three plasmids, pAgK84 (44 kb), pAtK84b (185 kb, previously called pNOC) and pAtK84c (388 kb, previously called pAgK434). The biovar 3 representative *A. vitis* S4 has seven replicons: two circular chromosomes (3.72 and 1.28 Mb) and five plasmids, pAtS4a (79 kb), pAtS4b (130 kb), pAtS4c (212 kb), pTiS4 (259 kb) and pAtS4e (631 kb).

The genomes of *A. tumefaciens* C58 and *A. vitis* S4 contain two true chromosomes, which we define as replicons containing both rRNA operons and genes essential for prototrophic growth. In both strains, the larger circular chromosome (chromosome I) contains an origin of replication that is similar to other chromosomal origins within the α-proteobacteria (Sibley et al., 2006) while chromosome II has a *repABC* origin of replication typical of the large plasmids within the Rhizobiaceae (MacLellan et al., 2006). These findings, in addition to similarities in gene content and the presence of significant syntenic blocks among the Rhizobiaceae described below, suggest that chromosome I is directly descended from the ancestral chromosome of the Rhizobiaceae, and that the large, secondary replicons are derived from a *repABC*-type plasmid as originally proposed (Goodner et al., 2001; Wood et al., 2001) and as discussed in more detail below.

Agrobacterium radiobacter K84, in contrast, has a single circular chromosome and a second, large (2.65 Mb) circular replicon (Slater et al., unpublished). Like the second chromosomes of C58 and S4, the 2.65 Mb replicon contains a plasmid-type origin of replication. However, it has no rRNA operon, and does not contain many essential genes. It does, however, have at least one gene that is likely to be essential, L-seryl-tRNA

Table 4.1 Genome features of *A. tumefaciens* C58, *A. vitis* S4 and *A. radiobacter* K84. Data for S4 and K84 are preliminary (Slater et al., unpublished data)

	C58	S4	K84
Genome size (bp)	5 674 259	6 320 946	7 273 300
% GC	59.0	57.5	59.9
Number of chromosomes	2	2	2
Number of plasmids	2	5	3
Protein coding genes	5267	5389	6698
Disrupted genes	30	90	60
rRNA operons	4	4	3
tRNAs	56	54	51
Other RNA (RNAse, tmRNA, misc RNA)	26	30	22
Genomic islands: number	38	20	59
Genomic islands: average size (kb)	23.3	33.0	28.2

selenium transferase (Arad7947). This gene is orthologous to proteins encoded by chromosome II of both C58 and S4 and is located in a region syntenic to regions in these secondary chromosomes.

Syntenic relationships of genomes

Genome alignments show a clear syntenic relationship among the primary chromosomes of all three *Agrobacterium* strains (Fig. 4.1 shows the alignment between C58 and S4). Remarkably, this synteny also extends to other members of the Rhizobiales. As an example, Fig. 4.2 shows the alignment between the circular chromosome of C58 and the large chromosome of *Ochrobactrum anthropi* (a soil bacterium that can cause opportunistic infections in humans and is most closely related to *Brucella*). Chromosome I of K84 shows a stronger relationship to the chromosomes of *R. etli* and *R. leguminosarum* than to the main chromosomes of either C58 or S4 (Fig. 4.3). These findings are consistent with 16S rRNA-based phylogenies (Fig. 4.4) and suggest that the K84 lineage diverged from the ancestral strain prior to the separation of C58 and S4.

While the overall backbone of the ancestral chromosome has been maintained to varying degrees across the Rhizobiales, this is not the case with the secondary large replicons of the three agrobacteria under study here (Figs. 4.5 and 4.6). No replicon-wide conservation of gene order is seen in these alignments; however smaller syntenic regions are observed. Gene clusters in small regions (sizes roughly between 5 and 10 kb) are conserved across several species, but most have undergone considerable shuffling (both intra-replicon and inter-replicon) with respect to one another. This had been observed earlier in a comparison between *A. tumefaciens* C58 and *S. meliloti* by Wong and Golding (Wong and Golding, 2003), who called these conserved regions 'patches'. It was also observed by Guerrero et al. (Guerrero et al., 2005), who analysed those two genomes plus *M. loti* and *Brucella melitensis* and called these regions 'microsyntenic'. One notable (and particularly large) example of such a region appears in the linear chromosome of C58 that is recognized as having originated from the ancestral chromosome (Goodner et al., 2001). Using orthologue alignments we find that the orthologous genes in this region appear, with order preserved, in the main chromosomes of nine other Rhizobiales species (Fig. 4.7). Such observations allow a more detailed investigation of the origins of these secondary large replicons as discussed below.

The origins of new chromosomes in *Agrobacterium* and other organisms

While the conservation of the ancestral chromosome is without question significant, there is growing evidence to support a common origin for chromosomes II of C58 and S4 along with the large 2.65 Mb replicon of K84. Of the 3364 genes shared by all three genomes, 290 are found on the chromosomes II of C58 and S4, and on the 2.65 Mb replicon of K84. This represents 16%, 27%, and 12% respectively of the total gene complement of these replicons. More specifically, several gene clusters found on the ancestral chromosome I, as modelled by the extant chromosomes of several sister taxa, are shared by chromosomes II of C58 and S4 and the 2.65 Mb replicon of K84. Some of these gene clusters are also found on plasmid p42e of *R. etli*, and plasmid pRL11 of *R. leguminosarum*. As detailed below, these findings indicate a common ancestry among these plasmid-based replicons and suggest that gene flow from chromosomes to a particular plasmid precursor has lead to the origin of second chromosomes in this group, and possibly other bacteria.

If intragenome transfer is a robust explanation for the origin of a new chromosome, then one would expect the 'transferred' genes to occur in clusters within which the synteny from the initial ancestral chromosome I is maintained. Homology-based comparisons across multiple protein-coding genes and multiple genomes were used to analyse the genomic history of the order Rhizobiales within the α-Proteobacteria. Within this order there are at least three clades with two chromosomes or other replicons that harbour intragenome transferred gene clusters. These include members of the genus *Brucella*, the genus *Sinorhizobium* and the mixed *Agrobacterium/Rhizobium* clade (Fig. 4.8). The most parsimonious explanation for these findings proposes a unichromosomal ancestor within the Rhizobiales that harboured one ancestral plasmid of interest

The Genomics of *Agrobacterium* | 99

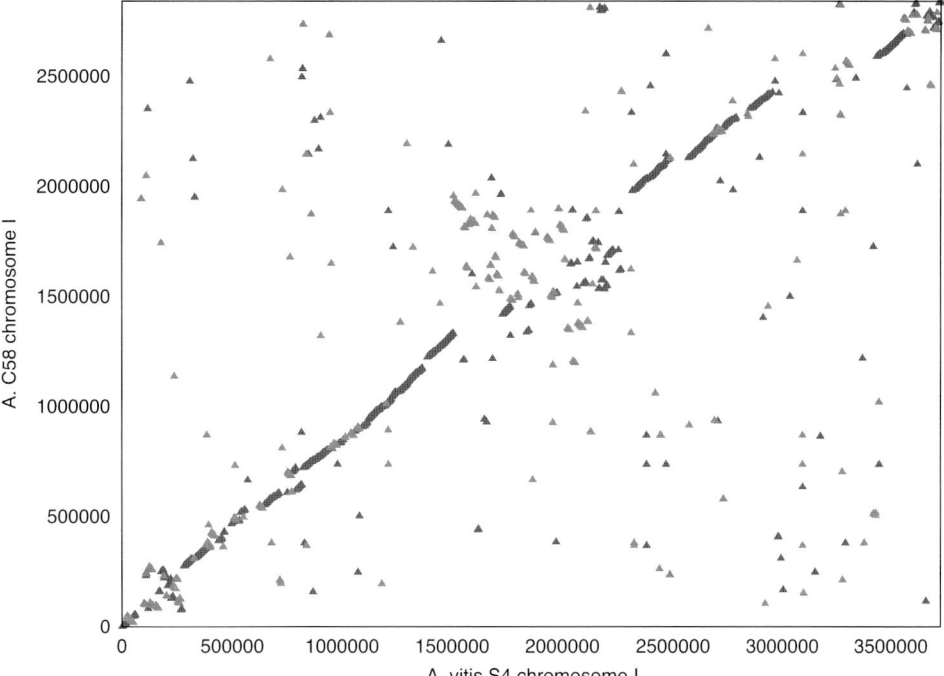

Figure 4.1 Alignment of the main chromosomes of *A. vitis* S4 and *A. tumefaciens* C58. Obtained with the program promer (Kurtz *et al.*, 2004).

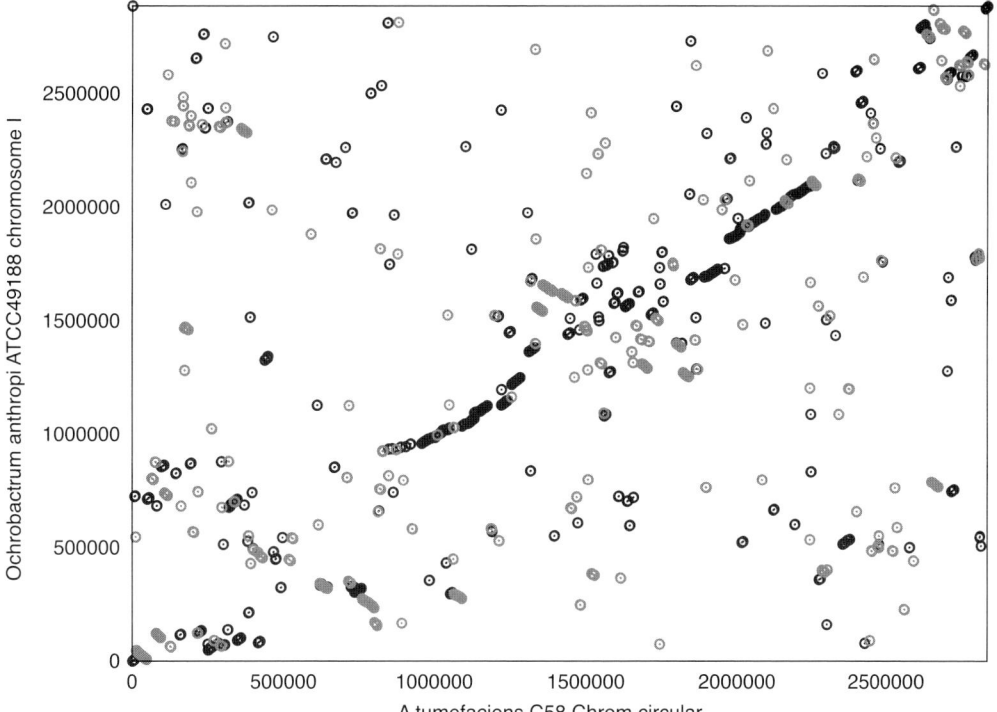

Figure 4.2 Alignment of the main chromosomes of *A. tumefaciens* C58 and *Ochrobactrum anthropi*. The *O. anthropi* sequence was reverse complemented and shifted to yield an easier-to-view alignment. Obtained with the program promer (Kurtz et al., 2004).

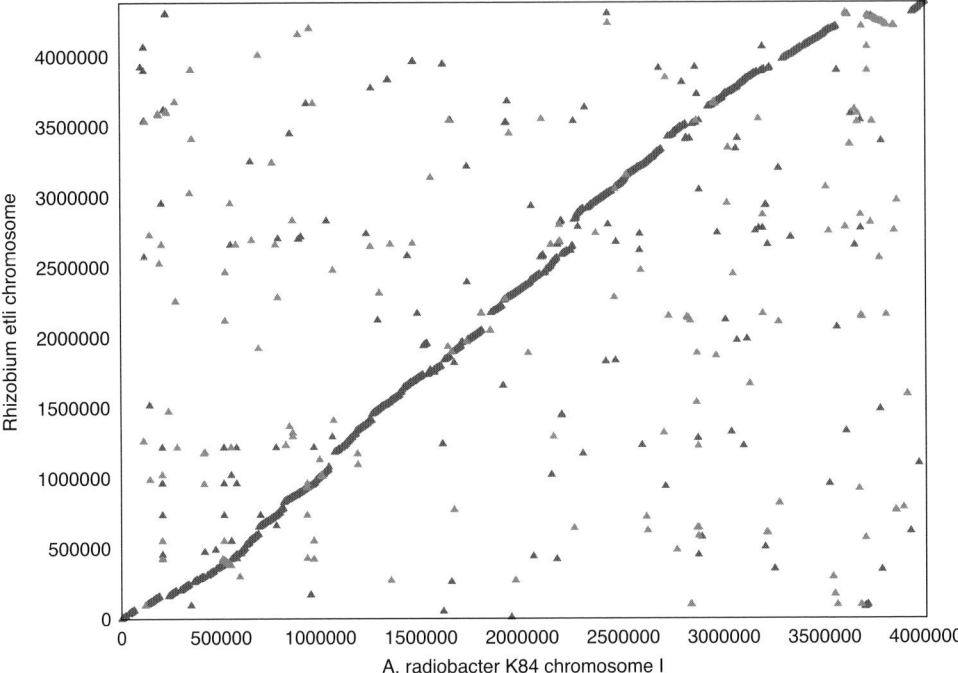

Figure 4.3 Alignment of the main chromosome of *A. radiobacter* K84 and *Rhizobium etli*. Obtained with the program promer (Kurtz et al., 2004).

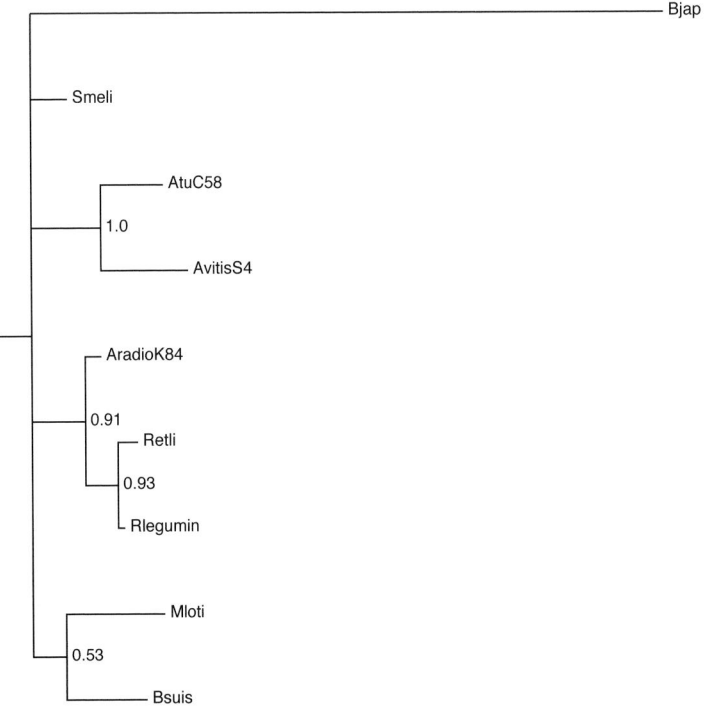

Figure 4.4 16S phylogenetic tree of selected Rhizobiaceae genomes. This tree was built using Muscle (Edgar, 2004) for the multiple alignment and MrBayes (Ronquist and Huelsenbeck, 2003) for the tree itself. The species are: Bjap, *Bradyrhizobium japonicum*; Smeli, *Sinorhizobium meliloti*; AtuC58, *A. tumefaciens* C58; AvitisS4, *A. vitis* S4; AradioK84, *A. radiobacter* K84; Retli, *Rhizobium etli*; Rlegumin: *Rhizobium leguminosarum*; Mloti, *Mesorhizobium loti* MAFF; and Bsuis, *Brucella suis* 1330. The numbers are bootstrap values for 1000 iterations.

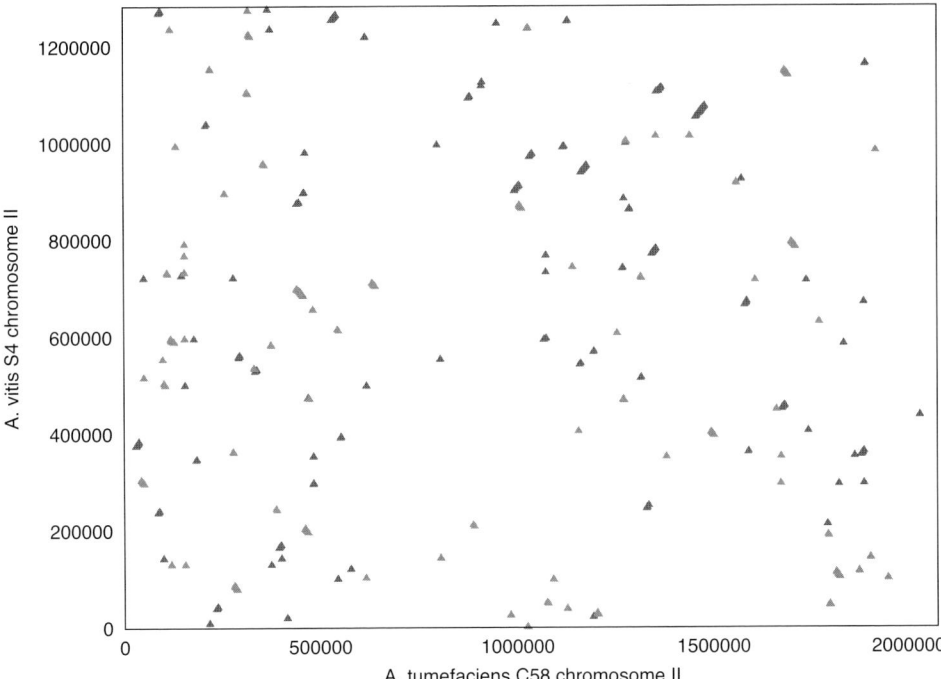

Figure 4.5 Alignment of the 2.65 Mb replicon of *A. radiobacter* K84 and chromosome II of *A. vitis* S4. Obtained with the program promer (Kurtz *et al.*, 2004).

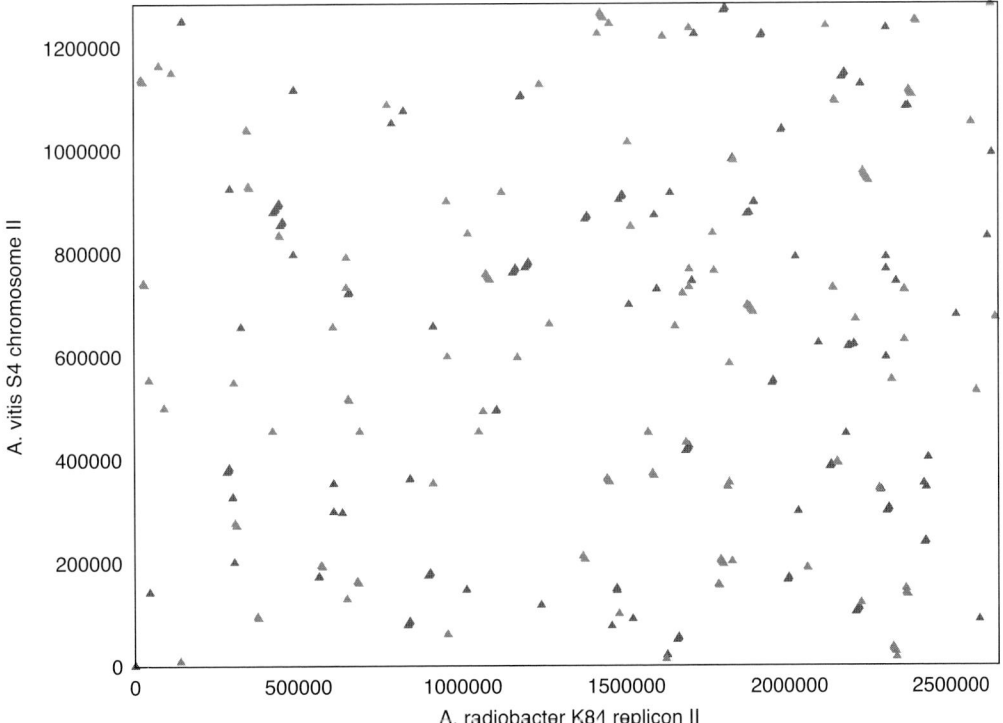

Figure 4.6 Alignment of the linear chromosome of *A. tumefaciens* C58 and chromosome II of *A. vitis* S4. Obtained with the program promer (Kurtz et al., 2004)

Ortholog alignment of *A. tumefaciens* C58

atu	avi	arad	RE	RL	SM	MLm / MLb	BH / BS	BJ	Description
atu3588	avi4002	arad4183	RE86359368	RL116254048	SM15966389	MLm13473004 MLb110635354	BS23502506 BH	BJ27376482	DNA-3-methyladenine glycosidase II
atu3589	avi4004	arad4184	RE86359369	RL	SM15966390	MLm13473005 MLb110635355	BS23502507 BH	BJ27376483	glutamyl-tRNA Synthetase
atu3590	avi4005	arad4186	RE86359370	RL116254049	SM15966391	MLm13473007 MLb110635357	BS23502826 BH49476272	BJ27376485	transporter
atu3591	avi4006	arad4187	RE86359371	RL116254050	SM15966392	MLm13473008 MLb110635358	BS23502825 BH	BJ	short chain dehydrogenase
atu3592	avi	arad	RE	RL	SM	MLm	BS	BJ	hypothetical protein
atu3593	avi4008	arad4189	RE86359373	RL116254052	SM15966394	MLm13473010 MLb110635360	BS23502823 BH	BJ27376487	conserved hypothetical protein
atu3594	avi	arad	RE86359374	RL116254053	SM15966395	MLm13473011 MLb110635361	BS BH	BJ	conserved hypothetical protein
atu3595	avi4010 / avi9247	arad12282 / arad4191	RE86359375 / RE86361135	RL116249185 / RL116254054	SM15966396 / SM16263215	MLm13473012 MLb110635362	BS23502819 BH49475943	BJ27376488	electron transfer flavoprotein beta subunit
atu3596	avi4012 / avi9248	arad12281 / arad4193	RE86359376 / RE86361134	RL116249186 / RL116254055	SM15966397 / SM16263214	MLm13473013 MLb110635363	BS23502818 BH49475944	BJ27376489	electron transfer flavoprotein alpha subunit
atu3597	avi4015	arad4195	RE86359377	RL116254056	SM15966398	MLm13473014 MLb110635364	BS23502817 BH	BJ27376490	3-hydroxybutyryl-CoA dehydrogenase
atu3598	avi4016	arad4196	RE86359378	RL116254057	SM15966399	MLm13473026 MLb110635366	BH49476270 BJ27376491		thiol:disulfide interchange protein
atu3599	avi4017	arad4197	RE86359379	RL116254058	SM15966400	MLm13473027 MLb110635367	BS23502829 BH	BJ27376492	argininosuccinate lyase
atu3600	avi4018	arad4198	RE	RL116254059	SM15966401	MLm	BS BH	BJ	hypothetical protein
atu3601	avi4020	arad4200	RE86359380	RL116254060	SM15966402	MLm13473029 MLb110635369	BS23502831 BH49476269	BJ27376494	diaminopimelate decarboxylase
atu3602	avi4021	arad4202	RE86359381	RL116254061	SM15966403	MLm13473032 MLb110635426	BS23502832 BH49476268	BJ27376497	conserved hypothetical protein
atu3603	avi4023	arad4203	RE86359382	RL116254062	SM15966405	MLm82791994 MLb110635423	BS BH	BJ27376499	two component response regulator
atu3604	avi4024	arad4205	RE86359383	RL116254063	SM15966406	MLm13473038 MLb110635422	BS23502833 BH	BJ	hypoxanthine phosphoribosyltransferase
atu3605	avi4026	arad4206	RE86359384	RL116254064	SM	MLm	BS BH	BJ	hypothetical protein
atu3606	avi4027	arad4207	RE86359385	RL116254065	SM15966408	MLm13473040 MLb110635420	BS23502845 BH49476197	BJ27376501	cell division ATP-binding protein
atu3607	avi4028	arad4208	RE86359386	RL116254066	SM15966409	MLm13473041 MLb110635419	BS23502844 BH49476198	BJ27376502	cell division protein
atu3608	avi4029	arad4209	RE86359387	RL116254067	SM15966410	MLm13473042 MLb110635418	BS23502843 BH49476199	BJ27376503	conserved hypothetical protein
atu3609	avi4030	arad4211	RE86359388	RL116254068	SM15966411	MLm13473046 MLb110635417	BS23502842 BH49476280	BJ27376504	1-acyl-sn-glycerol-3-phosphate acyltransferase
atu3610	avi4031	arad4212	RE86359389	RL116254069	SM15966412	MLm13473049 MLb110635416	BS23502839 BH	BJ27376505	transporter
atu3611	avi4033	arad4216	RE86359392	RL116254072	SM15966414	MLm13473052 MLb110635414	BS23502836 BH49476273	BJ27376507	prephenate dehydrogenase
atu3612	avi4034	arad4217	RE86359393	RL116254073	SM15966415	MLm13473053 MLb110635413	BS23502835 BH	BJ27376508	histidinol-phosphate aminotransferase
atu3613	avi4035	arad4218	RE86359394	RL116254074	SM15966416	MLm13473060 MLb110635410	BS23502834 BH49476267	BJ27375332	conserved hypothetical protein

Figure 4.7 Orthologue alignment of 11 Rhizobiales species, anchored by a region in the linear chromosome of *A. tumefaciens* C58. In this table, each cell represents a gene. Genes along a row belong to an orthologous family. Genes in a column belong to a specific genome. Genomes can be identified by the following identifiers: atu, *A. tumefaciens* C58; avi, *A. vitis* S4; arad, *A. radiobacter* K84; RE: *Rhizobium etli*; RL, *R. leguminosarum*; SM, *Sinorhizobium meliloti*; MLm, *Mesorhizobium loti* MAFF; MLb, *M. loti* BNC; BS, *Brucella suis* 1330; BH, *Bartonella henselae*; BJ, *Bradyrhizobium japonicum*. The numbers that appear next to the genome identifiers are gene identifiers. Generally these numbers follow physical location in the replicon, so consecutive numbers mean consecutive genes. As can thus be seen, except for BH, there is a large degree of gene conservation across all genomes. Except for the C58 genes, all others appearing on this table belong to the main chromosomes of their respective organisms. In the last column are given the protein descriptions of the C58 genes. The genes in boxes are putative paralogues. Orthologues were computed using OrthoMCL (Li *et al.*, 2003).

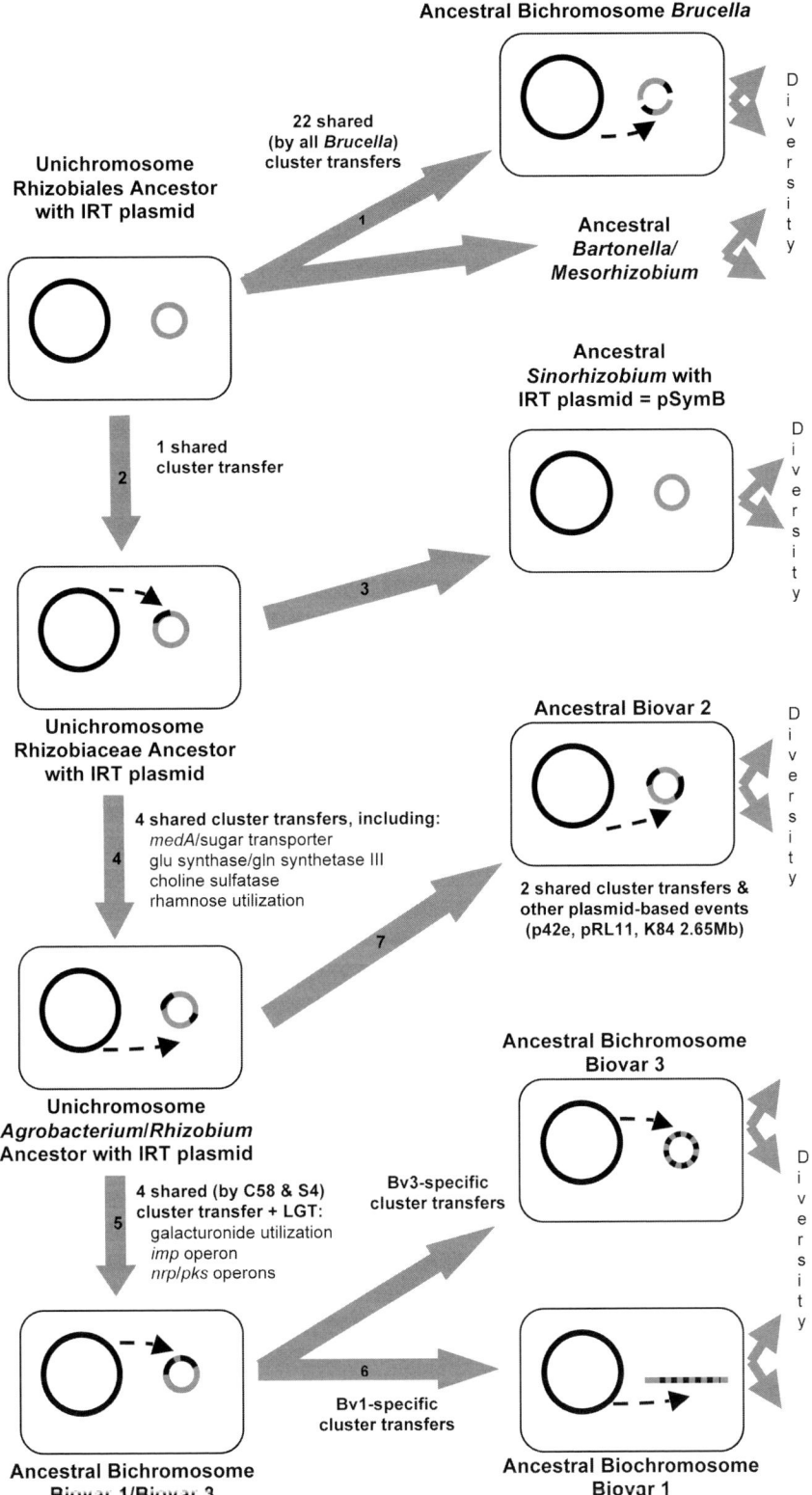

Figure 4.8 Proposed evolutionary path for origin of secondary chromosomes within the Rhizobiales. Key steps involving intrachromosomal gene transfers from the ancestral chromosome I to the Intragenomic Transfer Recipient (IRT) plasmid ancestor and its descendant replicons are numbered.

(labelled ITR for intragenomic transfer recipient). This ITR plasmid, with a *repABC*-type replication and partition system, carried three gene clusters that have been maintained on large secondary replicons throughout the subsequent evolution of this order, although it is not entirely clear whether these three gene clusters originated themselves from intragenomic transfer or by horizontal gene transfer or some combination of the two. Since the divergence of the *Brucella* clade away from that which gave rise to the family Rhizobiaceae, there have been 22 intragenomic transfers shared by all of the sequenced *Brucella* strains as the ancestral IRT plasmid evolved into chromosome II (step 1 in Fig. 4.8). Prior to the divergence of the *Sinorhizobium* clade (represented by *S. meliloti*) away from the *Agrobacterium/Rhizobium* clade there was then one shared intragenomic transfer event (step 2 in Fig. 4.8), and six subsequent transfers to the ancestral IRT plasmid as it evolved into pSymB (step 3 in Fig. 4.8).

Within the family Rhizobiaceae, there were four shared intragenomic gene clusters transferred to the ancestral IRT plasmid prior to the divergence of the clade that now contains *Agrobacterium* biovar 1+3 strains (*A. tumefaciens* C58 and *A. vitis* S4, respectively) from the clade that includes *Agrobacterium* biovar 2 strains (*A. radiobacter* K84, *Rhizobium etli*, and *R. leguminosarum*) (step 4 in Fig. 4.8).

Subsequent to this divergence four shared intragenomic transfers and two large lateral gene transfers to the IRT plasmid are apparent on chromosome II of *Agrobacterium* biovar 1+3 strains. These include genes encoding non-ribosomal peptide synthetases, polyketide synthases and a type VI secretion system (*imp* operon) along with putative effector proteins (step 5 in Fig. 4.8). After the separation of biovars 1 and 3, there have been multiple intragenomic transfers to chromosome II including seven large-scale gene transfer events, ranging from 10 kb to 220 kb, between the ancestral chromosome and chromosome II of biovar 1 that are not seen in biovar 3 as well as the linearization of chromosome II (step 6 in Fig. 4.8). In a separate but parallel track along the other branch of the *Agrobacterium/Rhizobium* clade, there are two intragenomic transfers shared by the IRT plasmid derivatives in *A. radiobacter* K84 (2.65 Mb replicon), *R. etli* (plasmid p42e) and *R. leguminosarum* (plasmid pRL11) (step 7 in Fig. 4.8).

Finally, although it is quite clear that chromosome II of C58 and S4 and the 2.65 Mb replicon of K84 have evolved from a common plasmid ancestor, the *repABC* operons involved in replication initiation, copy number control, and partitioning on all three molecules are distinct from one another and are more closely related to those of other *repABC* plasmids than they are to each other. These findings suggest that *repABC* operons, like many other genes, are being exchanged among replicons. This may reflect selective pressure to overcome replicon incompatibility allowing coexistence of several *repABC*-based elements in the genomes of these bacteria.

Importantly, this proposed mechanism of chromosome evolution does not appear to be restricted to members of the Rhizobiales. Multiple chromosomes and evolving plasmid replicons are also found in members of several other diverse genera across the Bacteria domain (see chapter by Sundin and Murillo). While the additional chromosomes of these genera did not arise via a single common event, and do not all share broad sequence homology, they do all appear to share the same mechanism just described within the Rhizobiales. There are no known examples of a chromosome in the Bacteria domain having more than one origin of replication [a few examples are now known in Archaea (Kelman and Kelman, 2004)], suggesting that simple breakage and reformation of two chromosomes from a single chromosome has not occurred. Rather, in these strains as in the Rhizobiales, plasmids have acted as nucleation centres for the formation of additional chromosomes through intragenome gene transfers. The two best examples of this are the *Vibrio–Photobacterium* clade and the *Burkholderia–Ralstonia* clade.

Photobacterium species were once considered part of the genus *Vibrio* and multiple lines of evidence support the fact that *Vibrio* and *Photobacterium* are closely related genera. Studies of *P. profundum* and four *Vibrio* species indicate that members of both genera have two chromosomes. Phylogenetic analysis of several conserved proteins showed that among the available sequenced genomes, *Aeromonas hydrophila* is the closest relative with a single chromosome. While chromosome II of *P. profundum* shares 136 protein-coding genes with chromosome II of the four *Vibrio* species, some of which are clearly plasmid-derived (e.g. plasmid-type *parAB*), there are six shared gene clusters that have been transferred from the ancestral chromosome I to the plasmid progenitor of chromosome II prior to the divergence of the *Photobacterium* and *Vibrio* clades (step 1 in Fig. 4.9). These shared gene clusters are found in the same relative order on chromosome II in the two genera, even though chromosome II in *Photobacterium* is roughly twice the size of that in *Vibrio*. Following the divergence of these two clades, there were seven additional shared gene cluster transfers to chromosome II from the common chromosomal ancestor in the ancestral *Vibrio* (step 2 in Fig. 4.9). On the *Photobacterium* side, there were 29 additional transfers including genes for the F_0/F_1 ATP synthase (step 3 in Fig. 4.9).

Fifteen years ago, the RNA homology group II of the γ-Proteobacterium genus *Pseudomonas* was re-designated as *Burkholderia* and placed within the β-Proteobacteria (Yabuuchi et al., 1992). A few years later, some members of *Burkholderia* along with some stragglers from other genera were reclassified into the genus *Ralstonia* (see chapters by Stavrinides and Brown). Several lines of evidence support a very close relationship between *Burkholderia* and *Ralstonia* and each consist of species with multiple chromosomes. The most closely related sequenced genomes with a single chromosome are those from the genus *Bordetella*, therefore *B. bronchiseptica* was used as the reference genome for the following analyses. Using chromosome II sequences from five *Burkholderia* species and three *Ralstonia* species, seven shared gene cluster transfers to the plasmid progenitor of chromosome II, including genes for cytochrome oxidase, can be identified (step 1 in Fig. 4.10). After the divergence of these two clades, there were 12 and 24 additional transfers unique to the *Burkholderia* bichromosome ancestor and *Ralstonia* bichromosome ancestor, respectively (steps 2 and 3 in Fig. 4.10). Making this example even more complex is the origin of a third chromosome within a subset of *Burkholderia* strains. In this case, the two chromosomes of *Burkholderia mallei* acted as comparison genomes. These data support at least five transfers from chromosome I to the plasmid progenitor of chromosome III (step 4 in Fig. 4.10).

It is apparent from these findings that a continued pattern of inter-replicon transfer exists among bacteria. These observed transfers appear to be widespread and readily explain the origins of second chromosomes as critical genes are transferred to plasmids such as the proposed IRT and lost from the ancestral chromosomes.

Linearization of the second chromosome in biovar 1 strains

A distinctive feature of the C58 genome, as compared to all other members of the Rhizobiaceae (and indeed the rest of α-Proteobacteria) sequenced to date, is the fact that its chromosome II is linear. Linear replicons are rare in prokaryotes, with other notable examples being *Streptomyces coelicolor*, a high GC Gram-positive species (Bentley et al., 2002) and *Borrelia burgdorferi*, a spirochaete (Casjens et al., 2000).

When the genome of *A. tumefaciens* C58 was published in 2001, no gene had been identified which could be linked to the linearity of its second chromosome. Slater et al. (unpublished) have identified in the genome the gene *telA* (Atu2523, located in the circular chromosome), coding for the enzyme protelomerase. TelA maintains the chromosome telomeres, which are covalently closed hairpin loops (Goodner et al., 2001). This gene was identified based on the similarity of its predicted product with protelomerases encoded by the *E. coli* bacteriophage N15 and *Klebsiella* phage φKO2 (Ravin et al., 2000; Ravin, 2003; Casjens et al., 2004). In closed hairpin telomeres, the end of one strand turns around and serves to form the second paired strand with no break in the phosphodiester backbone. The replication of such telomeres has been discussed previously (Casjens, 1999; Ravin, 2003; Huang et al., 2004). The functionality of the purified TelA protein has been demonstrated *in vitro*, and the telomere sequences, which were not in the originally published C58 linear chromosome sequences, were determined (Huang et al., unpublished).

The presence of linear chromosomes in other *Agrobacterium* strains was recently investigated

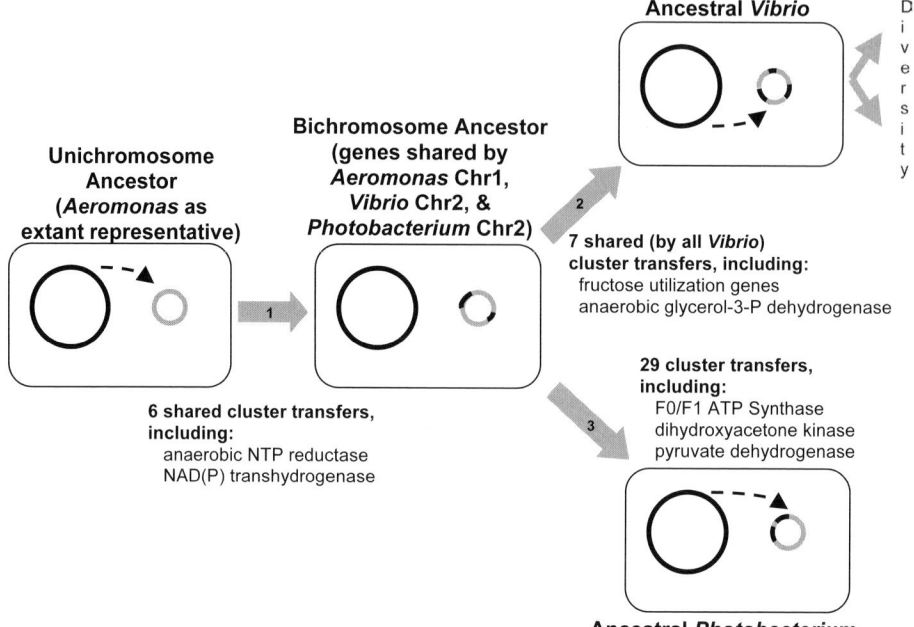

Figure 4.9 Proposed evolutionary path for origin of secondary chromosomes within the *Vibrio-Photobacterium* clade. Key steps involving intrachromosomal gene transfers from the ancestral chromosome I to a plasmid ancestor of chromosome II are numbered.

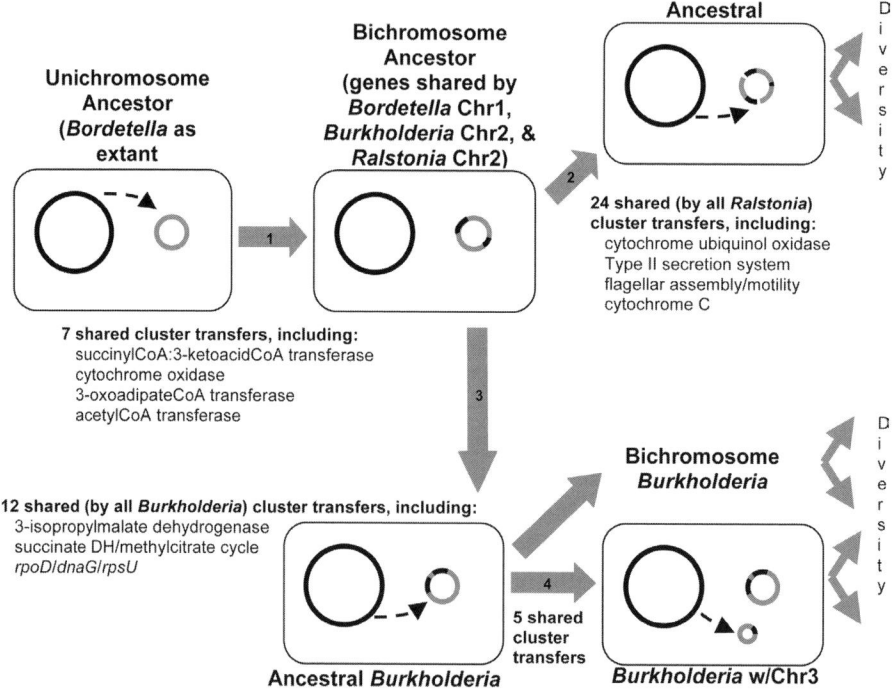

Figure 4.10 Proposed evolutionary path for origin of extra chromosomes within the *Ralstonia-Burkholderia* clade. Key steps involving intrachromosomal gene transfers from the ancestral chromosome I to the plasmid ancestors of chromosomes II and III are numbered.

by Slater et al. (unpublished data). Based on an examination of 25 strains representing all three *Agrobacterium* biovars, an absolute correlation was seen between the presence of the *telA* gene, as determined by PCR and Southern blots, and the presence of a linear chromosome, as determined by pulsed-field gel electrophoresis. Using low-stringency Southern blots, over 100 strains comprising all three biovars were examined for the presence of *telA*. These analyses showed that the *telA* gene was present only in genomes from *Agrobacterium* biovar 1 strains. They also determined that, in the strains examined, *telA* is always adjacent to the *lpiA-acvB* operon, which has been implicated in both pathogenesis and acid tolerance (Wirawan et al., 1993; Vinuesa et al., 2003; Reeve et al., 2006). The *acvB-telA* region appears to have been a region of divergence in biovars 1 and 3. A comparison of the biovar 1 (C58) and biovar 3 (*A. vitis* S4) genomes suggests that multiple recombination events in this region both incorporated *telA* and initiated two large transfers between chromosomes (Slater et al. unpublished).

Genome sequences are available for several representative nitrogen-fixing members of the Rhizobiaceae (Kaneko et al., 2000; Barnett et al., 2001; Capela et al., 2001; Galibert et al., 2001; Gonzalez et al., 2006; Young et al., 2006). All of these bacteria contain a single large, circular chromosome in addition to large plasmids that encode most of the genes necessary for symbiosis and nitrogen fixation. In 2003, Martínez et al. (2003) described several bacteria isolated from banana which they classified as *Rhizobium spp*. These strains produce small, ineffective nodules on *Acacia*, but have a strong positive effect (up to 60% increase) on growth when inoculated onto banana plants. Although they did not directly test nitrogen fixation by these strains, the authors reported that a plasmid harbouring *nifH* was present. Two of these isolates, F5H16b and F5R19 (Martínez et al., 2003), exhibited several traits in common with biovar 1 *Agrobacterium* strains, including very similar 16S rRNA genes (Martínez et al., 2003), growth patterns on biovar-specific differential media, and the ability to oxidize sucrose to 3-ketosucrose (Bernaerts and De Ley, 1952). These strains also contain at least two potential recombination junctions that are characteristic of transfers from the primordial chromosome to chromosome II in biovar 1 *Agrobacterium* (Slater et al., unpublished data).

An examination of the genome structure of these isolates using pulsed-field gels and PCR clearly demonstrated the presence of a linear chromosome, and a *telA* sequence very similar to that of *telA* from C58, in these mutualistic strains (Slater et al., unpublished data). Additional PCR analyses demonstrate that *telA* and *acvB* are adjacent to one-another on the chromosome, as in other biovar 1 strains. These results therefore show the intriguing fact that *A. tumefaciens* C58, a pathogen, is in a class of closely related organisms that includes mutualistic symbionts, leading to the speculation that the pathogenicity of biovar 1 isolates arose after their divergence from the biovar 3 branch.

Plasmid replication, transmission and unique functions

As described earlier, *A. radiobacter* strain K84 and *A. vitis* strain S4 harbour three and five plasmids respectively (see also chapter by Sundin and Murillo). All save the small, 44-kb plasmid pAgK84 of strain K84 encode *repABC*-type replication systems. Phylogenetic analyses of these systems suggest surprisingly diverse origins. For example, the *repABC* unit of pAtK84b is closely related to the replication proteins of the Ti plasmids while the two *repABC* units encoded on the pTiS4 plasmid are most closely related to those of the pRL7 plasmid from *R. leguminosarum* and the p42a plasmid from *R. etli*. Similar diversity exists among the replication proteins of the other *repABC* plasmids in the two strains.

All of the plasmids in both strains encode complete or partial conjugative transfer systems. Four of these replicons, pAtK84b of K84 and pTiS4, pAvS4b, and pAvS4c of S4 encode Ti plasmid-like transfer systems (Farrand et al., unpublished data). Interestingly, all also contain *traR-traI-traM*-type Ti plasmid quorum-sensing regulatory systems. In S4, pAvS4e codes for a Ti plasmid-like conjugative DNA metabolism system, but lacks a corresponding type IV secretion system. The smallest plasmid in S4, pAvS4a, encodes an apparent conjugative transfer system distantly related to that found on plasmid pAtC58 (termed 'Agrobacterium virulence homologue B', or AvhB) but closely related to transfer systems on pSmeSm11a of *Sinorhizobium meliloti* Rm1021 (Stiens et al., 2006) and *S. medicae* WSM419 (GenBank NC_009636). In K84, pAtK84c encodes a complete AvhB-like conjugative transfer system while pAgK84 encodes an AvhB-like DNA metabolism system, but lacks a type IV secretion system. Plasmid pAgK84 is transmissible by conjugation but likely uses a type IV secretion system encoded by one of the other replicons of K84. Among these systems only pAtK84b has been shown experimentally to mediate its own conjugation (Oger and Farrand, 2002). Remarkably, the chromosome of strain K84 carries an entire AvhB-like transfer system.

Most of the plasmids, including pAvS4e, pAvS4c, and pAvS4b of S4 and pAtK84c of K84, carry large numbers of genes coding for enzymes and ABC-type transporters associated with processing unknown substrates (Farrand et al. unpublished). Many of these genes, especially those of pAvS4b and pAvS4c, are not strongly conserved in the genomes of other members of the Rhizobiaceae. Such diversity resembles that seen in the *repABC*-type plasmids of *R. etli* and *R. leguminosarum* (Gonzalez et al., 2006; Young et al., 2006). There are, however, notable exceptions. pAvS4e carries a set of seven genes encoding polypeptide and polyketide synthetases (Avi7298, 7299, 7303, 7304, 7306, 7308, 7309, 7311) that is syntenic to a set of orthologous genes on the linear chromosome of strain C58 (Atu3683, 3681, 3677, 3675, 3673, 3672, 3670). More intriguingly, this plasmid encodes a set of genes that could be involved in nickel transport, and these are most closely related to orthologues in *Brucella* spp.

Conclusions

The recent sequencing of many genomes among the Rhizobiales has allowed us to examine the evolutionary history of these organisms and has shed significant light on the genes and systems involved in pathogenesis, biological control and symbiosis. One striking finding from these analyses has been the suggestion that chromosomes are evolving from plasmids in many of these bacteria. Another is that the diverse chromosomal structures in this group appear to be capable of supporting both symbiotic and pathogenic lifestyles. We expect that the stimulation of research facilitated by the availability of these genomes will continue to increase, resulting in substantial insights into the function and evolutionary history of this fascinating group of plant-associated microbes.

References

Almiron, M., Link, A.J., Furlong, D., and Kolter, R. (1992). A novel DNA-binding protein with regulatory and protective roles in starved *Escherichia coli*. Genes Dev. 6, 2646–2654.

Barnett, M.J., Fisher, R.F., Jones, T., Komp, C., Abola, A.P., Barloy-Hubler, F., Bowser, L., Capela, D., Galibert, F., Gouzy, J., Gurjal, M., Hong, A., Huizar, L., Hyman, R.W., Kahn, D., Kahn, M.L., Kalman, S., Keating, D.H., Palm, C., Peck, M.C., Surzycki, R., Wells, D.H., Yeh, K.C., Davis, R.W., Federspiel, N.A. and Long, S.R. (2001). Nucleotide sequence and predicted functions of the entire *Sinorhizobium meliloti* pSymA megaplasmid. Proc. Natl. Acad. Sci. USA 98, 9883–9888.

Bentley, S.D., Chater, K.F., Cerdeno-Tarraga, A.M., Challis, G.L., Thomson, N.R., James, K.D., Harris, D.E., Quail, M.A., Kieser, H., Harper, D., Bateman, A., Brown, S., Chandra, G., Chen, C.W., Collins, M., Cronin, A., Fraser, A., Goble, A., Hidalgo, J., Hornsby, T., Howarth, S., Huang, C.H., Kieser, T., Larke, L., Murphy, L., Oliver, K., O'Neil, S., Rabbinowitsch, E., Rajandream, M.A., Rutherford, K., Rutter, S., Seeger, K., Saunders, D., Sharp, S., Squares, R., Squares, S., Taylor, K., Warren, T., Wietzorrek, A., Woodward, J., Barrell, B.G., Parkhill, J. and Hopwood, D.A. (2002). Complete genome sequence of the model actinomycete *Streptomyces coelicolor* A3(2). Nature 417, 141–147.

Bernaerts, M., and De Ley, J. (1952). Microbiological decomposition of malto- and lacto-bionate. Nature 170, 713.

Binns, A.N., and Thomashow, M.F. (1988). Cell biology of *Agrobacterium* infection and transformation of plants. Annu. Rev. Microbiol. 42, 575–606.

Bouzar, H., and Moore, L.W. (1987). Isolation of Different *Agrobacterium* Biovars from a Natural Oak Savanna and Tallgrass Prairie. Appl. Environ. Microbiol. 53, 717–721.

Bouzar, H., Ouadah, D., Krimi, Z., Jones, J.B., Trovato, M., Petit, A., and Dessaux, Y. (1993). Correlative Association between Resident Plasmids and the Host Chromosome in a Diverse *Agrobacterium* Soil Population. Appl. Environ. Microbiol. 59, 1310–1317.

Burr, T.J., Bazzi, C., Süle, S., and Otten, L. (1998). Crown gall of grape: Biology of *Agrobacterium vitis* and the development of disease control strategies. Plant Dis. 82, 1288–1297.

Burr, T.J., Reid, C.L., Yoshimura, M., Momol, E.A., and Bazzi, C. (1995). Survival and tumorigenicity of *Agrobacterium vitis* in living and decaying grape roots and canes in soil. Plant Dis. 79, 677–682.

Capela, D., Barloy-Hubler, F., Gouzy, J., Bothe, G., Ampe, F., Batut, J., Boistard, P., Becker, A., Boutry, M., Cadieu, E., Dréano, S., Gloux, S., Godrie, T., Goffeau, A., Kahn, D., Kiss, E., Lelaure, V., Masuy, D., Pohl, T., Portetelle, D., Pühler, A., Purnelle, B., Ramsperger, U., Renard, C., Thébault, P., Vandenbol, M., Weidner, S. and Galibert, F. (2001). Analysis of the chromosome sequence of the legume symbiont *Sinorhizobium meliloti* strain 1021. Proc. Natl. Acad. Sci. USA 98, 9877–9882.

Carter, R.A., Worsley, P.S., Sawers, G., Challis, G.L., Dilworth, M.J., Carson, K.C., Lawrence, J.A., Wexler, M., Johnston, A.W., and Yeoman, K.H. (2002). The *vbs* genes that direct synthesis of the siderophore vicibactin in *Rhizobium leguminosarum*: their expression in other genera requires ECF sigma factor RpoI. Mol. Microbiol. 44, 1153–1166.

Casjens, S. (1999). Evolution of the linear DNA replicons of the *Borrelia* spirochetes. Curr. Opin. Microbiol. 2, 529–534.

Casjens, S., Palmer, N., van Vugt, R., Huang, W.M., Stevenson, B., Rosa, P., Lathigra, R., Sutton, G., Peterson, J., Dodson, R.J., Haft, D., Hickey, E., Gwinn, M., White, O. and Fraser, C.M. (2000). A bacterial genome in flux: the twelve linear and nine circular extrachromosomal DNAs in an infectious isolate of the Lyme disease spirochete *Borrelia burgdorferi*. Mol. Microbiol. 35, 490–516.

Casjens, S.R., Gilcrease, E.B., Huang, W.M., Bunny, K.L., Pedulla, M.L., Ford, M.E., Houtz, J.M., Hatfull, G.F., and Hendrix, R.W. (2004). The pKO2 linear plasmid prophage of *Klebsiella oxytoca*. J. Bacteriol. 186, 1818–1832.

Ceci, P., Ilari, A., Falvo, E., and Chiancone, E. (2003). The Dps protein of *Agrobacterium tumefaciens* does not bind to DNA but protects it toward oxidative cleavage. J. Biol. Chem. 278, 20319–20326.

Chevrot, R., Rosen, R., Haudecoeur, E., Cirou, A., Shelp, B.J., Ron, E., and Faure, D. (2006). GABA controls the level of quorum-sensing signal in *Agrobacterium tumefaciens*. Proc. Natl. Acad. Sci. USA 103, 7460–7464.

Chilton, W.S., Petit, A., Chilton, M.D., and Dessaux, Y. (2001). Structure and characterization of the crown gall opines heliopine, vitopine and ridéopine. Phytochemistry 58, 137–142.

Chuchue, T., Tanboon, W., Prapagdee, B., Dubbs, J.M., Vattanaviboon, P., and Mongkolsuk, S. (2006). *ohrR* and *ohr* are the primary sensor/regulator and protective genes against organic hydroperoxide stress in *Agrobacterium tumefaciens*. J. Bacteriol. 188, 842–851.

Citovsky, V., Kozlovsky, S.V., Lacroix, B., Zaltsman, A., Dafny-Yelin, M., Vyas, S., Tovkach, A., and Tzfira, T. (2007). Biological systems of the host cell involved in *Agrobacterium* infection. Cell. Microbiol. 9, 9–20.

Cornelis, G.R. (2006). The type III secretion injectisome. Nat Rev Microbiol 4, 811–825.

Cubero, J., Lastra, B., Salcedo, C.I., Piquer, J., and Lopez, M.M. (2006). Systemic movement of *Agrobacterium tumefaciens* in several plant species. J. Appl. Microbiol. 101, 412–421.

Das, S., and Choudhuri, K. (2003). Identification of a unique IAHP (IcmF-associated homologous proteins) cluster in *Vibrio cholerae* and other proteobacteria through in silico analysis. In Silico Biol. 3, 287–300.

De Cleene, M. (1979). Crown gall: economic importance and control. Zbl. Bakt. II Abt. 134, 551–554.

Dessaux, Y., Petit, A., Farrand, S.K., and Murphy, P.J. (1998). Opines and opine-like molecules involved in Plant-*Rhizobiaceae* interactions. In The Rhizobiaceae: Molecular Biology of Model Plant-Associated Bacteria, H.P. Spaink, A. Kondorosi, and P.J.J. Hooykaas, eds. (Dordrecht-Boston-London, Kluwer Academic Publisher), pp.173–197.

Edgar, R.C. (2004). MUSCLE: multiple sequence alignment with high accuracy and high throughput. Nucleic Acids Res. 32, 1792–1797.

Eiamphungporn, W., Nakjarung, K., Prapagdee, B., Vattanaviboon, P., and Mongkolsuk, S. (2003). Oxidant-inducible resistance to hydrogen peroxide killing in *Agrobacterium tumefaciens* requires the global peroxide sensor-regulator OxyR and KatA. FEMS Microbiol. Lett. 225, 167–172.

Escobar, M.A., and Dandekar, A.M. (2003). *Agrobacterium tumefaciens* as an agent of disease. Trends Plant Sci. 8, 380–386.

Galan, J.E., and Wolf-Watz, H. (2006). Protein delivery into eukaryotic cells by type III secretion machines. Nature 444, 567–573.

Galibert, F., Finan, T.M., Long, S.R., Puhler, A., Abola, P., Ampe, F., Barloy-Hubler, F., Barnett, M.J., Becker, A., Boistard, P., Bothe, G., Boutry, M., Bowser, L., Buhrmester, J., Cadieu, E., Capela, D., Chain, P., Cowie, A., Davis, R.W., Dreano, S., Federspiel, N.A., Fisher, R.F., Gloux, S., Godrie, T., Goffeau, A., Golding, B., Gouzy, J., Gurjal, M., Hernandez-Lucas, I., Hong, A., Huizar, L., Hyman, R.W., Jones, T., Kahn, D., Kahn, M.L., Kalman, S., Keating, D.H., Kiss, E., Komp, C., Lelaure, V., Masuy, D., Palm, C., Peck, M.C., Pohl, T.M., Portetelle, D., Purnelle, B., Ramsperger, U., Surzycki, R., Thebault, P., Vandenbol, M., Vorholter, F.J., Weidner, S., Wells, D.H., Wong, K., Yeh, K.C. and Batut, J. (2001). The Composite Genome of the Legume Symbiont *Sinorhizobium meliloti*. Science 293, 668–672.

Gelvin, S.B. (2000). *Agrobacterium* and Plant Genes Involved in T-DNA Transfer and Integration. Annu. Rev. Plant Physiol. Plant Mol. Biol. 51, 223–256.

Gelvin, S.B. (2003). *Agrobacterium*-mediated plant transformation: The biology behind the 'gene-jockeying' tool. Microbiol.Mol. Biol. Rev. 67, 16–37.

Gonzalez, V., Santamaria, R.I., Bustos, P., Hernandez-Gonzalez, I., Medrano-Soto, A., Moreno-Hagelsieb, G., Janga, S.C., Ramirez, M.A., Jimenez-Jacinto, V., Collado-Vides, J. and Davila, G. (2006). The partitioned *Rhizobium etli* genome: Genetic and metabolic redundancy in seven interacting replicons. Proc. Natl. Acad.Sci. USA 103, 3834–3839.

Goodner, B., Hinkle, G., Gattung, S., Miller, N., Blanchard, M., Qurollo, B., Goldman, B.S., Cao, Y., Askenazi, M., Halling, C., Mullin, L., Houmiel, K., Gordon, J., Vaudin, M., Iartchouk, O., Epp, A., Liu, F., Wollam, C., Allinger, M., Doughty, D., Scott, C.,

Lappas, C., Markelz, B., Flanagan, C., Crowell, C., Gurson, J., Lomo, C., Sear, C., Strub, G., Cielo, C. and Slater, S. (2001). Genome Sequence of the Plant Pathogen and Biotechnology Agent *Agrobacterium tumefaciens* C58. Science *294*, 2323–2328.

Grant, S.R., Fisher, E.J., Chang, J.H., Mole, B.M., and Dangl, J.L. (2006). Subterfuge and manipulation: type III effector proteins of phytopathogenic bacteria. Annu. Rev. Microbiol. *60*, 425–449.

Guerrero, G., Peralta, H., Aguilar, A., Diaz, R., Villalobos, M.A., Medrano-Soto, A., and Mora, J. (2005). Evolutionary, structural and functional relationships revealed by comparative analysis of syntenic genes in Rhizobiales. BMC Evol. Biol. *5*, 55.

Hao, G., and Burr, T.J. (2006). Regulation of long-chain N-acyl-homoserine lactones in *Agrobacterium vitis*. J. Bacteriol. *188*, 2173–2183.

Hao, G.X., Zhang, H.S., Zheng, D.S., and Burr, T.J. (2005). *luxR* Homolog *avhR* in *Agrobacterium vitis* affects the development of a grape-specific necrosis and a tobacco hypersensitive response. J. Bacteriol. *187*, 185–192.

Herlache, T.C., and Triplett, E.W. (2002). Expression of a crown gall biological control phenotype in an avirulent strain of *Agrobacterium vitis* by addition of the trifolitoxin production and resistance genes. BMC Biotechnol 2, 2.

Huang, W.M., Joss, L., Hsieh, T.T., and Casjens, S. (2004). Protelomerase uses a topoisomerase IB/Y-recombinase type mechanism to generate DNA hairpin ends. J. Mol. Biol. *337*, 77–92.

Kaneko, T., Nakamura, Y., Sato, S., Asamizu, E., Kato, T., Sasamoto, S., Watanabe, A., Idesawa, K., Ishikawa, A., Kawashima, K., Kimura, T., Kishida, Y., Kiyokawa, C., Kohara, M., Matsumoto, M., Matsuno, A., Mochizuki, Y., Nakayama, S., Nakazaki, N., Shimpo, S., Sugimoto, M., Takeuchi, C., Yamada, M. and Tabata, S. (2000). Complete genome structure of the nitrogen-fixing symbiotic bacterium *Mesorhizobium loti*. DNA Res. 7, 331–338.

Kelman, L.M., and Kelman, Z. (2004). Multiple origins of replication in archaea. Trends Microbiol. *12*, 399–401.

Kim, J.G., Park, B.K., Kim, S.U., Choi, D., Nahm, B.H., Moon, J.S., Reader, J.S., Farrand, S.K., and Hwang, I. (2006). Bases of biocontrol: sequence predicts synthesis and mode of action of agrocin 84, the Trojan horse antibiotic that controls crown gall. Proc. Natl. Acad. Sci. USA *103*, 8846–8851.

Krimi, Z., Petit, A., Mougel, C., Dessaux, Y., and Nesme, X. (2002). Seasonal fluctuations and long-term persistence of pathogenic populations of *Agrobacterium* spp. in soils. Appl. Environ. Microbiol. *68*, 3358–3365.

Kurtz, S., Phillippy, A., Delcher, A.L., Smoot, M., Shumway, M., Antonescu, C., and Salzberg, S.L. (2004). Versatile and open software for comparing large genomes. Genome Biol. *5*, R12.

Li, L., Stoeckert, C.J., Jr., and Roos, D.S. (2003). OrthoMCL: identification of ortholog groups for eukaryotic genomes. Genome Res. *13*, 2178–2189.

MacLellan, S.R., Zaheer, R., Sartor, A.L., MacLean, A.M., and Finan, T.M. (2006). Identification of a megaplasmid centromere reveals genetic structural diversity within the *repABC* family of basic replicons. Mol. Microbiol. *59*, 1559–1575.

McClure, N.C., Ahmadi, A.R., and Clare, B.G. (1998). Construction of a range of derivatives of the biological control strain *Agrobacterium rhizogenes* K84: a study of factors involved in biological control of crown gall disease. Appl. Environ. Microbiol. *64*, 3977–3982.

Marie, C., Broughton, W.J., and Deakin, W.J. (2001). *Rhizobium* type III secretion systems: legume charmers or alarmers? Curr. Opin. Plant Biol. *4*, 336–342.

Marketon, M.M., Gronquist, M.R., Eberhard, A., and Gonzalez, J.E. (2002). Characterization of the *Sinorhizobium meliloti sinR/sinI* locus and the production of novel N-acyl homoserine lactones. J. Bacteriol. *184*, 5686–5695.

Martínez, L., Caballero-Mellado, J., Orozco, J., and Martínez-Romero, E. (2003). Diazotrophic bacteria associated with banana (*Musa spp.*). Plant and Soil *257*, 35–47.

Matthysse, A.G., and Kijne, J.W. (1998). Attachment of *Rhizobiaceae* to plant cells. In The *Rhizobiaceae*: Molecular Biology of Model Plant-Associated Bacteria, H.P. Spaink, A. Kondorosi, and P.J.J. Hooykaas, eds. (Dordrecht/Boston/London, Kluwer Academic Publishers), pp. 235–249.

Mongkolsuk, S., Praituan, W., Loprasert, S., Fuangthong, M., and Chamnongpol, S. (1998). Identification and characterization of a new organic hydroperoxide resistance (*ohr*) gene with a novel pattern of oxidative stress regulation from *Xanthomonas campestris* pv. phaseoli. J. Bacteriol. *180*, 2636–2643.

Mota, L.J., and Cornelis, G.R. (2005). The bacterial injection kit: type III secretion systems. Ann. Med. *37*, 234–249.

Nair, G.R., Liu, Z.Y., and Binns, A.N. (2003). Reexamining the role of the accessory plasmid pAtC58 in the virulence of *Agrobacterium tumefaciens* strain C58. Plant Physiol. *133*, 989–999.

Nakjarung, K., Mongkolsuk, S., and Vattanaviboon, P. (2003). The *oxyR* from *Agrobacterium tumefaciens*: evaluation of its role in the regulation of catalase and peroxide responses. Biochem. Biophys. Res. Commun. *304*, 41–47.

Nester, E., Gordon, M.P., and Kerr, A. (2004). *Agrobacterium tumefaciens*: from Plant Pathology to Biotechnology (APS Press).

Oger, P., and Farrand, S.K. (2002). Two opines control conjugal transfer of an *Agrobacterium* plasmid by regulating expression of separate copies of the quorum-sensing activator gene *traR*. J. Bacteriol. *184*, 1121–1131.

Otten, L., and Schmidt, J. (1998). A T-DNA from the *Agrobacterium tumefaciens* limited-host-range strain AB2/73 contains a single oncogene. Mol. Plant–Microbe Interact. *11*, 335–342.

Penyalver, R., Oger, P., Lopez, M.M., and Farrand, S.K. (2001). Iron-binding compounds from *Agrobacterium* spp.: biological control strain *Agrobacterium rhizogenes* K84 produces a hydroxamate siderophore. Appl. Environ. Microbiol. *67*, 654–664.

Prapagdee, B., Vattanaviboon, P., and Mongkolsuk, S. (2004). The role of a bifunctional catalase-peroxidase

KatA in protection of *Agrobacterium tumefaciens* from menadione toxicity. FEMS Microbiol. Lett. *232*, 217–223.

Pueppke, S.G., and Broughton, W.J. (1999). *Rhizobium* sp. strain NGR234 and R. fredii USDA257 share exceptionally broad, nested host ranges. Mol. Plant–Microbe Interact. *12*, 293–318.

Pukatzki, S., Ma, A.T., Sturtevant, D., Krastins, B., Sarracino, D., Nelson, W.C., Heidelberg, J.F., and Mekalanos, J.J. (2006). Identification of a conserved bacterial protein secretion system in *Vibrio cholerae* using the *Dictyostelium* host model system. Proc. Natl. Acad. Sci. USA *103*, 1528–1533.

Ravin, N.V. (2003). Mechanisms of replication and telomere resolution of the linear plasmid prophage N15. FEMS Microbiol. Lett. *221*, 1–6.

Ravin, V., Ravin, N., Casjens, S., Ford, M.E., Hatfull, G.F., and Hendrix, R.W. (2000). Genomic sequence and analysis of the atypical temperate bacteriophage N15. J Mol. Biol. *299*, 53–73.

Reeve, W.G., Brau, L., Castelli, J., Garau, G., Sohlenkamp, C., Geiger, O., Dilworth, M.J., Glenn, A.R., Howieson, J.G., and Tiwari, R.P. (2006). The *Sinorhizobium medicae* WSM419 *lpiA* gene is transcriptionally activated by FsrR and required to enhance survival in lethal acid conditions. Microbiology *152*, 3049–3059.

Rondon, M.R., Ballering, K.S., and Thomas, M.G. (2004). Identification and analysis of a siderophore biosynthetic gene cluster from *Agrobacterium tumefaciens* C58. Microbiology *150*, 3857–3866.

Ronquist, F., and Huelsenbeck, J.P. (2003). MrBayes 3: Bayesian phylogenetic inference under mixed models. Bioinformatics *19*, 1572–1574.

Rosenberg, C., and Huguet, T. (1984). The pAtC58 plasmid of *Agrobacterium tumefaciens* is not essential for tumour induction. Mol. Gen. Genet. *196*, 533–536.

Saenkham, P., Eiamphungporn, W., Farrand, S.K., Vattanaviboon, P., and Mongkolsuk, S. (2007). Multiple superoxide dismutases in *Agrobacterium tumefaciens*: functional analysis, gene regulation, and influence on tumorigenesis. J. Bacteriol. *189*, 8807–8817.

Schell, M.A. (1993). Molecular biology of the LysR family of transcriptional regulators. Annu. Rev. Microbiol. *47*, 597–626.

Sibley, C.D., MacLellan, S.R., and Finan, T. (2006). The *Sinorhizobium meliloti* chromosomal origin of replication. Microbiology *152*, 443–455.

Skorpil, P., Saad, M.M., Boukli, N.M., Kobayashi, H., Ares-Orpel, F., Broughton, W.J., and Deakin, W.J. (2005). NopP, a phosphorylated effector of *Rhizobium* sp. strain NGR234, is a major determinant of nodulation of the tropical legumes *Flemingia congesta* and *Tephrosia vogelii*. Mol. Microbiol. *57*, 1304–1317.

Slater, S., Goodner, B., Setubal, J.C., Goldman, B.S., Wood, D., and Nester, E. (2008). The *Agrobacterium tumefaciens* C58 genome. In *Agrobacterium*: From Biology to Biotechnology, T. Tzfira, and V. Citovsky, eds. (New York, Springer), pp. 149–181.

Stiens, M., Schneiker, S., Keller, M., Kuhn, S., Pühler, A., and Schlüter, A. (2006). Sequence analysis of the 144-kilobase accessory plasmid pSmeSM11a, isolated from a dominant *Sinorhizobium meliloti* strain identified during a long-term field release experiment. Appl. Environ. Microbiol. *72*, 3662–3672.

Tzfira, T., and Citovsky, V. (2006). *Agrobacterium*-mediated genetic transformation of plants: biology and biotechnology. Curr. Opin. Biotechnol. *17*, 147–154.

Unger, L., Ziegler, S.F., Huffman, G.A., Knauf, V.C., Peet, R., Moore, L.W., Gordon, M.P., and Nester, E.W. (1985). New class of limited-host-range *Agrobacterium* mega-tumor-inducing plasmids lacking homology to the transferred DNA of a wide-host-range, tumor-inducing plasmid. J. Bacteriol. *164*, 723–730.

Vinuesa, P., Neumann-Silkow, F., Pacios-Bras, C., Spaink, H.P., Martinez-Romero, E., and Werner, D. (2003). Genetic analysis of a pH-regulated operon from *Rhizobium tropici* CIAT899 involved in acid tolerance and nodulation competitiveness. Mol. Plant–Microbe Interact. *16*, 159–168.

Wang, Y.H., Zhang, L.Q., Li, J.Y., Wang, J.H., and Wang, H.M. (2007). The quorum-sensing system AvsR-AvsI regulates both long-chain and short-chain acyl-homoserine lactones in *Agrobacterium vitis* E26. Antonie Van Leeuwenhoek *29*, 29.

Wirawan, I.G., Kang, H.W., and Kojima, M. (1993). Isolation and characterization of a new chromosomal virulence gene of *Agrobacterium tumefaciens*. J. Bacteriol. *175*, 3208–3212.

Wong, K., and Golding, G.B. (2003). A phylogenetic analysis of the pSymB replicon from the *Sinorhizobium meliloti* genome reveals a complex evolutionary history. Can. J. Microbiol. *49*, 269–280.

Wood, D.W., Setubal, J.C., Kaul, R., Monks, D.E., Kitajima, J.P., Okura, V.K., Zhou, Y., Chen, L., Wood, G.E., Almeida Jr., N.F., Woo, L., Chen, Y., Paulsen, I.T., Eisen, J.A., Karp, P.D., Bovee, D. Sr, Chapman, P., Clendenning, J., Deatherage, G., Gillet, W., Grant, C., Kutyavin, T., Levy, R., Li, M.J., McClelland, E., Palmieri, A., Raymond, C., Rouse, G., Saenphimmachak, C., Wu, Z., Romero, P., Gordon, D., Zhang, S., Yoo, H., Tao, Y., Biddle, P., Jung, M., Krespan, W., Perry, M., Gordon-Kamm, B., Liao, L., Kim, S., Hendrick, C., Zhao, Z.Y., Dolan, M., Chumley, F., Tingey, S.V., Tomb, J.F., Gordon, M.P., Olson, M.V. and Nester, E.W. (2001). The genome of the natural genetic engineer *Agrobacterium tumefaciens* C58. Science *294*, 2317–2323.

Xu, X.Q., Li, L.P., and Pan, S.Q. (2001). Feedback regulation of an *Agrobacterium* catalase gene *katA* involved in *Agrobacterium*-plant interaction. Mol. Microbiol. *42*, 645–657.

Yabuuchi, E., Kosako, Y., Oyaizu, H., Yano, I., Hotta, H., Hashimoto, Y., Ezaki, T., and Arakawa, M. (1992). Proposal of *Burkholderia* gen. nov. and transfer of seven species of the genus *Pseudomonas* homology group II to the new genus, with the type species *Burkholderia cepacia* (Palleroni and Holmes 1981) comb. nov. Microbiol. Immunol. *36*, 1251–1275.

Young, J.P., Crossman, L.C., Johnston, A.W., Thomson, N.R., Ghazoui, Z.F., Hull, K.H., Wexler, M., Curson, A.R., Todd, J.D., Poole, P.S., Mauchline, T.H., East, A.K., Quail, M.A., Churcher, C., Arrowsmith, C.,

Cherevach, I., Chillingworth, T., Clarke, K., Cronin, A., Davis, P., Fraser, A., Hance, Z., Hauser, H., Jagels, K., Moule, S., Mungall, K., Norbertczak, H., Rabbinowitsch, E., Sanders, M., Simmonds, M., Whitehead, S. and Parkhill, J. (2006). The genome of *Rhizobium leguminosarum* has recognizable core and accessory components. Genome Biol. 7, R34.

Yuan, Z.C., Edlind, M.P., Liu, P., Saenkham, P., Banta, L.M., Wise, A.A., Ronzone, E., Binns, A.N., Kerr, K., and Nester, E.W. (2007). The plant signal salicylic acid shuts down expression of the vir regulon and activates quormone-quenching genes in Agrobacterium. Proc. Natl. Acad. Sci. USA *104*, 11790–11795.

Zhao, G., Ceci, P., Ilari, A., Giangiacomo, L., Laue, T.M., Chiancone, E., and Chasteen, N.D. (2002). Iron and hydrogen peroxide detoxification properties of DNA-binding protein from starved cells. A ferritin-like DNA-binding protein of *Escherichia coli*. J. Biol. Chem. *277*, 27689–27696.

Zheng, D., Zhang, H.S., Carle, S., Hao, G., Holden, M.R., and Burr, T.J. (2003). A *luxR* homolog, *aviR*, in *Agrobacterium vitis* is associated with induction of necrosis on grape and a hypersensitive response on tobacco. Mol. Plant–Microbe Interact. *16*, 650–658.

# Common Genes and Genomic Breaks: a Detailed Case Study of the *Xylella fastidiosa* Genome Backbone and Evolutionary Insights

Alessandro M. Varani, Wanessa C. Lima, Leandro M. Moreira, Mariana C. de Oliveira, Rangel de Souza, Edwin Civerolo, Ana Tereza R. de Vasconcelos and Marie-Anne Van Sluys

Abstract

It has been more than 7 years since the first genome sequence of a plant pathogen, *Xylella fastidiosa* strain 9a5c, was published. At present, more than 10 genomes of the γ-Proteobacteria Xanthomonadales group are available for comparative genomics, and several studies related to functional genomics have been conducted, resulting in insights about the evolution, virulence and pathogenicity of this group of plant pathogens. The subject of this review is to explore the history of the chromosome backbone evolution and differentiation among four *X. fastidiosa* strains, and address the question of how and when those organisms became pathogenically competent. To address this question, three main lines of discussion are developed: (a) correlation of the bacterial life style with their genes involved in virulence and pathogenicity; (b) definition of the minimal core genome using a comparative approach; and (c) looking at the disruptions and rearrangements caused by mobile genetic elements. The discussion raised here allowed us to assess the differential evolutionary profiles inside the Xanthomonadales.

Introduction

In 2000, Simpson and co-workers published the first genome sequence of a plant pathogen, *Xylella fastidiosa* strain 9a5c, the causative agent of citrus variegated chlorosis (CVC). Since then, more than 20 plant pathogen genome sequences have become available (http://cmr.tigr.org/). As a result, researchers have gained new insights into gene content diversity, virulence factors, and environmental adaptation strategies. Of particular interest is the group of Xanthomonadales plant pathogens for which at least 10 genomes have been publicly released (Table 5.1) and an additional three will be available soon. The virulence components of the genome sequences and the genome organization, particularly of the fastidious plant pathogen *X. fastidiosa*, raised the question of how and when those organisms became pathogenically competent.

The 'California vine disease' (Pierce, 1892), now known as Pierce's disease (PD), caused by *X. fastidiosa*, was recognized in the late 1800s. However, the aetiology of the disease was not established until 1978 (Davis *et al.*, 1978). Although there was presumptive evidence for a bacterial aetiology of PD before this paper, Davis and colleagues (1978) described the isolation of a Gram-negative, catalase-positive, rod-shaped bacterium from grapevines affected with PD and fulfilled Koch's postulates. *X. fastidiosa* was only established as a new taxon in 1987 (Wells *et al.*, 1987). Subsequently, *X. fastidiosa* has been reported to cause diseases of a wide range of agronomic, horticultural, forest and landscape ornamental crops (Hopkins and Purcell, 2002).

Xylella belongs to the γ-Proteobacteria group and clusters with *Xanthomonas* and *Stenotrophomonas* at the very base of the corresponding phylogenetic branch. *Stenotrophomonas* is a free-living organism found mostly in soil, air and water, not particularly associated to plants and at least two distinct species are now being sequenced (Hauben *et al.*, 1999). To better understand the genetic relationships among the

Table 5.1 Xanthomonadales genomes with finished or in progress sequencing projects*

Organism	Size (Mb)	% GC	Chromosome no.	Plasmid no.	No. of ORFs	Status	Reference	GenBank accession no.	Project website
Stenotrophomonas maltophilia K279a	6.00	66.3				Draft		–	http://www.sanger.ac.uk/
Stenotrophomonas maltophilia R551-3	4.54	66.3			4032	Draft		NZ_AAVZ00000000	http://genome.jgi-psf.org/
Stenotrophomonas sp. SKA14	–	–				Incomplete		–	http://research.venterinstitute.org/
Xanthomonas albilineans GPE PC73	5.0	–				Incomplete		–	http://www.genoscope.cns.fr/
Xanthomonas fuscans subsp. *aurantifolii*	5.0	64.0				Incomplete		–	http://genoma2.fcav.unesp.br/
Xanthomonas citri subsp. *citri* str. 306	5.27	64.7	1	2	4312	Finished	da Silva et al. (2001)	AE008923.1	http://www.lbi.ic.unicamp.br/
Xanthomonas campestris pv. *campestris* str. 8004	5.15	65.0	1		4273	Finished	Qian et al. (2005)	CP000050.1	http://chgc.sh.cn/en/
Xanthomonas campestris pv. *campestris* str. ATCC 33913	5.08	65.1	1		4181	Finished	da Silva et al. (2001)	AE008922.1	http://www.lbi.ic.unicamp.br/
Xanthomonas campestris pv. *vesicatoria* str. 85-10	5.42	64.6	1	4	4487	Finished	Thieme et al. (2005)	AM039952.1	http://www.genetik.uni-bielefeld.de/

Organism	Size (Mb)	GC%	Chr	Genes	Status	Reference	Accession	URL
Xanthomonas oryzae pv. *oryzae* KACC10331	4.9	63.7	1	4147	Finished	Lee *et al.* (2005)	AE013598.1	http://www.niab.go.kr/
Xanthomonas oryzae pv. *oryzae* MAFF 311018	4.94	63.7	1	4372	Finished	Ochiai *et al.* (2005)	AP008229.1	http://www.nias.affrc.go.jp/
Xanthomonas oryzae pv. *oryzicola* BLS256	4.83	64.1			Draft		NZ_AAQN00000000	http://www.tigr.org/
Xylella fastidiosa 9a5c	2.73	52.6	2	2766	Finished	Simpson *et al.* (2000)	AE003849.1	http://www.xylella.lncc.br/
Xylella fastidiosa Ann-1	2.67	51.9		?	Draft	Bhattacharyya *et al* 2002b	NZ_AAAM00000000	http://www.jgi.doe.gov/re-annotation http://www.xylella.lncc.br/
Xylella fastidiosa Dixon	2.62	52.0		2358	Draft	Bhattacharyya *et a* (2002b)	NZ_AAAL00000000	http://www.jgi.doe.gov/re-annotation http://www.xylella.lncc.br/
Xylella fastidiosa M12	–	–			Incomplete		–	http://www.jgi.doe.gov/
Xylella fastidiosa M23	–	51.0			Incomplete		–	http://www.jgi.doe.gov/
Xylella fastidiosa Temecula-1	2.52	51.8	1	2034	Finished	Van Sluys *et al.* (2003)	AE009442.1	http://www.xylella.lncc.br/

*Data obtained from NCBI/GenBank and GOLD (Liolios *et al.*, 2006) databases.

Xanthomonadales, a ribosomal phylogenetic tree was made utilizing 4412 characters of the operon encompassing part of the 5S, 16S and 23S rDNA operon (Fig. 5.1). Of particular interest is the position of *Xylella* in the Xanthomonadales group. Although the three genera *Xanthomonas*, *Stenotrophomonas* and *Xylella* are closely related, *Xylella* is distinctly separated from the other two with a high bootstrap value. A closer look at the topology of the tree suggests that *X. fastidiosa* separated from the other two early within the Xanthomonadales clade and adapted to a specialized lifestyle independently. Thus, *Xylella* cannot be considered simply a parasitic form derived from *Xanthomonas* species as it has been sometimes considered. This analysis raises the issues of how and when these genomes diversified.

The purpose of this review is to explore the history of the chromosome backbone evolution and differentiation among four *X. fastidiosa* strains. Several questions that drove the study presented here are:

(a) Is it possible to determine when pathogenicity arose in the group?
(b) Is it a recent event?
(c) Which genes are shared among all Xanthomonadales?
(d) Can a minimal genome for the Xanthomonadales group be defined?

Stenotrophomonas maltophilia was included in this study because, although not strictly a plant pathogen, this species is found in the rhizosphere and on plant tissue surfaces as an epiphyte (Hauben et al., 1999). In fact, this bacterium is considered to be mostly a free living organism with no report of it causing plant diseases. However, it has been associated with some human infections (Senol, 2004; Looney, 2005).

In an attempt to answer the questions above, a *Xylella* comparative database has been developed and is available at www.xylella.lncc.br/comparative. This database contains the standardized and re-annotated genome sequences of four *X. fastidiosa* strains (9a5c, Temecula-1, Ann-1 and

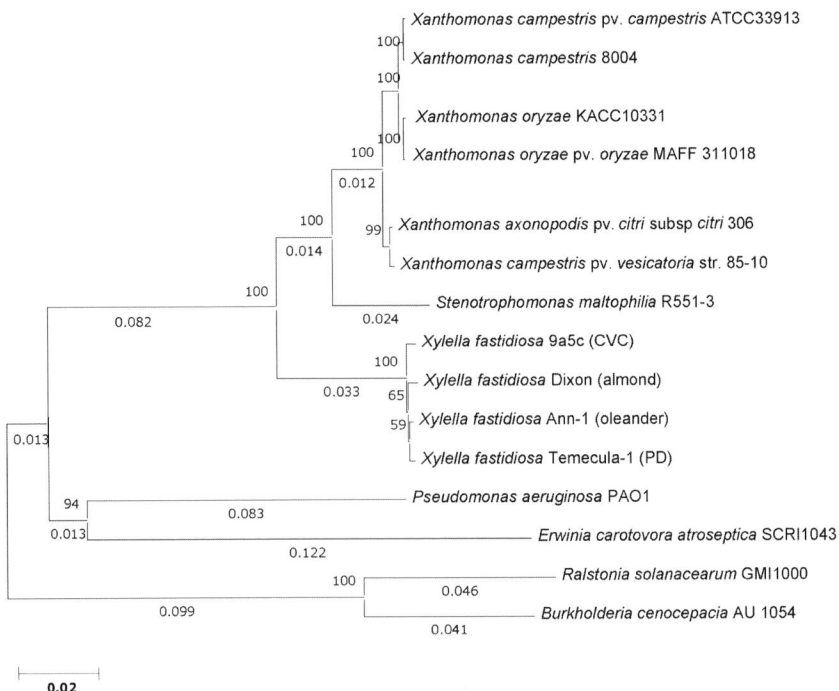

Figure 5.1 Distance tree of rRNA cluster from completely sequenced Xanthomonadales species. Values above branches indicate bootstrap support, and values below branches indicate branch length distance (values below 0.005 were excluded, for space reasons). Maximum parsimony trees show the same tree topology.

Dixon). To study the chromosome backbone history, this database also contains the genome information of other Xanthomonadales and plant-associated bacteria. To make this review compatible with previous published genome papers, *Xylella* and *Xanthomonas* strain names and abbreviations were kept as previously defined although taxonomic revisions have been published in which new taxons are defined for these prokaryote pathogens (see Rodrigues *et al.*, 2003; Jones *et al.*, 2004; Schaad *et al.*, 2006). Details on the bioinformatic pipeline are described at the end of this review.

Biology as depicted from the genome sequence and functional genomics outcomes

Xylella is environmentally restricted to the xylem vessels of host plants and certain insect vector tissue (Hopkins, 1989; Purcell and Hopkins, 1996). Thus, this pathogen needs to overcome environmental and nutritional limitations. These include continuous flow of nutritionally limited fluids in the insect vectors and host plant xylem vessels. Although *Xylella* can colonize both environments, it fails to cross the barriers of either the insect gut or the plant apoplastic tissue.

The first *X. fastidiosa* genome sequence made available (Simpson *et al.*, 2000) provided the basis for subsequent gene function analyses. For example, it was then possible to establish that the *Xylella*-plant interaction that leads to disease development is not dependent on effectors secreted through the type III secretion system as is the case in *Xanthomonas*. Also, the presence of genes encoding known animal virulence factors such as haemagglutinins and *vapD* were revealed. The other three *Xylella* genome projects that followed (Batthacharyya *et al.*, 2002a; Van Sluys *et al.*, 2003) confirmed the earlier observations since all the genomes shared those genes and diversity was mainly restricted to chromosome structure associated with the presence of large number of phage-related regions (Monteiro-Vitorello *et al.*, 2005; Moreira *et al.*, 2005).

In an attempt to summarize the functional data made available by several groups in the last 5 years, Fig. 5.2 was modified from the original one published in 2000 (Simpson *et al.*, 2000). A crucial step of the *Xylella* life cycle is attachment to a surface where bacterial cells can take advantage of fluid flow and initiate proliferation. At the bottom of Fig. 5.2, *Xylella* genes that were hypothesized to be involved in attachment are shown. The four *Xylella* strains harbour several genes related to fimbrial and afimbrial adhesins as will be detailed further below. *Xylella* needs to attach, colonize and move within xylem vessels, as well as attach to the insect vector foregut in order to complete the disease cycles. Colonization of the insect foregut is essential for dissemination of the bacteria within and between cropping systems, natural ecosystems and landscape plantings over time. Many Xf hosts have been identified (other than citrus) as being asymptomatic despite the fact that Xf strains in symptomless hosts are (probably) also disseminated by insect vectors. Thus, Xf pathogen–host interactions do not always result in disease development. Unfortunately, however, not much is known, if anything, about the nature of the interactions between Xf strains and hosts that do not develop 'disease' symptoms. Xylem feeding Cicadelinieae insects are known to be efficient vectors, but still little is known about the interaction between *Xylella* and this group of insects. The blue green sharpshooter and the more aggressive glassy-winged sharpshooter have been used in studies to characterize bacteria–vector interactions. Insects harbouring *Xylella* remain capable of transmitting the pathogen until they moult (Almeida and Purcell, 2003). After moulting, feeding on infected plants is needed for the insect vector to re-acquire the pathogen.

In both insect vector foreguts and plant xylem vessels, early adhesion of the bacteria to the surface is mediated by proteins known as adhesins. These sticky proteins are now known to be required for successful establishment of infection *in planta* and in insects. Based on genomic analyses, at least three distinct systems were predicted to be associated with this early infection event (Simpson *et al.*, 2000; Van Sluys *et al.*, 2002). All sequenced *Xylella* strains harbour type I and type IV pili, as well as haemagglutinin adhesins. The proteins encoded by these gene clusters are involved in attachment in both insects (Newman *et al.*, 2004; Li *et al.*, 2007) and plants (Guilhabert and Kirkpatrick, 2005; Meng *et al.*, 2005). Interestingly, Newman *et al* (2004) presented

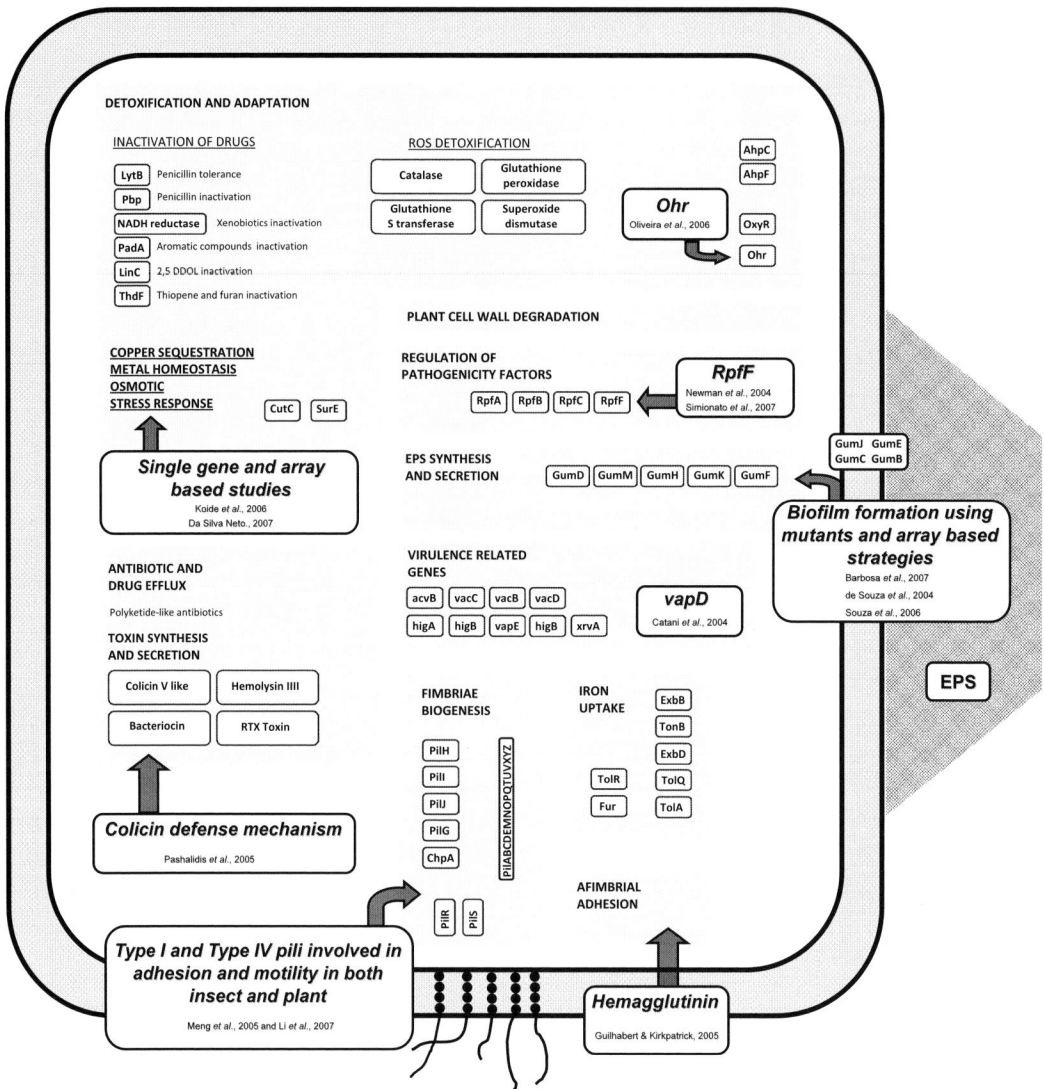

Figure 5.2 *Xylella* biology as depicted from the genome sequencing project (figure modified from Simpson *et al.*, 2000) and functional validation in subsequent studies. White boxes, with references included, depict the validation by experimentation (depicting specific references) based on the predictions from the available genome sequences.

evidence for a polar attachment of *Xylella* cells to insect vector foregut, while in plants a more disorganized mat is formed suggesting that different genes, protein structures or spatial construction of protein structures may be implicated in each system. Recently, Koide *et al.*, (2004) used a DNA microarray-based strategy to compare a non-pathogenic *Xylella* citrus strain J1a12 with the previously sequenced 9a5c strain and found that this non-pathogenic strain specifically lacks a fimbrial adhesin precursor of type I pilus which results in reduced cell adherence. Also, the work of Guilhabert and Kirkpatrick (2005) describing the attenuated impact of a disrupted haemagglutinin mutant strain on bacteria virulence led the authors to hypothesize that some of these genes may in fact act as anti-virulence components in the plant system providing the bacteria with the ability to colonize the vessels and be further transmitted by the vector. *Xylella* vessel-to-vessel movement is a critical step in systemic infection and disease development. As haemagglutinin proteins facilitate *Xylella* self-aggregation, and thereby reducing the rate of xylem vessel occlusion, disruption of haemagglutinin genes increases pathogen virulence by increasing the bacterial ability to translocate throughout the plant and, subsequently, colonize other vessels more rapidly than the wild-type cells (Fry and Milholland, 1994; Guilhabert and Kirkpatrick, 2005).

Koide et al (2004) used a DNA microarray-based strategy to compare a non-pathogenic *Xylella* citrus strain J1a12 with the previously sequenced Xf-9a5c and found that this non-pathogenic strain specifically lacks those genes which result in reduced cell adherence. Also, the work describing the impact of a disrupted haemagglutinin mutant strain on bacteria virulence led the authors to hypothesize that some of these genes may in fact act as anti-virulence components in the plant system providing the bacteria with the ability to colonize the vessels and be further transmitted by the vector.

As previously noted (Van Sluys *et al.* 2002; Moreira *et al.*, 2004), *Xylella* strains possess a larger number of genes involved in the production of sticky proteins associated with structures involved in adhesion and movement than *Xanthomonas* species. Based on bidirectional best hit (BBH) clustering performed across the Xanthomonadales, *Xylella* strains carry two clusters of *Tfp* pilus, while *Xanthomonas* and *Stenotrophomonas* harbour only one cluster. An exception is *fimT* that is present in three copies in *Xylella* while *Xanthomonas* species (XCV and XCC) each harbour two copies. Apart from the adhesins, gene products involved in biofilm formation and diffusible signal factor (DSF) are also critical for colonization of plant vessels and the insect gut (Fig. 5.2). From the studies described above and analysis of an *rpfF* mutant (Newman *et al.*, 2004), biofilm architecture is actively determined by *Xylella* gene expression rather than being a passive response to the environment. *Xylella* biofilms *in planta* are architecturally different from those *in insecta*. In addition, changes in gene expression profiles in *pilB* and *pilQ* mutants (possessing type I pili only) reported by Meng *et al.* (2005), indicate that *Xylella* express enhanced biofilm formation. Thus, it is apparent that *Xylella* cells utilize the expression of these genes to modulate their interactions with the two hosts.

This suggests these genes, while essential for establishing bacteria-plant interaction[s], may make the pathogen less efficient to be transmitted by insect vectors. Despite the absence of flagella, *Xylella* is capable of moving along a plant vessel against the flow of vascular fluid, as well as across vessels, as demonstrated by confocal studies using a GFP-expressing *Xylella* strain (Newman *et al.*, 2004) and microfluidic chamber devices (Meng *et al.*, 2005).

It is now apparent that *Xylella* cells may exist in two states within the plant vessels: a planktonic state and biofilm state. These two growth conditions have a distinct impact in the biology of *X. fastidiosa*. Planktonic cells mediate spread of the pathogen within the infected plant while cells in a biofilm state have an impact on the epidemiology of the disease. Biofilm formation has been recognized in several independent studies as an important mediator of the bacteria-host interaction, including those in infected, but asymptomatic, plants. De Souza and colleagues (2004; 2005) first described gene expression profile[s?] during biofilm formation *in vitro* and further compared those with gene expression *in planta*. The main conclusion from these studies is that adhesion-related genes are induced prior to biofilm formation and some adaptation-related genes are induced at later stages, as observed for *acrA*, *acrB* (multidrug efflux transporters) and *xpsE* (type II secretion system).

Also, of interest is a repressor protein, BigR (for biofilm growth-associated repressor; XF0767) that is predicted to regulate the transcription of the operon implicated in biofilm formation in both Xf and *Agrobacterium tumefaciens*. This gene is restricted to a few plant-associated bacteria and is located in an operon along with membrane proteins, and an unusual β-lactamase-like hydrolase (BLH) (Barbosa and Benedetti, 2007; Barbosa *et al.*, 2007). These authors elegantly demonstrated the BigR regulatory role upon the promoter region of the *blh* gene when fused to an EGFP reporter gene in *Escherichia coli* cells. In support of their results, is the fact that *bigR* is down-regulated under heat shock conditions while the neighbouring genes in the operon are expressed (Koide *et al.*, 2006).

Further studies focusing on Xf growth established that there is not only a difference in the expression profile of *Xylella* cells, but also that virulence is affected depending on whether cells are contained in a biofilm or are in a planktonic state (Newman *et al.*, 2004; Guilhabert and Kirkpatrick, 2005; Meng *et al.*, 2005; Souza *et al.*, 2006). The results suggest that in the biofilm state the cells are less virulent than the planktonic state, which is somewhat contradictory to the initial expectations where biofilm formation was concluded to be important in clogging the vessels. New data are needed to clearly define if *Xylella* strains can be found in two distinct 'physiological states' *in planta* i.e. one population of cells in a biofilm community that expresses anti-virulence genes (such *rpfF*, haemagglutinin, type I pili) resulting in latent infection without disease development, while planktonic cells move in all directions – acropetally, basipetally and laterally – resulting in less aggregation and disease development. Together these results point to the possibility that *Xylella* cells undergo a programmed variation in terms of lifestyle within the plant vessels and this could trigger the disease phenotype.

Minimal genome set

The chromosomal backbone of all genomes known to date is essentially structured as a sequence of genes that will, under different stimuli, be expressed to maintain the basic functions of life: metabolism and reproduction. The advent of whole genome sequencing first completed for *Haemophilus influenzae* in 1995 (Fleischmann *et al.*, 1995) established the basis for determining what the oldest common piece of DNA sequence in a chromosome of a given organism is in relation to other closely related organisms or clades. This led to the field of phylogenomic reconstructions (Sicheritz-Ponten and Andersson, 2001; Charlebois *et al.*, 2003; Eisen and Fraser, 2003; Snel *et al.*, 2005; Jeffroy *et al.*, 2006; Luo *et al.*, 2006) and to visualization of the dynamics of bacterial genome evolution. Also, comparative studies based on gene content and chromosomal gene order between completely sequenced genomes is a strategy to analyse whole genome evolution among closely related species (Moreira *et al.*, 2004; Zhou *et al.*, 2004; Vasconcelos *et al.*, 2005; Toth *et al.*, 2006). Results from these analyses and other studies (Pallen and Wren, 2007; Sirand-Pugnet *et al.*, 2007), revealed that most of these organisms harbour genetic novelties in stretches of the chromosome backbone most similar to unrelated species. These DNA stretches are considered to have resulted from horizontal gene transfer (HGT) events usually mediated by mobile genetic elements such as viruses, plasmids and transposable elements (Pallen and Wren, 2007). The genomes of *Xylella* and *Xanthomonas* include distinct pools of mobile elements, as previously described (Monteiro-Vitorello *et al.*, 2005). Lima *et al.* (2005) identified 40 regions of clustered gene islands in two sequenced *Xanthomonas* genomes (XAC and XCC) that have unusual best BLAST matches, which could be the result of HGT events. Of these islands, some are shared with *Xylella* and probably represent ancestral invasions that were kept in the group.

An alternate approach to study the history of the chromosome backbones of closely related species, such as the main forces driving genome differentiation (gene gain, gene loss or gene change), is to identify a common group of genes defined as the minimal genome set. To address the minimal genome content in *Xylella* regarding the Xanthomonadales and other selected bacteria, we undertook a BBH strategy in which individual genes from a group of organisms were compared to all other genes from the selected group. The strategy aims to establish the best bidirectional match among all genes to further organize them into clusters. Clustering was based on NCBI's BlastP program, and all the comparative analysis was performed using SABIA program (Almeida *et al.*, 2004). For the BBH approach, two genes – Xa of genome A and Xb of genome B – are considered as a BBH if, and only if, Xb is the best match for Xa (as shown by BLASTing Xa against genome B) and Xa is the best match for Xb (as shown by BLASTing Xb against genome A). All BLASTP searches considered an e-value cut-off E=1e-5 and query coverage of 60%. Once the BBH pairs were determined, we built the similarity clusters considering that a cluster is a set of genes in which every gene is a BBH with at least another one. There are two ways of checking the clustering database: either by checking *Search Genes in Cluster* or by checking *Clustering Statistics*. The two options are available under Comparative Genome on the *X. fastidiosa* genome project's website (http://www.xylella.lncc.br).

The start point of this analysis was the re-annotation of the *X. fastidiosa* strain 9a5c genome, and the exportation of the re-annotation results to the other three sequenced *Xylella* genomes (Xf-Temecula-1, Xf-Ann-1 and Xf-Dixon) in order to generate a complete and normalized annotation for the four strains. Start codons were manually inspected and adjusted for the genomes of all four strains using SWISSPROT information whenever available. Genes of the four *Xylella* strains found in the same cluster were defined as orthologues (Xf-9a5c, Xf-Ann-1, Xf-Dixon, and Xf-Temecula-1) with exceptions made for some particular gene families, such as integrases. Thirteen genomes were also downloaded and incorporated into the database. These were *Caulobacter crescentus* CB15 (Nierman et al., 2001), *Erwinia carotovora* subsp. *atroseptica* SCRI1043 (Bell et al., 2004), *E. coli* K12 (Blattner et al., 1997), *E. coli* O157:H7 (Hayashi et al., 2001), *Pseudomonas aeruginosa* PAO1 (Stover et al., 2000), *Pseudomonas syringae* pv. *tomato* str. DC3000 (Buell et al., 2003), *Ralstonia solanacearum* GMI1000 (Salanoubat et al., 2002), *S. maltophilia* R551-3, *Synechocystis* sp. PCC 6803 (Kaneko et al., 1995), *X. citri* subsp. *citri* str. 306 (da Silva et al., 2002), *Xanthomonas campestris* pv. *campestris* str. ATCC 33913 (da Silva et al., 2002), *X. campestris* pv. *vesicatoria* str. 85-10 (Thieme et al., 2005), and *X. oryzae* pv. *oryzae* KACC10331 (Lee et al., 2005). The selection of these organisms was based on their known phylogenetic relationship to the group because they are found in the rhizosphere of plants or because they are model systems, such as *E. coli* K12 and OH157:H7. All these genomes are available in GenBank.

The re-annotated Xf-9a5c genome database is composed of 2424 genes where the minimum size to annotate a CDS without database homologues was set as 70 bp. Of the total genes, 1876 presented a defined COG (77.39%), distributed in 25 possible functional categories. Of those, 1298 were found in all Xanthomonadales analysed in this study representing 52% of the Xf-9a5c putative coding sequences. Based on similar analyses outside the Xanthomonadales clade, *Synechocystis* PCC6803 contains only 24% of the genes in clusters supporting its distant phylogenetic position as a cyanobacterium. *Synechocystis* clustered genes are mainly related to basal metabolism. Accordingly, the free living, non-plant associated bacteria, *C. crescentus* and *E. coli* strains, follow *Synechocystis*, each with 35% and 39%, respectively. No particular deviation regarding clustering results was identified when the other plant pathogens outside from Xanthomonadales were compared.

The 2424 genes in *X. fastidiosa* strain 9a5c clustered into 1190 BBHs, thereby characterizing the minimal Xanthomonadales core genome composition based on the species focused on here. Luo et al (2006) clustered the genome of 13 γ-Proteobacteria in search of rare genomic markers and found that the Xf strain 9a5c genome possesses 1269 genes in clusters. However, they used a less stringent cut-off (E=1e-4 and query coverage of 40%) than was used here. The difference of 79 clusters observed between the two studies is most probably due to the distinct criteria used and the fact that CDSs smaller than 70 bp with no matches in GenBank were excluded from the work presented here.

Among the 1298 genes (grouped into 1190 BBHs) used to build the minimal Xanthomonadales genome, 555 were also found in other organisms such as *Erwinia*, *Pseudomonas*, *Escherichia*, *Caulobacter* and *Ralstonia* and those formed 490 BBH clusters. The most conserved functions among the above mentioned organisms are basal cellular processes related to translation, energy production and conversion as well as transport and metabolism of nucleotides, amino acids and lipids (Fig. 5.3). Thus, these shared clusters represent some of the ancestral bacteria functions.

Pallen and Wren (2007) suggest that there are three main forces driving genome innovation that ultimately lead to speciation and family definition. Innovation is dependent on shuffling the gene pool by gene gain (duplication and HGT), gene loss or gene change (SNPs and INDELs). The remaining 700 clusters shared by the Xanthomonadales is probably the result of these forces. The distribution of the *X. fastidiosa*

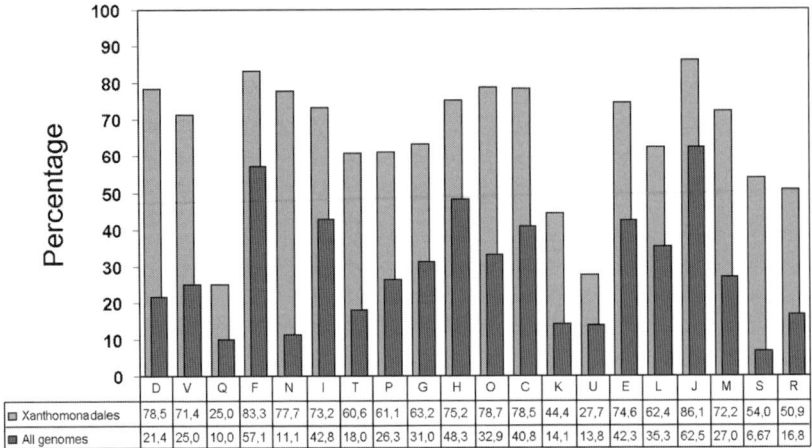

Figure 5.3 Minimal genome set defined by BBH clustering strategy among Xanthomonadales and other selected organisms. BBH clusters were categorized by COG using *Xylella fastidiosa* strain 9a5c as a reference.

strain 9a5c clustered genes in functional categories is presented in Fig. 5.3. Except for categories K, U and Q, at least 60% of the clustered genes are shared among all Xanthomonadales. The least conserved clusters (Q and U) correspond to biosynthesis, transport and catabolism of secondary metabolites and intracellular trafficking and secretion. These two categories probably reflect the ecologically distinctive niche occupied by three different lifestyles: free-living opportunists (*Stenotrophomonas*), free living pathogens (*Xanthomonas*), and insect gut and plant xylem restricted pathogens (*Xylella*). Interestingly, genes clustered in the cell wall/membrane biogenesis (M) and cell motility (N) categories are mostly shared by the three groups, suggesting that these are distinctive features in the group.

To gain insight about the chromosomal distribution of these clustered genes, a genome map was generated in which specific *Xylella* matching clusters (grey) are positioned as well as the Xanthomonadales (yellow) and other bacteria (red) matching clusters (Fig. 5.4). This survey revealed that the Xanthomonadales conserved genes identified in the present clustering analyses are found grouped, supporting their common ancestrality. Ten of these chromosomal regions have been identified as genomic islands in *Xanthomonas* and correspond to islands 4, 6, 12, 16–20, 22 and 25 as previously defined (Lima et al., 2008). These islands were defined based on their unusual GC content, codon bias and probably originated via HGT events into the ancestral genome of the group before speciation.

Careful inspection and categorization of the genes of the other syntenic regions was surveyed and some categories were found to be represented. The first one is related to genes involved in the metabolism of amino acids (mainly leucine, isoleucine and valine), nucleotides (almost exclusively purine biosynthesis) and cofactors (biotin, folic acid, coenzyme A and NAD). It is also worth noting the presence of several genes involved in the metabolism of LPS (lipopolysaccharide) and a large cluster of genes related to biogenesis of cytochromes. As expected, several genes are involved with the metabolism of nucleic acids and proteins, mainly with DNA repair and recombination, RNA modification and transcription, ribosomal proteins, and to a lesser extent related to protein modification. Finally, there are also some genes involved with detoxification and adaptation to stress including *sodA* (involved in detoxification of superoxide radicals), *ohr* (a gene first identified in *Xanthomonas*, coding for an organic hydroperoxide detoxification protein), *msrA* (responsible for the repair of proteins inactivated by oxidation) and *acrAB* (coding for a drug efflux protein) (the list is available upon request).

The organic hydroperoxide resistance proteins (Ohr) are enzymes with a thiol-dependent peroxidase activity and involved in detoxification of organic hydroperoxides. Although first described in *X. campestris* pv. *phaseoli* (Mongkolsuk *et al.*, 1998), this enzyme is present in all Xanthomonadales species, and its tridimensional structure was recently described for *X. fastidiosa* strain 9a5c (XF1562; PDB entries 1ZB9 and 1ZB8) (Oliveira *et al.*, 2006). These authors further demonstrated that, in *Xylella*, Ohr catalyses the following reaction: 2RSH+ROOH→RSSR+ROH+H2O, in the presence of dithiols.

Interestingly, only a few genes involved in uptake of small molecules from the environment are found in *Xylella* compared to *Xanthomonas* (Moreira *et al.*, 2005), where more than 60 genes are annotated as iron transporters and iron chelating molecules. In contrast, almost 25% of the 1126 genes in *Xylella* (but not in *Xanthomonas*) strains are located in phage-related regions with 7% located in phage remnants (category R). *X. fastidiosa* strain 9a5c harbours two interesting regions corresponding to 10% of the strain-specific genes. These are located in the previously described giCVC genomic island (Simpson *et al.*, 2000; Van Sluys *et al.*, 2002) and

the region between XF1746 and XF1800 genes, which most probably corresponds to a plasmid integration site, as are genes involved in type IV secretion system most closely related to *tra* genes and two copies of a DNA invertase gene are present.

Taken together over 60% of the *Xylella* genome is shared with other Xanthomonadales while 35% is shared among Bacteria (excluding *Synechocystis*). As predicted, the minimal genome set among *Xylella* and Bacteria correspond to common housekeeping functions such as energy metabolism, co-enzyme, amino acid and nucleotide metabolism and translation. These genes are mainly found scattered on the chromosome backbone with a few exceptions, such as the NADH complex, the RNA polymerase, EF-Tu and some ribosomal proteins. The minimal genome set for Xanthomonadales is more diverse with respect to gene function and gene order is mostly maintained within biosynthetic pathways, structural genes and some operons. At a higher level of chromosomal organization, these conserved blocks of DNA are shuffled basically by the presence of intervening sequences usually of mobile genetic element (MGE) origin.

tRNAs and phage-related regions locate at chromosomal break points

Xylella genomes carry several prophage regions and remnants of old insertions that are mostly rearranged within themselves among the sequenced genomes of the four strains while *Xanthomonas* genomes were invaded several times independently by a large group of transposons that are similar to IS elements. Expansion of any of these elements is dependent on the balance between amplification and cell viability.

Global chromosomal alignment was applied to the finished genome sequence of *X. fastidiosa* strain 9a5c and strain Temecula-1 and candidate genome molecules for strains Ann-1 and Dixon in order to identify syntenic regions and further evaluate the ancestral chromosome backbone. The results are presented schematically in Fig. 5.5. All *Xylella* strains share extensive synteny along the chromosome except for the phage-related regions (small black triangles) that are coincident with the chromosome breaks in the

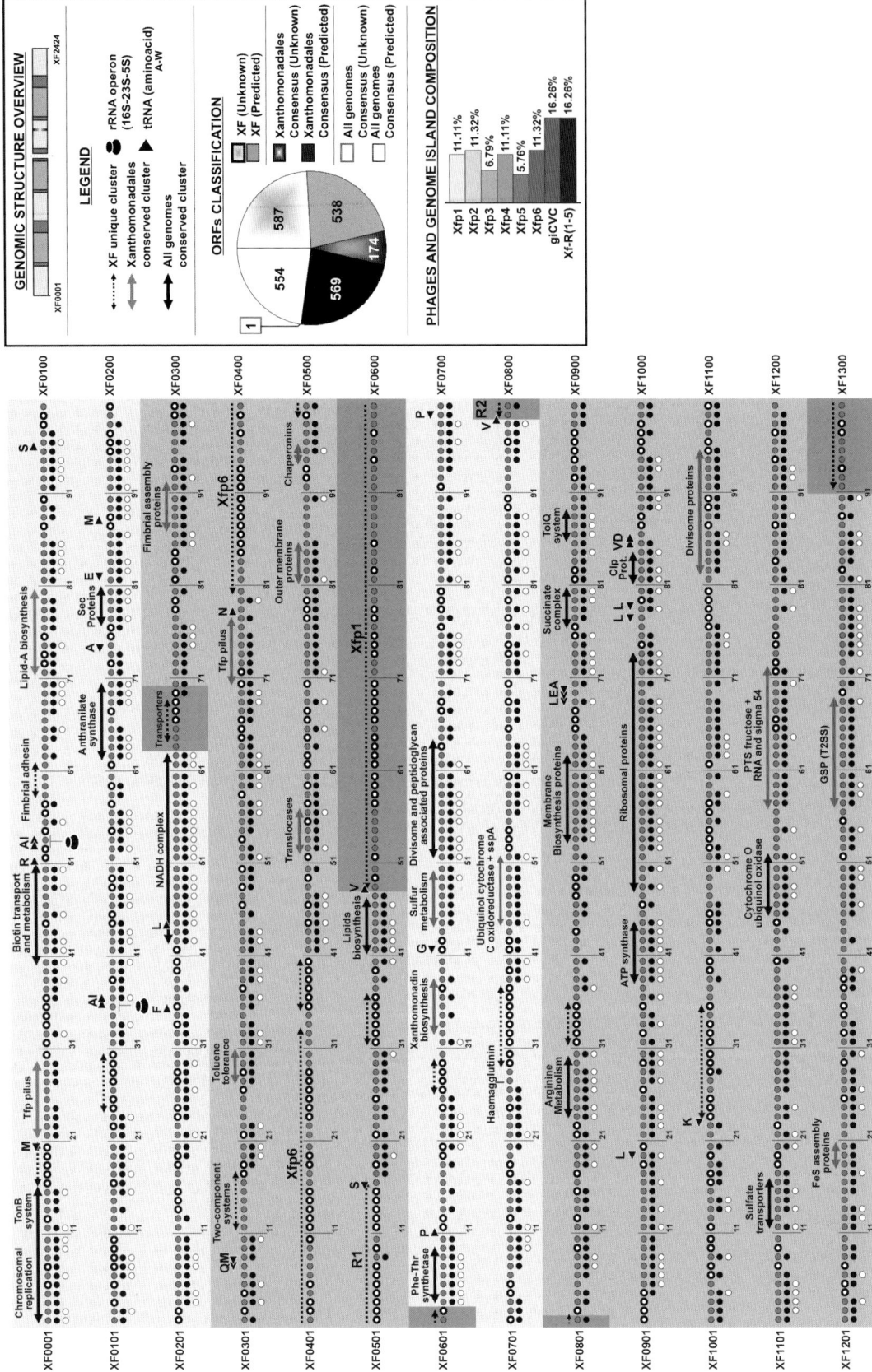

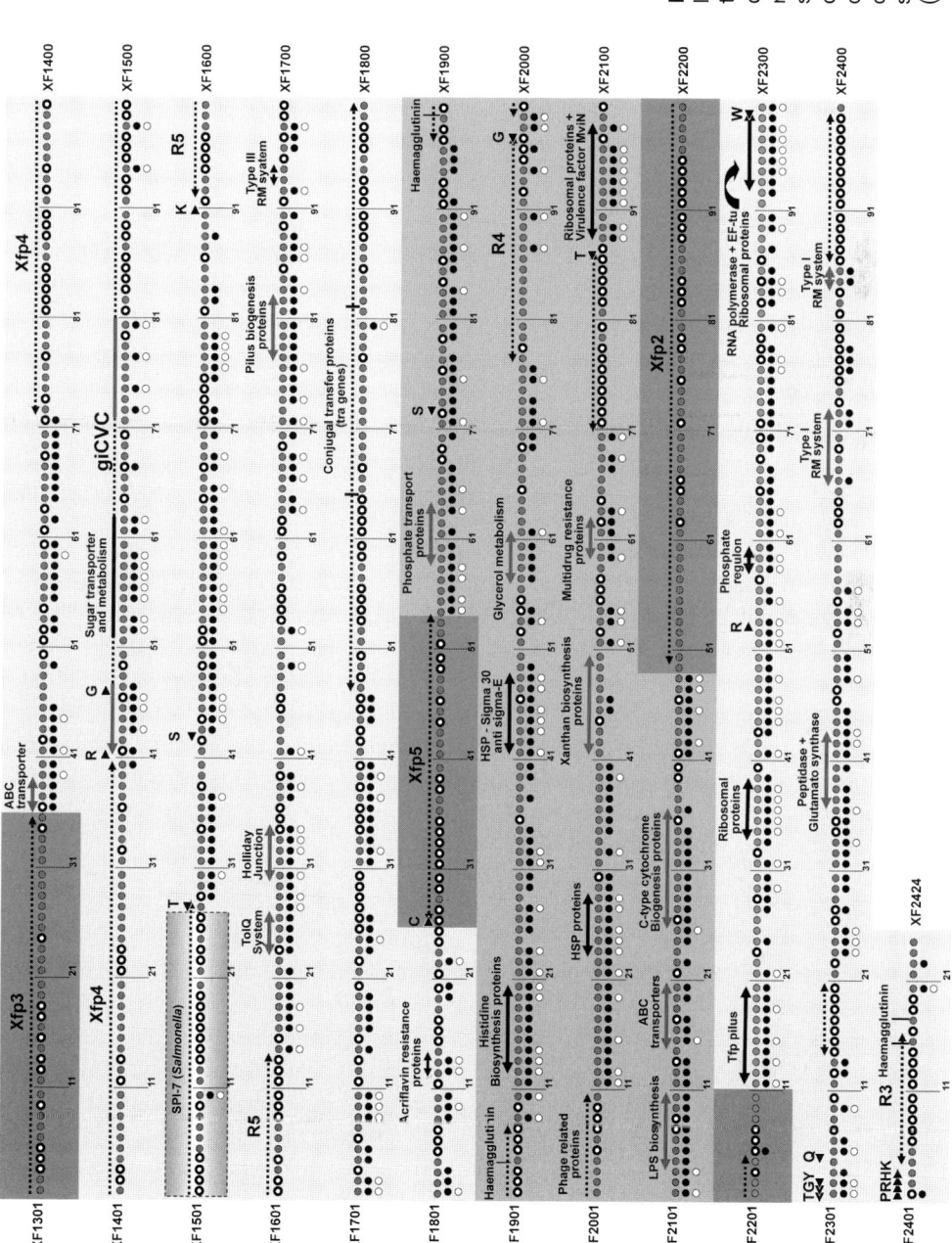

Figure 5.4 Schematic genome map illustrating the distribution of BBH clustered genes on the *Xylella fastidiosa* 9a5c chromosome backbone. *Xylella* unique regions (grey dots), Xanthomonadales shared genes (yellow dots) and genes conserved among all the organisms (red dots) studied here. Background colours correspond to the syntenic chromosomal regions as depicted previously (Van Sluys *et al.*, 2003).

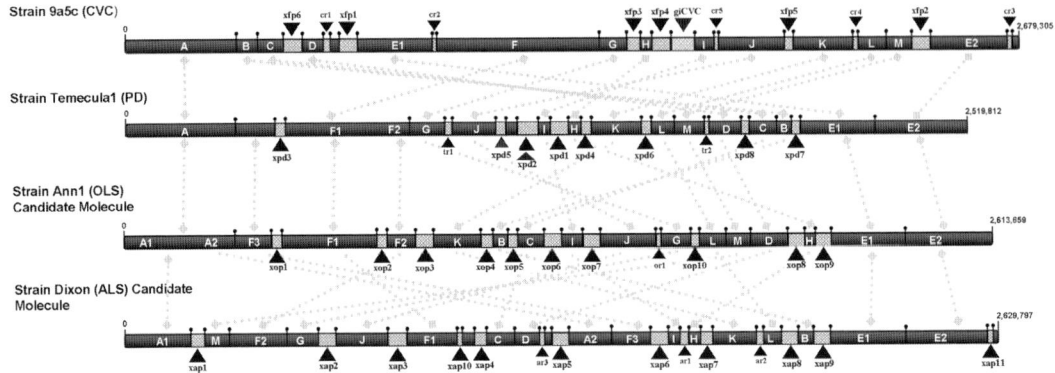

Figure 5.5 Schematic representation of chromosome alignment of the genomes of four *Xylella fastidiosa* strains from their predicted origin of replication. The letters (A-M) illustrated in the figure depict the chromosome backbone of the genomes, showing the relative position and size of collinear chromosome regions detected among the four strains. Large triangles illustrate the position of prophage-like and GI regions, and small black triangles illustrate the prophage remnants.

alignment. The four sequenced Xf genomes have 17 syntenic chromosomal regions (Table 5.2) and breaks/rearrangements generally map in association with phage-related regions. Altogether, phage-related regions correspond to between 5–10% of the *Xylella* chromosomes.

Mobile genetic elements (plasmids, phages and transposable elements) are considered important mechanisms in driving gene diversity through HGT events in prokaryotes. Except for plasmids, phages and transposons depend on enzymatic activity to fulfil integration. The actual DNA integration is guaranteed by enzymes able to generate breaks in the DNA molecule and depending on the enzymatic mechanisms these are usually named integrases, recombinases or transposases. Several reports describe the association of genomic islands with the presence of mobile elements and try to trace their origin applying phylogenetic tools on the integrative protein and the genes at their vicinity (Lima *et al*., 2005; Mantri and Williams, 2007). Accordingly, Monteiro-Vitorello and colleagues (2005) described the *Xylella* and *Xanthomonas* mobilomics and their association with the influx of new genes. *Xanthomonas* spp. have a profuse diversity of IS elements and almost no prophages. In addition, some of these elements border unique traits such as avr genes (Leach *et al*., 1992; da Silva *et al*., 2002; Hu *et al*., 2007). In contrast to the different xanthomonads, the *Xylella* chromosome was invaded by distinct bacteriophages from *Caudovirales* and these insertions possibly provided the substrate for recombination/rearrangement.

Bacteriophages are widely distributed among prokaryotic organisms, and influence genomic evolution and adaptive capabilities of their hosts. Prophages thus constitute, in many bacterial species, a substantial part of laterally acquired DNA. Fig. 5.5 schematically displays the four *Xylella* genomes illustrating the prophage-related regions and the phage remnants. *X. fastidiosa* strain 9a5c (Xf-CVC) contains six prophage-like regions and five phage remnant regions (six prophage-like regions, named xfp1 to xfp6, and xfp1 to xfp4 were previously reported by Simpson *et al*., 2000); the *X. fastidiosa* strain Temecula-1 (Xf-PD) contains eight prophage-like (called xpd1 to xpd8) regions and two small phage remnant regions (the eight prophage-like regions were previously reported by Van Sluys *et al*., 2003); the *X. fastidiosa* strain Ann-1 (Xf-OLS) candidate molecule contains ten prophage-like (called xop1-xop10) regions and just one phage remnant region; and the *X. fastidiosa* strain Dixon (Xf-ALS) candidate molecule contains eleven prophage-like (called xap1 -xap11) regions and three phage remnants.

Integration of bacteriophages is usually mediated by recombination events targeted to tRNAs through the activity of a phage-related integrase. These enzymes recognize conserved DNA sequences found in tRNAs. Further analyses were performed based on the tRNA

Table 5.2 Syntenic blocks amongst the genomes of four *Xylella fastidiosa* strains. IDs as defined in Simpson *et al.* (2000)

Syntenic bloc	Initial ORF	Final ORF	Chromosome disruption caused by phages?
A1	XF0001	XF0169	xap1
A2	XF0171	XF0324	
B	XF0338	XF0384	xap9, xop5, xpd7
C	XF0387	XF0478	xop6, xfp6
D	XF0549	XF0677	xap5, xop8, xpd8, xfp1
E1	XF0736	XF0967	
E2	XF2534	XF2780	
F1	XF0973	XF1082	xap10, xap4, xop2
F2	XF1084	XF1304	xop3,
F3	XF1310	XF1486	xap6, xop1, xpd3,
G	XF1487	XF1554	xap2, xop10, xfp3
H	XF1601	XF1641	xap7, xop9, xpd4, xfp4
I	XF1794	XF1849	xop7, xpd1,
J	XF1886	XF2106	xap3, xpd5, xfp5
K	XF2133	XF2287	xop4, xpd6,
L	XF2324	XF2395	xap8,
M	XF2416	XF2476	xfp2

content and similarity across four genomes, two *Xylella* genomes (Xf-CVC, Xf-PD) and two *Xanthomonas* genomes (XAC and XCV). All sequenced *Xylella* genomes posses 49 tRNAs (Simpson *et al.*, 2000; Bhattacharyya *et al.*, 2002a; Van Sluys *et al.*, 2003) while XAC has 54 (da Silva *et al.*, 2002) and XCV has 55 (Thieme *et al.*, 2005). Nucleotide sequence divergence ranges from 0 to 12% and seven unique tRNAs were identified in *Xanthomonas* genomes while *Xylella* genomes have two. None of these unique tRNAs map on the SMA genome. Furthermore, most of the prophage-like regions (72%) found in *Xylella* genomes are associated with tRNAs as insertion sites. Thus, tRNA sequences are interesting tools for studying the impact of phage on the chromosomal genome evolution.

The chromosomal location and nucleotide identity of *Xylella* tRNAs, expressed in % towards XAC and XCV, are listed in Table 5.3. Within the genus, tRNA identity was 100%, except in four cases where the identity is 97–98% between Xf-PD and Xf-CVC. In contrast, there are 28 tRNA sequences displaying >95% nucleotide identity between the most divergent *Xylella* and the two xanthomonads (XAC and XCV) analysed. There are 13 sequences with identities ranging from 90% to 94.99% and the remaining 5 tRNAs are 63.53% (Leu), 83.64% (Ser), 88.16% (Lys), 88.89% (Thr) and 89.47% (Met), each representing the existing redundancy of the genetic code for a given amino acid. It is not known if these latter comparisons represent decaying tRNAs or fast evolving sequences. Alternatively, these could be indicative of distinct evolutionary pieces of the chromosome backbone. Interestingly, three of these less conserved tRNAs are adjacent to a phage-remnant region (Xf-remnant 1 and 5) or the island similar to SPI-7 from *Salmonella* described previously (Pickard *et al.*, 2003; Moreira *et al.*, 2004). The most conserved tRNAs lie embedded in the 'all genome' BBH containing regions as can be observed in the map representation (Fig. 5.3). Other than the gene conservation found by the BBH approach and the minimal genome set, this review extends the conservation concept to a common ancestral origin of most of the chromosome backbone in Xanthomonadales. Forces driving genome variation in this group are highly determined by gene gain and gene loss rather then gene variation. MGEs are major players in both gene gain and

Table 5.3 tRNAs in Xf-CVC/Xf-PD and XCC/XCV genomes, their position on the corresponding genome and the identity percentage amongst Xf-CVC/PD and XCC/XCV tRNAs

tRNA	Xf-CVC	Xf-PD	XCC	XCV	% identity
Met-CAT1	20945	20804	577886	590861	89.47
Arg-CCG	62407	61898	462225	459479	94.80
Ala-TGC1	68209	67692	4559959	4664466	98.68
Ala-TGC2	–	–	4669357	1229693	–
Ile-GAT1	68298	67781	4559865	4664372	100.00
Ser-GCT	123996	122451	2004911	1999469	94.62
Ala-TGC3	173925	172775	5009201	5008645	98.68
Ile-GAT2	174014	172864	5009107	5008551	100.00
Ala-CGC	224513	223245	–	–	
Glu-CTC	236114	234838	2002391	2006870	93.33
Met-CAT2	242842	241438	3110387	3221642	100.00
Phe-GAA*	308643	307078	3009111	3118799	93.42
Leu-GAG	319058	317497	3115578	3226853	97.65
Gln-CTG	405834	1994587	577757	590732	93.33
Met-CAT3	405956	1994465	4445163	4440110	98.70
Asn-GTT	486925	1881069	3003839	3221684	100.00
Ser-TGA	615973	1772646	1998540	1992626	83.64
Val-CAC	649104	1778264	1224101	1221512	96.00
Pro-GGG	699322	2229563	3007985	3118701	96.10
Gly CCC	732147	2220084	3887789	3991347	98.65
Pro-CGG	816600	2114945	2773586	2882478	100.00
Ala-CGC	-	–	2888302	2999185	
Val-GAC*	926649	2000502	3000979	3118758	98.67
Leu-CAA	1007939	431266	2443460	2559134	96.55
Glu-TTC1	1008144	431472	2444461	2554496	96.05
Ala-GGC1	1008273	431601	2444584	2554619	90.79
Glu-TTC2	–	–	2445418	2555444	
Ala-GGC2	–	–	2445540	2555567	
Leu-CAG1	1008913	506276	1661012	1666635	90.59
Leu-CAG2	–	–	1889221	1886615	
Leu-TAA	1111319	558377	2448927	2771396	91.95
Leu-TAG	1115824	562897	1221193	1222077	63.53
Val-TAC*	1112842	569915	1228015	1228900	98.67
Asp-GTC1	1112947	570020	1228118	1229003	100.00
Asp-GTC2	–	–	1228289	1229174	
Lys-TTT2	1111317	1112638	–	–	
Arg-CCT	1661333	1227175	2003089	2004889	100.00
Gly GCC1	1668644	1229860	2443657	2661647	90.14
Thr-TGT	1777580	1225939	4776230	4773337	88.89
Ser-GGA	1770323	1223216	1229673	1227426	100.00
Lys-TTT1	1776435	617940	3669881	3776101	88.16
Cys-GCA	2004768	953053	2443502	2661492	91.89

Table 5.3 *continued*

tRNA	Xf-CVC	Xf-PD	XCC	XCV	% identity
Ser-CGA	2 007 093	1 447 310	1 226 290	1 226 946	95.24
Gly GCC2	2 114 005	1 556 021	2 443 358	2 553 428	98.68
Thr-CGT	2 225 229	1 660 178	1 445 449	1 441 062	92.11
Arg-ACG1	2 449 885	2 333 865	2 001 560	2 006 127	97.40
Arg-ACG2	–	–	2 001 732	2 006 305	
Trp-CCA	2 554 186	2 338 491	1 117 364	1 116 198	98.61
Thr-GGT	2 555 616	2 339 922	1 115 941	1 114 775	97.37
Gly TCC	2 555 732	2 330 038	1 115 834	1 114 668	98.65
Tyr-GTA	2 555 851	2 330 157	1 115 719	1 114 553	98.84
Gln-TTG	2 559 679	2 333 982	1 111 817	1 110 651	94.81
Pro-TGG*	2 662 606	2 449 453	1 229 044	1 229 693	98.70
Arg-TCT	2 662 744	2 449 593	1 229 163	1 229 812	92.21
His-GTG	2 662 862	2 449 711	1 229 272	1 229 921	97.40
Lys-CTT	2 663 049	2 449 898	1 229 179	1 220 091	100.00

*tRNAs not 100% identical among PD and CVC strains.

loss through site-specific recombination, and illegitimate recombination events. Changes in gene regulatory networks are less considered in these analyses, but may be crucial determinants of phenotypic variation among closely related species. Identification of regulatory networks are now coming into the genome scene and demand that the work developed by de Souza *et al.* (2004, 2005), Koide *et al.* (2006) and da Silva Neto *et al.* (2007a,b) be consolidated to provide the basis for identifying differences related to pathogenesis and host specificity among *X. fastidiosa* strains. Nunes *et al.* (2003) provided evidence that *X fastidiosa* CVC-inducing strains diversify mainly at their phage-related regions both in gene content as well as in their expression profile.

Final comments

The analysis of syntenic regions and core backbone in Xanthomonadales genomes enabled our ability to extend conservation beyond genes and the striking differences in genome size and genome structure among the genera does not help to draw a clear evolutionary history for the group. Based on the rDNA tree, it is worth noting the relative recent speciation inside each genus/genera, although the core chromosome backbone is already shared by the ancestral species. It seems that, after the first speciation inside the Xanthomonadales group, it took a long time for further speciation to occur or that the ancestral species was exposed to a recurrent highly infectious environment composed of diverse MGEs. Several lines of evidence described in this review support the hypothesis that the current differences between Xanthomonadales chromosome backbone occurred recently in time and is the result of bacterial lines that remained susceptible to phage infections (*Xylellas*' genomes) while other lines gained immunity to phage infection and became 'obese' by capturing new genes associated with IS elements and conjugative plasmids (*Xanthomonas*' genomes). It is tempting to speculate that several of the previously described virulence genes were already present in the ancestral species and that *Xylella* and *Xanthomonas* acquired their plant pathogenicity capability mainly through distinct gene acquisition events.

Acknowledgements

The authors would like to thank the financial support from FAPESP grant to M.A.V.S., CNPq grants to A.T.R.V. and M.A.V.S. and USDA-ARS grant to M.C.O. and M.A.V.S. A.M.V is the recipient of a CAPES fellowship and W.L. of a FAPESP fellowship.

References

Almeida, L.G., Paixão, R., Souza, R.C., Costa, G.C., Barrientos, F.J., Santos, M.T., Almeida, D.F. and Vasconcelos, A.T. (2004). A System for Automated Bacterial (genome) Integrated Annotation – SABIA. Bioinformatics 20, 2832–3.

Almeida, R.P., and Purcell, A.H. (2003). Transmission of *Xylella fastidiosa* to grapevines by *Homalodisca coagulata* (Hemiptera: Cicadellidae). J. Econ. Entomol. 96, 264–271.

Barbosa, R.L., and Benedetti, C.E. (2007). BigR, a transcriptional repressor from plant-associated bacteria, regulates an operon implicated in biofilm growth. J. Bacteriol. 189, 6185–6194.

Barbosa, R.L., Rinaldi, F.C., Guimaraes, B.G., and Benedetti, C.E. (2007). Crystallization and preliminary X-ray analysis of BigR, a transcription repressor from *Xylella fastidiosa* involved in biofilm formation. Acta Crystallogr. Sect. F Struct. Biol. Cryst. Commun. 63, 596–598.

Bell, K.S., Sebaihia, M., Pritchard, L., Holden, M.T., Hyman, L.J., Holeva, M.C., Thomson, N.R., Bentley, S.D., Churcher, L.J., Mungall, K., Atkin, R., Bason, N., Brooks, K., Chillingworth, T., Clark, K., Doggett, J., Fraser, A., Hance, Z., Hauser, H., Jagels, K., Moule, S., Norbertczak, H., Ormond, D., Price, C., Quail, M.A., Sanders, M., Walker, D., Whitehead, S., Salmond, G.P., Birch, P.R., Parkhill, J., Toth, I.K. (2004). Genome sequence of the enterobacterial phytopathogen *Erwinia carotovora* subsp. *atroseptica* and characterization of virulence factors. Proc. Natl. Acad. Sci. USA 101, 11105–10.

Bhattacharyya, A., Stilwagen, S., Ivanova, N., D'Souza, M., Bernal, A., Lykidis, A., Kapatral, V., Anderson, I., Larsen, N., Los, T., Reznik, G., Selkov, E. Jr., Walunas, T.L., Feil, H., Feil, W.S., Purcell, A., Lassez, J.L., Hawkins, T.L., Haselkorn, R., Overbeek, R., Predki, P.F., Kyrpides, N.C. (2002a). Whole-genome comparative analysis of three phytopathogenic *Xylella fastidiosa* strains. Proc. Natl. Acad. Sci. U.S.A. 99, 12403–12408.

Bhattacharyya, A., Stilwagen, S., Reznik, G., Feil, H., Feil, W.S., Anderson, I., Bernal, A., D'Souza, M., Ivanova, N., Kapatral, V., Larsen, N., Los, T., Lykidis, A., Selkov, E. Jr., Walunas, T.L., Purcell, A., Edwards, R.A., Hawkins, T., Haselkorn, R., Overbeek, R., Kyrpides, N.C., Predki, P.F. (2002b). Draft sequencing and comparative genomics of *Xylella fastidiosa* strains reveal novel biological insights. Genome Res. 12, 1556–1563.

Blattner, F.R., Plunkett, G. 3rd, Bloch, C.A., Perna, N.T., Burland, V., Riley, M., Collado-Vides, J., Glasner, J.D., Rode, C.K., Mayhew, G.F., Gregor, J., Davis, N.W., Kirkpatrick, H.A., Goeden, M.A., Rose, D.J., Mau, B., Shao, Y. (1997). The complete genome sequence of *Escherichia coli* K-12. Science 277, 1453–74.

Buell, C.R., Joardar, V., Lindeberg, M., Selengut, J., Paulsen, I.T., Gwinn, M.L., Dodson, R.J., Deboy, R.T., Durkin, A.S., Kolonay, J.F., Madupu, R., Daugherty, S., Brinkac, L., Beanan, M.J., Haft, D.H., Nelson, W.C., Davidsen, T., Zafar, N., Zhou, L., Liu, J., Yuan, Q., Khouri, H., Fedorova, N., Tran, B., Russell, D., Berry, K., Utterback, T., Van Aken, S.E., Feldblyum, T.V., D'Ascenzo, M., Deng, W.L., Ramos, A.R., Alfano, J.R., Cartinhour, S., Chatterjee, A.K., Delaney, T.P., Lazarowitz, S.G., Martin, G.B., Schneider, D.J., Tang, X., Bender, C.L., White, O., Fraser, C.M., Collmer, A.. (2003). The complete genome sequence of the Arabidopsis and tomato pathogen *Pseudomonas syringae* pv. *tomato* DC3000. Proc. Natl. Acad. Sci. U.S.A. 100, 10181–6.

Catani, C.F., Azzoni, A.R., Paula, D.P., Tada, S.F., Rosselli, L.K., de Souza, A.P., and Yano, T. (2004). Cloning, expression, and purification of the virulence-associated protein D from *Xylella fastidiosa*. Protein Expr. Purif. 37, 320–6.

Charlebois, R.L., Beiko, R.G., and Ragan, M.A. (2003). Microbial phylogenomics: branching out. Nature 421, 217.

da Silva, A.C., Ferro, J.A., Reinach, F.C., Farah, C.S., Furlan, L.R., Quaggio, R.B., Monteiro-Vitorello, C.B., Van Sluys, M.A., Almeida, N.F., Alves, L.M., do Amaral, A.M., Bertolini, M.C., Camargo, L.E., Camarotte, G., Cannavan, F., Cardozo, J., Chambergo, F., Ciapina, L.P., Cicarelli, R.M., Coutinho, L.L., Cursino-Santos, J.R., El-Dorry, H., Faria, J.B., Ferreira, A.J., Ferreira, R.C., Ferro, M.I., Formighieri, E.F., Franco, M.C., Greggio, C.C., Gruber, A., Katsuyama, A.M., Kishi, L.T., Leite, R.P., Lemos, E.G., Lemos, M.V., Locali, E.C., Machado, M.A., Madeira, A.M., Martinez-Rossi, N.M., Martins, E.C., Meidanis, J., Menck, C.F., Miyaki, C.Y., Moon, D.H., Moreira, L.M., Novo, M.T., Okura, V.K., Oliveira, M.C., Oliveira, V.R., Pereira, H.A., Rossi, A., Sena, J.A., Silva, C., de Souza, R.F., Spinola, L.A., Takita, M.A., Tamura, R.E., Teixeira, E.C., Tezza, R.I., Trindade, S.M., Truffi, D., Tsai, S.M., White, F.F., Setubal, J.C., Kitajima, J.P. (2002). Comparison of the genomes of two *Xanthomonas* pathogens with differing host specificities. Nature 417, 459–463.

da Silva Neto, J.F., Koide, T., Abe, C.M., Gomes, S.L., and Marques, M.V. (2007). Role of sigma (54) in the regulation of genes involved in type I and type IV pili biogenesis in *Xylella fastidiosa*. Arch Microbiol, in press.

da Silva Neto, J.F., Koide, T., Gomes, S.L., and Marques, M.V. (2007). The single extracytoplasmic-function sigma factor of *Xylella fastidiosa* is involved in the heat shock response and presents an unusual regulatory mechanism. J. Bacteriol. 189, 551–60.

Davis, M.J., Purcell, A.H., and Thomson, S.V. (1978). Pierce's disease of grapevines: isolation of the causal bacterium. Science 199, 75–77.

de Souza, A.A., Takita, M.A., Coletta-Filho, H.D., Caldana, C., Yanai., G.M., Muto, N.H., de Oliveira, R.C., Nunes, L.R., and Machado, M.A. (2004). Gene expression profile of the plant pathogen *Xylella fastidiosa* during biofilm formation in vitro. FEMS Microbiol. Lett. 237, 341–53.

de Souza, A.A., Takita, M.A., Pereira, E.O., Coletta-Filho, H.D., and Machado, M.A. (2005). Expression of pathogenicity-related genes of *Xylella fastidiosa* in vitro and in planta. Curr. Microbiol. 50, 223–8.

Eisen, J.A., and Fraser, C.M. (2003). Phylogenomics: intersection of evolution and genomics. Science 300, 1706–1707.

Fleischmann, R.D., Adams, M.D., White, O., Clayton, R.A., Kirkness, E.F., Kerlavage, A.R., Bult, C.J., Tomb, J.F., Dougherty, B.A., Merrick, J.M., et al. (1995). Whole-genome random sequencing and assembly of *Haemophilus influenzae* Rd. Science 269, 496–512.

Fry, S.M., Huang, J-S., and Milholland, R.D. (1994). Isolation and preliminary characterization of extracellular proteases produced by strains of *Xylella fastidiosa* from grapevines. Phytopathology 84, 357–363.

Guilhabert, M.R., and Kirkpatrick, B.C. (2005). Identification of *Xylella fastidiosa* antivirulence genes: hemagglutinin adhesins contribute a biofilm maturation to *X. fastidiosa* and colonization and attenuate virulence. Mol. Plant–Microbe Interact. 18, 856–868.

Hauben, L., Vauterin, L., Moore, E.R., Hoste, B., and Swings, J. (1999). Genomic diversity of the genus *Stenotrophomonas*. Int. J. Syst. Bacteriol. 49, 1749–1760.

Hayashi, T., Makino, K., Ohnishi, M., Kurokawa, K., Ishii, K., Yokoyama, K., Han, C. G., Ohtsubo, E., Nakayama, K., Murata, T., Tanaka, M., Tobe, T., Iida, T., Takami, H., Honda, T., Sasakawa, C., Ogasawara, N., Yasunaga, T., Kuhara, S., Shiba, T., Hattori, M., and Shinagawa, H. (2001). Complete genome sequence of enterohemorrhagic *Escherichia coli* O157:H7 and genomic comparison with a laboratory strain K-12. DNA Res. 28, 8:11–22.

Hopkins, D.L., and Purcell, A.H. (2002). *Xylella fastidiosa*: cause of Pierce's disease of grapevine and other emergent diseases. Plant Disease 86, 1056–1066.

Hopkins, D.L. (1989). *Xylella fastidiosa*: xylem-limited bacterial pathogen of plants. Ann. Rev. Phytopathol. 27, 271–290.

Hu, J., Zhang, Y., Qian, W., and He, C. (2007). Avirulence gene and insertion element-based RFLP as well as RAPD markers reveal high levels of genomic polymorphism in the rice pathogen *Xanthomonas oryzae* pv. *oryzae*. Syst. Appl. Microbiol. 30, 587–600.

Jeffroy, O., Brinkmann, H., Delsuc, F., and Philippe, H. (2006). Phylogenomics: the beginning of incongruence? Trends Genet. 22, 225–231.

Jones, J.B., Lacy, G.H., Bouzar, H., Stall, R.E., and Schaad, N.W. (2004). Reclassification of the xanthomonads associated with bacterial spot disease of tomato and pepper. Syst. Appl. Microbiol. 27, 755–762.

Kaneko, T., Nakamura, Y., Sasamoto, S., Watanabe, A., Kohara, M., Matsumoto, M., Shimpo, S., Yamada, M., and Tabata, S. (2003). Structural analysis of four large plasmids harboring in a unicellular cyanobacterium, *Synechocystis* sp. PCC 6803. DNA Res. 10, 221–8.

Koide, T., Vencio, R.Z., and Gomes, S.L. (2006). Global gene expression analysis of the heat shock response in the phytopathogen *Xylella fastidiosa*. J. Bacteriol. 188, 5821–30.

Koide, T., Zaini, P.A., Moreira, L.M., Vencio, R.Z., Matsukuma, A.Y., Durham, A.M., Teixeira, D.C., El-Dorry, H., Monteiro, P.B., da Silva, A.C., Gomes, S.L., (2004). DNA microarray-based genome comparison of a pathogenic and a nonpathogenic strain of *Xylella fastidiosa* delineates genes important for bacterial virulence. J. Bacteriol. 186, 5442–5449.

Leach, J.E., Rhoads, M.L., Vera Cruz, C.M., White, F.F., Mew, T.W., and Leung, H. (1992). Assessment of genetic diversity and population structure of *Xanthomonas oryzae* pv. *oryzae* with a repetitive DNA element. Appl. Environ. Microbiol. 58, 2188–95.

Lee, B.M., Park, Y.J., Park, D.S., Kang, H.W., Kim, J.G., Song, E.S., Park, I.C., Yoon, U.H., Hahn, J.H., Koo, B.S., Lee, G.B., Kim, H., Park, H.S., Yoon, K.O., Kim, J.H., Jung, C.H., Koh, N.H., Seo, J.S., Go, S.J.. (2005). The genome sequence of *Xanthomonas oryzae* pathovar *oryzae* KACC10331, the bacterial blight pathogen of rice. Nucleic Acids Res. 33, 577–86.

Li, Y., Hao, G., Galvani, C.D., Meng, Y., De La Fuente, L., Hoch, H.C., and Burr, T.J. (2007). Type I and type IV pili of *Xylella fastidiosa* affect twitching motility, biofilm formation and cell–cell aggregation. Microbiology 153, 719–726.

Lima, W.C., Van Sluys, M.A., and Menck, C.F. (2005). Non-gamma-proteobacteria gene islands contribute to the *Xanthomonas* genome. OMICS 9, 160–172.

Lima, W.C., Paquola, A., Varani, A.M., Van-Sluys, M.A., and Menck, C.F. (2008). Laterally transferred genomic islands in Xanthomonadales related to pathogenicity and primary metabolism. FEMS Microbiol. Lett., in press.

Liolios, K., Tavernarakis, N., Hugenholtz, P., and Kyrpides, N.C. (2006). The Genomes On Line Database (GOLD) v.2, a monitor of genome projects worldwide. Nucleic Acid Res. 34, D332–334.

Looney, W.J. (2005). Role of *Stenotrophomonas maltophilia* in hospital-acquired infection. Br. J. Biomed. Sci. 62, 145–54.

Luo, Y., Fu, C., Zhang, D.Y., and Lin, K. (2006). Overlapping genes as rare genomic markers: the phylogeny of gamma-Proteobacteria as a case study. Trends Genet. 22, 593–596.

Mantri, Y., and Williams, K.P. (2004). Islander: a database of integrative islands in prokaryotic genomes, the associated integrases and their DNA site specificities. Nucleic Acids Res. 32, D55–8.

Meng, Y., Li, Y., Galvani, C.D., Hao, G., Turner, J.N., Burr, T.J., Hoch, H.C. (2005). Upstream migration of *Xylella fastidiosa* via pilus-driven twitching motility. J. Bacteriol. 187, 5560–5567.

Mongkolsuk, S., Praituan, W., Loprasert, S., Fuangthong, M., and Chamnongpol, S. (1998). Identification and characterization of a new organic hydroperoxide resistance (ohr) gene with a novel pattern of oxidative stress regulation from *Xanthomonas campestris* pv. *phaseoli*. J. Bacteriol. 180, 2636–2643.

Monteiro-Vitorello, C.B., de Oliveira, M.C., Zerillo, M.M., Varani, A.M., Civerolo, E., and Van Sluys, M.A. (2005). *Xylella* and *Xanthomonas* Mobil'omics. OMICS 9, 146–159.

Moreira, L.M., de Souza, R.F., Almeida, N.F. Jr., Setubal, J.C., Oliveira, J.C., Furlan, L.R., Ferro, J.A., and da Silva, A.C. (2004). Comparative genomics analyses of citrus-associated bacteria. Annu. Rev. Phytopathol. 42, 163–184.

Moreira, L.M., De Souza, R.F., Digiampietri, L.A., Da Silva, A.C., and Setubal, J.C. (2005). Comparative analyses of *Xanthomonas* and *Xylella* complete genomes. OMICS 9, 43–76.

Newman, K.L., Almeida, R.P., Purcell, A.H., and Lindow, S.E. (2004). Cell–cell signaling controls *Xylella fastidiosa* interactions with both insects and plants. Proc. Natl. Acad. Sci. U.S.A. 101, 1737–1742.

Nierman, W.C., Feldblyum, T.V., Laub, M.T., Paulsen, I.T., Nelson, K.E., Eisen, J.A., Heidelberg, J.F., Alley, M.R., Ohta, N., Maddock, J.R., Potocka, I., Nelson, W.C., Newton, A., Stephens, C., Phadke, N.D., Ely, B., DeBoy, R.T., Dodson, R.J., Durkin, A.S., Gwinn, M.L., Haft, D.H., Kolonay, J.F., Smit, J., Craven, M.B., Khouri, H., Shetty, J., Berry, K., Utterback, T., Tran, K., Wolf, A., Vamathevan, J., Ermolaeva, M., White, O., Salzberg, S.L., Venter, J.C., Shapiro, L., Fraser, C.M.. (2001). Complete genome sequence of *Caulobacter crescentus*. Proc. Natl. Acad. Sci. U.S.A. 98, 4136–41.

Nunes, L.R., Rosato, Y.B., Muto, N.H., Yanai, G.M., da Silva, V.S., Leite, D.B., Gonçalves, E.R., de Souza, A.A., Coletta-Filho, H.D., Machado, M.A., Lopes, S.A, de Oliveira, R.C. (2003). Microarray analyses of *Xylella fastidiosa* provide evidence of coordinated transcription control of laterally transferred elements. Genome Res. 13, 570–8.

Ochiai, H., Inoue, Y., Takeya, M., Sasaki, A., and Kaku, H. (2005). Genome sequence of *Xanthomonas oryzae* pv. *oryzae* suggests contribution of large numbers of effector genes and insertion sequences to its race diversity. Jpn. Agric. Res. Q. 39, 275–287.

Oliveira, M.A., Guimaraes, B.G., Cussiol, J.R., Medrano, F.J., Gozzo, F.C., and Netto, L.E. (2006). Structural insights into enzyme-substrate interaction and characterization of enzymatic intermediates of organic hydroperoxide resistance protein from *Xylella fastidiosa*. J. Mol. Biol. 359, 433–445.

Pallen, M.J., and Wren, B.W. (2007). Bacterial pathogenomics. Nature 449, 835–42.

Pashalidis, S., Moreira, L.M., Zaini, P.A., Campanharo, J.C., Alves, L.M., Ciapina, L.P., Vencio, R.Z., Lemos, E.G., Da Silva, A.M., and Da Silva, A.C. (2005). Whole-genome expression profiling of *Xylella fastidiosa* in response to growth on glucose. OMICS 9, 77–90.

Pierce, N.B. (1892). The California vine disease. Veg. Pathol. Bull., 2, 1–222.

Pickard, D., Wain, J., Baker, S., Line, A., Chohan, S., Fookes, M., Barron, A., Gaora, P.O., Chabalgoity, J.A., Thanky, N., Scholes, C., Thomson, N., Quail, M., Parkhill, J., Dougan, G.(2003). Composition, acquisition, and distribution of the Vi exopolysaccharide-encoding *Salmonella enterica* pathogenicity island SPI-7. J. Bacteriol. 185, 5055–5065.

Purcell, A.H., and Hopkins, D.L. (1996). Fastidious xylem-limited bacterial plant pathogens. Annu. Rev. Phytopathol. 34, 131–151.

Qian, W., Jia, Y., Ren, S.X., He, Y.Q., Feng, J.X., Lu, L.F., Sun, Q., Ying, G., Tang, D.J., Tang, H., Wu, W., Hao, P., Wang, L., Jiang, B.L., Zeng, S., Gu, W.Y., Lu, G., Rong, L., Tian, Y., Yao, Z., Fu, G., Chen, B., Fang, R., Qiang, B., Chen, Z., Zhao, G.P., Tang, J.L., He, C. (2005). Comparative and functional genomic analyses of the pathogenicity of phytopathogen *Xanthomonas campestris* pv. *campestris*. Genome Res. 15, 757–67.

Rodrigues, J.L., Silva-Stenico, M.E., Gomes, J.E., Lopes, J.R., and Tsai, S.M. (2003). Detection and diversity assessment of *Xylella fastidiosa* in field-collected plant and insect samples by using 16S rRNA and gyrB sequences. Appl. Environ. Microbiol. 69, 4249–55.

Salanoubat, M., Genin, S., Artiguenave, F., Gouzy, J., Mangenot, S., Arlat, M., Billault, A., Brottier, P., Camus, J.C., Cattolico, L., Chandler, M., Choisne, N., Claudel-Renard, C., Cunnac, S., Demange, N., Gaspin, C., Lavie, M., Moisan, A., Robert, C., Saurin, W., Schiex, T., Siguier, P., Thébault, P., Whalen, M., Wincker, P., Levy, M., Weissenbach, J., Boucher, C.A. (2002). Genome sequence of the plant pathogen *Ralstonia solanacearum*. Nature 415, 497–502.

Schaad, N.W., Pstnikova, E., Lacy, G., Sechler, A., Agarkova, I., Stromberg, P.E., Stromberg, V.K., and Vidaver, A.K. (2006). Emended classification of xanthomonad pathogens on citrus. Syst. Appl. Microbiol. 29, 690–695

Senol, E. (2004). *Stenotrophomonas maltophilia*: the significance and role as a nosocomial pathogen. J. Hosp. Infect. 57, 1–7.

Sicheritz-Ponten, T., and Andersson, S.G. (2001). A phylogenomic approach to microbial evolution. Nucleic Acids Res. 29, 545–552.

Simpson, A.J., Reinach, F.C., Arruda, P., Abreu, F.A., Acencio, M., Alvarenga, R., Alves, L.M., Araya, J.E., Baia, G.S., Baptista, C.S., Barros, M.H., Bonaccorsi, E.D., Bordin, S., Bové, J.M., Briones, M.R., Bueno, M.R., Camargo, A.A., Camargo, L.E., Carraro, D.M., Carrer, H., Colauto, N.B., Colombo, C., Costa, F.F., Costa, M.C., Costa-Neto, C.M., Coutinho, L.L., Cristofani, M., Dias-Neto, E., Docena, C., El-Dorry, H., Facincani, A.P., Ferreira, A.J., Ferreira, V.C., Ferro, J.A., Fraga, J.S., França, S.C., Franco, M.C., Frohme, M., Furlan, L.R., Garnier, M., Goldman, G.H., Goldman, M.H., Gomes, S.L., Gruber, A., Ho, P.L., Hoheisel, J.D., Junqueira, M.L., Kemper, E.L., Kitajima, J.P., Krieger, J.E., Kuramae, E.E., Laigret, F., Lambais, M.R., Leite, L.C., Lemos, E.G., Lemos, M.V., Lopes, S.A., Lopes, C.R., Machado, J.A., Machado, M.A., Madeira, A.M., Madeira, H.M., Marino, C.L., Marques, M.V., Martins, E.A., Martins, E.M., Matsukuma, A.Y., Menck, C.F., Miracca, E.C., Miyaki, C.Y., Monteiro-Vitorello, C.B., Moon, D.H., Nagai, M.A., Nascimento, A.L., Netto, L.E., Nhani, A. Jr., Nobrega, F.G., Nunes, L.R., Oliveira, M.A., de Oliveira, M.C., de Oliveira, R.C., Palmieri, D.A., Paris, A., Peixoto, B.R., Pereira, G.A., Pereira, H.A Jr., Pesquero, J.B., Quaggio, R.B., Roberto, P.G., Rodrigues, V., Rosa, A.J., Rosa, V.E Jr., Sá, R.G., Santelli, R.V., Sawasaki, H.E., Silva, A.C., Silva, A.M., Silva, F.R., Silva, W.A., Silveira, J.F., Silvestri, M.L., Siqueira, W.J., Souza, A.A., Souza, A.P., Terenzi, M.F., Truffi, D., Tsai, S.M., Tsuhako, M.H., Vallada, H., Van Sluys, M.A., Verjovski-Almeida, S., Vettore, A.L., Zago, M.A., Zatz, M., Meidanis, J., Setubal, J.C. (2000). The genome sequence of the plant pathogen *Xylella fastidiosa*. The *Xylella fastidiosa* Consortium of the Organization for Nucleotide Sequencing and Analysis. Nature 406, 151–159.

Sirand-Pugnet, P., Lartigue, C., Marenda, M., Jacob, D., Barré, A., Barbe, V., Schenowitz, C., Mangenot, S., Couloux, A., Segurens, B., Daruvar, A., Blanchard, A., Citti, C., (2007). Being pathogenic, plastic, and sexual while living with a nearly minimal bacterial genome. PLoS Genet. 3, e75.

Snel, B., Huynen, M.A., and Dutilh, B.E. (2005). Genome trees and the nature of genome evolution. Annu. Rev. Microbiol. 59, 191–209.

Souza, L.C., Wulff, N.A., Gaurivaud, P., Mariano, A.G., Virgílio, A.C., Azevedo, J.L., and Monteiro, P.B. (2006). Disruption of *Xylella fastidiosa* CVC gumB and gumF genes affects biofilm formation without a detectable influence on exopolysaccharide production. FEMS Microbiol. Lett. 257, 236–42.

Stover, C.K., Pham, X.Q., Erwin, A.L., Mizoguchi, S.D., Warrener, P., Hickey, M.J., Brinkman, F.S., Hufnagle, W.O., Kowalik, D.J., Lagrou, M., Garber, R.L., Goltry, L., Tolentino, E., Westbrock-Wadman, S., Yuan, Y., Brody, L.L., Coulter, S.N., Folger, K.R., Kas, A., Larbig, K., Lim, R., Smith, K., Spencer, D., Wong, G.K., Wu, Z., Paulsen, I.T., Reizer, J., Saier, M.H., Hancock, R.E., Lory, S., Olson, M.V. (2000). Complete genome sequence of *Pseudomonas aeruginosa* PA01, an opportunistic pathogen. Nature 406, 959–64.

Thieme, F., Koebnik, R., Bekel, T., Berger, C., Boch, J., Buttner, D., Caldana, C., Gaigalat, L., Goesmann, A., Kay, S., Kirchner, O., Lanz, C., Linke, B., McHardy, A.C., Meyer, F., Mittenhuber, G., Nies, D.H., Niesbach-Klösgen, U., Patschkowski, T., Rückert, C., Rupp, O., Schneiker, S., Schuster, S.C., Vorhölter, F.J., Weber, E., Pühler, A., Bonas, U., Bartels, D., Kaiser, O. (2005). Insights into genome plasticity and pathogenicity of the plant pathogenic bacterium *Xanthomonas campestris* pv. *vesicatoria* revealed by the complete genome sequence. J. Bacteriol. 187, 7254–7266.

Toth, I.K., Pritchard, L., and Birch, P.R. (2006). Comparative genomics reveals what makes an enterobacterial plant pathogen. Annu. Rev. Phytopathol. 44, 305–36.

Van Sluys, M.A., de Oliveira, M.C., Monteiro-Vitorello, C.B., Miyaki, C.Y., Furlan, L.R., Camargo, L.E., da Silva, A.C., Moon, D.H., Takita, M.A., Lemos, E.G., El-Dorry, H., Tsai, S.M., Carrer, H., Carraro, D.M., de Oliveira, R.C., Nunes, L.R., Siqueira, W.J., Coutinho, L.L., Kimura, E.T., Ferro, E.S., Harakava, R., Kuramae, E.E., Marino, C.L., Giglioti, E., Abreu, I.L., Alves, L.M., do Amaral, A.M., Baia, G.S., Blanco, S.R., Brito, M.S., Cannavan, F.S., Celestino, A.V., da Cunha, A.F., Fenille, R.C., Ferro, J.A., Formighieri, E.F., Kishi, L.T., Leoni, S.G., Oliveira, A.R., Rosa, V.E., Sassaki, F.T., Sena, J.A., de Souza, A.A., Truffi, D., Tsukumo, F., Yanai, G.M., Zaros, L.G., Civerolo, E.L., Simpson, A.J., Almeida, N.F., Setubal, J.C., Kitajima, J.P. (2003). Comparative analyses of the complete genome sequences of Pierce's disease and citrus variegated chlorosis strains of *Xylella fastidiosa*. J. Bacteriol. 185, 1018–1026.

Van Sluys, M.A., Monteiro-Vitorello, C.B., Camargo, L.E., Menck, C.F., Da Silva, A.C., Ferro, J.A., Oliveira, M.C., Setubal, J.C., Kitajima, J.P., and Simpson, A.J. (2002). Comparative genomic analysis of plant-associated bacteria. Annu. Rev. Phytopathol. 40, 169–189.

Vasconcelos, A.T., Ferreira, H.B., Bizarro, C.V., Bonatto, S.L., Carvalho, M.O., Pinto, P.M., Almeida, D.F., Almeida, L.G., Almeida, R., Alves-Filho, L., Assunção, E.N., Azevedo, V.A., Bogo, M.R., Brigido, M.M., Brocchi, M., Burity, H.A., Camargo, A.A., Camargo, S.S., Carepo, M.S., Carraro, D.M., de Mattos Cascardo, J.C., Castro, L.A., Cavalcanti, G., Chemale, G., Collevatti, R.G., Cunha, C.W., Dallagiovanna, B., Dambrós, B.P., Dellagostin, O.A., Falcão, C., Fantinatti-Garboggini, F., Felipe, M.S., Fiorentin, L., Franco, G.R., Freitas, N.S., Frías, D., Grangeiro, T.B., Grisard, E.C., Guimarães, C.T., Hungria, M., Jardim, S.N., Krieger, M.A., Laurino, J.P., Lima, L.F., Lopes, M.I., Loreto, E.L., Madeira, H.M., Manfio, G.P., Maranhão, A.Q., Martinkovics, C.T., Medeiros, S.R., Moreira, M.A., Neiva, M., Ramalho-Neto, C.E., Nicolás, M.F., Oliveira, S.C., Paixão, R.F., Pedrosa, F.O., Pena, S.D., Pereira, M., Pereira-Ferrari, L., Piffer, I., Pinto, L.S., Potrich, D.P., Salim, A.C., Santos, F.R., Schmitt, R., Schneider, M.P., Schrank, A., Schrank, I.S., Schuck, A.F., Seuanez, H.N., Silva, D.W., Silva, R., Silva, S.C., Soares, C.M., Souza, K.R., Souza, R.C., Staats, C.C., Steffens, M.B., Teixeira, S.M., Urmenyi, T.P., Vainstein, M.H., Zuccherato, L.W., Simpson, A.J., Zaha, A. (2005). Swine and poultry pathogens: the complete genome sequences of two strains of *Mycoplasma hyopneumoniae* and a strain of *Mycoplasma synoviae*. J. Bacteriol. 187, 5568–77.

Wells, J.M., Raju, B.C., Jung, H.Y., Weisburg, W.G., Mandelco-Paul, L., and Brenner, D.J. (1987). *Xylella fastidiosa*, new-genus, new-species gram negative xylem-limited fastidious plant bacteria related to *Xanthomonas* spp. Int. J. Syst. Bacteriol. 37, 136–143.

Zhou, D., Han, Y., Song, Y., Huang, P., and Yang, R. (2004). Comparative and evolutionary genomics of *Yersinia pestis*. Microbes Infect. 6, 1226–34.

Genome Sequence-based Insights into the Biology of the Sugarcane Pathogen *Leifsonia xyli* subsp. *xyli*

Claudia B. Monteiro-Vitorello, Marcelo Marques Zerillo, Marie-Anne Van Sluys and Luis Eduardo Aranha Camargo

Abstract

Leifsonia xyli subsp. *xyli* (*Lxx*) causes ratoon stunting disease (RSD), a major worldwide disease of sugarcane. Formerly classified as *Clavibacter xyli* subsp. *xyli*, *Lxx* is a fastidious member of the GC-rich Actinomycetales, a taxonomic order that comprises two other genera of plant pathogens of great agricultural impact. In this review we present some interesting features of the genome of *Lxx* with emphasis on pathogenicity. Most striking is the observation that *Lxx* has a relatively large number of pseudogenes suggestive of an ongoing process of genome decay. It has been proposed that *Lxx* was once a free-living bacterium that is now restricted to the xylem as a consequence or cause of the accumulation of pseudogenes. This point stems from the observation that although *Lxx* has only been detected inhabiting the xylem of sugarcane, it carries several genes typical of free-living organism. In this review we also discuss the relevance of lateral gene transfer in the acquisition of a few genes associated with pathogenicity and the contribution of mobile genetic elements.

Introduction

Leifsonia xyli subsp. *xyli* (*Lxx*) is a small, fastidious, Gram-positive, coryneform bacterium that causes the ratoon stunting disease (RSD) of sugarcane. *Lxx* belongs to the phylum Actinobacteria, which is the largest taxonomic unit within the domain of Bacteria. Members of this group inhabit a variety of environments and have different lifestyles (Ventura *et al.*, 2007). Species of the genus *Mycobacterium*, for instance, are the most studied due to their importance as human and animal pathogens (Marri *et al.*, 2006) and various virulence factors are well known. However, the same is not true for a group of species of this phylum that are plant pathogens, such as *Lxx*, and some species of *Clavibacter* (Gartemann *et al.*, 2003), *Curtobacterium* (Davis and Vidaver, 2001), and *Streptomyces* (Loria *et al.*, 2006). Despite the significant yield losses related to their attack of crops such as beans, beets, potato, tomato, soybean and sugarcane, little is known about their biology and the mechanisms of interaction with their plant hosts.

Biological studies with *Lxx* have mostly been impaired by its 'misbehaviour' in the laboratory. *Lxx* is renowned for its fastidious growth habit and nutritional requirements, taking 4 weeks for single colonies to be visible on solid medium and 2 weeks for reasonable growth in liquid medium. There is no defined minimal medium and although transposon mutagenesis has been accomplished, it was with great difficulty (Brumbley *et al.*, 2002). As a result of the effort to gather genomic information on different plant related Actinobacteria, complete genome sequencing projects were undertaken or are in progress for *Lxx* (Monteiro-Vitorello *et al.*, 2004), *Clavibacter michiganensis* subsp. *michiganensis* (https://www.genetik.uni-bielefeld.de/GenoMik/partner/bi_eichen.html); *Clavibacter michiganensis* subsp. *sepedonicus* (http://www.sanger.ac.uk/Projects/C_michiganensis/), and *Streptomyces scabies* (http://www.sanger.ac.uk/Projects/S_scabies/). *Lxx* is the only finished and annotated plant associated Actinobacterial genome at the time of writing. It is interesting to note that out of the more than two hundred Actinobacterial genomes sequenced or in the process of being sequenced (http://www.genomesonline.org/), only four are plant pathogens and another four are plant-associated bacteria. We expect that the availability of the genome sequence of *Lxx* will be of help to identify virulence factors as well as key genes that allow *Lxx* to live in the xylem of sugarcane, which is a particularly inhospitable environment compared to other existent niches within a plant host. This chapter reviews some of the inferences and hypotheses on the biology of this organism made possible by 'reading' its genome sequence.

A brief history of *Leifsonia* as a pathogen

Formerly classified as *Clavibacter xyli* subsp. *xyli* (Davis *et al.*, 1984), *Lxx* was removed from the genus *Clavibacter* to create the genus *Leifsonia*, together with *L. poae*, found in *Poa annua* root galls, and *L. aquatica*, a free-living bacterium (Evtushenko *et al.* 2000). So far, *Lxx* has been reported as only infecting sugarcane. The disease it provokes (RSD) is found in most sugarcane growing areas of the world and can cause yield losses of up to 30% in susceptible varieties (Gillaspie and Teakle, 1989). Since the bacterium is present in the liquids of infected plants, it can be mechanically disseminated after contamination of the cane knives used in harvesting. Thus, the incidence of infected plants increases during successive ratoon crops. For this reason and for quite some time, cumulative losses due to RSD have probably been greater than the losses caused by any other sugarcane disease (Gillaspie and Teakle, 1989). Curiously, despite being a worldwide presence in commercial fields, *Lxx* has never been found affecting wild clones of *Saccharum officinarum* in its centre of diversity (Magarey *et al.*, 2002), Papua New Guinea, which suggests it has recently evolved to infect its host. The disease was primarily observed in the 1940s, after the production of the first hybrids, based on breeding of *S. officinarum* and *S. spontaneum*. Therefore, RSD may be favoured by the establishment of modern worldwide commercial crops of sugarcane (Brumbley *et al.*, 2006). *Lxx* is believed to have evolved from a single pathogenic clone, since no genetic variation was found among different isolates of distinct countries and cultivars (Young *et al.*, 2006).

Lxx may be regarded as a stealth pathogen since the symptoms it causes are subtle compared to the ones caused by necrogenic bacteria and may be confounded with symptoms caused by other biotic and abiotic stresses. Moreover, symptoms are highly dependent on the genetic background of the varieties and on environmental conditions. Because of this, the spread of RSD through planting material has beset most sugarcane growing areas in the world. Infected plants show reduced cane diameter and shortening of the internodes, i.e., stunting. Discolouration of vascular bundles of mature stalks may be seen in the form of discrete rosy dots or streaks just below the internodes where the bundles branch into the leaf sheath (Gillaspie and Teakle, 1989), but this is of little diagnostic value as infection by other pathogens may cause the same symptom. Internally, *Lxx* colonizes the lumen and the pits of the xylem cells, but not the phloem or parenchyma (Weaver *et al.*, 1977). Thus, *Lxx* does not access the valuable carbon source accumulated by sugarcane but rather lives in the relative nutritionally poor environment of the xylem. No general tissue disorganization or necrosis results from the colonization of the xylem vessels by *Lxx*, although the xylem conduits may be blocked with a mucilaginous substance probably produced by the host. This blockage is reported to reduce sap flow up to 34% (Teakle and Appleton, 1978; Kao and Damann, 1980) which in its turn may result in wilting but only during extensive drought (James, 1996).

Diagnosis is based on the detection of the bacteria by a number of different techniques, including phase contrast microscopy (Steindl, 1976), serology (Gillaspie, 1978) and PCR (Pan *et al.*, 1998). The successful growth of axenic cultures of *Lxx* on specific media (Davis *et al.*, 1980) has made it possible to produce specific antibodies useful for diagnosis.

Since sugarcane is vegetatively propagated and due to the nature of the transmission of *Lxx*, control measures rely primarily on using healthy stalks as planting material. These are raised in special disease-free nurseries in which all cutting instruments are disinfected either by heating in a flame or by dipping in a chemical solution. Moreover, before planting in the nurseries, these 'seed' stalks are dipped in hot water so as to eliminate the bacteria.

General features of the genome of *Lxx*

The genome of *Lxx* is 2 584 158 bp in length and is characterized by a high content of G and C bases (67.7%). The total number of predicted genes is 2351, which were divided into two classes: 2044 intact protein-coding genes and 307 pseudogenes (Monteiro-Vitorello et al., 2004). Pseudogenes were considered as those disrupted by one or more authentic frameshifts and/or a point mutation that introduced a stop codon in frame, or genes that are partially represented based on BLAST results. There is only one copy of the ribosomal genes, 45 tRNAs representing all amino acids, and one tmRNA. A striking feature is the presence of a large number of insertion sequences (IS elements) some of which are within genes, probably impairing their function. Fifty-one genes are predicted to encode transposases located within copies of six IS elements (IS*Lxx*1–6) (Monteiro-Vitorello et al., 2004; Zerillo et al., submitted) distributed along the chromosome (Fig. 6.1). The IS elements belong to the major families IS*110*, IS*21*, IS*481*, IS*30* and IS5, as described by Mahillon and Chandler (1998). Two IS elements had a clear recent expansion since all copies are nearly identical (IS*Lxx*4 – 26 copies; and IS*Lxx*5 – 15 copies). Nine elements were found inserted within genes and 14 were in close vicinity of housekeeping pseudogenes. Another 46 transposases were not classified into one of the six IS elements described and were primarily associated with genomic islands (see discussion below). Sixteen transposases were found clustered in one putative genomic island (*Lxx*GI3), 11 of which are uncharacterized elements of families IS256, IS3, IS*481* and IS*110*, and the other five belong to copies of IS*Lxx*1, 2, 4, 5 and 6.

The theory of a recent niche conversion and genome decay

It has been proposed that the accumulation of pseudogenes adversely affects the ability of some obligate parasites and symbionts to colonize diverse niches (Cole et al., 2001; Babu, 2003). As mentioned before, although *Lxx* is cultivable but fastidious, its niche is very restricted since it colonizes only the xylem vessels of sugarcane. A central hypothesis that our research group has proposed after analyzing the *Lxx* genome sequence is that *Lxx* was once a free-living bacterium that is now confined to the xylem and the consequent confinement may be the cause or the result (or probably both) of the accumulation of pseudogenes. This reduction process due to gene loss and its biological consequences, known as genome decay, would be similar to that proposed for the related species *Mycobacterium leprae* (Cole et al., 2001) and for several other obligate pathogens and endosymbionts (Thomson et al., 2003). According to our hypothesis, the loss of functionality of important genes would have restricted the

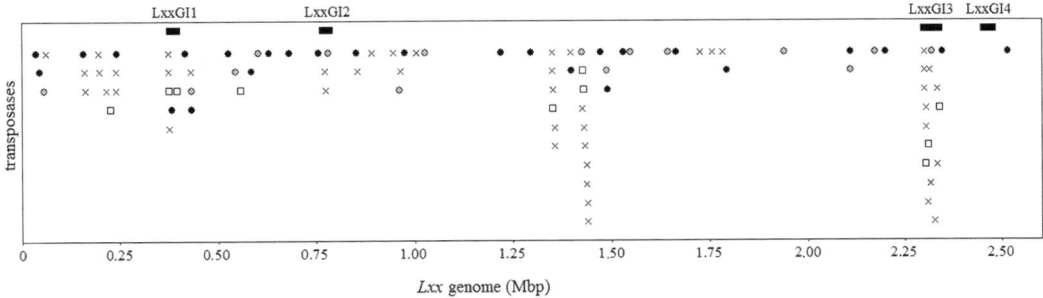

Figure 6.1 Localization of transposase-encoding genes represented on the complete nucleotide sequence of the *Lxx* genome. Genes separated by less than 20 kbp are shown on different levels for ease of visualization. The most abundant IS elements (IS*Lxx*4 and IS*Lxx*5) are represented as black and grey circles, respectively. Other characterized IS elements are represented as white squares; and transposases of uncharacterized IS elements are indicated by 'X's. The genomic islands (*Lxx*GI1–4) are shown as thick black lines.

ability of *Lxx* to live as a free-living bacterium as other species of *Leifsonia* do. This initial niche restriction by its turn would then have resulted in a neutral selection pressure against mutations on genes that were no longer necessary for the free-living lifestyle, thus resulting in the accumulation of pseudogenes (Lawrence et al., 2001). Clues of this ancient free-living lifestyle of *Lxx*, can still be found in the genome of *Lxx* as genes that code for proteins that would apparently be necessary mostly for a free-living bacteria such as a light inducible photolyase, a (non-functional) flagellar operon, a threalose synthase, and genes that code for mechanosensitive channels and transporters, including ones for glycine and betaine, which are typically involved in tolerance to osmotic stress. In addition, *Lxx* has a strikingly large repertoire of ABC transporters compared with *Xylella fastidiosa* (42 as opposed to 26) (Ren et al., 2007) (Fig. 6.2), another xylem-limited plant pathogen, and a complete phosphoenolpyruvate (PEP)-dependent phosphotransferase system (PTS), which is absent in *X. fastidiosa*. Together, these systems would enable *Lxx* to uptake a range of carbohydrates. However, the presence of this arsenal of sugar transporters is in sharp contrast to the carbon-poor content of the xylem sap of sugarcane (Dong et al., 1997).

Perhaps the most striking suggestion of an early free-living habit of *Lxx* is the finding that as much as 6.9% of all genes are predicted to be transcriptional regulators based on the presence of regulatory motifs (Monteiro-Vitorello et al., 2004). This high number is typical of free-living bacteria, as organisms that live in complex environments need to respond to diverse stimuli compared to those organisms restricted to a more uniform environment (Stover et al., 2000). For instance, in *Pseudomonas aeruginosa* and in *Streptomyces coelicolor*, which are free-living, this percentage is 12% and 9.8%, respectively, whereas in *X. fastidiosa* (which has a genome size equivalent to that of *Lxx*) it is only 3.7%. This same percentage is found in the obligate parasite *M. leprae*. Counting genes associated with transcription as predicted by COG, the differences mentioned were also detected (Fig. 6.3). Among the 173 total regulators found in *Lxx*, 10 are predicted to be non-functional, which is also consistent with our theory of a genome decay

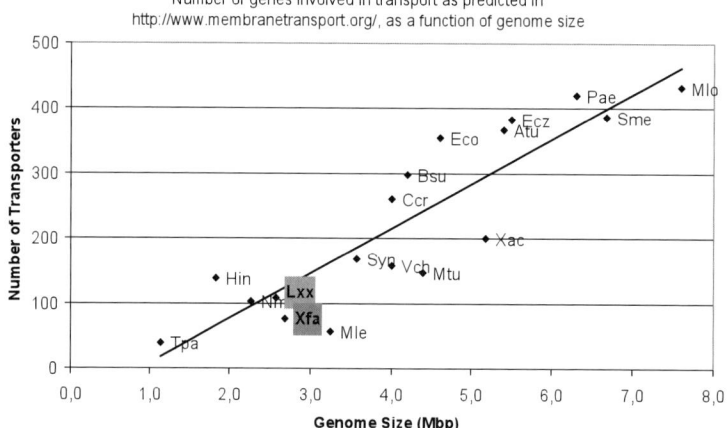

Figure 6.2 Number of genes associated with membrane transport as found in TransporterDB (http://www.membranetransport.org/index.html) (Ren et al., 2007) as a function of genome size. Among more than five hundred genomes completely sequenced, we chose to represent those organisms with different genome sizes and lifestyle, to emphasize a general tendency of bacterial genomes and number of genes associated with membrane transport and lifestyles. Below are the abbreviations used to represent each of the organisms: Tpa, *Treponema pallidum* Nichols; Hin, *Haemophilus influenzae* KW20; Nme, *Neisseria meningitidis* MC58; Lxx, *Leifsonia xyli* subsp. *xyli* CTCB07; Xfa, *Xylella fastidiosa* 9a5c; Mle, *Mycobacterium leprae* TN; Syn, *Synechocystis* sp. PCC6803; Vch, *Vibrio cholerae* El Tor 16961; Ccr, *Caulobacter crescentus* CB15; Bsu, *Bacillus subtilis* 168; Mtu, *Mycobacterium tuberculosis* H37Rv; Eco, *Escherichia coli* K12-MG1655; Xac, *Xanthomonas axonopodis* pv. *citri* 306; Atu, *Agrobacterium tumefaciens* C58; Ecz, *Escherichia coli* O157:H7 EDL933; Pae, *Pseudomonas aeruginosa* PAO1; Sme, *Sinorhizobium meliloti* 1021; Mlo, *Mesorhizobium loti* MAFF303099.

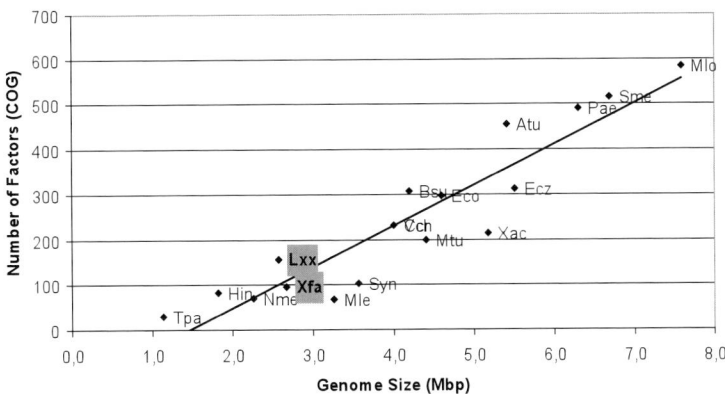

Figure 6.3 Number of genes involved in transcription as predicted by COG (cluster of orthologous groups), as a function of genome size. Among the more than five hundred genomes completely sequenced, we chose to represent those organisms with different genome sizes and lifestyle, to emphasize a general tendency of bacterial genomes and number of genes associated with transcription and lifestyles. Below are the abbreviations used to represent each of the organisms: Tpa, *Treponema pallidum* Nichols; Hin, *Haemophilus influenzae* KW20; Nme, *Neisseria meningitidis* MC58; Lxx, *Leifsonia xyli* subsp. *xyli* CTCB07; Xfa, *Xylella fastidiosa* 9a5c; Mle, *Mycobacterium leprae* TN; Syn, *Synechocystis* sp. PCC6803; Vch, *Vibrio cholerae* El Tor 16961; Ccr, *Caulobacter crescentus* CB15; Bsu, *Bacillus subtilis* 168; Mtu, *Mycobacterium tuberculosis* H37Rv; Eco, *Escherichia coli* K12-MG1655; Xac, *Xanthomonas axonopodis* pv. *citri* 306; Atu, *Agrobacterium tumefaciens* C58; Ecz, *Escherichia coli* O157:H7 EDL933; Pae, *Pseudomonas aeruginosa* PAO1; Sme, *Sinorhizobium meliloti* 1021; Mlo, *Mesorhizobium loti* MAFF303099.

process and niche restriction. Analyzing *Lxx* and *X. fastidiosa*, the over representations of transcriptional regulators in *Lxx* were mainly found in LuxR, ROK, GntR, LacI and TetR regulator families (Monteiro-Vitorello et al., 2004).

Other truncated proteins which were considered to be non-functional are involved in the synthesis of methionine and cysteine. Although cysteine biosynthesis could occur via an atypical pathway similar to that of *Bifidobacterium longum*, which includes cystathionine γ- and β-synthases and cystathionine γ-lyase (Schell et al., 2002), the more common sulphite/sulphate pathway is affected by the inactivation of cysteine synthase (*cysK*). This explains the need for cysteine in the standard *Lxx* culture medium developed by Davis and collaborators (1980). Also genes *metE* and *metF*, both involved in the folate branch of the methionine biosynthetic pathway, are truncated. In *Streptomyces lividans*, disruption of *metF* led to methionine auxotrophy (Blanco et al., 1998) and this might be the case of *Lxx* since its growth is greatly improved in liquid medium by the addition of this amino acid (Monteiro-Vitorello et al., 2004).

Lateral transfer and pathogenicity

Analysis of codon bias, GC composition and dinucleotide signatures indicated the existence of anomalous regions in the genome of *Lxx* that is suggestive of lateral transfer. Four regions were defined as genomic islands and named *Lxx*GI1-*Lxx*GI4 (Table 6.1) (Monteiro-Vitorello et al., 2004). Regions *Lxx*GI2, *Lxx*GI3 and *Lxx*GI4 contain genes known to be involved in pathogenicity in other organisms. *Lxx*GI1 and *Lxx*GI4 are probably of phage origin, due to the presence of phage-related genes. In both cases we defined the site of insertion and the duplicated sequence upon insertion (Fig. 6.4).

Genomic island *Lxx*GI2 could be the result of a plasmid integration event, since it harbours a relaxase/mobilization (Rlx) protein required for the horizontal transfer of plasmids during bacterial conjugation (Parker et al., 2005). In islands *Lxx*GI2 and *Lxx*GI3, genes were found coding for a pectinase and a cellulase, respectively, enzymes which are present in variable number of copies in bacterial plant pathogens (Van Sluys et al., 2002). It has been speculated that these genes

Table 6.1 Genomic Islands described in *Leifsonia xyli* subsp. *xyli* CTC B07

Genomic island	5'–3' ends[1]	Size (bp)	GC %	Gene content[2]
*Lxx*GI1	370 529–404 065	33 537	62.4	phage-related genes, IS*Lxx*1, IS*Lxx*2, IS*Lxx*4
*Lxx*GI2	762 248–791 148	28 901	62.5	plasmid-related genes, IS*Lxx*5, pectinase
*Lxx*GI3	2 291 270–2 347 438	50 169	63.0	IS*Lxx*1, IS*Lxx*2, IS*Lxx*4, IS*Lxx*5, IS*Lxx*6, *celA* homologue and *desA* homologue
*Lxx*GI4	2 447 539–2 485 053	37 515	62.7	Phage-related genes, *pat-1* homologue

[1]Position in base pairs (bp) relative to origin of replication.
[2]All the islands described contain hypothetical genes.

play a primary colonization and adaptation role similar to that predicted for *Xylella fastidiosa*, since they would both enable these pathogens to migrate through the xylem vessels and to use these polysaccharides as a source of carbon (Simpson *et al.*, 2000). Another interesting finding on genomic island *Lxx*GI3 is a gene, *desA*, that codes for a δ-fatty acid desaturase family. Another working hypothesis developed by our group is that this gene is involved in the synthesis of absicic acid (ABA) or an ABA analogue since it is reported in the literature that the synthesis of this plant hormone could derive from an indirect pathway associated with the degradation of β-carotene and other intermediates (Bartley and Scolnik, 1995; Kende and Zeevaart, 1997; Lee and Schmidt-Dannert, 2002). The link between the carotenoid biosynthetic pathways to the intermediates leading to ABA synthesis would depend on a desaturase activity. The identification of a putative desaturase (DesA) in *Lxx* sharing 35% of amino acid similarity with a desaturase from *Synechocystis* is a possible candidate for this function. Thus the putative DesA from *Lxx* could divert the putative carotenoid biosynthetic pathway present in *Lxx* to the synthesis of an abscisic acid-like molecule. We believe this to be an important hypothesis since the production of ABA-like molecule by *Lxx* could be related to the stunting symptom, as ABA is a known plant growth inhibitor. Most important to our hypothesis, however, is that recent findings indicate that besides playing a regulatory and signalling role in plants against abiotic stresses, ABA also appears to play such roles during pathogen attack. Audenaert *et al.* (2002) demonstrated that an ABA deficient tomato mutant is more resistant to the fungus *Botrytis cinerea* than the wild-type and that this can be reversed by exogenous application of ABA. Exogenous treatment with this hormone also rendered *Arabidopsis* plants susceptible to *Pseudomonas syringae* pv. *tomato* and this was related to suppression of the accumulation of disease-resistance related compounds

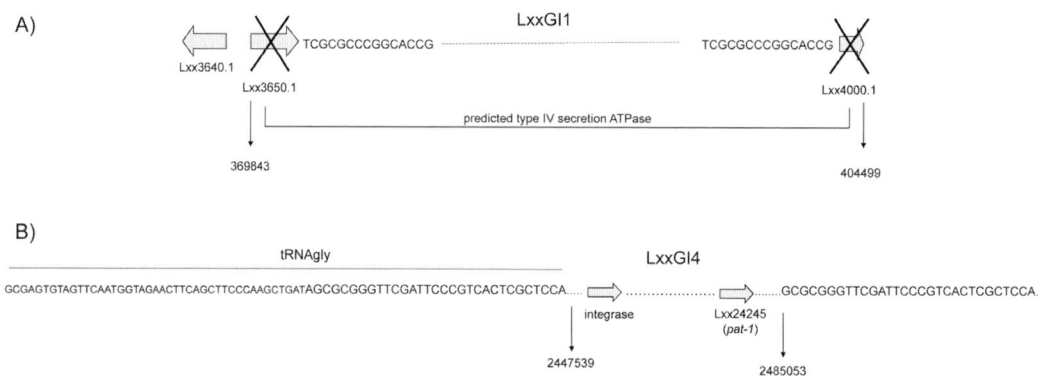

Figure 6.4 Analyses of the *Lxx*GI1 (A) and *Lxx*GI4 (B) insertion sites. Nucleotides represent the target duplication sites for each genomic island insertion. Arrows indicate predicted genes within the regions. Crossed arrows represent the genes truncated upon *Lxx*GI1 insertion. *Lxx*GI4 is inserted within a tRNA Gly. Numbers on both sides indicate the coordinates of each genome.

(Mohr and Cahill, 2007). The putative carotenoid biosynthetic pathway of *Lxx* is represented by five essential genes that would enable this bacterium to produce β-carotene from geranyl pyrophosphate (GPP) and isopentenyl pyrophosphate (Armstrong, 1997). These genes are closely related to the *crt* operon of *Brevibacterium linens* (Krubasik and Sandmann, 2000) and *Corynebacterium glutamicum* (Krubasik *et al.*, 2001). We are currently testing the functionality of this gene cluster by heterologous expression in non-carotenogenic bacteria, such as *Escherichia coli*.

*Lxx*GI4 harbours a gene similar to the *pat-1* gene of *Clavibacter michiganensis* subsp. *michiganensis*, a pathogen of tomatoes. In this organism, this plasmid-encoded gene plays a decisive role in causing plant wilting (Dreier *et al.*, 1997). Upon transformation, mildly virulent strains which lack this gene and do not cause wilting, acquire the wilt-inducing phenotype. Also, this phenotype is attenuated if a repetitive sequence positioned downstream of the gene (*pat-1rep*) is deleted. Thus it may be regarded as a major determinant of virulence of this closely related bacterium. In *Lxx*, the *pat-1* homologue (e-value 5e-15) is present in two copies, although only the one present in *Lxx*GI4 may be functional. This copy has the consensus motif of the trypsin family of serine proteases characteristic of *pat-1*, but lacks the downstream repetitive sequence. The absence of *pat-1rep* in *Lxx* might be the reason why this pathogen causes wilting only in highly susceptible sugarcane varieties under very particular environmental conditions.

Given the primary roles of the genes discussed above in the pathogenicity of various plant pathogens, they would be suitable candidates for further functional studies aiming to determine the virulence mechanisms of *Lxx*. Also, from this analysis, it is clear that lateral gene transfer has played a central role in shaping the pathogenicity of *Lxx* CTCB07.

In addition to potential virulence genes located in genomic islands, *Lxx* encodes genes implicated in pathogenicity in other bacterial pathogens distributed throughout the genomic sequence. These include haemolysins similar to the *tly*A and *tly*C widely present in pathogens as well as a haemolysin III homologue. The type III secretory system is absent, as expected for a Gram-positive bacterium. Also absent is the cytolysin-mediated translocation system, which is predicted to be the counterpart of this system in Gram-positive bacteria (Madden *et al.*, 2001). Noteworthy is the presence of a flagellar operon comprising 17 genes encoding proteins of the basal body (MotA, FlgC, F, FliE, G, O), a sigma-like transcriptional factor (FliA), hook-assembly proteins (FlgD, E, K, L, FliK), flagellar filament proteins (FliC, D, FlhA), and a flagellum-specific ATP-synthase (FliI). However, notably absent or predicted to be non-functional are four genes involved in the assembly of the flagellar filament (FlgK, FliD, FliK) and exportation of flagellin (FlhA). Consistent with these findings, electron microscopy indicates that *Lxx* lacks flagella (Davis *et al.*, 1984). Since flagellar proteins have been shown to play important roles in host-parasite interactions such as adhesion and transfer of virulence proteins to animal host cells (Arora *et al.*, 1998) and induction of rapid cell death in non-host plants (Taguchi *et al.*, 2003), it would be interesting to determine whether intact operons occur in other strains of *Lxx* or related species and to establish the role of such an intact cellular apparatus, if any, in pathogenicity.

Occlusions of xylem vessels are seen in infected sugarcane (Teakle and Appleton, 1978; Kao and Damann, 1980) but these are probably of host origin since electron microscopic studies of xylem extracts indicated that cells of *Lxx* lack an external capsule (Weaver *et al.*, 1977). Moreover, gum producing genes such as the ones found in *Xylella* (Simpson *et al.*, 2000) are missing. The only two clusters of genes found (*xan*A and *xan*B and the *rml*A/B/C/D) that could be associated with EPS and/or LPS production remain to be characterized in more detail.

As mentioned before, *Lxx* is a stealth pathogen due to the subtle and non-typical symptoms of the disease it causes. One reason for this could be the small number of putative pathogenicity genes compared to other bacterial plant pathogens. For instance, only 85 intact genes were categorized as pathogenicity genes in *Lxx* (Monteiro-Vitorello *et al.*, 2004), as opposed to 287 in the necrogenic, 'brute-force' and rotting pathogen of cauliflower and cabbages *Xanthomonas campestris* pv. *campestris* (da Silva *et al.*, 2002). This number is small even when compared to *X. fastidiosa* 9a5C and Temecula 1 strains (134 and 159 genes, respectively). Although artificial co-inoculations of *Lxx* and *X. albilineans* were never attempted, the two bacteria can co-exist in the same plant as indicated by PCR assays with species-specific primers carried out in our lab (unpublished results).

Insights to the general biology of *Lxx*

Energy metabolism

The genome of *Lxx* encodes none of the 14 proteins that compose complex I, which is the first oxidoreductase complex in the respiratory chain. This absence is also evident in *Bacillus*, *Haemophilus* and *Vibrio cholerae* (COG- NCBI). All theses organism may transfer the hydrogen from NADH to Ubiquinone using a type II NADH dehydrogenase (EC 1.6.99.3) or a Na^+-transporting NADH:ubiquinone oxidoreductase. The *Lxx* genome has only one type II NADH dehydrogenase (ORF 17150.1), which is highly similar to a *Streptomyces* enzyme. Other plant pathogens such as *Xanthomonas*, *Ralstonia* and *Xylella* have both the type I and II NADH dehydrogenase. Complex II and III are present and complete. Regarding complex IV, *Lxx* has more options than *Xylella* (Simpson *et al.*, 2000), since it harbours cytochrome C oxidase and cytochrome D Ubiquinol oxidase while *Xylella* has only cytochrome O Ubiquinol oxidase. However, both are less complete than *Xanthomonas* that has the three cytochrome oxidases (da Silva *et al.*, 2002). These differences might be related to more versatility needed by *Xanthomonas* due to it being an epiphytic organism, while *Xylella* and *Lxx* are limited to the xylem of their plant hosts. It has been pointed out before that the restricted respiratory complex could be an interesting target for drug design (Bhattacharyya *et al.*, 2002) and might also be related with the fastidious behaviour of xylem-limited bacteria.

Secretion systems

Several genes encoding proteins of the general secretion pathway were identified including *secA* (translocation motor), *secD/F*, *secY*, *secE*, *secG* and *yajC* (translocation channel). Although the SecB chaperone is not represented in the genome, gene products such as Ffh and FtsY may substitute for the SecB function. A type I signal peptidase is present, but a type II peptidase was not identified (Monteiro-Vitorello *et al.*, 2004). For secretion of folded cofactor-containing proteins across the membrane, *Lxx* probably accomplishes it with the twin arginine-motif translocation (TAT) pathway as TAT genes (A–E) are present.

Adaptation to the host

Albicidin is the major component of a complex of toxins produced by *Xanthomonas albilineans* (Birch and Patil, 1985; 1987ab), another sugarcane bacterial pathogen which inhabits the xylem vessels causing leaf scald disease (Rott *et al.*, 1996). In sugarcane, albicidin inhibits plastid DNA replication, resulting in blocked chloroplast differentiation and the chlorotic streaks that are characteristic of the disease (Birch and Patil, 1985; Birch and Patil, 1987a,b; Birch *et al.*, 1990). *X. albilineans* strains unable to produce albicidin do not induce leaf scald symptoms in sugarcane.

Albicidin is also bactericidal to a large range of Gram positive and Gram negative bacteria pathogenic to humans, animals and plants, and may be important during colonization of sugarcane by inhibiting growth of other xylem invading bacteria. Genes associated with albicidin detoxification are known to be present in *Klebsiella oxytoca* (Walker *et al.*, 1988), *Alcaligenes denitrificans* (Basnayake *et al.*, 1995), *Pantoea dispersa* (syn. *Erwinia herbicola*) (Zhang & Birch, 1997), and also in *X. albilineans*. Four predicted genes similar to those encoding multidrug resistance proteins, and one similar to the *albF* of *X. albilineans* are found in *Lxx*. We speculate that one or more of these genes could be associated with *Lxx* albicidin resistance.

Concluding remarks

Although genomics does not provide immediate solutions, the science does offer the opportunity to create hypotheses and to design precise experiments. The 'reading' of the *Lxx* genome sequence revealed a wide range of biological issues that can be further addressed. For example, considering the lifestyles of *X. fastidiosa* and *Lxx* even though both are xylem-limited fastidious plant pathogens and their presence is associated with clogging of xylem vessels, comparative analysis of their genomes revealed that the pool of genes and strategies used by them are probably quite different. *Xylella* seems to be much more specialized to this niche than *Lxx*. Particularly intriguing is the presence of a variable number of transcriptional regulators and transporters. Pathogens with restricted niches appear to have reduced transport capabilities and regulatory potential when compared to bacteria that can be found in a variety of environments. This is in agreement with what was found for *Xylella*, but not what was found for *Lxx*. The theory of a recent event of niche restriction fits well within this context, along with the large number of pseudogenes still recognized by BLAST searches. A closer analysis of the origin of those pseudogenes may help to understand the evolution of the interaction of *Lxx* and sugarcane. Within the same context, the comparative genomic analysis with *L. xyli* subsp. *cynodontis* (*Lxc*) genome, a bacterium that can be inoculated in the xylem of sugarcane causing none of the RSD disease symptoms would provide valuable new insights on *Lxx* biology. Currently, we are investigating the presence of IS elements associated with specific or rearranged fragments between *Lxx* and *Lxc* genomes (Zerillo *et al.*, unpublished).

As yet, the relationship of *Lxx* and sugarcane is intriguing, as the mechanisms of colonization and symptom-induction are not completely understood. The discrete symptoms of the disease even at high bacterial titres of plant-infection was tentatively explained by the limited number of pathogenicity genes (Monteiro-Vitorello *et al.*, 2004), which made *Lxx* be regarded as a near-perfect pathogen (Metzler *et al.*, 1997). The evolutionary forces of lateral gene transfer have also played an important role in shaping the *Lxx* genome by restricting the niche as well as bringing in new pathogenicity genes. It is also worthwhile mentioning the presence of *desA* in *Lxx*, suggesting the production of an ABA-like molecule that could help to explain the symptoms of stunting associated with RSD.

Lxx used to be part of the *Clavibacter* genus and indeed *Lxx* and *Clavibacter michiganensis* are closer to each other than to any other Actinobacterium with a completely sequenced genome. Preliminary analyses using draft sequences of *C. michiganensis* subsp. *michiganensis* (*Cmm*) and *C. michiganensis* subsp. *sepedonicus* (*Cms*) have shown that none of the *Lxx* genomic islands are present, even though some similarity exists among most of the transposases found in *Lxx*GI3 and the genome of *Cms*. The cluster of *Lxx* genes associated with flagella assembly is not found in either of the two *Clavibacter* genomes. However, the desaturase *desA* gene is present in the *Clavibacter* genomes and is thus a strong candidate for further experimental analysis.

Some questions about *Lxx* biology have been answered, but many other questions have been raised since finishing the *Lxx* genome sequence. We have presented the information of what is known in terms of crop management of the disease and genomic analysis in this review so other questions can be raised, leading to development of new hypotheses and more studies be devoted to understanding the interaction of *Lxx* and sugarcane.

Acknowledgements

This work was supported by grants from Fundação de Amparo à Pesquisa do Estado de São Paulo (FAPESP) and Conselho Nacional de Desenvolvimento Científico e Tecnológico (CNPq) to C.B.M.V., to M.A.V.S. and to L.E.A.C., and a scholarship to M.M.Z.

References

Armstrong, G.A. (1997). Genetics of eubacterial carotenoid biosynthesis: a colorful tale. Annu. Rev. Microbiol. 51, 629–659.

Arora, S.K., Ritchings, B.W., Almira, E.C., Lory, S., and Ramphal, R. (1998). The *Pseudomonas aeruginosa* flagellar cap protein, FliD, is responsible for mucin adhesion. Infect. Immun. 66, 1000–1007.

Audenaert, K., De Meyer, G.B., and Höfte, M.M. (2002). Abscisic acid determines basal susceptibility of tomato to *Brotytis cinerea* and suppresses salicylic acid-dependent signaling mechanisms. Plant. Physiol. 128, 491–501.

Babu, M.M. (2003). Did the loss of sigma factors initiate pseudogene accumulation in *M. leprae*? Trends Microbiol. 11, 59–61.

Bartley, G.E., and Scolnik, P.A. (1995). Plant carotenoids: pigments for photoprotection, visual attraction, and human health. Plant Cell. 7, 1027–1038.

Basnayake, W.V., Birch, R.G. (1995). A gene from *Alcaligenes denitrificans* that confers albicidin resistance by reversible antibiotic binding. Microbiology. 141, 551–560.

Bhattacharyya, A., Stilwagen, S., Ivanova, N., D'Souza, M., Bernal, A., Lykidis, A., Kapatral, V., Anderson, I., Larsen, N., Los, T., Reznik, G., Selkov Jr., E., Walunas, T.L., Feil, H., Feil, W.S., Purcell, A., Lassez, J.L., Hawkins, T.L., Haselkorn, R., Overbeek, R., Predki, P.F., and Kyrpides, N.C. (2002). Whole-genome comparative analysis of three phytopathogenic *Xylella fastidiosa* strains. Proc. Natl. Acad. Sci. USA. 99, 12403–12408.

Birch, R.G., and Patil, S.S. (1985). Preliminary characterization of an antibiotic produced by *Xanthomonas albilineans* which inhibits DNA synthesis in *Escherichia coli*. J. Gen. Microbiol. 131, 1069–1075.

Birch, R.G., and Patil, S.S. (1987a). Correlation between albicidin production and chlorosis induction by *Xanthomonas albilineans*, the sugarcane leaf scald pathogen. Physiol. Mol. Plant Pathol. 30, 199–206.

Birch, R.G., and Patil, S.S. (1987b). Evidence that an albicidin-like phytotoxin induces chlorosis in sugarcane leaf scald disease by blocking plastid DNA replication. Physiol. Mol. Plant. Pathol. 30, 207–214.

Birch, R.G., Pemberton, J.M., and Basnayake, W.V. (1990). Stable albicidin resistance in *Escherichia coli* involves an altered outermembrane nucleoside uptake system. J. Gen. Microbiol. *136*, **51–58.**

Blanco, J., Coque, J.J., and Martin, J.F. (1998). The folate branch of the methionine biosynthesis pathway in *Streptomyces lividans*: disruption of the 5,10-methylenetetrahydrofolate reductase gene leads to methionine auxotrophy. J. Bacteriol. 180, 1586–1591.

Brumbley, S.M., Petrasovits, L.A., Birch, R.G., and Taylor, P.W.J. (2002). Transformation and transposon mutagenesis of *Leifsonia xyli* subsp. *xyli*, causal organism of Ratoon Stunting Disease of sugarcane. Mol. Plant. Microbe Interact. 15, 262–268.

Brumbley, S.M., Petrasovits, L.A., Hermann, S.R., Young, A.J., and Croft, B.J. (2006). Recent advances in the molecular biology of *Leifsonia xyli* subsp. *xyli*, causal organism of ratoon stunting disease. Austr. Plant. Pathol. 35, 681–689.

Cole, S.T., Eiglmeier, K., Parkhill, J., James, K.D., Thomson, N.R., Wheeler, P.R., Honoré, N., Garnier, T., Churcher, C., Harris, D., Mungall, K., Basham, D., Brown, D., Chillingworth, T., Connor, R., Davies, R.M., Devlin, K., Duthoy, S., Feltwell, T., Fraser, A., Hamlin, N., Holroyd, S., Hornsby, T., Jagels, K., Lacroix, C., Maclean, J., Moule, S., Murphy, L., Oliver, K., Quail, M.A., Rajandream, M.A., Rutherford, K.M., Rutter, S., Seeger, K., Simon, S., Simmonds, M., Skelton, J., Squares, R., Squares, S., Stevens, K., Taylor, K., Whitehead, S., Woodward, J.R., Barrell, B.G. (2001). Massive gene decay in the leprosy bacillus. Nature. 409, 1007–1011.

Davis, M.J., Gillaspi.e. Jr., A.G., Harris, R.W., and Lawson, R.H. (1980). Ratoon stunting disease of sugarcane – isolation of the causal bacterium. Science. 210, 1365–1367.

Davis, M.J., Gillaspie, A.G., Vidaver, A.K., and Harris, R.W. (1984). *Clavibacter*: A new genus containing some phytopathogenic coryneform bacteria, including *Clavibacter xyli* subsp. *xyli* sp. nov., subsp. nov. and *Clavibacter xyli* subsp. *cynodontis* subsp. nov., pathogens that cause ratoon stunting disease of sugarcane and Bermuda grass stunting disease. Int. J. Syst. Bacteriol. 34, 107–117.

Davis, M.J. and Vidaver, A.K. (2001). Coryneform plant pathogens. In Laboratory Guide for Identification of Plant Pathogenic Bacteria, N.W. Schaad, J.B. Jones and W. Chun, 3rd ed. (Minnesota, USA: APS Press), pp. 218–235.

Dong, Z., Mc Cully, M.E., and, Canny, M.J. (1997). Does *Acetobacter diazotrophicus* live and move in the xylem of sugarcane stems? Anatomical and physiological data. Ann. Bot. 80, 147–158.

Dreier, J., Meletzus, D., Eichenlaub, R. (1997). Characterization of the plasmid encoded virulence region *pat-1* of phytopathogenic *Clavibacter michiganensis* subsp. *michiganensis*. Mol. Plant Microbe Interact. 10, 195–206.

Evtushenko, L.I., Dorofeeva, L.V., Subbotin, S.A., Cole, J.R., and Tiedje, J.M. (2000). *Leifsonia poae* gen. nov., sp. nov., isolated from nematode galls on *Poa annua*, and reclassification of '*Corynebacterium aquaticum*' Leifson 1962 as *Leifsonia aquatica* (ex Leifson 1962) gen. nov., nom. rev., comb. nov. and *Clavibacter xyli* Davis et al. 1984 with two subspecies as *Leifsonia xyli* (Davis et al. 1984) gen. nov., comb. nov. Int. J. of Syst. and Evol. Microb. 50, 371–380.

Gartemann, K.H., Kirchner, O., Engemann, J., Grafen, I., Eichenlaub, R., and Burger, A. (2003). *Clavibacter michiganensis* subsp. *michiganensis*: first steps in the understanding of virulence of a Gram-positive phytopathogenic bacterium. J. Biotechnol. 106, 179–191.

Gillaspie, A.G. (1978). Ratoon stunting disease of sugarcane – serology. Phytopathology. 68, 529–532.

Gillaspie, A.G., and Teakle, D.S. (1989) Diseases of Sugarcane: Major Diseases, Ricaud et al. ed. (Amsterdam, ND: Elsiever Science Publishers), pp. 399.

James, G. (1996). A review of ratoon stunting disease. Int. Sugar J. 98, 532–541.

Kende, H., and Zeevaart, J. (1997). The Five 'Classical' Plant Hormones. Plant Cell. 9, 1197–1210.

Kao, J., and Damann, K.E. Jr., (1980) In situ localisation and morphology of the bacterium associated with ratoon stunting disease of sugarcane. Canadian Journal of Botany. 58, 310–315.

Krubasik, P., and Sandmann, G. (2000). A carotenogenic gene cluster from *Brevibacterium linens* with novel lycopene cyclase genes involved in the synthesis of aromatic carotenoids. Mol. Gen. Genet. 263, 423–432.

Krubasik, P., Takaichi, S., Maoka, T., Kobayashi, M., Masamoto, K., Sandmann, G. (2001). Detailed biosynthetic pathway to decaprenoxanthin diglucoside in *Corynebacterium glutamicum* and identification of novel intermediates. Arch. Microbiol. 176, 217–223.

Lawrence, J.G., Hendrix, R.W., and Casjens, S. (2001). Where are the pseudogenes in bacterial genomes? Trends Microbiol. 9, 535–540.

Lee, P.C., and Schmidt-Dannert, C. (2002). Metabolic engineering towards biotechnological production of carotenoids in microorganisms. Appl. Microbiol. Biotechnol. 60, 1–11.

Loria, R., Kers, J., and Joshi, M. (2006). Evolution of Plant Pathogenicity in *Streptomyces*. Annu. Rev. Phytopathol. 44, 469–487.

Madden, J.C., Ruiz, N., Caparon, M. (2001). Cytolysin-mediated translocation (CMT): a functional equivalent of type III secretion in gram-positive bacteria. Cell. 104, 143–152.

Magarey, R.C., Suma, S., Irawan, Kuniata, L.S., and Allsopp, P.G. (2002). Sik na binatang bilong suka — diseases and pests encountered during a survey of *Saccharum* germplasm 'in the wild' in Papua New Guinea. Proc. of the Aust. Soc. of Sugar Cane Technologists. 24, 219–227.

Mahillon, J., and Chandler, M. (1998). Insertion sequences. Microbiol. Mol. Biol. R. 62, 725–774.

Marri, P.R., Bannantine, J.P., and Golding, G.B. (2006). Comparative genomics of metabolic pathways in *Mycobacterium* species: gene duplication, gene decay and lateral gene transfer. FEMS Microbiol. Rev. 30, 906–925.

Metzler, M.C., Laine, M.J., and De Boer, S.H. (1997). The status of molecular biological research on the plant pathogenic genus *Clavibacter*. FEMS Microbiol. Lett. 150, 1–8.

Mohr, P.G., and Cahill, D.M. (2007). Suppresion by ABA of salicylic acid and lignin accumulation and the expression of multiple genes, in Arabidopsis infected with *Pseudomonas sytingae* pv. tomato. Funct. Integr. Genomics. 7, 181–191.

Monteiro-Vitorello, C.B., Camargo, L.E.A., Van Sluys, M.A., Kitajima, J.P., Truffi, D., do Amaral, A.M., Harakava, R., de Oliveira, J.C., Wood, D., de Oliveira, M.C., Miyaki, C., Takita, M.A., da Silva, A.C., Furlan, L.R., Carraro, D.M., Camarotte, G., Almeida, N.F.Jr, Carrer, H., Coutinho, L.L., El-Dorry, H.A., Ferro, M.I., Gagliardi, P.R., Giglioti, E., Goldman, M.H., Goldman, G.H., Kimura, E.T., Ferro, E.S., Kuramae, E.E., Lemos, E.G., Lemos, M.V., Mauro, S.M., Machado, M.A., Marino, C.L., Menck, C.F., Nunes, L.R., Oliveira, R.C., Pereira, G.G., Siqueira, W., de Souza, A.A., Tsai, S.M., Zanca, A.S., Simpson, A.J., Brumbley, S.M., Setúbal, J.C. (2004). The genome sequence of the Gram-positive sugarcane pathogen *Leifsonia xyli* subsp. *xyli*. Mol. Plant Microbe Interact. 17, 827–836.

Pan, Y.B., Grisham, M.P., Burner, D.M., Damann, K.E., and Wei, Q. (1998). A Polimerase Chain Reaction protocol for the detection of *Clavibacter xyli* subsp. *xyli*, the causal bacterium of Ratoon Stuntig Disease. Plant Dis. 82, 285–290.

Parker, C., Becker, E., Zhangm X., Jandle, S., and Meyer, R. (2005). Elements in the co-evolution of relaxases and their origins of transfer. Plasmid. 53, 113–118.

Ren, Q., Chen, K., and Paulsen, I.T. (2007). TransportDB: a comprehensive database resource for cytoplasmic membrane transport systems and outer membrane channels. Nucleic Acids Res. 35, 274–279.

Rott, P.C., Costet, L., Davis, M.J., Frutos, R., and Gabriel, D.W. (1996). At least two separate gene clusters are involved in albicidin production by *Xanthomonas albilineans*. J. Bacteriol. 178, 4590–4596.

Schell, M.A., Karmirantzou, M., Snel, B., Vilanova, D., and Berger, B. (2002). The genome sequence of *Bifidobacterium longum* reflects its adaptation to the human gastrointestinal tract. Proc. Natl. Acad. Sci. USA. 99, 14422–14427.

da Silva, A.C., Ferro, J.A., Reinach, F.C., Farah, C.S., Furlan, L.R., Quaggio, R.B., Monteiro-Vitorello, C.B., Van Sluys, M.A., Almeida, N.F., Alves, L.M., do Amaral, A.M., Bertolini, M.C., Camargo, L.E., Camarotte, G., Cannavan, F., Cardozo, J., Chambergo, F., Ciapina, L.P., Cicarelli, R.M., Coutinho, L.L., Cursino-Santos, J.R., El-Dorry, H., Faria, J.B., Ferreira, A.J., Ferreira, R.C., Ferro, M.I., Formighieri, E.F., Franco, M.C., Greggio, C.C., Gruber, A., Katsuyama, A.M., Kishi, L.T., Leite, R.P., Lemos, E.G., Lemos, M.V., Locali, E.C., Machado, M.A., Madeira, A.M., Martinez-Rossi, N.M., Martins, E.C., Meidanis, J., Menck, C.F., Miyaki, C.Y., Moon, D.H., Moreira, L.M., Novo, M.T., Okura, V.K., Oliveira, M.C., Oliveira, V.R., Pereira, H.A., Rossi, A., Sena, J.A., Silva, C., de Souza, R.F., Spinola, L.A., Takita, M.A., Tamura, R.E., Teixeira, E.C., Tezza, R.I., Trindade dos Santos, M., Truffi, D., Tsai, S.M., White, F.F., Setubal, J.C., Kitajima, J.P. (2002). Comparison of the genomes of two *Xanthomonas* pathogens with differing host specificities. Nature. 417, 459–63.

Simpson, A.J., Reinach, F.C., Arruda, P., Abreu, F.A., Acencio, M., Alvarenga, R., Alves, L.M., Araya, J.E., Baia, G.S., Baptista, C.S., Barros, M.H., Bonaccorsi, E.D., Bordin, S., Bové, J.M., Briones, M.R., Bueno, M.R., Camargo, A.A., Camargo, L.E., Carraro, D.M., Carrer, H., Colauto, N.B., Colombo, C., Costa, F.F., Costa, M.C., Costa-Neto, C.M., Coutinho, L.L., Cristofani, M., Dias-Neto, E., Docena, C., El-Dorry, H., Facincani, A.P., Ferreira, A.J., Ferreira, V.C., Ferro, J.A., Fraga, J.S., França, S.C., Franco, M.C., Frohme, M., Furlan, L.R., Garnier, M., Goldman, G.H., Goldman, M.H., Gomes, S.L., Gruber, A., Ho, P.L., Hoheisel, J.D., Junqueira, M.L., Kemper, E.L., Kitajima, J.P., Krieger, J.E., Kuramae, E.E., Laigret, F., Lambais, M.R., Leite, L.C., Lemos, E.G., Lemos, M.V., Lopes, S.A., Lopes, C.R., Machado, J.A., Machado, M.A., Madeira, A.M., Madeira, H.M., Marino, C.L., Marques, M.V., Martins, E.A., Martins, E.M., Matsukuma, A.Y., Menck, C.F., Miracca, E.C., Miyaki, C.Y., Monteiro-Vitorello, C.B., Moon, D.H., Nagai, M.A., Nascimento, A.L., Netto, L.E., Nhani, A.Jr., Nobrega, F.G., Nunes, L.R., Oliveira, M.A., de Oliveira, M.C., de Oliveira, R.C., Palmieri, D.A., Paris, A., Peixoto, B.R., Pereira, G.A., Pereira, H.A.Jr., Pesquero, J.B., Quaggio, R.B., Roberto, P.G., Rodrigues, V. M., Rosa, A.J., de Rosa, V.E.Jr, de Sá, R.G., Santelli, R.V., Sawasaki, H.E., da Silva, A.C., da Silva, A.M., da Silva, F.R., da Silva, W.A.Jr., da Silveira, J.F., Silvestri, M.L., Siqueira, W.J., de Souza, A.A., de Souza, A.P., Terenzi, M.F., Truffi, D., Tsai, S.M., Tsuhako, M.H., Vallada, H., Van Sluys, M.A., Verjovski-Almeida, S., Vettore, A.L., Zago, M.A., Zatz, M., Meidanis, J., Setubal, J.C. (2000). The genome sequence of the plant pathogen *Xylella fastidiosa*. The *Xylella fastidiosa* Consortium of the Organization for Nucleotide Sequencing and Analysis. Nature. *406*, 151–159.

Steindl, D.R.L. (1976). The use of Phase Contrast Microscopy in the identification of Ratoon Stunting Disease. Proc. Queensl. Soc. Sugarcane Technol. *43*, 71–77.

Stover, C.K., Pham, X.Q., Erwin, A.L., Mizoguchi, S.D., Warrener, P., Hickey, M.J., Brinkman, F.S., Hufnagle, W.O., Kowalik, D.J., Lagrou, M., Garber, R.L., Goltry, L., Tolentino, E., Westbrock-Wadman, S., Yuan, Y., Brody, L.L., Coulter, S.N., Folger, K.R., Kas, A., Larbig, K., Lim, R., Smith, K., Spencer, D., Wong, G.K., Wu, Z., Paulsen, I.T., Reizer, J., Saier, M.H., Hancock, R.E., Lory, S., Olson, M.V. (2000). Complete genome sequence of *Pseudomonas aeruginosa* PA01, an opportunistic pathogen. Nature. *406*, 959–964.

Taguchi, F., Shimizu, R., Inagaki, Y., Toyoda, K., Shiraishi, T., and Ichinose, Y. (2003) Post-translational modification of flagellin determines the specificity of HR induction. Plant Cell Physiol. *44*, 342–349.

Teakle, D.S., Appleton, J.M., and Steindl, D.R.L. (1978). Anatomical basis for resistance of sugar-cane to ratoon stunting disease. Physiological. Plant. Pathol. *12*, 83–91.

Thomson, N., Bentley, S., Holden, M., and Parkhill, J. (2003). Fitting the niche by genomic adaptation. Nat. Rev. Microbiol. *1*:92–93.

Van Sluys, M.A., Monteiro-Vitorello, C.B., Camargo, L.E.A., Menck, C.F., da Silva, C.R., Ferro, J.A., Oliveira, M.C., Setubal, J.C., Kitajima, J.P., Simpson, A.J. (2002). Comparative genomic analysis of plant-associated bacteria. Annu. Rev. Phytopathol. *40*, 169–189.

Ventura, M., Canchaya, C., Tauch, A., Chandra, G., Fitzgerald, G.F., Chater, K.F., and van Sinderen, D. (2007). Genomics of Actinobacteria: tracing the evolutionary history of an ancient phylum. Microbiol. Mol. Biol. Rev. *71*, 495–548.

Walker, M.J., Birch, R.G., and Pemberton, J.M. (1988). Cloning and characterization of an albicidin resistance gene from *Klebsiella oxytoca*. Mol. Microbiol. *2*, 443–454.

Weaver, L., Teakle, D.S., and Hayward, A.C. (1977). Ultrastructural studies on bacterium associated with ratoon stunting disease of sugar-cane. Aust. J. Agric. Res. *28*, 843–852.

Young, A.J., Petrasovits, L.A., Croft, B.J., Gillings, D.M., and Brumbley, S.M. (2006). Genetic uniformity of international isolates of *Leifsonia xyli* subsp. *xyli*, causal agent of ratoon stunting disease of sugarcane. Austr. Plant. Pathol. *35*, 503–511.

Zhang, L., and Birch, R.G. (1997). The gene for albicidin detoxificationfrom *Pantoea dispersa* encodes an esterase and attenuates pathogenicity of *Xanthomonas albilineans* to sugarcane. Proc. Natl. Acad. Sci. USA. *94*, 9984–9989.

Genomics-driven Advances in *Xanthomonas* Biology

Damien F. Meyer and Adam J. Bogdanove

Abstract

The genus *Xanthomonas* consists of 20 plant-associated species, many of which cause important diseases of crops and ornamentals. Individual species comprise multiple pathovars, characterized by distinctive host specificity or mode of infection. Genomics is at the centre of a revolution in *Xanthomonas* biology. Complete genome sequences are available for nine *Xanthomonas* strains, representing three species and five pathovars, including vascular and non-vascular pathogens of the important models for plant biology, *Arabidopsis thaliana* and rice. With the diversity of complete and pending *Xanthomonas* genome sequences, the genus has become a superb model for understanding functional, regulatory, epidemiological, and evolutionary aspects of host- and tissue-specific plant pathogenesis. In this chapter, we review structural, functional, and comparative genomics studies that are driving rapid advances in our understanding of this important group of bacteria.

Introduction

Xanthomonas is a genus in the gamma subdivision of the Proteobacteria that contains a large number of plant pathogens. Complete genome sequences are available for several, diverse *Xanthomonas* strains, and more are pending. This chapter presents an overview of the genus, a review of functional, structural, and comparative genomics studies to date, and prospects for continued rapid progress toward understanding mechanistic, regulatory, epidemiological, and evolutionary aspects of host- and tissue-specific plant pathogenesis through genome-enabled studies of this important group of bacteria.

Xanthomonas, a uniquely important genus for genomic analysis

The genus *Xanthomonas* (from Gk. *xanthos* 'yellow' and *monas* 'entity') consists of 20 plant-associated species, many of which cause important diseases of crops and ornamentals. Individual species comprise multiple pathogenic variants (pathovars, pv.). Collectively, members of the genus cause disease on at least 124 monocot species and 268 dicot species, including fruit and nut trees, solanaceous and brassicaceous plants, and cereals (Hayward, 1993). They cause a variety of symptoms including necrosis, cankers, spots, and blight, and they affect a variety of plant parts including leaves, stems, and fruits (Leyns et al., 1984). The collectively broad host range of the genus contrasts strikingly with the typically narrow host range of individual species and pathovars (Vauterin et al., 1995), which also typically exhibit a marked tissue-specificity, infecting either through stomata to colonize the intercellular spaces of the mesophyll parenchyma, or via hydathodes (water pores at the leaf margin) or wounds to spread systemically through the vascular system.

The pathogenic diversity of *Xanthomonas* contrasts with a characteristic uniformity with regard to morphology and physiology. Cells are rod-shaped, round-ended, and vary in length from approximately 0.7 µm to 2.0 µm and in width from 0.4 µm to 0.7 µm. They are motile by a single polar flagellum. *Xanthomonas* species are catalase positive, unable to reduce nitrate, and weak producers of acids from carbohydrates (Bradbury, 1984). The vast majority are also yellow pigmented, due to the production of photoprotector carotenoids called xanthomonadins (Rajagopal *et al.*, 1997). Colonies of *Xanthomonas* are also typically highly mucoid on sugar-rich media, due to production of the exopolysaccharide xanthan (Sutherland, 1993). The xanthan of *X. campestris* pv. *campestris* (Xcc) is used extensively in food, cosmetic, and oil-producing industries (Sutherland, 1993; Becker *et al.*, 1998). Xanthan often plays an important role in pathogenicity but is also believed to be important in survival of the bacterium, protecting against UV light, freezing, and desiccation (Bretschneider *et al.*, 1989; Dow and Daniels, 1994; Rajeshwari and Sonti, 2000; Dow *et al.*, 2003; Crossman and Dow, 2004).

The morphological and physiological uniformity within *Xanthomonas* hampered the genesis of a stable classification in the genus for a long time. Early on, a taxonomy based on pathogenicity was attempted. Each *Xanthomonas* variant presenting a different host range or inducing different disease symptoms was considered as a full species based on a 'new host-new species' concept (Starr, 1981). This led to a complex genus with 100 species otherwise indistinguishable with the phenotypic tests available. The later adoption of the infrasubspecific designation of pathovar (Young *et al.*, 1978), led to the assignment of 140 variants as pathovars of *X. campestris* (Hayward, 1993). Despite its practical utility, this classification, based essentially on a single feature, was insufficient to represent both the diversity and the evolutionary relationships within the genus.

With the rapid expansion of molecular genetics, a proposal emerged for a new classification of *Xanthomonas* using a polyphasic approach that integrates phenotypic and genotypic data (Vandamme *et al.*, 1996; for further examples, see chapters by Stavrinides, and Bull and Vinatzer). As molecular genotyping techniques such as DNA-DNA hybridization, PCR based on repeated sequences (rep-PCR), and genotyping by amplified fragment-length polymorphism (AFLP) were developed and applied, the taxonomy of *Xanthomonas* came to reflect a more precise phylogenetic classification (Vauterin *et al.*, 1995; Rademaker *et al.*, 2000; Vauterin *et al.*, 2000). Thus, the current taxonomy, proposed by Vauterin *et al.* (1995) using polyphasic analysis including DNA-DNA hybridization, and later substantiated and refined by Rademaker *et al.* (2005) using rep-PCR, recognizes 20 species (genomic groups). A majority of these are composed of former *X. campestris* pathovars. *X. campestris* has contracted to six pathovars: *aberrans*, *armoraciae*, *barbareae*, *campestris*, *incanae*, and *raphani*. In the Rademaker *et al.* study, six distinct subgroups of *X. axonopodis* (genomic group 9) were proposed as candidates for further characterization to determine whether they should be elevated to distinct species. At the same time, less dramatic but robust distinctions were made among different pathovars and in some cases even among particular strains of a pathovar, by rep-PCR based clustering. Finally, examples were found both of apparent convergent evolution with regard to pathogenic traits, e.g. isolates in different genomic groups that infect the same host(s), as well as divergent evolution, e.g. isolates in the same genomic group that infect different hosts, or the same host differently (Rademaker *et al.*, 2005).

The current taxonomy represents a solid framework in which poorly studied and future isolates can be evaluated, and to which new species can be added. Also, by establishing molecular genetic relationships among the more than 140 known members of the genus with distinctive pathogenic characteristics, it sets the stage for informed comparisons directed toward identifying unique determinants of host and tissue specificity as well as pathogenicity factors that are universally important in plant disease. Recently completed and ongoing sequencing of several *Xanthomonas* genomes provides an unprecedented opportunity to carry out these comparisons in a comprehensive and conclusive way.

Xanthomonas genome sequences

Complete genome sequences of nine *Xanthomonas* strains, representing three species and five pathovars, have been determined and made available (Table 7.1). These include strain

Table 7.1 Sequenced *Xanthomonas* genomes[1]

Organism	Disease	Abbreviation	Size (Mb)	Components	% G+C	% coding	Genes	% genes assigned a role category	GenBank accession(s)	Reference
X. axonopodis pv. *citri* 306	Citrus canker	Xac	5.27	Circular chromosome (5175554 bp), plasmids pXAC64 (64920 bp), pXAC33 (33700 bp)	64.8	90.3	5809	48.0	NC_003919, NC_003922, NC_003921	da Silva et al. (2002)
X. axonopodis pv. *vesicatoria* (*X. campestris* pv. *vesicatoria*) 85-10	Bacterial spot disease of pepper and tomato	Xav	5.42	Circular chromosome (5178466 bp), plasmids pXCV183 (182572 bp), pXCV38 (38116 bp), pXCV19 (19146 bp), pXCV2 (1852 bp)	64.6	86.6	5229	73.2	NC_007508, NC_007507, NC_007506, NC_007505, NC_007504	Thieme et al. (2005)
X. campestris pv. *campestris* 8004	Black rot of crucifers	Xcc8	5.15	Circular chromosome	65.0	87.2	5079	67.0	NC_007086	Qian et al. (2005)
X. campestris pv. *campestris* ATCC33913	Black rot of crucifers	Xcc	5.08	Circular chromosome	65.1	90.1	5832	48.1	NC_003902	da Silva et al. (2002)
X. campestris pv. *armoraciae* 756C	Leaf spot disease of crucifers	Xca	4.94	Circular chromosome	65.3	85.3	4598	69.2	Pending[2]	A.J. Bogdanove, unpublished
X. oryzae pv. *oryzae* KACC10331	Bacterial blight of rice	XooK	4.94	Circular chromosome	63.7	87.6	5805	71.1	NC_006834	Lee et al. (2005)
X. oryzae pv. *oryzae* MAFF311018	Bacterial blight of rice	XooM	4.94	Circular chromosome	63.7	83.9	5091	55.0	NC_007705	Ochiai et al. (2005)
X. oryzae pv. *oryzae* PXO99^A	Bacterial blight of rice	XooP	5.24	Circular chromosome	63.6	83.0	5083	55.0	CP000967	Salzberg et al. (2008)
X. oryzae pv. *oryzicola* BLS256	Bacterial leaf streak of rice	Xoc	4.83	Circular chromosome	64.0	86.0	4686	72.2	AAQN01000001[3]	A.J. Bogdanove, unpublished

[1]To allow for direct comparison, genome statistics were derived from the TIGR automated annotation for each genome available through the Comprehensive Microbial Resource (CMR; http://cmr.jcvi.org), since methods used in authors' original annotation may differ across genomes.

[2]Finished sequence and draft annotation are available through the CMR.

[3]Finished sequence only. Finished sequence and draft annotation are available through the CMR.

306 of *X. axonopodis* pv. *citri* (Xac), which causes citrus canker (da Silva *et al.*, 2002); strain 85-10 of *X. axonopodis* pv. *vesicatoria* (Xav), the bacterial spot pathogen of pepper and tomato, formerly a pathovar of *X. campestris*, and now a proposed new species, *X. euvesicatoria* (Jones *et al.*, 2004; Thieme *et al.*, 2005); strains 8004 and ATCC33913 of *X. campestris* pv. *campestris* (Xcc8 and XccA), the causal agent of black rot in crucifers, including the model plant *Arabidopsis thaliana* (da Silva *et al.*, 2002; Qian *et al.*, 2005); strain 756C of *X. campestris* pv. *armoraciae* (Xca), which causes bacterial spot disease of crucifers (A. J. Bogdanove, unpublished); strains KACC10331, MAFF311018, and PXO99^A of *X. oryzae* pv. *oryzae* (XooK, XooM, and XooP), which is responsible for bacterial blight of rice (Lee *et al.*, 2005; Ochiai *et al.*, 2005; Salzberg *et al.*, 2008); and strain BLS256 of *X. oryzae* pv. *oryzicola* (Xoc), which is the causal agent of bacterial leaf streak of rice (A. J. Bogdanove, unpublished). A partial sequence is also available for Xcc strain B100 (Vorholter *et al.*, 2003). The completely sequenced genomes are similar in general characteristics: sizes range from 4.83 million base pairs (Mb) to 5.42 Mb, G+C contents from 63.6% to 65.3%, and numbers of genes from 4598 to 5809. Gene content is largely conserved, but whole genome alignments reveal numerous inversions, indels, and rearrangements in the genomes relative to one another (Thieme *et al.*, 2005; Fig. 7.1). The structural variation among these genomes suggests a high degree of genome plasticity within the genus overall, consistent with the molecular genotyping studies cited above.

A striking feature shared by the *Xanthomonas* genomes is an abundance of insertion sequence (IS) elements, which are postulated to be important drivers of *Xanthomonas* genome evolution (Monteiro-Vitorello *et al.*, 2005). In addition to serving as vectors for lateral gene transfer, IS elements can generate other types of genome modifications, including rearrangements, inversions, and deletions, any of which can lead to acquisition, modification, or loss of gene content. The *X. oryzae* genomes have the greatest number and diversity of IS elements. For example, there are about 700 IS elements or element fragments in the XooP genome (Salzberg *et al.*, 2008). Not surprisingly, alignment of the Xoo genomes shows at least ten major rearrangements and indels in XooP relative to the XooK and XooM genomes, and seven of these are indeed associated with IS elements (Salzberg *et al.*, 2008). Phage-related sequences are also prevalent among the genomes, suggesting that interactions with phage have contributed to genome variation within the genus (da Silva *et al.*, 2002; Lee *et al.*, 2005; Ochiai *et al.*, 2005; Qian *et al.*, 2005; Thieme *et al.*, 2005; A. J. Bogdanove, unpublished; Salzberg *et al.*, 2008).

The sequenced *Xanthomonas* genomes represent a diversity of pathogen classes. Phylogeny of the group based on ribosomal RNA operon sequences is presented in Fig. 7.2, and is consistent with the current taxonomy (Vauterin *et al.*, 1995; Rademaker *et al.*, 2005). Relationships among these strains suggest comparisons of particular interest. For example, despite being pathogens of markedly distinct hosts, i.e. citrus trees vs. tomato and pepper plants, Xac and

Figure 7.1 (on facing page) Alignments of nine *Xanthomonas* genomes generated using MAUVE (Darling *et al.*, 2004, http://gel.ahabs.wisc.edu/mauve/). Locally collinear blocks (LCBs), shown as rounded rectangles, represent regions without rearrangement of homologous sequence across genomes. The orientation of the LCBs, forward or reverse, is indicated by their position above or below the line, respectively. Lines between genomes trace orthologous LCBs. Across all the genomes, using default parameters, resulting in a minimum LCB weight of 83, there are 154 LCBs. The LCB weight sets the minimum number of matching nucleotides in a collinear region for it to be considered homologous across genomes and not the result of a spurious match. Regions outside LCBs were too divergent in at least one genome to be aligned successfully. Inside each LCB, a similarity profile of the genome sequence is represented by vertical bars. The height of each bar corresponds to the average level of conservation in that region of the genome sequence. Xac, *X. axonopodis* pv. *citri* 306; Xav, *X. axonopodis* pv. *vesicatoria* (*X. campestris* pv. *vesicatoria*) 85-10; Xcc8, *X. campestris* pv. *campestris* 8004; XccA, *X. campestris* pv. *campestris* ATCC33913; Xca, *X. campestris* pv. *armoraciae* 756C; XooK, *X. oryzae* pv. *oryzae* KACC10331; XooM, *X. oryzae* pv. *oryzae* MAFF311018; XooP, *X. oryzae* pv. *oryzae* PXO99^A; Xoc, *X. oryzae* pv. *oryzicola* BLS256.

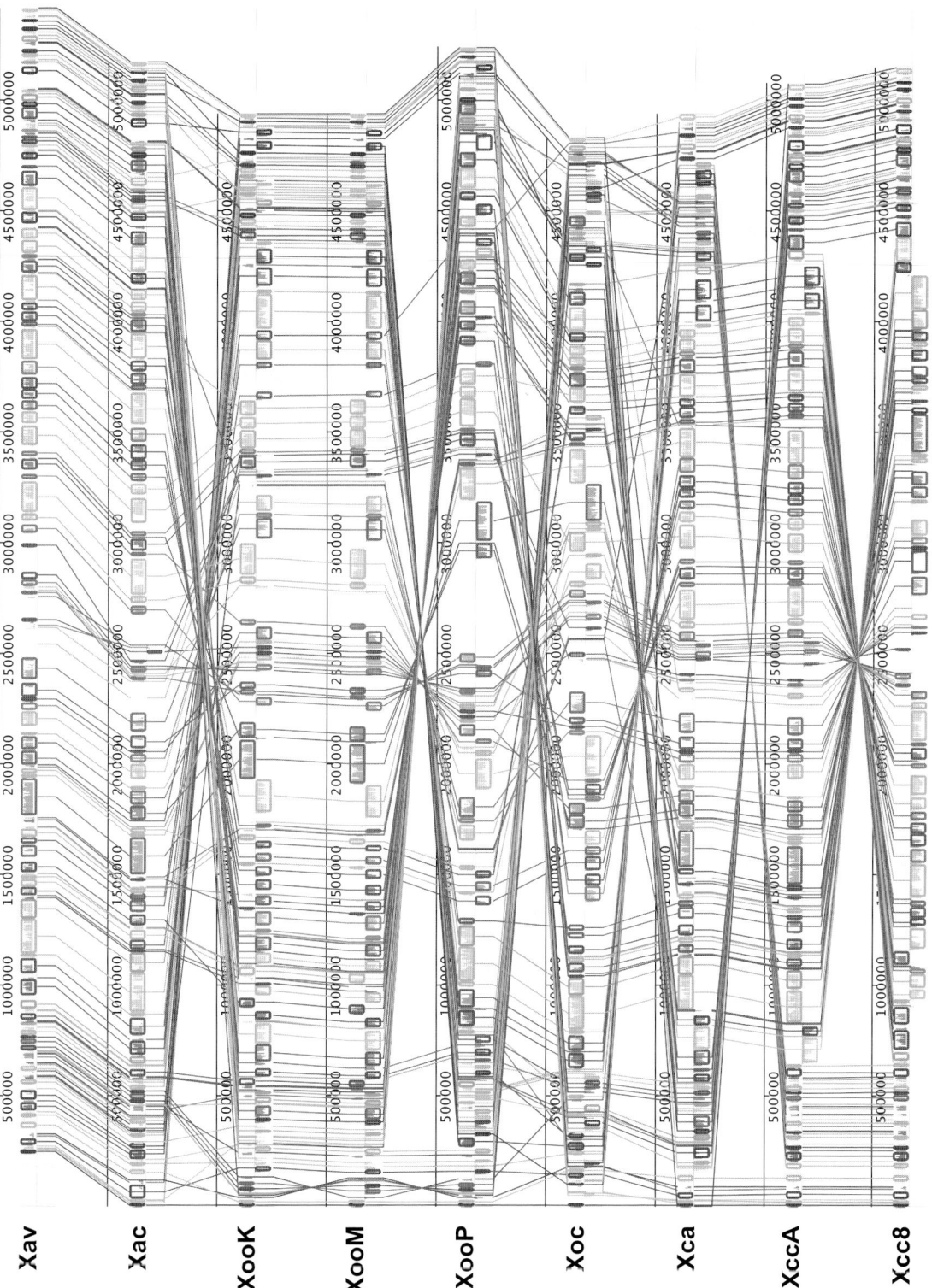

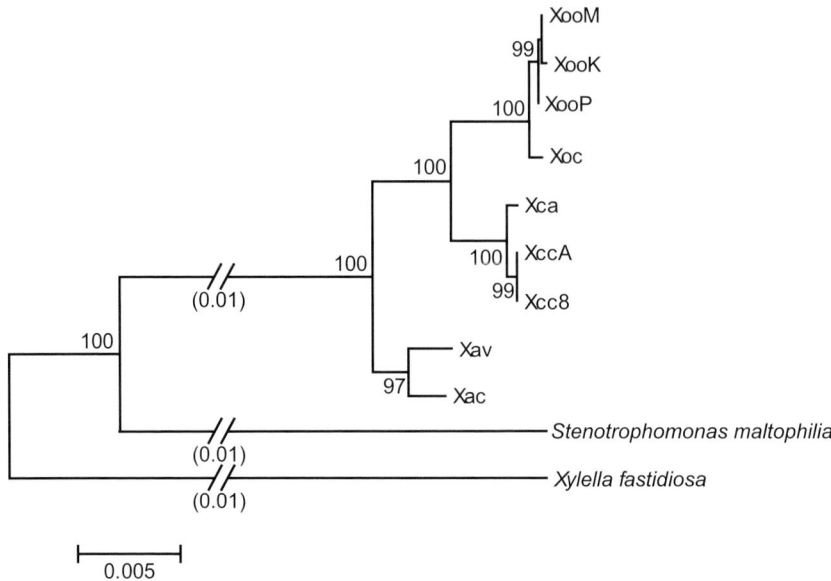

Figure 7.2 Phylogeny of the sequenced *Xanthomonas* strains, as well as *Xylella fastidiosa* strain 9a5c (citrus pathogen) and *Strenotrophomonas maltophilia* strain K279a, based on alignment of ribosomal RNA operon sequences. The alignment was carried out using ClustalW (Chenna et al., 2003). The tree was drawn using Phylip (http://evolution.genetics.washington.edu/phylip.html). Numbers above branches represent the bootstrap values for 1000 replicates. Xac, *X. axonopodis* pv. *citri* 306; Xav, *X. axonopodis* pv. *vesicatoria* (*X. campestris* pv. *vesicatoria*) 85-10; Xcc8, *X. campestris* pv. *campestris* 8004; XccA, *X. campestris* pv. *campestris* ATCC33913; Xca, *X. campestris* pv. *armoraciae* 756C; XooK, *X. oryzae* pv. *oryzae* KACC10331; XooM, *X. oryzae* pv. *oryzae* MAFF311018; XooP, *X. oryzae* pv. *oryzae* PXO99[A]; Xoc, *X. oryzae* pv. *oryzicola* BLS256.

Xav are closely related, and basal to the group overall. Thus, sequences representing candidate adaptations toward the respective hosts of these pathogens might be relatively straightforward to identify through a comparative genomic approach. Likewise, the Xcc and Xca genomes, and the Xoo and Xoc genomes, respectively, represent vascular and non-vascular pathogen pairs that infect leading models for dicot and monocot biology, namely, *A. thaliana* and rice. These genomes especially are a valuable starting point for comparative analysis toward identifying candidate determinants both of host and tissue specificity, and for functional characterization *in planta*. Comparative analysis across all the genomes has potential for identifying genes fundamental to pathogenesis, irrespective of the host and tissue infected. Complete or near complete genome data available for other members of the Xanthomonadaceae, namely *Xylella fastidiosa*, a highly adapted plant pathogen (Simpson et al., 2000; Bhattacharyya et al., 2002; Van Sluys et al., 2003), and *Stenotrophomonas maltophilia* (*S. maltophilia* K279a Sequencing Group at the Sanger Institute, ftp://ftp.sanger.ac.uk/pub/pathogens/sma/; The Joint Genome Institute, US Department of Energy, http://genome.jgi-psf.org/draft_microbes/stema/stema.download.ftp), which is not pathogenic to plants, empower this approach.

X. fastidiosa is a xylem-limited, insect-vectored plant pathogen with a genome roughly half the size of a typical *Xanthomonas* genome (van Sluys et al., 2002; see chapter by van Sluys et al.), leading some to postulate that it evolved from an ancestor shared with *Xanthomonas* by genome reduction during adaptation to life within its hosts (Dow and Daniels, 2000). *X. fastidiosa* strains collectively cause disease on diverse hosts, with some specificity (Purcell and Hopkins, 1996). The genomes of four strains, respectively pathogenic to citrus, almond, oleander, and grape, have been sequenced (see chapter by Varani et al.). *S. maltophilia* includes free-living as well as endophytic isolates, and opportunistic human pathogens (Berg et al., 1999). It represents an excellent backdrop against which adaptations in *Xanthomonas* for plant pathogenesis might be identified. In addition to the sequences described, several more *Xanthomonas* complete genomes are pending, including those of *X. albilineans*, *X. axonopodis* pv. *phaseoli*, and additional strains of Xcc and Xoo (M. Arlat, P. Rott, V. Verdier, and A. Puhler, personal communication). *X. albilineans* is an interesting species that has a genome roughly two-thirds the size of the sequenced *Xanthomonas* genomes, and depends largely on a single toxin for pathogenesis of its host, sugarcane. The *X. albilineans* genome may represent an ancestral genome that lacks many of the adaptations present in other *Xanthomonas* strains, or as postulated for *Xylella*, a reduced and highly adapted genome with a minimal complement of genes needed for survival within a plant. Comparisons with the genomes of other strains promise to shed light on this interesting question.

Altogether, with the multiple complete and pending *Xanthomonas* genome sequences, and a growing number of contextual and reference genome sequences, the genus is rapidly coalescing as a superb model for understanding functional, regulatory, epidemiological, and evolutionary aspects of host- and tissue- specific bacterial plant pathogenesis through comparative and functional genomic studies.

Functional genomics based on random mutagenesis, heterologous expression, and other genome-wide screening approaches

The beginning of the genomics era is generally associated with the release of the first bacterial whole genome sequence in 1995 (Fleischmann *et al.*, 1995). But genome-scale analyses were already yielding results prior to that time. In *Xanthomonas*, genome-wide mutagenesis and heterologous expression of whole genome libraries early on led to the identification of genes for essential pathogenicity factors. The clusters of *hrp* genes encoding the type III secretion system in Xcc and Xav were discovered and characterized by random, transposon-based mutagenesis (Arlat *et al.*, 1991; Bonas *et al.*, 1991). Subsequent studies, largely in Xav, further elucidated the functions of individual genes in the cluster, including the master *hrp* regulatory genes, *hrpG*

and *hrpX* (Schulte and Bonas, 1992; Wengelnik *et al.*, 1996; Rossier *et al.*, 1999; Wengelnik *et al.*, 1999; Rossier *et al.*, 2000). Transposon mutagenesis also allowed the cloning of regulatory and biosynthetic genes for the production of extracellular enzymes and polysaccharides of Xcc and Xoo (Hotte *et al.*, 1990; Tang *et al.*, 1990; Wilson *et al.*, 1998; Dharmapuri and Sonti, 1999). Lipopolysaccharide biosynthesis and its regulation have also been elucidated both by random and directed mutagenesis, coupled with biochemical analysis, in Xcc and *X. campestris* pv. *citrumelo* (Koplin *et al.*, 1992; Kingsley *et al.*, 1993; Dow *et al.*, 1995; Vorholter *et al.*, 2001). Whole genome chemical mutagenesis and complementation led to the discovery of the *aroE* gene involved in xanthomonadin production and virulence of Xoo (Goel *et al.*, 2001). Several type III secreted avirulence proteins were also identified prior to the availability of whole genome sequences by using heterologous expression of genomic libraries to screen for clones that restrict host range (Bonas *et al.*, 1989; Minsavage *et al.*, 1990; Parker *et al.*, 1993; Yang *et al.*, 1994; Swords *et al.*, 1996; Huguet and Bonas, 1997; Astua-Monge *et al.*, 2000; Yang *et al.*, 2000). Whole genome cosmid libraries of different strains were used in a kind of rudimentary comparative genomics approach to assess functional conservation of the *avrBs2* gene of Xav in two other pathovars (Kearney and Staskawicz, 1990).

Other, more targeted, genome-scale functional analyses have been carried out prior to whole genome sequence availability for the strains being tested. For example, a cDNA-AFLP-based analysis in Xav compared gene expression in a wild type genetic background and in a strain expressing a constitutively active allele of the *hrpG* regulatory gene, and allowed the identification of Hrp-regulated genes, including several new candidate type III effector genes (Noel *et al.*, 2001). Candidate effectors were then tested using an *in vitro* secretion assay, and several (XopA, B, C, D, and J) were confirmed (Noel *et al.*, 2002; Noel *et al.*, 2003). Similarly, a reporter-based genome-wide screen for effectors in Xav led to the discovery of other new type III effectors, XopD and XopN (Roden *et al.*, 2004a). Altogether, these studies led to the identification of 14 type III effectors in Xav, several of which were observed to be conserved in the two published genomes at the time (XccA and Xac) and were suggested as a 'core' group of effectors (Buttner and Bonas, 2003; Roden et al., 2004a; Roden et al., 2004b). This study was a harbinger of more good things to come. The completion of several whole genome sequences quickly brought about a new era of genome sequence-enabled and comparative genomics-driven biology.

Genome-sequence enabled functional genomics

With a complete genome sequence, the cloning and characterization of genes identified through traditional approaches such as random mutagenesis is greatly accelerated. A snippet of sequence generated using a transposon-specific primer is enough to map the location of the insertion in a mutant of interest and identify the disrupted gene. Cloning the gene is a simple matter of amplifying it by PCR, using the genome sequence to design appropriate primers. Genome sequence also drives non-experimental approaches to gene discovery, for example, mining for genes with specific *cis*-regulatory elements, eukaryotic features, evidence of horizontal acquisition, or other characteristics. Finally, the sequence enables transcript profiling and proteomic studies.

The first outstanding example of genome-sequence enabled functional genomics in *Xanthomonas* occurred with the release of the Xcc8 genome. The authors were able to screen and map, in a high-throughput fashion, transposon insertions that impaired pathogenesis, leading to the identification of several candidate virulence factors (Qian et al., 2005). Moreover, the authors were able to assess whether candidates were present in the previously sequenced XccA genome. In addition to identifying known virulence factors, the study implicated genes not previously associated with virulence in *Xanthomonas*, including certain metabolic genes, genes for type IV secretion, and cell signalling genes. Three candidates were unique to Xcc8 relative to XccA, implying that virulence of different strains can depend in part on different genes. The same genome-enabled approach in Xoc similarly identified both known and new virulence factors in this pathogen, including a lipopolysaccharide O-chain synthesis gene, genes for twitching motility, and candidate genes for the production of cell-surface-associated carbohydrates (Wang et al., 2007). A good example of genome mining for gene discovery is provided by consecutive studies in Xoo. Tsuge et al. (2005) characterized the effect of base substitutions in the plant inducible promoter (PIP) box sequence that is targeted by the positive regulator HrpX. In this way, the authors refined the consensus sequence necessary for function and also identified several variants with complete or partial activity. Using this information, Furutani et al. (2006) mined the XooM genome sequence for open reading frames preceded by potentially active PIP box variants, and identified several genes in the Hrp regulon, including, in addition to type III secretion structural genes, candidate genes for effectors, genes encoding conserved hypothetical proteins, a phosphatase gene, and an ABC transporter gene. The authors confirmed many of these using a reporter assay. A number of type III effector genes have been identified in several of the sequenced *Xanthomonas* genomes by sequence similarity to known effectors of other pathogens (Gurlebeck et al., 2006). Thieme et al (2007) recently characterized a set of *Xanthomonas* type III effectors mined from the genome sequences and selected based on the presence of a predicted N-terminal myristylation motif. Cloning and characterization of the unique class of *Xanthomonas* transcription activator-like (TAL) effector genes from Xoc and Xoo has been facilitated greatly by genome sequence availability (D. O. Nino-Liu and A. J. Bogdanove, unpublished). TAL effector genes are characterized by large numbers of 102 bp repeats in their central coding region, and there are 28 family members in the Xoc genome. Availability of complete genome sequence has made it a simple matter to identify, by mapping, a minimal set of cosmids that contains the complete inventory of these genes, and to quickly map insertions in TAL genes in a mutant library by end sequencing of large rescue clones. In Xcc, the genome sequences have enabled the generation of oligonucleotide microarrays, which have been used to decipher regulation by the diffusible signal factor (DSF), which plays an important role in pathogenesis (He et al., 2006). The Xcc genome sequences have also made possible proteomic studies for identifying proteins involved in interactions with plants (Watt *et al.*, 2005).

By combining genome enabled functional genomics with comparative genomics, Blanvillain *et al.* (2007) provided insight into the functions of the large number of TonB-dependent receptors (TBDRs) present in Xcc. They mined the XccA genome sequence to identify the TBDR genes, and demonstrated a role for the proteins in bacterial adaptation to plants, involving scavenging plant carbohydrates. The authors also analysed the distribution of TBDRs in the sequenced genomes of 226 Gram-negative bacteria and demonstrated conservation among *Xanthomonas* spp. and aquatic bacteria of several carbohydrate utilization loci that depend on TBDRs for the active transport of plant molecules across the outer membrane. They postulated that, as plant compounds are widespread in nature, TBDRs might play a key role in bacterial survival in diverse environments. This work highlights the potential for combined functional and comparative genomics in *Xanthomonas* to advance not only phytopathology, but also biotechnology, and our fundamental understanding of bacterial evolution and adaptation.

Comparative genomics

A search for genetic adaptations that are responsible for certain characteristics of a particular strain can be aided by determining what genes are shared (or not shared) in other, related genomes. Functional genomics thus can be driven by comparative genomics, which gives rise to hypotheses and helps to prioritize candidates for experimental characterization. By comparing genomes of different *Xanthomonas* species and pathovars, we can address fundamental questions in phytobacteriology, including what genes are important for host-species specificity, what genes are required for invasion of different tissues, and what genes are required for pathogenesis generally, irrespective of host and tissue. Moreover, comparative genomics can refine taxonomy and contribute to a better understanding of genome evolution, including horizontal gene transfer (HGT).

Annotation of the first completed *Xanthomonas* genomes, Xac and XccA, increased considerably the known inventory of genes potentially involved in interactions of these pathogens with their plant hosts. Annotation relies on comparison of sequences to known genes, and inference of shared function. Thus, based on comparison to sequences in other bacteria, several genes or gene classes potentially involved in pathogenesis, and therefore targets for functional characterization, were identified. These include genes required for adhesion, motility, oxidative stress resistance, plant cell wall degradation, phytohormone production, synthesis and injection of effectors into the host cell, and others (da Silva *et al.*, 2002).

The comparison of genomes of closely related strains that differ in their pathogenicity may be the best approach to identify specific genetic determinants of that difference. Yet, a comparison of gene content in two *Xylella fastidiosa* strains respectively pathogenic to grape and citrus failed to detect significant differences in gene content, suggesting that the two strains rely on similar mechanisms to infect their hosts (Van Sluys *et al.*, 2003; see chapter by van Sluys *et al.*). Inversely, comparative genomics between the citrus strain of *Xylella fastidiosa* with Xac did not reveal strong candidates for shared genes representing adaptations to the same host (Moreira *et al.*, 2004). The emerging theme is that adaptations in closely related genomes are likely to involve differences in gene structure, such as point mutations and small indels, rather than differences in gene content, and that less closely related organisms may harbour distinct adaptations that result in similar characteristics. Recent results of ours are consistent with this idea. We have compared across the sequenced *Xanthomonas* genomes four clusters of genes associated with pathogenesis to look for correlations of gene content and structure with host or tissue specificity. The clusters are the *hrp* gene cluster for type III secretion, the *gum* cluster for extracellular polysaccharide production, the *rpf* cluster for DSF-dependent signalling, and the *xps* cluster for type II secretion. We found that (1) the clusters are present in each genome and (2) phylogenetic relationships based on concatenated predicted amino acid sequences of the clusters each reflected lineage based on a rooted tree for the *rrnA* operon. Our results indicate that each of these clusters was present in the ancestral genome, indicating that they all are likely fundamentally important. Further, the results suggest that shared tissue specificity, which does not coincide with phylogeny, resulted either from coincident changes in the structure of specific gene sequences within the clusters, in regulatory sequences, and/or among genes regulated or encoding proteins secreted by products of the gene clusters, or from changes elsewhere in the genomes (A. J. Bogdanove, unpublished).

An expanded comparison of *Xanthomonas* and *Xylella* strains, including Xac, Xcc, and both the citrus and grape strains of *Xylella*, however, did in fact bring to the surface an apparently horizontally acquired gene cluster unique to the citrus pathogens, which is present also in *Salmonella* spp., and carries genes for exopolysaccharide synthesis, type IV pilus production, and conjugal transfer (Moreira et al., 2005). Although functional characterization of this cluster has not been carried out in either of the xanthomonads, the observation is consistent with the notion that such 'genomic islands' could in fact account for shared characteristics between divergent pathogens.

As discussed earlier in this chapter, *Xanthomonas* genomes are replete with IS elements, and many of these elements border islands (Monteiro-Vitorello et al., 2005). Comparative genomics among two *Xanthomonas* genomes and one *Xylella fastidiosa* genome, as well as 17 other genomes representing each of the three subdivisions (alpha, beta, and gamma) of the Proteobacteria, allowed Comas et al. (2006) to carefully analyse HGT events in the Xanthomonadaceae and to define the events as ancient or recent. The authors observed that the Xanthomonadaceae genomes represent extreme mosaics. They also presented evidence that genes associated with ancient HGT associate equally with sequences in the three subdivisions of the Proteobacteria, while recently acquired genes appear to be predominantly from other xanthomonads, providing novel insight into the role of HGT in shaping *Xanthomonas* genomes.

Xanthomonas genomics into the future

Recent advances in DNA sequencing technology (Marusina, 2006; see chapter by MacLean and Studholme) are bringing high-throughput whole genome sequencing closer to being a practical reality. The typically short read lengths of the new techniques probably limit our ability to obtain accurate *Xanthomonas* genome assemblies without some relatively more costly and time-consuming Sanger sequencing, due to the prevalence in *Xanthomonas* genomes of IS elements and other relatively long repeat sequences. Nevertheless, these technologies are rapidly improving, and current methods are certainly capable of generating high-quality draft genome sequences for mining gene content and comparing strains by large scale multi locus sequence typing (MLST). The prospects for addressing questions in *Xanthomonas* genome evolution and pinpointing adaptations to environments in and outside of plants, and for developing highly refined diagnostic and epidemiological tools are exciting. It will be imperative to sequence both broadly, across species and pathovars, and deeply, within pathovars and across strains that differ in geographical origin, virulence, or other characteristics. A large collection of sequences will also permit the construction of a robust pan-genomic microarray for *Xanthomonas*, which will be useful in typing strains by DNA content and in exploring their biology by gene expression profiling.

Even with the currently available genome sequences, there are ample opportunities for discovery through targeted comparative genomics. Fig. 7.3 illustrates a combinatorial approach to comparing just the nine completely sequenced *Xanthomonas* genomes, in order to extract genes unique to a particular genome or group of genomes. Twenty-seven theoretical classes of genes (with some classes overlapping) result. Such genes would be of interest as candidate determinants of traits specific to the corresponding strains. Conceptually straightforward, but computationally intensive, this example underscores the importance of bioinformatics resources and databases to generate and serve comparative data. Clearly, as more genome sequences become available, the value of correlations of gene content with pathogenic characteristics grows, enabling better targeting of genes for functional characterization. But the complexity of the analyses and the volume of data grow as well. Moreover, integration of comparative and functional genomic data with transcript profiling, proteomic,

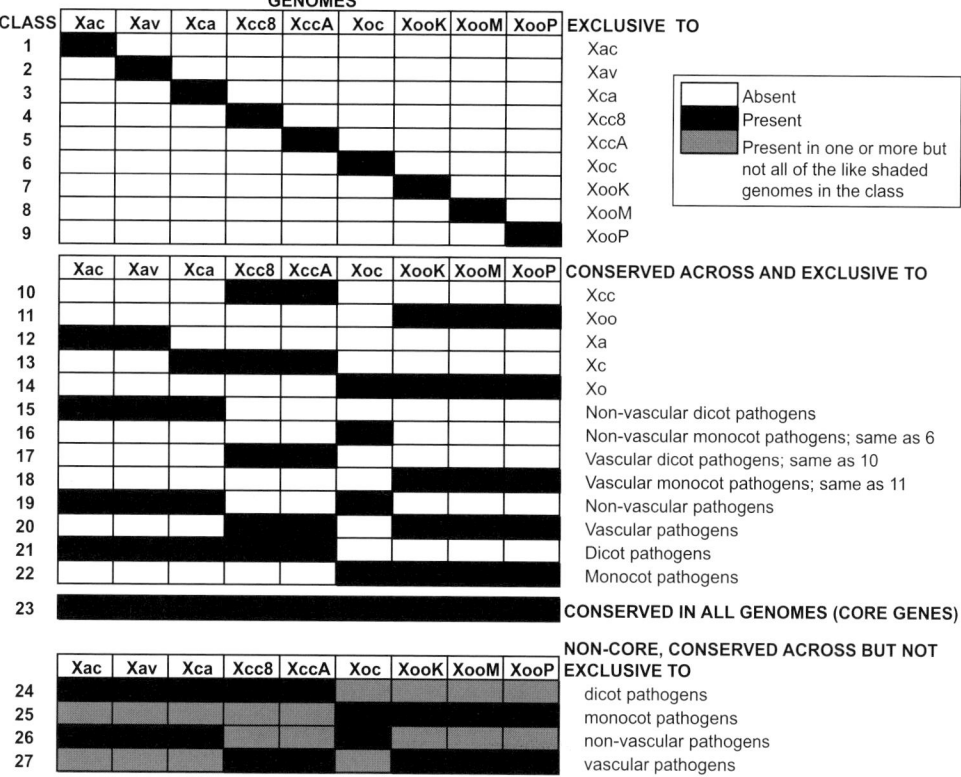

Figure 7.3 Classification of Xanthomonas genes based on comparative genomics of nine sequenced strains toward discovery of determinants of host and tissue specificity. Twenty-seven hypothetical classes of genes are shown, based on their uniqueness to individual genomes or groups of genomes representing species and pathovars with particular host and tissue specificities. Xac, *X. axonopodis* pv. *citri* 306; Xav, *X. axonopodis* pv. *vesicatoria* (*X. campestris* pv. *vesicatoria*) 85-10; Xcc8, *X. campestris* pv. *campestris* 8004; XccA, *X. campestris* pv. *campestris* ATCC33913; Xca, *X. campestris* pv. *armoraciae* 756C; XooK, *X. oryzae* pv. *oryzae* KACC10331; XooM, *X. oryzae* pv. *oryzae* MAFF311018; XooP, *X. oryzae* pv. *oryzae* PXO99[A]; Xoc, *X. oryzae* pv. *oryzicola* BLS256.

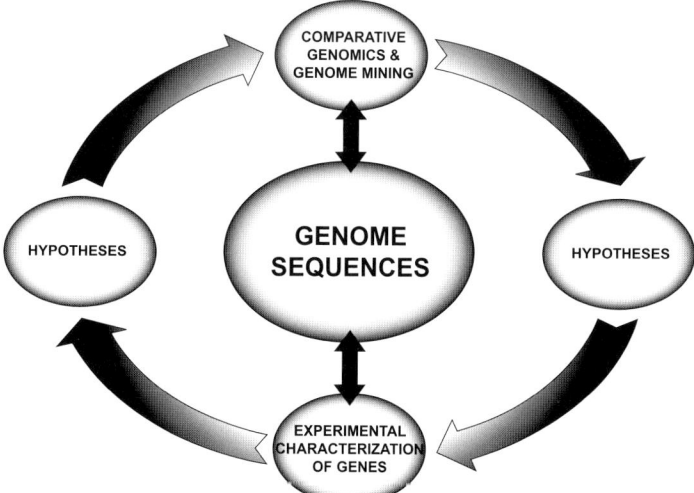

Figure 7.4 Genome sequence-driven revolution in Xanthomonas biology. Comparative genomics at the level of genes, operons, and whole genomes, and genome mining, generate novel hypotheses regarding gene function. Genes are characterized experimentally to test these hypotheses. New knowledge leads to additional hypotheses. Predictions based on these hypotheses are then tested or extended by further comparative genome mining. High-quality sequence and annotation drive this cycle of discovery.

and metabolomic data will be important. Thus databases must be versatile and expandable.

Genomics is at the centre of a revolution in *Xanthomonas* biology (Fig. 7.4). Genomic data are beginning to create a sort of molecular Rosetta stone, by which, through comparison of the various iterations of the *Xanthomonas* genome, coupled with genome sequence enabled functional analysis, we will be able to extract the determinants of the defining characteristics of the genus, as well as the differences that define its diversity. As genomics continues to drive advances in the fundamental biology of *Xanthomonas*, it will also drive the development of better ways to control the many important plant pathogens that the genus encompasses, and potentially other plant and even animal pathogenic bacteria as well.

Acknowledgements

The authors thank Prabhu Patil, Ramesh Sonti, and Marie-Anne Van Sluys for enlightening discussion. D.F.M. was supported by grants to A.J.B. from the National Science Foundation (award 0227357) and the United States Department of Agriculture Cooperative State Research, Education, and Extension Service (Award 2004-35600-15022).

References

Arlat, M., Gough, C.L., Barber, C.E., Boucher, C., and Daniels, M.J. (1991). *Xanthomonas campestris* contains a cluster of *hrp* genes related to the larger *hrp* cluster of *Pseudomonas solanacearum*. Mol. Plant Microbe Interact. 4, 593–601.

Astua-Monge, G., Minsavage, G.V., Stall, R.E., Vallejos, C.E., Davis, M.J., and Jones, J.B. (2000). *Xv4-avrxv4*: a new gene-for-gene interaction identified between *Xanthomonas campestris* pv. *vesicatoria* race T3 and wild tomato relative *Lycopersicon pennellii*. Mol. Plant–Microbe Interact. 13, 1346–1355.

Becker, A., Katzen, F., Puhler, A., and Ielpi, L. (1998). Xanthan gum biosynthesis and application: a biochemical/genetic perspective. Appl. Microbiol. Biotechnol. 50, 145–152.

Berg, G., Roskot, N., and Smalla, K. (1999). Genotypic and phenotypic relationships between clinical and environmental isolates of *Stenotrophomonas maltophilia*. J. Clin. Microbiol. 37, 3594–3600.

Bhattacharyya, A., Stilwagen, S., Ivanova, N., D'Souza, M., Bernal, A., Lykidis, A., Kapatral, V., Anderson, I., Larsen, N., Los, T., et al. (2002). Whole-genome comparative analysis of three phytopathogenic *Xylella fastidiosa* strains. Proc. Natl. Acad. Sci. U.S.A. 99, 12403–12408.

Blanvillain, S., Meyer, D., Boulanger, A., Lautier, M., Guynet, C., Denance, N., Vasse, J., Lauber, E., and Arlat, M. (2007). Plant carbohydrate scavenging through *tonB*-dependent receptors: a feature shared by phytopathogenic and aquatic bacteria. PLoS ONE 2, e224.

Bonas, U., Schulte, R., Fenselau, S., Minsavage, G.V., Staskawicz, B.J., and Stall, R.E. (1991). Isolation of a gene cluster from *Xanthomonas campestris* pv. *vesicatoria* that determines pathogenicity and the hypersensitive response on pepper and tomato. Mol. Plant Microbe Interact. 4, 81–88.

Bonas, U., Stall, R.E., and Staskawicz, B. (1989). Genetic and structural characterization of the avirulence gene *avrBs3* from *Xanthomonas campestris* pv. *vesicatoria*. Mol. Gen. Genet. 218, 127–136.

Bradbury, J.F. (1984). Genus II. *Xanthomonas* Dowson. In *Bergey's Manual of Systematic Bacteriology*, N.R. Krieg, and J.G. Holt, eds. (Baltimore, Williams and Wilkins), pp. 199–210.

Bretschneider, K.E., Gonella, M.G., and Robeson, D.J. (1989). A comparative light- and electron microscopical study of compatible and incompatible interactions between *Xanthomonas campestris* pv. *campestris* and cabbage (*Brassica oleracea*). Physiol. Mol. Plant Pathol. 34, 285–297.

Buttner, D., and Bonas, U. (2003). Common infection strategies of plant and animal pathogenic bacteria. Curr. Opin. Plant Biol. 6, 312–319.

Chenna, R., Sugawara, H., Koike, T., Lopez, R., Gibson, T.J., Higgins, D.G., and Thompson, J.D. (2003). Multiple sequence alignment with the Clustal series of programs. Nucleic Acids Res. 31, 3497–3500.

Comas, I., Moya, A., Azad, R.K., Lawrence, J.G., and Gonzalez-Candelas, F. (2006). The evolutionary origin of Xanthomonadales genomes and the nature of the horizontal gene transfer process. Mol. Biol. Evol. 23, 2049–2057.

Crossman, L., and Dow, J.M. (2004). Biofilm formation and dispersal in *Xanthomonas campestris*. Microbes Infect. 6, 623–629.

da Silva, A.C., Ferro, J.A., Reinach, F.C., Farah, C.S., Furlan, L.R., Quaggio, R.B., Monteiro-Vitorello, C.B., Van Sluys, M.A., Almeida, N.F., Alves, L.M., et al. (2002). Comparison of the genomes of two *Xanthomonas* pathogens with differing host specificities. Nature 417, 459–463.

Darling, A.C., Mau, B., Blattner, F.R., and Perna, N.T. (2004). Mauve: multiple alignment of conserved genomic sequence with rearrangements. Genome Res. 14, 1394–1403.

Dharmapuri, S., and Sonti, R.V. (1999). A transposon insertion in the *gumG* homologue of *Xanthomonas oryzae* pv. *oryzae* causes loss of extracellular polysaccharide production and virulence. FEMS Microbiol. Lett. 179, 53–59.

Dow, J.M., Crossman, L., Findlay, K., He, Y.Q., Feng, J.X., and Tang, J.L. (2003). Biofilm dispersal in *Xanthomonas campestris* is controlled by cell–cell signaling and is required for full virulence to plants. Proc. Natl. Acad. Sci. USA 100, 10995–11000.

Dow, J.M., and Daniels, M.J. (1994). Pathogenicity determinants and global regulation of pathogenicity of *Xanthomonas campestris* pv. *campestris*. Curr. Top. Microbiol. Immunol. *192*, 29–41.

Dow, J.M., and Daniels, M.J. (2000). *Xylella* genomics and bacterial pathogenicity to plants. Yeast *17*, 263–271.

Dow, J.M., Osbourn, A.E., Wilson, T.J., and Daniels, M.J. (1995). A locus determining pathogenicity of *Xanthomonas campestris* is involved in lipopolysaccharide biosynthesis. Mol. Plant–Microbe Interact. *8*, 768–777.

Fleischmann, R.D., Adams, M.D., White, O., Clayton, R.A., Kirkness, E.F., Kerlavage, A.R., Bult, C.J., Tomb, J.F., Dougherty, B.A., Merrick, J.M., *et al.* (1995). Whole-genome random sequencing and assembly of *Haemophilus influenzae* Rd. Science (New York, N.Y *269*, 496–512.

Furutani, A., Nakayama, T., Ochiai, H., Kaku, H., Kubo, Y., and Tsuge, S. (2006). Identification of novel HrpXo regulons preceded by two cis-acting elements, a plant-inducible promoter box and a -10 box-like sequence, from the genome database of *Xanthomonas oryzae* pv. *oryzae*. FEMS Microbiol. Lett. *259*, 133–141.

Goel, A.K., Rajagopal, L., and Sonti, R.V. (2001). Pigment and virulence deficiencies associated with mutations in the *aroE* gene of *Xanthomonas oryzae* pv. *oryzae*. Appl. Environ. Microbiol. *67*, 245–250.

Gurlebeck, D., Thieme, F., and Bonas, U. (2006). Type III effector proteins from the plant pathogen *Xanthomonas* and their role in the interaction with the host plant. J. Plant Physiol. *163*, 233–255.

Hayward, A.C. (1993). The hosts of *Xanthomonas*. In *Xanthomonas*, J.G. Swings, and E.L. Civerolo, eds. (London, United Kingdom, Chapman and Hall), pp. 1–119.

He, Y.-W., Xu, M., Lin, K., Ng, Y.-J.A., Wen, C.-M., Wang, L.-H., Liu, Z.-D., Zhang, H.-B., Dong, Y.-H., Dow, J.M., *et al.* (2006). Genome scale analysis of diffusible signal factor regulon in *Xanthomonas campestris* pv. *campestris*: identification of novel cell–cell communication-dependent genes and functions. Mol. Microbiol. *59*, 610–622.

Hotte, B., Rath-Arnold, I., Puhler, A., and Simon, R. (1990). Cloning and analysis of a 35.3-kilobase DNA region involved in exopolysaccharide production by *Xanthomonas campestris* pv. *campestris*. J. Bacteriol. *172*, 2804–2807.

Huguet, E., and Bonas, U. (1997). hrpF of *Xanthomonas campestris* pv. *vesicatoria* encodes an 87-kDa protein with homology to NoIX of *Rhizobium fredii*. Mol. Plant Microbe Interact. *10*, 488–498.

Jones, J.B., Lacy, G.H., Bouzar, H., Stall, R.E., and Schaad, N.W. (2004). Reclassification of the xanthomonads associated with bacterial spot disease of tomato and pepper. Syst. Appl. Microbiol. *27*, 755–762.

Kearney, B., and Staskawicz, B.J. (1990). Widespread distribution and fitness contribution of *Xanthomonas campestris* avirulence gene *avrBs2*. Nature *346*, 385–386.

Kingsley, M.T., Gabriel, D.W., Marlow, G.C., and Roberts, P.D. (1993). The *opsX* locus of *Xanthomonas campestris* affects host range and biosynthesis of lipopolysaccharide and extracellular polysaccharide. J. Bacteriol. *175*, 5839–5850.

Koplin, R., Arnold, W., Hotte, B., Simon, R., Wang, G., and Puhler, A. (1992). Genetics of xanthan production in *Xanthomonas campestris*: the *xanA* and *xanB* genes are involved in UDP-glucose and GDP-mannose biosynthesis. J. Bacteriol. *174*, 191–199.

Lee, B.M., Park, Y.J., Park, D.S., Kang, H.W., Kim, J.G., Song, E.S., Park, I.C., Yoon, U.H., Hahn, J.H., Koo, B.S., *et al.* (2005). The genome sequence of *Xanthomonas oryzae* pathovar *oryzae* KACC10331, the bacterial blight pathogen of rice. Nucleic Acids Res. *33*, 577–586.

Leyns, F., De Cleene, M., Swings, J., and De Ley, J. (1984). The host range of the genus *Xanthomonas*. Botanical Review *50*, 308.

Marusina, K. (2006). The next generation of DNA sequencing. Genetic Engineering Biotechnol. News *26*, www.genengnews.com/articles/chitem.aspx?aid=1946.

Minsavage, G.V., Dahlbeck, D., Whalen, M.C., Kearney, B., Bonas, U., Staskawicz, B.J., and Stall, R.E. (1990). Gene-for-gene relationships specifying disease resistance in *Xanthomonas campestris* pv. *vesicatoria*-pepper interactions. Mol. Plant Microbe Interact. *3*, 41–47.

Monteiro-Vitorello, C.B., de Oliveira, M.C., Zerillo, M.M., Varani, A.M., Civerolo, E., and Van Sluys, M.A. (2005). *Xylella* and *Xanthomonas* Mobil'omics. OMICS *9*, 146–159.

Moreira, L.M., de Souza, R.F., Almeida Jr, N.F., Setubal, J.C., Oliveira, J.C.F., Furlan, L.R., Ferro, J.A., and da Silva, A.C.R. (2004). Comparative genomics analyses of citrus-associated bacteria. Annu. Rev. Phytopathol. *42*, 163–184.

Moreira, L.M., De Souza, R.F., Digiampietri, L.A., Da Silva, A.C., and Setubal, J.C. (2005). Comparative analyses of *Xanthomonas* and *Xylella* complete genomes. OMICS *9*, 43–76.

Noel, L., Thieme, F., Gabler, J., Buttner, D., and Bonas, U. (2003). XopC and XopJ, two novel Type III effector proteins from *Xanthomonas campestris* pv. *vesicatoria*. J. Bacteriol. *185*, 7092–7102.

Noel, L., Thieme, F., Nennstiel, D., and Bonas, U. (2001). cDNA-AFLP analysis unravels a genome-wide hrpG-regulon in the plant pathogen *Xanthomonas campestris* pv. *vesicatoria*. Mol. Microbiol. *41*, 1271–1281.

Noel, L., Thieme, F., Nennstiel, D., and Bonas, U. (2002). Two novel type III-secreted proteins of *Xanthomonas campestris* pv. vesicatoria are encoded within the *hrp* pathogenicity island. J. Bacteriol. *184*, 1340–1348.

Ochiai, H., Inoue, Y., Takeya, M., Sasaki, A., and Kaku, H. (2005). Genome sequence of *Xanthomonas oryzae* pv. *oryzae* suggests contribution of large numbers of effector genes and insertion sequences to its race diversity. Jpn Agric. Res. Q. *39*, 275–287.

Parker, J.E., Barber, C.E., Fan, M.J., and Daniels, M.J. (1993). Interaction of *Xanthomonas campestris* with *Arabidopsis thaliana*: characterization of a gene from *X. c.* pv. *raphani* that confers avirulence to most *A.*

thaliana accessions. Mol. Plant Microbe Interact. 6, 216–224.

Purcell, A.H., and Hopkins, D.L. (1996). Fastidious xylem-limited bacterial plant pathogens. Annu. Rev. Phytopathol. 34, 131–151.

Qian, W., Jia, Y., Ren, S.-X., He, Y.-Q., Feng, J.-X., Lu, L.-F., Sun, Q., Ying, G., Tang, D.-J., Tang, H., et al. (2005). Comparative and functional genomic analyses of the pathogenicity of phytopathogen Xanthomonas campestris pv. campestris. Genome Res. 15, 757–767.

Rademaker, J.L.W., Hoste, B., Louws, F.J., Kersters, K., Swings, J., Vauterin, L., Vauterin, P., and Bruijn, F.J.d. (2000). Comparison of AFLP and rep-PCR genomic fingerprinting with DNA-DNA homology studies: Xanthomonas as a model system. In Int. J. Syst. Evol. Microbiol., pp. 665–677.

Rademaker, J.L.W., Louws, F.J., Schultz, M.H., Rossbach, U., Vauterin, L., Swings, J., and de Bruijn, F.J. (2005). A comprehensive species to strain taxonomic framework for Xanthomonas. Phytopathology 95, 1098–1111.

Rajagopal, L., Sundari, C.S., Balasubramanian, D., and Sonti, R.V. (1997). The bacterial pigment xanthomonadin offers protection against photodamage. FEBS Lett. 415, 125–128.

Rajeshwari, R., and Sonti, R.V. (2000). Stationary-phase variation due to transposition of novel insertion elements in Xanthomonas oryzae pv. oryzae. J. Bacteriol. 182, 4797–4802.

Roden, J., Eardley, L., Hotson, A., Cao, Y., and Mudgett, M.B. (2004a). Characterization of the Xanthomonas AvrXv4 effector, a SUMO protease translocated into plant cells. Mol. Plant–Microbe Interact. 17, 633–643.

Roden, J.A., Belt, B., Ross, J.B., Tachibana, T., Vargas, J., and Mudgett, M.B. (2004b). A genetic screen to isolate type III effectors translocated into pepper cells during Xanthomonas infection. Proc. Natl. Acad. Sci. USA 101, 16624–16629.

Rossier, O., Van den Ackerveken, G., and Bonas, U. (2000). HrpB2 and HrpF from Xanthomonas are type III-secreted proteins and essential for pathogenicity and recognition by the host plant. Mol. Microbiol. 38, 828–838.

Rossier, O., Wengelnik, K., Hahn, K., and Bonas, U. (1999). The Xanthomonas Hrp type III system secretes proteins from plant and mammalian bacterial pathogens. Proc. Natl. Acad. Sci. USA 96, 9368–9373.

Salzberg, S.L., Sommer, D.D., Schatz, M.C., Phillippy, A.M., Rabinowicz, P.D., Tsuge, S., Furutani, A., Ochiai, H., Delcher, A.L., Kelley, D., et al. (2008). Genome sequence and rapid evolution of the rice pathogen Xanthomonas oryzae pv. oryzae PXO99A. BMC Genomics 9, 204.

Schulte, R., and Bonas, U. (1992). Expression of the Xanthomonas campestris pv. vesicatoria hrp gene cluster, which determines pathogenicity and hypersensitivity on pepper and tomato, is plant inducible. J. Bacteriol. 174, 815–823.

Simpson, A.J., Reinach, F.C., Arruda, P., Abreu, F.A., Acencio, M., Alvarenga, R., Alves, L.M., Araya, J.E., Baia, G.S., Baptista, C.S., et al. (2000). The genome sequence of the plant pathogen Xylella fastidiosa. Nature 406, 151–159.

Starr, M.P. (1981). The genus Xanthomonas. In The Prokaryotes, M.P. Starr, H. Stolp, H.G. Truper, A. Balows, and H.G. Schlegel, eds. (Berlin, Springer Verlag), pp. 742–763.

Sutherland, I.W. (1993). Xanthan. In Xanthomonas, J.G. Swings, and E.L. Civerolo, eds. (London, United Kingdom, Chapman & Hall), pp. 363–388.

Swords, K.M., Dahlbeck, D., Kearney, B., Roy, M., and Staskawicz, B.J. (1996). Spontaneous and induced mutations in a single open reading frame alter both virulence and avirulence in Xanthomonas campestris pv. vesicatoria avrBs2. J. Bacteriol. 178, 4661–4669.

Tang, J.L., Gough, C.L., and Daniels, M.J. (1990). Cloning of genes involved in negative regulation of production of extracellular enzymes and polysaccharide of Xanthomonas campestris pathovar campestris. Mol. Gen. Genet. 222, 157–160.

Thieme, F., Koebnik, R., Bekel, T., Berger, C., Boch, J., Buttner, D., Caldana, C., Gaigalat, L., Goesmann, A., Kay, S., et al. (2005). Insights into genome plasticity and pathogenicity of the plant pathogenic bacterium Xanthomonas campestris pv. vesicatoria revealed by the complete genome sequence. J. Bacteriol. 187, 7254–7266.

Thieme, F., Szczesny, R., Urban, A., Kirchner, O., Hause, G., and Bonas, U. (2007). New type III effectors from Xanthomonas campestris pv. vesicatoria trigger plant reactions dependent on a conserved N-myristoylation motif. Mol. Plant Microbe Interact. 20, 1250–1261.

Tsuge, S., Terashima, S., Furutani, A., Ochiai, H., Oku, T., Tsuno, K., Kaku, H., and Kubo, Y. (2005). Effects on promoter activity of base substitutions in the cis-acting regulatory element of HrpXo regulons in Xanthomonas oryzae pv. oryzae. J. Bacteriol. 187, 2308–2314.

Van Sluys, M.A., de Oliveira, M.C., Monteiro-Vitorello, C.B., Miyaki, C.Y., Furlan, L.R., Camargo, L.E., da Silva, A.C., Moon, D.H., Takita, M.A., Lemos, E.G., et al. (2003). Comparative analyses of the complete genome sequences of Pierce's disease and citrus variegated chlorosis strains of Xylella fastidiosa. J. Bacteriol. 185, 1018–1026.

Van Sluys, M.A., Monteiro-Vitorello, C.B., Camargo, L.E.A., Menck, C.F.M., da Silva, A.C.R., Ferro, J.A., Oliveira, M.C., Setubal, J.C., Kitajima, J.P., and Simpson, A.J. (2002). Comparative genomic analysis of plant-associated bacteria. Annu. Rev. Phytopathol. 40, 169–189.

Vandamme, P., Pot, B., Gillis, M., de Vos, P., Kersters, K., and Swings, J. (1996). Polyphasic taxonomy, a consensus approach to bacterial systematics. Microbiol. Rev. 60, 407–438.

Vauterin, L., Hoste, B., Kersters, K., and Swings, J. (1995). Reclassification of Xanthomonas. Int. J. Syst. Bacteriol. 45, 472–489.

Vauterin, L., Rademaker, J., and Swings, J. (2000). Synopsis on the taxonomy of the genus Xanthomonas. In Phytopathology, pp. 677–682.

Vorholter, F.J., Niehaus, K., and Puhler, A. (2001). Lipopolysaccharide biosynthesis in Xanthomonas campestris pv. campestris: a cluster of 15 genes is involved in the biosynthesis of the LPS O-antigen and the LPS core. Mol. Genet. Genomics 266, 79–95.

Vorholter, F.J., Thias, T., Meyer, F., Bekel, T., Kaiser, O., Puhler, A., and Niehaus, K. (2003). Comparison of two *Xanthomonas campestris* pathovar campestris genomes revealed differences in their gene composition. J. Biotechnol. 106, 193–202.

Wang, L., Makino, S., Subedee, A., and Bogdanove, A.J. (2007). Novel candidate virulence factors in rice pathogen *Xanthomonas oryzae* pv. oryzicola as revealed by mutational analysis. Appl. Environ. Microbiol. 73, 8023–8027.

Watt, S.A., Wilke, A., Patschkowski, T., and Niehaus, K. (2005). Comprehensive analysis of the extracellular proteins from *Xanthomonas campestris* pv. *campestris* B100. Proteomics 5, 153–167.

Wengelnik, K., Rossier, O., and Bonas, U. (1999). Mutations in the regulatory gene *hrp*G of *Xanthomonas campestris* pv. vesicatoria result in constitutive expression of all *hrp* genes. J. Bacteriol. 181, 6828–6831.

Wengelnik, K., Van den Ackerveken, G., and Bonas, U. (1996). HrpG, a key *hrp* regulatory protein of *Xanthomonas campestris* pv. vesicatoria is homologous to two-component response regulators. Mol. Plant–Microbe Interact. 9, 704–712.

Wilson, T.J., Bertrand, N., Tang, J.L., Feng, J.X., Pan, M.Q., Barber, C.E., Dow, J.M., and Daniels, M.J. (1998). The *rpfA* gene of *Xanthomonas campestris* pathovar *campestris*, which is involved in the regulation of pathogenicity factor production, encodes an aconitase. Mol. Microbiol. 28, 961–970.

Yang, B., Zhu, W., Johnson, L.B., and White, F.F. (2000). The virulence factor AvrXa7 of *Xanthomonas oryzae* pv. *oryzae* is a type III secretion pathwaydependent nuclear-localized double-stranded DNA-binding protein. Proc. Natl. Acad. Sci. USA 97, 9807–9812.

Yang, Y., Feyter, R.d., and Gabriel, D.W. (1994). Host-specific symptoms and increased release of *Xanthomonas citri* and *X. campestris* pv. *malvacearum* from leaves are determined by the 102-bp tandem repeats of *pthA* and *avrb6*, respectively. Mol. Plant Microbe Interact. 7, 345–355.

Young, J.M., Dye, D.W., Bradbury, J.F., Panagopoulos, C.G., and Robbs, C.F. (1978). A proposed nomenclature and classification for plant pathogenic bacteria. *New Zealand Journal of Agricultural Research*, pp. 153–177.

Genomics of the Enterobacterial Plant Pathogens

Ian Toth, Leighton Pritchard, Paul Birch and Hui Liu

Abstract

The enterobacterial plant pathogens are an important group of bacteria responsible for disease on a wide range of plant species in geographically diverse regions. These pathogens entered the postgenomics age in 2004, with the completion of the *Pectobacterium atrosepticum* SCRI1043 genome sequence. Since then, other enterobacterial plant pathogens have been sequenced, and an array of computational and functional genomics tools have developed in parallel with these sequences. Genomic analyses are helping to identify many shared pathogenicity and lifestyle mechanisms among plant (and animal) pathogens, helping to forge new collaborations between scientists working on different pathosystems. These analyses are also uncovering pathogen-specific features, often acquired through horizontal gene transfer, which offer new insights into how individual pathogens infect particular hosts, their modes of pathogenesis or their life in the wider environment. This chapter describes current progress in genomic and post-genomic research on enterobacterial plant pathogens, together with some of the most recent findings.

The family Enterobacteriaceae falls within the γ-Proteobacteria, and contains bacteria with a wide range of biological functions and niches. The best known of these bacteria are the human pathogens, including *Escherichia coli*, *Salmonella enterica* and *Yersinia pestis*. However, enterobacterial pathogens also cause disease on plants, and these are currently divided into the six genera *Erwinia*, *Pantoea*, *Pectobacterium*, *Dickeya*, *Brenneria* and *Enterobacter*. These genera include species that cause disease on a wide variety of plants across a range of geographical regions, the most economically important of which include: *Erwinia amylovora* causing fire blight on apples and pears; *Pantoea stewartii* subsp. *stewartii* causing Stewart's wilt and blight of sweet corn and maize; and the soft rot pathogens *Pectobacterium atrosepticum* (formerly *Erwinia carotovora* subsp. *atroseptica*) causing blackleg on potato, and *P. carotovorum* and *Dickeya* species (the latter formerly *Erwinia chrysanthemi*), causing soft rot disease on a wide range of crop and ornamental plants, including potato (Hauben *et al.*, 1998; Samson *et al.*, 2005). The genomes of many enterobacterial animal pathogens have now been fully sequenced, and functional and comparative genomics technologies have been developed. Recent years have also seen progress in sequencing, and functional and comparative genomics of the enterobacterial plant pathogens (EPPs).

Completed genome sequences

To date, the genome sequence of only one EPP (*P. atrosepticum* strain SCRI1043 [*Pba*1043]) has been published (Bell *et al.*, 2004). However, at least one other (*Dickeya dadantii* strain 3937 [*Dda*3937], and now one of six named *Dickeya* species), has been fully sequenced and annotated and is publicly available (http://asap.ahabs.wisc.edu/annotation/php/ASAP1.htm), while several other genome sequences are in progress (Table 8.1). The *Pba*1043 and *Dda*3937 genomes were sequenced using the Sanger sequencing approach (Sanger and Coulson, 1975) but more recent draft genome sequences have been obtained using

Table 8.1 Enterobacterial plant pathogen sequencing projects. Eight enterobacterial plant pathogen genome sequencing projects, spanning four genera of EPPs, are either completed or under way. Four of these projects use the traditional Sanger sequencing method. More recent projects are being carried out using the high-throughput 454 sequencing approach

Organism	Sequencing centre	Project co-ordinator(s)
Pectobacterium atrosepticum (Pba) SCRI1043	Sanger Institute, Cambridge, UK	SCRI, Dundee, UK (Bell et al., 2004).
Dickeya dadantii (Dda) 3937	The Institute for Genomic Research (TIGR) and University of Wisconsin	University of Wisconsin, USA (ASAP database – see below)
Erwinia amylovora (Ea) 273	Sanger Institute, Cambridge, UK	Universities of Cornell and Wisconsin, USA
Pantoea stewartii subsp. stewartii DC283	Baylor College of Medicine Human Genome Sequencing Center	Universities of Connecticut, Ohio State and Wisconsin, USA
Pantoea ananatis LMG20103	454 draft sequencing – commercial	University of Pretoria, South Africa
Pectobacterium carotovorum subsp. carotovorum WPP14	454 draft sequencing – commercial	Universities of Wisconsin and North Carolina, USA
Pectobacterium carotovorum subsp. brasiliensis WPP1692	454 draft sequencing – commercial	Universities of Wisconsin and North Carolina, USA
Dickeya dianthicola (Ddi) NCPPB3534	454 draft sequencing – commercial	SCRI, Dundee, UK

the high-throughput pyro-sequencing method developed by 454 Life Sciences (Margulies et al., 2005). Pyro-sequencing generates short read lengths of individual sequences, making assembly of repeats problematic, and draft genome sequences may remain in several fragmentary contigs. However, the technology offers a rapid and inexpensive means of obtaining large amounts of highly informative genome sequence data, and is ideally suited for comparative analyses between strains and species, especially where an existing complete genome sequence is available as a framework for assembly.

Genome comparisons, databases and software tools

A major goal of genome sequencing of the pathogens is to gain as broad an understanding as possible of the capacity of that organism to colonize certain niches, including an ability to cause disease. The genome sequences of both Pba1043 and Dda3937 have revealed a complement of potential virulence-related determinants, and provided many new targets for functional analysis. Mutants disabled in some of these genes, e.g. those encoding polyketide phytotoxin coronafacoyl conjugates and a putative type IV secretion system, have been tested on plants and found to be reduced in virulence (Bell et al., 2004). More surprising is evidence of genes that are involved in pathogenesis on aphids (Grenier et al., 2006). Unexpected aspects of these organisms' biochemistry have also been revealed, including genes that contribute to the pathogen's life in association with plants but not directly related to disease processes, e.g. those involved in nitrogen fixation, opine catabolism and root binding (Bell et al., 2004).

In addition to facilitating the discovery of new virulence and lifestyle determinants, these genome sequences provide 'blueprints' to assist in the development of other genomics approaches, and the analysis of data derived from them, e.g. comparative genomics, the location and modelling of regulatory binding sites, gene expression studies, proteomics, and mutation libraries. However, there are limits to some of these comparative techniques, as bacterial regulatory and metabolic networks, for example, may evolve in terms of their structure as well as in terms of the sequences of their components (Price et al., 2007). Nevertheless, genomic technologies provide information on such things as the location and context of a particular gene within the genome (perhaps as part of a gene cluster and/or on a plasmid or pathogenicity island); whether it has functionally similar counterparts (orthologues) in other genomes; is expressed under defined conditions or life stages (such as during pathogenesis); is part of a known, novel or predicted regulatory network; or produces a protein product that is secreted or intracellular.

The rate of generation of sequence data continues to increase rapidly, too rapidly for manual, expert annotation, and it is thus essential that this information is stored in a robust and easily accessible way so that automated annotation methods may be applied (Stothard and Wishart, 2006; Baumgartner et al., 2007; see also chapter by MacLean and Studholme). Public databases such as GenBank and EMBL provide such a facility for raw sequence data and curated annotations. However, specialized databases that focus on particular groups of organisms are also available (for enterobacteria these include ASAP and coliBase), which offer additional information and analysis tools. ASAP is a database primarily intended for community annotation of bacterial genome sequences, with strong emphasis on the enterobacteria [including *Pba*1043 and *Dda*3937; Glasner et al. (2003) – https://asap.ahabs.wisc.edu/annotation/php/ASAP1.htm]. coliBASE is similarly focused on the enterobacteria, and provides comparative data such as whole genome alignments and sets of putative orthologues. Like ASAP, it also has a graphical interface [including an Artemis applet (Rutherford et al., 2000), analytical tools and links to other on-line resources (Chaudhuri et al., 2004) – http://xbase.bham.ac.uk/colibase]. The PathwayTools, EcoCyc, BioCyc and MetaCyc systems provide local and network solutions for genomic and metabolic annotation, with organism-specific databases and annotations for *Pba* and many other sequenced enterobacteria (Caspi et al., 2008). The availability of genome databases and genome analysis tools for prokaryotic genomes is ably summarized in a recent review by Vinatzer and Yan (2008).

A number of methods are available for aligning and visualizing whole genome sequence data sets, including Mauve (http://gel.ahabs.wisc.edu/mauve/), Artemis (http://www.sanger.ac.uk/Software/Artemis/), ACT (Artemis Comparison Tool – http://www.sanger.ac.uk/Software/ACT/) and GenomeDiagram (http://bioinf.scri.ac.uk/lp/programs.php) (Rutherford et al., 2000; Darling et al., 2004; Carver et al., 2005; Pritchard et al., 2006). Mauve produces multiple alignments of conserved genomic DNA in the presence of rearrangements and insertions. It generates diagrammatic representations of the relationships between multiple sequences, and allows users to interact by zooming in on regions of structural interest (Darling et al., 2004). Artemis is an annotation and genome visualization tool that permits viewing of a single genome at any one time, and ACT may be considered as a comparative genomic variant of Artemis that also allows interactive visualization of comparisons between genome sequences, using comparison data generated by one or more of a number of sequence alignment methods, such as BLAST or FASTA. Like Mauve, it is able to visualize sequence comparisons in a range from whole genomes to base pairs and, as it is based on the Artemis program, has powerful search and analysis tools (Carver et al., 2005). Unlike Mauve and ACT, GenomeDiagram is not interactive but does allow visualization of very large-scale comparisons of genomic data, representing a comparison between a query sequence and potentially hundreds of complete genome sequences. Bell et al. (2004) used a combination of both GenomeDiagram and ACT to visualize and target genomic regions of interest following the complete sequencing of *Pba*1043. GenomeDiagram was used to investigate the *Pba*1043 genome in a broad comparative context, and to confirm the position of 17 putative horizontally acquired islands (HAIs) (Fig. 8.1). ACT was then used to investigate synteny between genomes in a more targeted investigation of pathogenicity-related genes within these islands. More recently, Toth et al. (2006) used GenomeDiagram to identify the presence of genes in *Pba*1043 that appeared to have greater sequence identity to plant-associated bacteria than to those of the more closely related animal pathogenic enterobacteria, and so revealed potentially niche-adaptive functional islands. Over 10% of the *Pba*1043 genome was found to contain such sequences, compared to a much smaller number in the animal pathogen *S. enterica* Typhi, consistent with the notion that such sequences

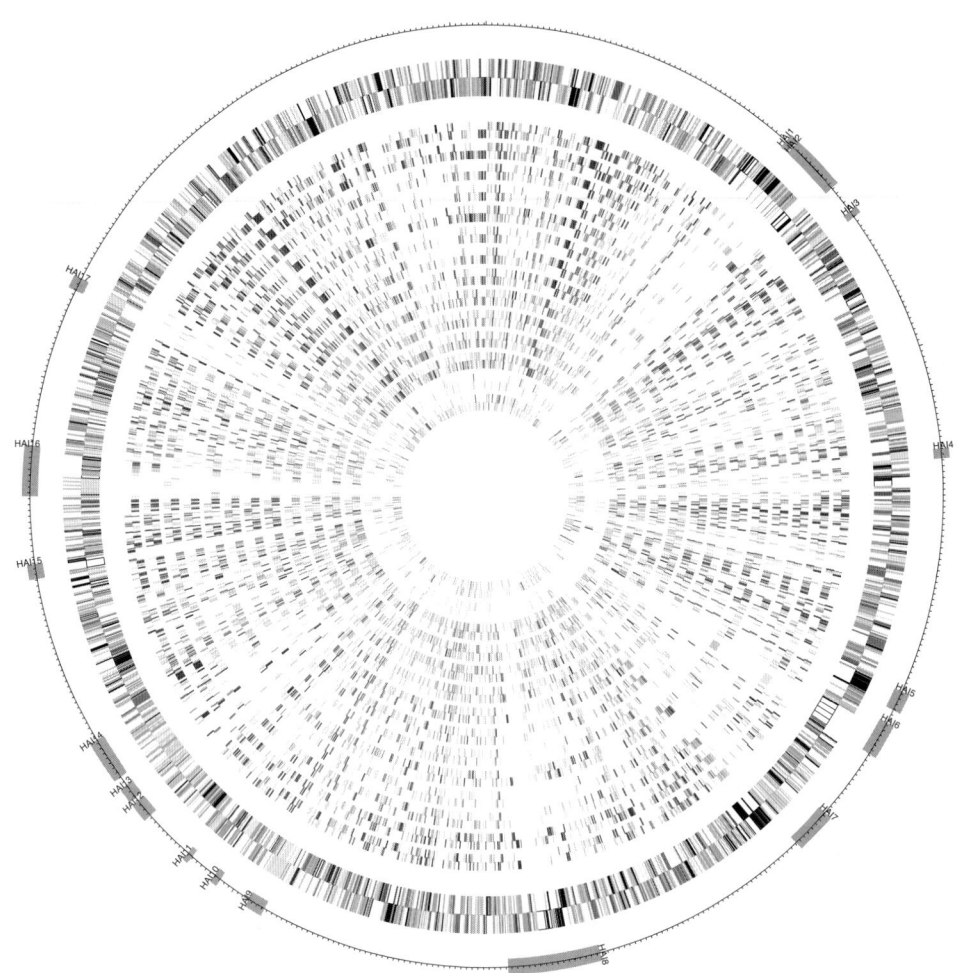

Figure 8.1 Circular representation (produced by GenomeDiagram) of the genome sequence of *Pectobacterium atrosepticum* strain SCRI1043 (*Pba*1043) compared with the genomes of other plant-associated bacteria. From the outer edge, successive rings represent: locations of horizontally acquired islands (HAI); genes from the *Pba*1043 reference sequence in forward and reverse directions; genes similar (>30% identity over 80% of its length) to those of *Pba*1043 from 14 other plant-associated bacteria. Genes are coloured individually on a scale from 30% (*lighter*) to 100% (*darker*) amino acid identity.

may have been acquired during the adaptation of *Pba* to a plant-associated lifestyle.

The majority of genes similar to other plant-associated bacteria were found within the 17 HAIs, clearly implicating horizontal gene transfer in the evolution of *Pba*1043 towards a plant-associated (pathogenic) life style. Some of these HAIs (or genomic islands) contain genes encoding known key pathogenicity and other lifestyle determinants, including the *cfa* gene cluster on HAI2 involved in the synthesis of the phytotoxin coronafacic acid, the *hrp* gene cluster on HAI8 encoding the type III secretion system (T3SS) and its putative helper and effector proteins, and the *nif* gene cluster on HAI14 with a putative role in nitrogen fixation. At least five of the HAIs represent integrated plasmids (HAIs 2, 7 and 13) and prophages (HAIs 9 and 17). Not surprisingly, therefore, determinants found within the islands are, in other isolates or species, often or mostly found on plasmids, e.g. the *cfa* cluster in *P. syringae* (Alarcon-Chaidez et al., 1999), the *hrp* cluster in *Enterobacter agglomerans* [formerly *Erwinia herbicola* – (Nizan *et al.*, 1997)] and the *nif* cluster in *Sinorhizobium meliloti* (Barnett *et al.*, 2001). Indeed, through genome sequencing and comparative genomics, the importance of plasmids, bacteriophages and genomic islands in the evolution and shared function of many plant (and animal) pathogens is becoming apparent (Wagner and Waldor, 2002; Arnold *et al.*, 2003; Sundin, 2007; see also chapters by Stavrinides, and Sundin and Murillo).

Laboratory-based approaches to genome comparison have also been undertaken. An analysis by Triplett *et al.* (2006) used PCR-based suppression subtractive hybridization (SSH) to isolate sequences present in *E. amylovora* Ea110 that were absent in other closely related strains from different hosts. With SSH, pools of genomic DNA from a bacterial strain of interest (tester) are mixed with an excess of DNA from another closely related strain (driver), removing common DNA fragments from the reaction. PCR-enriched tester-specific sequences are then cloned for further investigation. Using this approach, a large number of strain-specific sequences were identified, including components of the T3SS with homology to genes from the insect endosymbiont *Sodalis glossinidius* and the *Yersinia pestis* effector protein YopH, as well as hypothetical membrane and ATP-binding proteins (Triplett *et al.*, 2006).

A complete genome sequence or, ideally, multiple sequences, enable genome-wide searches for the presence and structure of regulatory binding sites, potentially linking many seemingly unrelated functions through a common regulatory pathway(s). This approach was used by Rodionov *et al.* (2004) to build on existing knowledge of the pectin degradation pathway and KdgR regulon found in enterobacteria. Information based on known KdgR DNA binding sites from *Dda*3937 was used to identify further potential KdgR-regulated genes and operons amongst eight fully sequenced enterobacteria, including both plant and animal pathogens. Again, the caveats that bacterial regulatory and metabolic networks may evolve in terms of their structure as well as the sequences of their components also apply here (Price *et al.*, 2007). In the Rodionov *et al.* (2004) study, orthologues of KdgR and those of potential KdgR-controlled genes from the different genomes were subsequently compared using a range of protein alignment and orthologue-searching software tools. Gene expression of key novel genes was then tested using the *uidA* gene reporter system during growth in pectin and other carbon sources, culminating in the reconstruction of the pectin catabolic pathway in this group of bacteria (Rodionov *et al.*, 2004).

Where the sequence of a DNA binding site is not known, large-scale *in vitro* selection methods have been used. One such method called SELEX (systematic evolution of ligands by exponential enrichment) was used to identify the consensus binding site of the virulence-related transcriptional repressor PecS in *Dda*3937 (Rouanet *et al.*, 2004). Many millions of random oligonucleotides were generated flanked by defined primer binding sites, which were then screened for their ability to bind to PecS using band shift assays. Bound DNA was recovered and, using the defined primer binding sites, was PCR amplified prior to cloning and sequencing. Optimal binding sequences were then determined following site-directed mutagenesis of the consensus sequence. Scanning the *Dda*3937 genome sequence revealed the presence of strong PecS binding sites in the intergenic regions within, e.g. the flagella biosynthesis gene (*fli*) cluster, indicating repression of motility during pathogenesis (Rouanet *et al.*, 2004).

Computational approaches

Whole-genome sequence comparisons between enterobacterial plant and animal pathogens first became possible on sequencing of the *Pba*1043 genome. Bell *et al.* (2004) made reciprocal best hit comparisons of the *Pba*1043 genome with all other available sequenced bacteria to identify islands of putative horizontal gene transfer. This analysis was subsequently extended to identify those genes that are more conserved between *Pba* and plant associated bacteria, than between *Pba* and animal pathogenic enterobacteria, and thus may indicate adaptation to its plant niche (Toth *et al.*, 2006). The *Pba* genome sequence has since been incorporated into other computational studies including the identification of regions of putative horizontal gene transfer in *Salmonella* species (Vernikos *et al.*, 2007), and optimization of annotation techniques (Chen et al., 2008). In the Chen et al. study, the authors identified 49 of 1254 hypothetical ORFs in Pba1043 as non-coding, and were able to assign putative function to a further 427 hypothetical ORFs from the original annotation.

In their landmark analysis of the phylogeny of microbial species, Kunin et al. (2005) proposed that a 'net of life' was more appropriate than a 'tree of life' for the representation and reconstruction of bacterial lineages that involve a large number of horizontal gene transfer events. In such nets, the bulk of horizontal gene transfer was evident as multiple small connections between individual species. However, the existence of several highly connected hubs in the network was observed. These hubs were proposed to facilitate rapid gene propagation between microbial species, acting as microbial gene 'banks', providing a medium to acquire and redistribute genes in the microbial environment. The five major hubs identified included the plant-associated bacteria *Bradyrhizobium japonicum* and *Pba*. Both of these species were found robustly to be amongst the top three most highly connected hub organisms (Kunin et al., 2005). This predicted role is consistent with the observations made from sequence analyses that *Pba* contains many virulence determinants with orthologues in distantly related bacteria, including fungal and oomycete plant pathogens, e.g. NEP1-like proteins (Mattinen et al., 2004; Pemberton and Salmond, 2004; Pemberton et al., 2005). It remains to be seen whether other EPPs are similarly facilitative of horizontal gene transfer in these networks, but supportive evidence is available, such as the observation that the *Pcc* phytotoxin carotovoricin (Ctv) appears to have evolved from a common ancestor with *S. enterica* Typhi (Yamada et al., 2006).

Transcriptomics approaches

Reporter gene and hybridization strategies for a single or a small number of genes continue to be at the centre of bacterial gene expression studies. However, high-throughput systems are playing an increasingly large role. cDNA-Amplified Fragment Length Polymorphism (cDNA-AFLP) has identified important differentially expressed genes from eukaryotes (Bachem et al., 1996), but has also been used to study gene expression in the prokaryotes (Dellagi et al., 2000). Dellagi and co-workers used short 11-mer primers that anneal preferentially to conserved sequences within the 3′ region of enterobacterial genes, producing first strand cDNA from mRNA after growth in different conditions. Following ligation of adaptor sequences to the ends of the cDNA fragments, pre-amplification of the cDNA by PCR and end-labelling with ^{33}P, the fragments were electrophoresed through a polyacrylamide gel to reveal cDNA profiles. Differentially amplified products were then excised from the gel, re-amplified and sequenced. In this case, growth of *Pba* and *Pcc* strains in minimal medium (MM) with polygalacturonic acid (PGA), a known inducer of virulence factors such as plant cell wall degrading enzymes (PCWDEs), was compared to that in Luria broth and a number of differentially expressed genes identified. One of these genes was similar to a putative avirulence gene from *Xanthomonas campestris* pv. *raphani*, (now named *svx*) and has since been shown to be type II secreted, quorum sensing regulated and required for full virulence in *Pba*1043 (Corbett et al., 2005).

The IVET (*in vivo expression technology*) method searches the entire genome for genes that are expressed in one or more environmental conditions. Random fragments of chromosomal DNA are inserted into a plasmid between an antibiotic resistance marker (for plasmid selection) and a reporter gene, e.g. *uidA* encoding β-glucuronidase or *gfp* encoding green fluorescent protein. Gene expression from promoters present on these DNA fragments is then compared following, for example, growth in laboratory medium and *in planta*. Clones that show reporter activity *in planta* but not *in vitro* are considered to contain potentially important genes in the plant interaction and are further analysed (Jackson and Giddens, 2006). IVET has been used to study *in planta* gene expression in both *Dda*3937 and *Ea*110 (Yang et al., 2004; Zhao et al., 2005). From 10 000 *Dda*3937 clones, Yang et al. (2004) identified 61 clones that were up-regulated *in planta* and putatively involved in functions such as metabolism, information transfer, regulation, transport, cell processing, transposition, as well as 30% unknown or hypothetical proteins. Candidate clones then end-sequenced and compared in the ASAP database to the *Dda*3937 genome sequence, and full length sequences obtained. Tn5 mutations were then made in ten plant up-regulated genes to reveal a number of virulence candidates, including genes involved in type III secretion. Zhao et al. (2005) carried out a similar study in *Ea*110 to reveal 498 from 19 200 clones that were differentially expressed on pear disks compared to laboratory medium. Putative functional groups included host–microbe interactions, stress, regulation, cell surface, transport, mobile elements and phages, metabolism, nutrient acquisition and synthesis, with 19.8% unknown or hypothetical proteins. Again, genes involved in type III secretion were among the differentially expressed genes, together with those encoding components of the type II secretion system, a polygalacturonase (*peh* – a known substrate of this system), and a haemagglutinin family adhesin (*hecA*), all of which have been associated with virulence in one or more of the soft rot EPPs (Toth et al., 2003). A mutation in the murein transglycosylase gene *mltE* (a known virulence factor in *P. syringae* – Boch et al., 2002) was also shown to reduce virulence on pears (Zhao et al., 2005).

Microarray technologies, in a similar way to IVET, produce expression data for many genes simultaneously. However, unlike IVET, which requires the presence of a promoter sequence and reporter gene, microarrays are able to report altered transcript accumulation directly by comparing different conditions following hybridization of cDNA to gene probes. Such an approach is of use where the global effects of gene expression are of interest, e.g. on exposure to changing environmental conditions or between a wild type and regulatory mutant. Okinaka et al. (2002) were the first to attempt microarray analysis in the EPPs, using a spotted array consisting of 5000 randomly selected 3-kb clones from *Dda*3937. The bacteria were grown in their host plant African violet (*Saintpaulia ionantha* var. Katja) and gene expression compared with growth in laboratory medium. To avoid potential contamination from plant RNA, which when converted to cDNA can bind non-specifically to probes on the microarray and/or mask small amounts of bacterial target cDNA, bacterial cells were separated by centrifugation from the inoculated plant leaves prior to bacterial RNA extraction and cDNA synthesis. The analysis revealed that 15 and 74 clones were down- and up-regulated, respectively, in plant extract. Selected clones were then end-sequenced and the resulting sequences compared to those from DNA databases. In some cases the entire clone was sequenced by sequential primer walking, as many clones contained more than one gene (usually between 2 and 4). To identify which of the gene(s) were changed in expression, it was necessary to PCR amplify individual genes from the clones, re-array them, and repeat the hybridization. In general, plant down-regulated genes were involved in house-keeping functions, while up-regulated genes included putative virulence factors, such as a polygalacturonase, protease/peptidase, and peptide synthase genes. Transposase genes were also found to be up-regulated in the plant compared to the laboratory medium, indicating a stress response under these conditions and, potentially, the movement of DNA within or between individual bacterial genomes. Finally, transposon (Tn5) insertion and plant inoculation was then used to identify three mutants in genes homologous to a peptide synthase, *purU* and *pheC*, which were all affected in virulence (Okinaka et al., 2002).

More recently, Venkatesh et al. (2006) published data derived from the first whole-genome cDNA microarrays developed for EPPs. These arrays, manufactured by Agilent Technologies as part of an SCRI-led functional genomics project, are based on the complete genome sequence of *Dda*3937 (microarrays are also available for *Pba*1043 – http://www.scri.ac.uk/research/genetics/genesanddevelopment/sequencingandmicroarray). The *Dda* arrays contain 11 000 60mer cDNA probes, with 6000 distinct probe sequences representing the majority of genes in the completed genome, and include additional commercial controls (Lucidea Universal Score Card, Amersham Biosciences). The microarrays were used to characterize the regulon of the master regulator PhoQ in *Dda*3937, following hybridization of cDNA from both a *phoQ* mutant and wild-type strain grown in minimal medium. Around 40 genes showed statistically different levels of expression (>1.5-fold) between the strains, relating to iron assimilation, membrane transport, stress, toxin (cytolytic delta-endotoxin) production, and at least five additional transcriptional regulators, some of which have been shown to affect virulence. Quantitative real-time (qRT)-PCR was then used to validate the microarray data. qRT-PCR is a powerful method of monitoring bacterial gene expression in infected host tissues, but relies on the use of internal standards for accurate normalization of data. A recent study by Takle et al. (2007) identified a number of different reference genes in Pba1043, including recA and ffh, whose expression remained constant during growth in laboratory media and in planta. In a similar microarray study, Ravirala et al. (2007) used the same whole-genome Dda3937 microarrays to demonstrate that salicylic acid and its precursors activate expression of genes relating to multi-drug efflux pumps (including acrA, emrA), oxidative stress (osmC and gst) and antibiotic/detergent resistance (ompX), which enhance survival of the pathogen in the presence of the plant and plant-derived antimicrobial chemicals.

Proteomic approaches

Proteomic analyses have been conducted on a number of EPPs, including Dda3937, Pba1043, PccATTn10 and Ea273. 2D polyacrylamide gel electrophoresis (PGE) and, more recently, 2D difference gel electrophoresis (DIGE) have been used to profile changes in the proteome between different growth conditions or strains. Cells are grown to mid/late log or stationary phase in laboratory media before concentrating proteins from the culture supernatant (soluble fraction) and/or the intracellular fractions (which may include both soluble and insoluble fractions) to obtain sufficient protein for analysis. Unlike the more traditional 2D-PGE, which uses Coomassie or silver staining and compares samples between gels, DIGE labels each of two protein samples with a different fluorescent dye (Cy3 or Cy5), which are then electrophoresed and compared on a single gel. In some cases a third dye (Cy2) is used to label pooled biological samples for use as an internal standard (Alban et al., 2003). The fluorescent signal from each sample can then be accurately measured to allow a quantitative comparison of proteins, while eliminating between gel variation (Unlu et al., 1997). A number of groups have used protemics to study the biology of the EPPs, while Karp and coworkers used the soluble proteins of Pba1043 and PccATTn10 as a simple biological system for studying improvements in proteomics methodologies (Karp et al., 2005; Karp and Lilley, 2005).

Corbett et al. (2005) used a combination of 2D-PGE and DIGE to analyse secreted proteins from Pba1043 grown in MM containing PGA. As well as seeing changes in spots relating to the PCWDEs, two adjacent proteins of low abundance were also observed and, following mass spectroscopy, were both found to be homologous to a X. campestris pv. raphani putative avirulence protein. Subsequent proteomic analyses, comparing the wild-type strain with type II secretion ($outD^-$) and quorum sensing ($expI^-$) mutants, showed that this protein was both type II secreted and under QS control. Potato tuber and stem inoculation assays then revealed that it played a role in virulence, and the gene was named svx (secreted virulence factor from Xanthomonas) (see above). Coulthurst et al. (2006) used DIGE to compare protein profiles from both intracellular and secreted fractions of wild type and luxS mutant strains of Pba1043 and PccATTn10. LuxS is associated with quorum sensing through the production of autoinducer 2 (AI-2) in a number of bacteria (Vendeville et al., 2005). Differences in protein profiles between the wild type and luxS mutant strains indicated major strain-specific differences, although similar changes in PCWDEs were seen. The luxS mutation in both strains led to a reduction in virulence (Coulthurst et al., 2006).

Other researchers with an interest in the identification of virulence factors in Pba1043 used 2D-PGE to identify 50–60 protein spots from a stationary phase culture of Pba1043 following growth in MM supplemented with potato (host) tuber extract (Mattinen et al., 2007). These spots included known virulence factors, such as the PCWDEs and Svx, proteins involved in transport and utilization of nutrients, and a number of hypothetical proteins. A combination of qRT-PCR and Northern blot analyses revealed that the majority of these proteins were also differentially regulated at the transcriptional level. Amongst the proteins identified were those similar to VgrG, a large protein of unknown function, and four homologues of Hcp (haemolysin co-regulated protein). In *Vibrio cholerae* and *Pseudomonas aeruginosa*, these proteins are secreted by the recently described type VI secretion system, which has been implicated in virulence (Mougous *et al.*, 2006; Pukatzki *et al.*, 2006). Although the modes of action of these proteins have not yet been elucidated, it was shown that over-expression of Hcp1 led to an increase in virulence and may thus be a new virulence factor in *Pba*1043 (Mattinen *et al.*, 2007).

Proteome analysis has also been used to identify virulence-related proteins in *Dda*3937 and *Ea*273. Bouchart *et al.* (2007) investigated the role of osmoregulated *Dda*3937 periplasmic glucans in virulence using LB-grown cells. Although these glucans are known to affect a number of different phenotypes, such as motility and the secretion of PCWDEs, the analysis also revealed proteins suggestive of a role in environmental perception during infection. Nissinen *et al.* (2007) compared wild type and *hrp* secretion and regulatory mutants of *Ea*273 to identify novel type III secreted proteins. Following growth of cells in *hrp*-inducing minimal medium, the supernatant from the wild type grown cells contained protein spots that were absent in supernatant from the *hrp* mutants. These included spots similar to the putative harpins HrpN and HrpW, and effector DspA/E. Others showed similarity to potential helper and effector proteins not previously associated with this pathogen, including Eop1, similar to a YopJ/AvrRxv/HopZ family cysteine protease effector; Eop2, similar to a type III helper HopAK1 of *P. syringae*; and Eop3, similar to an AvrPphE (HopX) family type III effector protein. However, possible roles for these proteins in virulence remain to be determined.

High-throughput mutant screening

As more genomic information becomes available and potential gene targets for analysis are revealed, there is an increasing need for high-throughput mutant generation and screening. One approach to achieving the former is by the generation of transposon-based mutant libraries. Holeva *et al.* (2004) generated an arrayed set of Tn5 insertional mutants for *Pba*1043 that were pooled for rapid isolation of mutants for any gene of interest using PCR-based screening. Using standard methods, around 4500 mutants were constructed (the same number as genes in the genome and therefore equivalent to a one-fold mutation library), pooled in five dimensions and DNA extracted to yield 31 DNA pools. Screening for a mutation in any given gene of interest was then carried out on the 31 pooled samples by PCR using a combination of gene- and transposon-specific primers. Using this approach, a number of mutants affected in T3SS were identified and their role in disease determined *in planta*.

Conclusions

Genome sequences provide a blueprint of all the biochemical and cellular functions required for life. They reveal the capacity of an organism to colonize environmental niches and, through comparisons with other bacterial genomes, they reveal the plasticity in genome organization and evolution that leads to such dramatic, and rapid, adaptation. Given the constantly decreasing costs of sequencing, bacterial genomics, although still in its infancy, seems set to boom. This is particularly important in an age where, with climate change and increased global travel and trade, new pathogens are rapidly emerging to successfully colonize new niches. Sequencing the genomes of these organisms provides the fastest, most cost-effective and comprehensive route to understanding the molecular processes underlying their successful emergence. Through comparing genome sequences of the EPPs we are beginning to fully appreciate the considerable influence of horizontal gene transfer as a driver in their evolutionary change. Our observations are starting to reveal the influence of this gene transfer in directing the adaptation of both the enterobacterial plant and animal pathogens to their specific hosts, while revealing a previously unsuspected capacity to colonize wider environmental niches. The experimental tractability of enterobacteria allows gene functional experiments to be conducted in a high-throughput manner, facilitating rapid exploitation of the observations yielded through transcriptome and proteome profiling, to fully understand the biology of these organisms. Still, bottlenecks remain, both in terms of designing bioassays to determine gene function in an appropriate biological context, and in following such studies with detailed biochemical analyses. Nevertheless, the power of comparative genomics, and the increasing accessibility of genome sequencing, suggests that the use of genomics for studying the EPP will continue to grow.

References

Alarcon-Chaidez, F.J., Penaloza-Vazquez, A., Ullrich, M., and Bender, C.L. (1999). Characterization of plasmids encoding the phytotoxin coronatine in *Pseudomonas syringae*. Plasmid 42, 210–220.

Alban, A., David, S.O., Bjorkesten, L., Andersson, C., Sloge, E., Lewis, S., and Currie, I. (2003). A novel experimental design for comparative two-dimensional gel analysis: two-dimensional difference gel electrophoresis incorporating a pooled internal standard. Proteomics 3, 36–44.

Arnold, D.L., Pitman, A., and Jackson, R.W. (2003). Pathogenicity and other genomic islands in plant pathogenic bacteria. Mol. Plant Pathol. 4, 407–420.

Bachem, C.W. B., van der Hoeven, R.S., de Bruijn, S.M., Vreugdenhil, D., Zabeau, M., and Visser, R.G. F. (1996). Visualization of differential gene expression using a novel method of RNA fingerprinting based on AFLP: Analysis of gene expression during potato tuber development. Plant J. 9, 745–753.

Barnett, M.J., Fisher, R.F., Jones, T., Komp, C., Abola, A.P., Barloy-Hubler, F., Bowser, L., Capela, D., Galibert, F., Gouzy, J., Gurjal, M., Hong, A., Huizar, L., Hyman, R.W., Kahn, D., Kahn, M.L., Kalman, S., Keating, D.H., Palm, C., Peck, M.C., Surzycki, R., Wells, D.H., Yeh, K.C., Davis, R.W., Federspiel, N.A. and Long, S.R. (2001). Nucleotide sequence and predicted functions of the entire *Sinorhizobium meliloti* pSymA megaplasmid. Proc. Natl. Acad. Sci. USA 98, 9883–9888.

Baumgartner, W.A. Jr., Cohen, K.B., Fox, L.M., Acquaah-Mensah, G., and Hunter, L. (2007). Manual curation is not sufficient for annotation of genomic databases. Bioinformatics 23, i41–48.

Bell, K.S., Sebaihia, M., Pritchard, L., Holden, M.T., Hyman, L.J., Holeva, M.C., Thomson, N.R., Bentley, S.D., Churcher, L.J., Mungall, K., Atkin, R., Bason, N., Brooks, K., Chillingworth, T., Clark, K., Doggett, J., Fraser, A., Hance, Z., Hauser, H., Jagels, K., Moule, S., Norbertczak, H., Ormond, D., Price, C., Quail, M.A., Sanders, M., Walker, D., Whitehead, S., Salmond, G.P., Birch, P.R.J., Parkhill, J. and Toth, I.K. (2004). Genome sequence of the enterobacterial phytopathogen *Erwinia carotovora* subsp. *atroseptica* and characterization of virulence factors. Proc. Natl. Acad. Sci. USA 101, 11105–11110.

Boch, J., Joardar, v., Gao, L., Roberson, T.L., Lim, M., and Kunkel, B.N. (2002). Identification of *Pseudomonas syringae* pv. *tomato* genes induced during infection of *Arabidopsis thaliana*. Mol. Microbiol. 44, 73–88.

Bouchart, F., Delangle, A., Lemoine, J., Bohin, J.P., and Lacroix, J.M. (2007). Proteomic analysis of a non-virulent mutant of the phytopathogenic bacterium *Erwinia chrysanthemi* deficient in osmoregulated periplasmic glucans: change in protein expression is not restricted to the envelope, but affects general metabolism. Microbiology 153, 760–767.

Carver, T.J., Rutherford, K.M., Berriman, M., Rajandream, M.A., Barrell, B.G., and Parkhill, J. (2005). ACT: the Artemis Comparison Tool. Bioinformatics 21, 3422–3423.

Caspi, R., Foerster, H., Fulcher, C.A., Kaipa, P., Krummenacker, M., Latendresse, M., Paley, S., Rhee, S.Y., Shearer, A.G., Tissier, C., Walk, T.C., Zhang, P. and Karp, P.D. (2008). The MetaCyc Database of metabolic pathways and enzymes and the BioCyc collection of Pathway/Genome Databases. Nucleic Acids Res. 36, D623–631.

Chaudhuri, R.R., Khan, A.M., and Pallen, M.J. (2004). coliBASE: an online database for *Escherichia coli*, *Shigella* and *Salmonella* comparative genomics. Nucleic Acids Res. 32, D296–299.

Chen, L.L., Ma, B.G., and Gao, N. (2008). Reannotation of hypothetical ORFs in plant pathogen *Erwinia carotovora* subsp. *atroseptica* SCRI1043. FEBS J. 275, 198–206.

Corbett, M., Virtue, S., Bell, K., Birch, P., Burr, T., Hyman, L., Lilley, K., Poock, S., Toth, I., and Salmond, G. (2005). Identification of a new quorum-sensing-controlled virulence factor in *Erwinia carotovora* subsp. *atroseptica* secreted via the type II targeting pathway. Mol. Plant–Microbe Interact. 18, 334–342.

Coulthurst, S.J., Lilley, K.S., and Salmond, G.P.C. (2006). Genetic and proteomic analysis of the role of *luxS* in the enteric phytopathogen, *Erwinia carotovora*. Mol. Plant Pathol. 7, 31–45.

Darling, A.C., Mau, B., Blattner, F.R., and Perna, N.T. (2004). Mauve: multiple alignment of conserved genomic sequence with rearrangements. Genome Res. 14, 1394–1403.

Dellagi, A., Birch, P.R.J., Heilbronn, J., Lyon, G.D., and Toth, I.K. (2000). cDNA-AFLP analysis of differential gene expression in the prokaryotic plant pathogen *Erwinia carotovora*. Microbiology 146, 165–171.

Glasner, J.D., Liss, P., Plunkett, G., 3rd, Darling, A., Prasad, T., Rusch, M., Byrnes, A., Gilson, M., Biehl, B., Blattner, F.R., and Perna, N.T. (2003). ASAP, a systematic annotation package for community analysis of genomes. Nucleic Acids Res. 31, 147–151.

Grenier, A.M., Duport, G., Pages, S., Condemine, G., and Rahbe, Y. (2006). The phytopathogen *Dickeya dadantii* (*Erwinia chrysanthemi* 3937) is a pathogen of the pea aphid. Appl. Environ. Microbiol. 72, 1956–1965.

Hauben, L., Moore, E.R., Vauterin, L., Steenackers, M., Mergaert, J., Verdonck, L., and Swings, J. (1998). Phylogenetic position of phytopathogens within the Enterobacteriaceae. Syst. Appl. Microbiol. 21, 384–397.

Holeva, M.C., Bell, K.S., Hyman, L.J., Avrova, A.O., Whisson, S.C., Birch, P.R.J., and Toth, I.K. (2004). Use of a pooled transposon mutation grid to demonstrate roles in disease development for *Erwinia carotovora* subsp. *atroseptica* putative type III secreted effector (DspE/A) and helper (HrpN) proteins. Mol. Plant–Microbe Interact. 17, 943–950.

Jackson, R.W., and Giddens, S.R. (2006). Development and application of *in vivo* expression technology (IVET) for analysing microbial gene expression in complex environments. Infect. Disord. Drug Targets 6, 207–240.

Karp, N.A., Griffin, J.L., and Lilley, K.S. (2005). Application of partial least squares discriminant analysis to two-dimensional difference gel studies in expression proteomics. Proteomics 5, 81–90.

Karp, N.A., and Lilley, K.S. (2005). Maximising sensitivity for detecting changes in protein expression: experimental design using minimal CyDyes. Proteomics 5, 3105–3115.

Kunin, V., Goldovsky, L., Darzentas, N., and Ouzounis, C.A. (2005). The net of life: reconstructing the microbial phylogenetic network. Genome Res. 15, 954–959.

Margulies, M., Egholm, M., Altman, W.E., Attiya, S., Bader, J.S., Bemben, L.A., Berka, J., Braverman, M.S., Chen, Y.J., Chen, Z., Dewell, S.B., Du, L., Fierro, J.M., Gomes, X.V., Godwin, B.C., He, W., Helgesen, S., Ho, C.H., Irzyk, G.P., Jando, S.C., Alenquer, M.L., Jarvie, T.P., Jirage, K.B., Kim, J.B., Knight, J.R., Lanza, J.R., Leamon, J.H., Lefkowitz, S.M., Lei, M., Li, J., Lohman, K.L., Lu, H., Makhijani, V.B., McDade, K.E., McKenna, M.P., Myers, E.W., Nickerson, E., Nobile, J.R., Plant, R., Puc, B.P., Ronan, M.T., Roth, G.T., Sarkis, G.J., Simons, J.F., Simpson, J.W., Srinivasan, M., Tartaro, K.R., Tomasz, A., Vogt, K.A., Volkmer, G.A., Wang, S.H., Wang, Y., Weiner, M.P., Yu, P., Begley, R.F. and Rothberg, J.M. (2005). Genome sequencing in microfabricated high-density picolitre reactors. Nature 437, 376–380.

Mattinen, L., Nissinen, R., Riipi, T., Kalkkinen, N., and Pirhonen, M. (2007). Host-extract induced changes in the secretome of the plant pathogenic bacterium *Pectobacterium atrosepticum*. Proteomics 7, 3527–3537.

Mattinen, L., Tshuikina, M., Mae, A., and Pirhonen, M. (2004). Identification and characterization of Nip, necrosis-inducing virulence protein of *Erwinia carotovora* subsp. *carotovora*. Mol. Plant–Microbe Interact. 17, 1366–1375.

Mougous, J.D., Cuff, M.E., Raunser, S., Shen, A., Zhou, M., Gifford, C.A., Goodman, A.L., Joachimiak, G., Ordonez, C.L., Lory, S., Walz, T., Joachimiak, A., and Mekalanos, J.J. (2006). A virulence locus of *Pseudomonas aeruginosa* encodes a protein secretion apparatus. Science 312, 1526–1530.

Nissinen, R.M., Ytterberg, A.J., Bogdanove, A.J., van Wijk, K.J., and Beer, S.V. (2007). Analyses of the secretomes of *Erwinia amylovora* and selected hrp mutants reveal novel type III secreted proteins and an effect of HrpJ on extracellular harpin levels. Mol. Plant Pathol. 8, 55–67.

Nizan, R., Barash, I., Valinsky, L., Lichter, A., and Manulis, S. (1997). The presence of *hrp* genes on the pathogenicity-associated plasmid of the tumorigenic bacterium *Erwinia herbicola* pv. *gypsophilae*. Mol. Plant–Microbe Interact. 10, 677–682.

Okinaka, Y., Yang, C.H., Perna, N.T., and Keen, N.T. (2002). Microarray profiling of *Erwinia chrysanthemi* 3937 genes that are regulated during plant infection. Mol. Plant–Microbe Interact. 15, 619–629.

Pemberton, C.L., and Salmond, G.P. C. (2004). The Nep1-like proteins-a growing family of microbial elicitors of plant necrosis. Mol. Plant Pathol. 5, 353–359.

Pemberton, C.L., Whitehead, N.A., Sebaihia, M., Bell, K.S., Hyman, L.J., Harris, S.J., Matlin, A.J., Robson, N.D., Birch, P.R.J., Carr, J.P., Toth, I.K. and Salmond, G.P. (2005). Novel quorum-sensing-controlled genes in *Erwinia carotovora* subsp. *carotovora*: identification of a fungal elicitor homologue in a soft-rotting bacterium. Mol. Plant–Microbe Interact. 18, 343–353.

Price, M.N., Dehal, P.S., and Arkin, A.P. (2007). Orthologous transcription factors in bacteria have different functions and regulate different genes. PLoS Comput. Biol. 3, 1739–1750.

Pritchard, L., White, J.A., Birch, P.R.J., and Toth, I.K. (2006). GenomeDiagram: a python package for the visualization of large-scale genomic data. Bioinformatics 22, 616–617.

Pukatzki, S., Ma, A.T., Sturtevant, D., Krastins, B., Sarracino, D., Nelson, W.C., Heidelberg, J.F., and Mekalanos, J.J. (2006). Identification of a conserved bacterial protein secretion system in *Vibrio cholerae* using the *Dictyostelium* host model system. Proc. Natl. Acad. Sci. USA 103, 1528–1533.

Ravirala, R.S., Barabote, R.D., Wheeler, D.M., Reverchon, S., Tatum, O., Malouf, J., Liu, H., Pritchard, L., Hedley, P.E., Birch, P.R., Toth, I.K., Payton, P. and San Francisco, M.J. (2007). Efflux pump gene expression in *Erwinia chrysanthemi* is induced by exposure to phenolic acids. Mol. Plant–Microbe Interact. 20, 313–320.

Rodionov, D.A., Gelfand, M.S., and Hugouvieux-Cotte-Pattat, N. (2004). Comparative genomics of the KdgR regulon in *Erwinia chrysanthemi* 3937 and other gamma-proteobacteria. Microbiology 150, 3571–3590.

Rouanet, C., Reverchon, S., Rodionov, D.A., and Nasser, W. (2004). Definition of a consensus DNA-binding site for PecS, a global regulator of virulence gene expression in *Erwinia chrysanthemi* and identification of new members of the PecS regulon. J. Biol. Chem. 279, 30158–30167.

Rutherford, K., Parkhill, J., Crook, J., Horsnell, T., Rice, P., Rajandream, M.A., and Barrell, B. (2000). Artemis: sequence visualization and annotation. Bioinformatics 16, 944–945.

Samson, R., Legendre, J.B., Christen, R., Saux, M.F., Achouak, W., and Gardan, L. (2005). Transfer of *Pectobacterium chrysanthemi* (Burkholder et al. 1953) Brenner et al. 1973 and *Brenneria paradisiaca* to the genus *Dickeya* gen. nov. as *Dickeya chrysanthemi* comb. nov. and *Dickeya paradisiaca* comb. nov. and delineation of four novel species, *Dickeya dadantii* sp. nov., *Dickeya dianthicola* sp. nov., *Dickeya dieffenbachiae* sp. nov. and *Dickeya zeae* sp. nov. Int. J. Syst. Evol. Microbiol. 55, 1415–1427.

Sanger, F., and Coulson, A.R. (1975). A rapid method for determining sequences in DNA by primed synthesis with DNA polymerase. J. Mol. Biol. 94, 441–448.

Stothard, P., and Wishart, D.S. (2006). Automated bacterial genome analysis and annotation. Curr. Opin. Microbiol. 9, 505–510.

Sundin, G.W. (2007). Genomic insights into the contribution of phytopathogenic bacterial plasmids to the evolutionary history of their hosts. Annu. Rev. Phytopathol. 45, 129–151.

Takle, G.W., Toth, I.K., and Brurberg, M.B. (2007). Evaluation of reference genes for real-time RT-PCR expression studies in the plant pathogen *Pectobacterium atrosepticum*. BMC Plant Biol. 7, 50.

Toth, I.K., Bell, K.S., Holeva, M.C., and Birch, P.R.J. (2003). Soft rot erwiniae: from genes to genomes. Mol. Plant Pathol. 4, 17–30.

Toth, I.K., Pritchard, L., and Birch, P.R.J. (2006). Comparative genomics reveals what makes an enterobacterial plant pathogen. Annu. Rev. Phytopathol. 44, 305–336.

Triplett, L.R., Zhao, Y., and Sundin, G.W. (2006). Genetic differences between blight-causing *Erwinia* species with differing host specificities, identified by suppression subtractive hybridization. Appl. Environ. Microbiol. 72, 7359–7364.

Unlu, M., Morgan, M.E., and Minden, J.S. (1997). Difference gel electrophoresis: a single gel method for detecting changes in protein extracts. Electrophoresis 18, 2071–2077.

Vendeville, A., Winzer, K., Heurlier, K., Tang, C.M., and Hardie, K.R. (2005). Making 'sense' of metabolism: autoinducer-2, LuxS and pathogenic bacteria. Nat. Rev. Microbiol. 3, 383–396.

Venkatesh, B., Babujee, L., Liu, H., Hedley, P., Fujikawa, T., Birch, P., Toth, I., and Tsuyumu, S. (2006). The *Erwinia chrysanthemi* 3937 PhoQ sensor kinase regulates several virulence determinants. J. Bacteriol. 188, 3088–3098.

Vernikos, G.S., Thomson, N.R., and Parkhill, J. (2007). Genetic flux over time in the *Salmonella* lineage. Genome Biol. 8, R100.

Vinatzer, B.A., and Yan, S.-C. (2008). Mining the genomes of plant pathogenic bacteria: how not to drown in gigabases of sequence. Mol. Plant Pathol. 9, 105–118.

Wagner, P.L., and Waldor, M.K. (2002). Bacteriophage control of bacterial virulence. Infect. Immun. 70, 3985–3993.

Yamada, K., Hirota, M., Niimi, Y., Nguyen, H.A., Takahara, Y., Kamio, Y., and Kaneko, J. (2006). Nucleotide sequences and organization of the genes for carotovoricin (Ctv) from *Erwinia carotovora* indicate that Ctv evolved from the same ancestor as *Salmonella typhi* prophage. Biosci. Biotechnol. Biochem. 70, 2236–2247.

Yang, S., Perna, N.T., Cooksey, D.A., Okinaka, Y., Lindow, S.E., Ibekwe, A.M., Keen, N.T., and Yang, C.H. (2004). Genome-wide identification of plant-up-regulated genes of *Erwinia chrysanthemi* 3937 using a GFP-based IVET leaf array. Mol. Plant–Microbe Interact. 17, 999–1008.

Zhao, Y., Blumer, S.E., and Sundin, G.W. (2005). Identification of *Erwinia amylovora* genes induced during infection of immature pear tissue. J. Bacteriol. 187, 8088–8103.

Ralstonia solanacearum and Bacterial Wilt in the Post-genomics Era

Darby Brown

Abstract

The first *Ralstonia solanacearum* GMI1000 genome sequence was completed in 2002. This work paved the way for an enhanced understanding of the molecular genetics of this important plant pathogen. The availability of a fully annotated genome sequence made it possible to further our understanding of gene expression *in planta* and led to the discovery of important nuances in virulence gene regulatory cascades. Moreover, analysis of the more recently sequenced race 3 biovar 2 strain, UW551, made comparative genomics possible. Other recent advances stemming from the genomic profile of *R. solanacearum* has led to an appreciation of this pathogen as a species complex, with a rich evolutionary history as well as the discovery and molecular dissection of a lysogenic bacteriophage that contributed to virulence on tobacco. Now that the genome is available, progress towards a more comprehensive study of the life cycle of this successful soil inhabitant and deadly plant pathogen can be made.

R. solanacearum is a long-known and potent plant pathogen

The history of *Ralstonia solanacearum* as a plant pathogen begins with its description as *Bacillus solanacearum* in 1896 by Erwin Fink Smith (Smith, 1896). For the last 111 years, this soil-borne bacterium that causes lethal wilting of its host plant has been known as *Pseudomonas solanacearum*, *Burkholderia solanacearum* and finally, *Ralstonia solanacearum*. Its designation into the genus, *Ralstonia*, took place in 1995 as a result of more specific phylogenetic categorization methods (Yabuuchi *et al.*, 1995). *R. solanacearum* is a member of the β-proteobacteria (along with *Neisseria* and *Burkholderia* spp.), and like many plant pathogenic bacteria, *R. solanacearum* exists as a Gram-negative rod with polar flagella.

The importance of *R. solanacearum* as a serious plant pest cannot be overemphasized. Not only does infection with this bacterium very often result in host plant death but once established, *R. solanacearum* can survive in soils for many years. It is thought that such prolonged survival in soils is due to the ability of the bacterium to form latent infections within certain indigenous weeds and non-food crops (Tusiime *et al.*, 1998; Wenneker *et al.*, 1999; Pradhanang *et al.*, 2000; Janse *et al.*, 2004). A serious problem has recently emerged whereby cold-tolerant race 3 strains of *R. solanacearum* have been introduced to Europe and North America via Africa and Central America on geranium cuttings (Janse *et al.*, 2004; Swanson, 2005). This poses a major threat to many of the crops grown in these regions, particularly potatoes. Furthermore, members of the *R. solanacearum* species complex can infect over 200 different plant species, many of which are economically important food crops (Hayward, 1991). Furthermore, increasing globalization of commodity markets poses significant challenges in controlling this plant pathogen, especially if imported crop plants are latently infected. In a recent survey of solanaceous crops in Ethiopia, researchers identified both *R. solanacearum* biovar I and biovar II strains. The authors correlated their findings with the observation that bacterial wilt incidence had been steadily increasing and that peppers, which were not previously susceptible, were succumbing to newly introduced biovar I strains. In fact, a previous survey of *R. solanacearum* incidence in Ethiopia showed that only biovar II strains were present and that within an 18-year window, biovar I strains had been introduced (likely from imported crops) and proved to be successful pathogens in the Ethiopian soils (Lamessa and Zeller, 2007).

Ralstonia solanacearum: a taxonomically diverse species complex

Before the genomic revolution, much effort was expended describing and organizing the many different strains of *R. solanacearum* (see also chapter by Vinatzer and Bull). More than 40 years ago, Buddenhagen and colleagues (1962) developed the Race system which was based on the host plants from which different strains were isolated. There were five proposed races: race 1 (the solanaceous race); race 2, (the Musa race); race 3, (the potato race); race 4 (the mulberry race); and race 5 (the ginger race). Later, researchers quickly realized that *R. solanacearum* can infect a vast number of hosts, so a better way to classify *R. solanacearum* isolates was required. This realization led to an effort to define additional classification methods. As a result, the biovar and biotype systems were developed. These systems are based on the intrinsic metabolic activities of different *R. solanacearum* isolates. The biovar test organizes *R. solanacearum* isolates for the utilization of three hexose alcohols and three disaccharides (cellobiose, lactose, maltose, dulcitol, mannitol, and sorbitol), while the biotype test provides a metabolic profile using maltose, mannitol, malonate, trehalose, inositol, and hippurate (Hayward, 1964, 1991).

Of the five known races, race 1 is composed of strains derived from a variety of hosts. Members of this group can infect ornamental and food crops such as petunia, zinnia, tomato, potato, eggplant, and peanut (Hayward, 1991). Race 2 is associated specifically with members of the banana family. Race 3 strains are commonly isolated from potato, tomato and, more recently, geranium, bitter nightshade and stinging nettle. This race is also unique in that it has the ability to tolerate cooler soil temperatures. This unique characteristic is believed to be the underlying cause of recent outbreaks and long-term infestation in northern Europe. Moreover, race 3 strains are not only a quarantine pathogen worldwide, but are listed as a bioagroterrorism agent in the United States (Swanson, 2005). Races 4 and 5 are primarily associated with ginger and mulberry, respectively, and are the least well characterized of all *R. solanacearum* strains.

Currently, the rapid advances in genomic technology make it possible to carefully study the taxonomy of the *R. solanacearum* species complex at the molecular level. Such studies have broadly divided *R. solanacearum* isolates into four phylotypes (genetic groups). Each phylotype is assigned based on the sequence of the 16S-23S rRNA internal transcribed spacer (ITS) (Prior and Fegan, 2005) (Fig. 9.1). Phylotype I, also known as Asiaticum, comprises *R. solanacearum* isolates from Asia and Australia while phylotype II, or Americanum, comprises bacterial isolates from South and Central America. The third phylotype, Africanum, accounts for strains originating in Africa. The last (fourth) phylotype represents strains originating from Indonesia. Members of each phylotype can be further divided into subgroups called sequevars. Each sequevar group is determined by the nucleotide sequences of highly conserved regions of the *hrpB*, *mutS*, and endonuclease genes (Prior and Fegan, 2005). There are currently 23 sequevars representing over 140 *R. solanacearum* strains (see Table 9.1 for a summary) and as additional genetic data become available, the sequevar groups can be confirmed and further refined.

However, it should be noted, that with all classification schemes, they are subject to change. A case in point is the recent report of a new pathogenic variant of *R. solanacearum* isolated from the French West Indies (Wicker et al., 2007). Furthermore, a phylotype analysis of *R. solanacearum* isolates isolated from the 1980s was compared to a large collection of isolates obtained from 1989–2003. The results were surprising in that a particular phylotype II strain collected in 1988 from tomato belonged to the phylotype II/sequevar 4 (II/4). The phylogenetic trees from

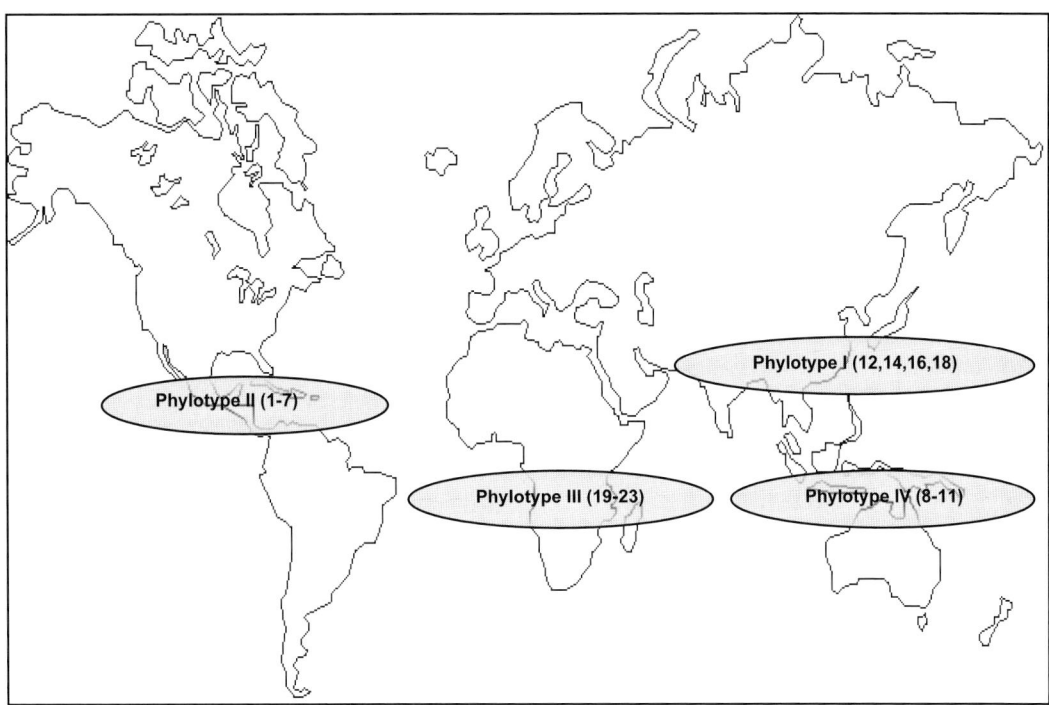

Figure 9.1 Geographical distribution of R. solanacearum phylotypes.

Table 9.1 Comparison of race/biovar and phylotype/sequevar/biotype schemes (adapted from Prior and Fegan, 2005)

Strain	Isolated	Host (race)	Biovar	Phylotype	Sequevar	Biotype
K60	USA	Tomato (1)	1	II	7	7
GMI1000	Fr. Guyana	Tomato (1)	3	I	12	8
UW551	Kenya	Geranium (3)	2	II	ND	ND
UW20	Honduras	Banana (2)	1	II	6	6
UW127	Peru	Plantain (2)	1	II	4	4
UW11	Costa Rica	Heliconia (2)	1	II	3	3
CIP239	Brazil	Potato (3)	1	II	5	5
BP385	Netherlands	Potato (3)	2	II	1	1
CIP309	Colombia	Potato (3)	2	II	2	1
J25	Kenya	Potato (3)	2	III	22	9
R288	China	M. Alba (?)	5	I	18	8
CFBP765	Japan	Tomato (1)	4	I	ND	8
PS107	Indonesia	Tomato (1)	2	IV	8	1
ACH0732	Australia	Tomato (1)	2	IV	7	1

ND, not determined.

this study show that these isolates cluster with the strains that are commonly found on banana (causing Moko disease), yet they are not pathogenic on banana. Instead, these newly emergent strains are highly pathogenic on solanaceous, cucurbit, and ornamental plants, including the 'resistant' tomato line Hawaii7996. This study highlights the sensitivity of the phylotype scheme and its use as a diagnostic tool. Moreover, with large databases full of genomic information, it is possible to quickly make phylogenetic comparisons and discover novel bacterial isolates.

The phylotype/sequevar/biotype regime was proposed to supersede the race/biovar system due to its greater specificity, ease of use and because it can successfully encompass the diversity of the *R. solanacearum* species complex (Fegan and Prior, 2005). However, the race/biovar and biotype systems serve as the international standard for *R. solanacearum* taxonomy (Fegan and Prior, 2005). In fact, the biotype test can reliably predict the phylotype/sequevar of a given isolate. The reader is referred to Table 9.1, which shows the relationship between the phylotype/sequevar/biotype scheme and the race/biovar scheme for a representative group of *R. solanacearum* strains.

Understanding bacterial wilt in the genomics era

The entire genome sequence of *R. solanacearum* strain GMI1000 (race 1, biovar 3) was released in 2002 while in 2005 the second *R. solanacearum* (strain UW551; race 3 biovar 2) sequence was published (Salanoubat *et al.*, 2002, Gabriel *et al.*, 2006). The availability of these genome sequences allow for a more comprehensive understanding of bacterial wilt, but also gives rise to more sophisticated studies that include detailed mechanisms of pathogenesis, and also makes it possible to ask questions about the non-pathogenic or saprophytic role of *R. solanacearum* in the soil.

The genome of *R. solanacearum* is divided into two replicons: the chromosome (3.7 Mb) and the megaplasmid (2.1 Mb). The chromosome harbours a greater majority of the usual 'housekeeping' genes, while the megaplasmid harbours hallmark features of a plasmid (such as a *repA* gene flanking the origin of replication) (Genin and Boucher, 2004). Importantly, however, the megaplasmid does not function as a plasmid (it is not curable), but rather, is akin to a second chromosome. For example, in GMI1000, the megaplasmid harbours 55 genes that are involved in the biosynthesis of amino acids, nucleotides, and cofactors, some of which are not present on the chromosome and are probably required for growth in nutritionally poor environments such as minimal medium and inside the host plant.

The G+C content of the *R. solanacearum* genome is approximately 67% (64.5% for UW551); however, there are regions that show either increased or decreased G + C content and alternate codon usage regions denoted as ACURs. These ACURs include tRNA sequences, bacteriophage, insertion elements and many known and putative type III effectors. The genes in and surrounding the ACURs are hypothesized to be the result of recent lateral gene transfer (LGT) events (Salanoubat *et al.*, 2002; Lavie *et al.*, 2004).

Genomic mining has revealed that, like other soil-borne bacteria, such as *Pseudomonas putida* and *P. aeruginosa*, *R. solanacearum* possesses catabolic pathways that can break down organic matter found in soil. Moreover, in GMI1000, there are 83 ABC transporters that likely contribute to the ability of *R. solanacearum* to survive on myriad carbon sources in both plant and soil environments (Genin and Boucher, 2004). This observation is consistent with the understanding that *R. solanacearum* can survive for long periods of time in various soil habitats and can infect an enormous range of host plants, each representing a unique nutritional challenge. Furthermore, UW551 is lacking a 22-kb region that contains the genes required for catabolism of three sugars used in the biovar tests: mannitol, sorbitol and dulcitol, while strain GMI1000 possesses these genes (Gabriel *et al.*, 2006). These metabolic differences have long been known, but it was only through the comparison of the two genome sequences that the specific genes could be identified.

Analysis of the *R. solanacearum* genome led to the 'discovery' of many genes that likely contribute to virulence based on homology to other systems (Salanoubat *et al.*, 2002). However, a role for such genes in virulence awaits experimental evidence. Much work is now required to determine which, if any, of these hypothetical virulence factors are indeed sufficient to allow *R. solanacearum* to wreak havoc on its host plant.

Comparative analysis of the genomes of several plant pathogens (representing distinct phylogenetic clusters) for which the genome sequences are available has revealed, as expected, both conserved and divergent genomic loci/modules (Van Sluys et al., 2002). BLASTP (Altschul et al., 1997) comparisons were performed to identify orthologous genes in the genomes of six plant-associated bacteria (*Xanthomonas campestris* pv. *campetris*, *Ralstonia solanacearum*, *Mezorhizobium loti*, *Sinorhizobium meliloti*, *Xylella fastidiosa*, *Xanthomonas axonopodis* pv. *citri*). Nineteen common genes were identified. Ten of these are proteins associated with the bacterial cell surface, which is a significant discovery since membrane- or cell wall- associated proteins are at the crux of the plant–bacterium interaction (Manoil and Beckwith, 1985). It should be noted however, that the comparison across genomes is at best an imperfect science. With little experimental evidence, we can only make predictions based on sequence similarity, and as a result we should exercise great caution when interpreting the results of genome annotations and broad comparisons to avoid the fallacy of what Wassenaar and Gaastra refer to as 'putativism' (Wassenaar and Gaastra, 2001). However, until we have experimental data to confirm the true function of a given gene product, such annotations provide a framework from which to begin.

The second *R. solanacearum* genome sequence: UW551

UW551 is a race 3 biovar 2 (R3B2) strain that was isolated from an infected Geranium plant in Wisconsin (Swanson et al., 2007). This strain originated from Kenya, where the ornamental geranium plants were vegetatively propagated. It is unknown how the bacteria were introduced to the plants, but either the mother plant from which the new plants were obtained was latently infected or that in the process of propagation, bacteria were introduced and remained dormant in the plant tissue until conditions favoured the rapid growth of the host plant. Because the R3B2 strains are capable of infecting potato and other closely related solanaceous plants and they are able to cause disease at cooler temperatures than other *R. solanacearum* isolates, the R3B2 strains are listed by the USDA as a select agent and they are quarantine pathogens in Canada and Europe. Thus, it is of great economic interest to fully understand the nature of this *R. solanacearum* group with the hope of eradicating, or controlling its spread and virulence. As mentioned above, R3B2 strains have the ability to cause disease in relatively cool temperatures (about 16°C) compared with other *R. solanacearum* isolates which require warmer temperatures (24–28°C) to fully cause disease. In order to better understand this so-called 'cold tolerance' and to reveal additional unique characteristics of this *R. solanacearum* phylotype, the genome of UW551 was sequenced (Gabriel et al., 2006). The sequencing project revealed that UW551 possesses approximately 402 unique open reading frames (ORFs) that are absent from the GMI1000 genome. Furthermore, UW551 and other R3B2 strains all possess a 38 gene region with similarity to bacteriophage genes. Another interesting feature of the R3B2 strains is the possible difference in expression of a major virulence determinant, AvrA (Robertson et al., 2004). R3B2 strains have a more narrow host range than the Race 1 (like GMI1000) strains; one reason for this narrow host range could be due to differences between the expression of the *avrA* genes. AvrA, when present in Race 1 strains, elicits an HR on tobacco, but R3B2 strains are not virulent on tobacco, nor do they cause an HR on this plant host. Analysis of the UW551 genome revealed that there is an insertion sequence located upstream of the putative *avrA* locus; this insertion may render the gene inactive (Gabriel et al., 2006). Unfortunately, the genome sequence analysis did not reveal any obvious genes that would allow for the 'cold tolerance' of UW551 and other R3B2 strains. Nevertheless, the additional genomic information is helpful for anyone wishing to better understand the virulence and evolution of *R. solanacearum* at the molecular level. The more genome sequences present, the better able we are to tease apart the specific genomic differences, which will ultimately lead to more informed hypotheses about the role these different genes play during the course of infection.

The *R. solanacearum* life cycle: from the soil to the stem and back again

Bacterial wilt is typically characterized by a progressive wilting of plant foliage due to impaired xylem vessel function. Wilting is also preceded or accompanied by stunting, chlorosis, enhanced adventitious root growth (a symptom of water stress), and browning of infected stem tissue (Schell, 2000). Once a susceptible plant exhibits the first signs of wilt, the prognosis for recovery is grim with death the final outcome.

Vasse and coworkers (1995) described a three-stage process for *R. solanacearum* infection. First, *R. solanacearum* gains entry to the host plant through the roots. The bacteria initially colonize the root exterior at the elongation zone (near the tip of the growing root) and near sites of secondary (branching) root emergence. Next, once bacteria enter the root tissue, the intercellular spaces of the root cortex and the vascular parenchyma (living cells near the xylem vessels that function in starch storage) are targets of further bacterial ingress. The final destination of the invading bacterial population is the xylem tissue. Once inside the xylem, the bacteria multiply rapidly and spread throughout the plant; bacterial population densities can approach 10^{10} cfu/gram stem tissue.

A recent report described the effects of *R. solanacearum* strain GMI1000 and Hrp mutants (see below) on the root growth of petunia plants. Interestingly, both wild type and the mutants lacking a functional type three secretion system were able to inhibit lateral root elongation and swelling of the root tips, but only the wild type strain could stimulate the growth of new lateral roots (Zolobowska and Van Gijsegem, 2006). Thus, it would appear that *R. solanacearum* is able to manipulate its host plant even before it gains entry to the root tissue and that while the type three secretion system is involved, there are other unknown mechanisms that also promote the bacterial-plant interaction.

Despite the fact that bacterial numbers can reach astronomical levels inside the tiny space of the plant xylem vessel, it has been observed that once infected plants develop wilt symptoms, the internal bacterial numbers significantly decline (C. Allen, unpublished observations). This phenomenon is something of an enigma; are the bacteria killed by some toxic compound produced by the plant as it begins to wilt or is it possible that bacteria are capable of moving freely out of the host and into the soil during the course of the disease? Recent evidence suggests the latter scenario occurs whereby bacteria are continually shed from the roots preceding the death of the infected plant (Swanson *et al.*, 2005). The reason for these dynamic population changes is unclear, and may be due to sloughing off of the dead plant cells, or the bacteria may utilize an active escape mechanism driven by chemotaxis and motility. The plant could also produce toxic compounds as it wilts, which acts as an escape signal to the bacterial population.

R. solanacearum is a very destructive plant pathogen

Since *R. solanacearum* can infect such a large number of host plants, it is a very difficult pathogen to control in the field. The most common methods for controlling bacterial wilt include breeding for plant resistance and crop rotation. Fumigation, which involves the application of toxic chemicals (i.e. methyl bromide) to the soil, is not generally useful because it is economically and environmentally disadvantageous. Thus, these methods are met with little success due to the worldwide distribution and diversity of *R. solanacearum*. Furthermore, the ability of *R. solanacearum* to survive in soil, water, and in the rhizosphere of weedy and native non-host plants undermines many control regimens because of the many factors that must be eliminated to ensure eradication or prevention (Jenkins *et al.*, 1966; Jackson and Gonzalez, 1981). The economic impact of bacterial wilt is difficult to assess, since many regions plagued with bacterial wilt are devoted to sustenance farming rather than commercial agriculture, which monitors crop yields more closely. Moreover, crop losses are rampant in developing nations (where this disease is often endemic) because there is little support financially, and sometimes socially, for the implementation of hygienic agricultural practices.

So far, the most effective control measures employed against *R. solanacearum* involves the development of resistant plants. Some progress has been made in developing highly resistant tomato varieties by crossing wild tomato relatives that grow naturally in infested regions with cultivated varieties. Several resistant tomato cultivars have been developed, the best of which is Hawaii 7996. This tomato line was derived from crosses between cultivated tomato and the wild tomato relative, *Lycopersicon pimpinellifolium* (Grimault et al., 1995). In a recent survey of *R. solanacearum* growth in different resistant tomatoes, Hawaii 7996 showed the greatest level of resistance (Nakaho, 2004). While the entire repertoire of plant genes required for resistance is unknown, it is hypothesized that the resistance phenotype originates from multiple quantitative trait loci (QTL) present on chromosome six (Mangin et al., 1999). Resistance to *R. solanacearum* in these tomato lines is not due to the pathogen's inability to infect the roots, but to the inability of the pathogen to fully colonize the xylem and stem tissue. Using immunofluorescence microscopy, McGarvey and coworkers (McGarvey et al., 1999) showed that bacteria were able to spread throughout the xylem and intercellular spaces of the stem in susceptible plants, but in resistant plants, bacteria were confined only to the xylem vessels. In another resistant tomato line, LS-89, it was observed that *R. solanacearum* is restricted to the primary xylem vessels, and bacterial growth within the xylem tissue is suppressed (Nakaho et al., 2004). Moreover, in both resistant plant lines, electron-dense thickening in the pit membranes surrounded the bacteria, a response that is thought to reduce bacterial spread (Nakaho et al., 2000). Also, it was observed that resistant tomato roots accumulated polyphenolic cell wall components (Vasse et al., 2005), further suggesting that the plant cell wall undergoes structural changes to limit bacterial spread. This cell wall mediated defence response, best studied in *Arabidopsis thaliana*, is characterized by cell wall thickening via deposition of callose, phenolics, and proline-rich glycoproteins, and has been observed in response to several phytopathogenic bacteria (Hauck et al., 2003). Moreover, RP-HPLC analysis of the liquid culture medium collected from resistant or susceptible tomato varieties grown in hydroponic solution in the presence and absence of *R. solanacearum* revealed that aromatic molecules of low molecular weight were present only in the inoculated culture medium of the resistant tomato line, Hawaii 7996 (Vasse et al., 2005). Recently, Feng and coworkers (Feng et al., 2003) identified a novel antimicrobial (and anti-*Ralstonia*) peptide produced in the leaves of the wilt-resistant potato line MS-42.3. These results suggest that resistant plants may react to *R. solanacearum* infection by producing antimicrobial compounds.

Another means to control *R. solanacearum* might come from what is learned about the non-host plant habitats where *R. solanacearum* is found. It is known that *R. solanacearum* can survive in sterile water over a very long period of time, but the survival dynamics of *R. solanacearum* in non-sterile environments is poorly understood. In a recent study carried out by Alvarez et al. (2007), it was found that survival rates of *R. solanacearum* in river water microcosms decreased in non-sterile river water, while populations remained constant if the river water was first sterilized before introducing *R. solanacearum*. In the non-sterile microcosms, it was observed that temperature altered the population dynamics such that the cooler the temperatures, the less well *R. solanacearum* survived. Additional biotic factors contributed to the decline of *R. solanacearum* numbers as well, including lytic bacteriophage and predators and competitors such as protozoa and other bacteria, respectively. Such studies are extremely useful in understanding which ecological factors contribute to the fitness of *R. solanacearum* in different environments. For many years, these kinds of studies were hampered by the lack of available technology, but now it is possible to use genomics to monitor population changes in natural and laboratory ecosystems. Complex ecological experiments are facilitated through the availability of specific genomic tools to monitor specific strains of *R. solanacearum* as well as other members in an ecosystem community.

R. solanacearum can also infect the model plant *Arabidopsis thaliana* (Deslandes et al., 1998; Yang and Ho, 1998). In this pathosystem, it was observed that several *R. solanacearum* isolates could infect and cause wilt symptoms on the Col-5 accession of *A. thaliana*, while the Nd-1 accession was resistant to the bacterium (Deslandes et al., 1998). This work also identified the resistance locus, RRS1 on chromosome V. RRS1 is a recessive allele in the Nd-1 accession, but in transgenic plants, it functions like a dominant allele conferring *R. solanacearum* resistance to the transgenic plant (Deslandes et al., 2002) which suggests that it may be useful in developing transgenic crop plants containing such resistance genes. Further work to elucidate the RRS1 gene product (a protein that contains the common R protein domain, TIR-NBS-LRR and a WRKY motif common to transcriptional activators) and its interaction with R. solanacearum effector proteins revealed that the effector PopP2 (but not PopP1) could interact with RRS1 and direct it to the plant host nucleus (Deslandes et al., 2003). Understanding the specific genes involved in host recognition and response to bacterial pathogens in both susceptible and resistant hosts is greatly facilitated by the availability of full genome sequences.

Furthermore, because R. solanacearum infects a large number of leguminous plants such as peanut (Arachis hypogea) and common bean (Phaseolus vulgaris) it is of interest to the scientific community to develop a model pathosystem to elucidate the R. solanacearum–legume interaction. Recently, Vailleau and colleagues have described an initial study using the model legume, Medicago truncatula (Vailleau et al., 2007). Using two M. truncatula lines, (Jemalong A17 and F83005.5) and a diverse collection of legume-isolated R. solanacearum, the authors showed that two strains of R. solanacearum could cause severe wilt symptoms on M. truncatula. The most virulent R. solanacearum isolate was UW377, a peanut isolate from China. UW377 caused severe wilt symptoms on both the resistant (F83005.5) and susceptible plants (A17). GMI1000, an isolate from tomato, was also capable of infecting M. truncatula A17 and the ability to cause disease was *hrp* dependent: inactivation of the *hrp*-dependent type III secretion system rendered the bacterial strain avirulent on the susceptible M. truncatula lines.

Genomic plasticity drives heterogeneity in the *R. solanacearum* species complex

One reason why R. solanacearum may be able to infect a high number of diverse host plants is the fact that the genome of this bacterium is prone to rapid genetic diversification (Grover et al., 2006). That is, when the genomes of R. solanacearum strains isolated from the same field were subjected to random amplified polymorphic DNA (RAPD) analysis, 95% of the 44 isolates analysed differed by as much as 70% at the genomic level, which is the typical cut-off value for a species definition. These 44 isolates are believed to have been derived from an initial population that was introduced to the field site nearly 50 years ago. This much diversity seems astonishing considering the relatively short time span. Even more astonishing is the fact that Grover and coworkers (2006) showed that a single clonal isolate required only nine generations to reach an average genomic similarity of only 65%. This genomic study highlights the profound ability of a single bacterium to undergo genetic change in a very short span of time; and compared to the life span of a plant host it seems that the bacteria are always going to be ahead in the so-called 'arms race' between host resistance and susceptibility.

Moreover, this descriptive study leads to additional questions about the underlying mechanisms that drive this intrinsic genomic variability, and paves the way for 'real time' evolutionary studies to better understand the role that different selective pressures play on the R. solanacearum saprophytic fitness and pathogenic flexibility. How does the host contribute to this variability? What impact do mechanisms that generate genomic plasticity have on the 'core genome' (see below) and what mechanism specifically limits variation in the well-conserved genes?

The core genome of the *R. solanacearum* species complex

What set of genes make R. solanacearum such a successful plant pathogen? This important question has been the topic of many research projects during the last 5 years since the R. solanacearum genome of strain GMI1000 was annotated and published. Despite the large genetic and phenotypic (metabolic) variability observed between members of the R. solanacearum species complex (see above), the disease phenotype of each strain is always the same: rapid growth in plant xylem tissue that almost always results in lethal wilting of the host plant. Thus, it would seem logical to hypothesize that within each strain, there exists a set of well-conserved core genes that govern virulence.

To address this question, Guidot et al. (2007) performed comparative genomic hybridization (CGH) using a microarray containing the genome of the fully sequenced phylotype I strain, GMI1000. CGH experiments were carried out for a total of 17 R. solanacearum isolates, representing the phenotypic biodiversity of the R. solanacearum species complex. The results showed that 53% of all R. solanacearum genes are conserved, or present in the GMI1000 genome. Thus, about half of all R. solanacearum genes comprise the 'core-genome'. Within this core genome are 152 genes that are known or predicted to play an important role in R. solanacearum pathogenicity. Included in this set of genes are eight core effector genes that are predicted to play a part in the type three secretion system (see Table 9.2 for a description of these effectors). The remaining 50% of the genes make up the 'variable genome'. Interestingly, a large proportion of the core genome is located on the chromosome while the so-called variable genome is over-represented on the megaplasmid (described below). Such distribution further confirms that the chromosome of R. solanacearum preceded the megaplasmid, albeit a long time ago in evolutionary history. Further analysis of the variable genome revealed two classes of genomic islands. The first class is characterized by mobile genetic elements with an altered GC content or alternate codon usage. The second class of genes found in the variable genome is more similar to the core genome in terms of genetic composition despite their variable presence in different R. solanacearum isolates. Thus, it is possible that these genes are the result of various evolutionary paths from a common ancestor.

Not surprisingly, the CGH analysis showed that many of the major virulence genes are widely conserved among the test isolates, with the exception of two gene classes: the type III secretion system effectors, and the genes encoding haemagglutinin-related proteins. Yet, despite such variability, there were nine effector genes conserved in all R. solanacearum isolates tested.

Lastly, the CGH experiments were able to assign a level of relatedness (per cent identity between each isolate and GMI1000) between each of the strains tested. Moreover, these results matched the phylotype designation remarkably well, thus providing additional data to support the phylotype-based classification.

Such broad-reaching experiments have been possible because of the genomic revolution. Now that sequencing costs are decreasing, the future holds exciting possibilities for dissecting the genomic differences among R. solanacearum isolates in the same and different phylotypes.

Table 9.2 Effector genes identified in the R. solanacearum ancestral effector core putative effectors present in all Ralstonia solanacearum genomes included in the study (adapted from Guidot et al., 2007)

Effector (GMI1000 names are used)	Relevant features
RSc1475	Transmembrane domain
RSc3272	Hypothetical protein
RSp0099	ripA/AWR family
RSp0845	AWR family
RSp0846	AWR family
RSp0882	Hypothetical protein
RSp1218	Pseudogene
RSp1281	AvrE/DspA family

The evolutionary history of *R. solanacearum*

To assess the evolutionary history of *R. solanacearum*, Castillo and Greenburg (2007) utilized a variety of genomics tools including multilocus sequence typing (MLST) to analyse housekeeping and virulence-associated genes (see also chapter by Vinatzer and Bull). Not surprisingly, the authors concluded that the core genome (housekeeping genes) was under purifying selection while certain virulence genes were subject to diversifying selection. However, two well-studied virulence genes, *hrpB* and *fliC* are probably under purifying selection. The key finding of this study was that since both the chromosome and megaplasmid contain essential genes that are of similar genetic composition, they have co-evolved over a long period of time, despite the fact that the megaplasmid exhibits greater genetic diversity compared to the genome. While this study is congruous with what has been known about the variation between *R. solanacearum* isolates and corroborates the findings of Coenye and Vandamme (2003), it further substantiated the phylotype taxonomy (described above) and took into consideration the role that geographic isolation played during *R. solanacearum* evolution. But how will the information from this study guide future genomic explorations? Can it be used as a framework from which scientists a century from now can determine how field isolates have changed? Can laboratory experiments be designed to test the hypothesis that the *R. solanacearum* core and accessory genomes are truly under different selective pressures? Lastly this study exemplifies the fact that genomics has revolutionized comparative taxonomic studies. It is now quite easy to evaluate traits across large groups of related organisms using standard PCR and genome sequencing methods.

Mining the genome for genomic islands

Genomic islands (GI) are thought to arise via horizontal gene transfer (HGT) which is accepted as a common evolutionary mechanism by which bacterial genes are exchanged between different bacterial isolates (see also chapters by Arnold *et al.* and Monteiro-Vitorello *et al.*). HGT can result in the creation of new bacterial species or at the very least, significantly alter the genotype to the extent that the there is a measurable phenotype. Specifically, genomic islands consist of long DNA segments and they are notorious for harbouring genes that improve the virulence of the pathogen. At the genetic level, GIs are typically flanked by tRNA genes and may contain transposase and/or integrase genes. Another hallmark feature of a GI is an altered GC content of the DNA, which was noted in the annotation of the GMI1000 genome (Salanoubat *et al.*, 2002). Recently, Chen (2006) screened strain GMI1000 for the presence of genomic islands and found that there were six GIs on the chromosome and two GIs on the megaplasmid. It is not surprising that *R. solanacearum* possesses a number of GIs throughout its genome, given its history as a plant pathogen. Further genomic analyses (algorithms) should be developed to assess the evolutionary history of these elements and novel ways to test the function and stability of GIs in pathogenic bacteria will help to elucidate their role in shaping the pathogenic identity of *R. solanacearum*. See Table 9.3 for a summary of these findings.

Pathogens of the pathogen increase pathogenicity!

Like other bacteria, *R. solanacearum* is susceptible to bacteriophage infection. The underlying biological mechanisms underlying the *R. solanacearum*-phage interaction has recently been more closely examined thanks to the availability of large genomic databases. A consortium of Japanese researchers has been actively studying the phage of *R. solanacearum* since the early 1990s, and recently two very interesting papers were published on the subject of *R. solanacearum* phages (Kawasaki *et al.*, 2007; Yamada *et al.*, 2007). Yamada *et al.* (2007) reported the identification of four new phages from Japanese soil samples that show variable infection profiles. Two myovirus-type phage (φRSA1 and φRSL1) and two Ff-type (Inovirus) phage (φRSM1 and φRSS1) were characterized. The phage sensitivity of 15 different *R. solanacearum* strains exhibited a varied pattern with all strains sensitive to φRSA1, while φRSS1 could only infect

Table 9.3 Location of genomic islands (GI) in *Ralstonia solanacearum* GMI1000 (adapted from Chen 2006) The important features of each GI are given

Genomic island	Location (kb) (number of genes)	tRNA genes at 3′ junction?	Transposase? (locus)	Integrase?	Other features?
Chromosome					
CGI-1	202–238 (37)	No	Yes (Rsc0208)	No	Altered codon usage
CGI-2	1545–1589 (43)	No	Yes (Rsc1484)	No	Altered codon usage
CGI-3	1621–1672 (38)	Yes	Yes (RS05235, Rsc1549, Rsc1550)	Yes (Rs05241)	Altered codon usage
CGI-4	2019–2064 (42)	Yes	Yes (RS04283, RS05589, Rsc1864)	Yes (RS03431)	Altered codon usage
CGI-5	2218–2252 (37)	Yes	No	No	Altered codon usage
CGI-6	2348–2411 (61)	Yes	Yes (Rsc2176)	No	Altered codon usage
Megaplasmid			No	No	
PGI-1	45–108 (61)	No	Yes (Rsp0041, Rsp0042, Rsp0073, Rsp0093, Rsp0094)	Yes (Rsp0090)	Altered codon usage
PGI-2	1393–1414 (18)	No	No	Yes (Rsp1102)	Drug efflux genes (Rsp1112 and Rsp1114), altered codon usage

four *R. solanacearum* isolates. Overall the myoviruses could infect a larger number of strains tested including those belonging to different races. Furthermore, the more virulent φRSAI behaved as a lysogenic phage, and hybridization studies revealed that all 15 *R. solanacearum* strains showed multiple signals when probed with φRSAI (Yamada et al., 2007). The Ff-type viruses were restricted in their host range (infecting only race I *R. solanacearum* strains) and it was observed that bacterial strains that were sensitive to RSS1 were resistant to RSM1 and vice versa. Of the four phage recently described, the lysogenic RSS1 is the most interesting because its presence was shown to increase the virulence of *R. solanacearum* strain C319 in tobacco soil soak experiments. Moreover, the genomes of all 15 bacterial strains examined could hybridize to RSS1 sequences, and the banding patterns corresponded to the different taxa of *R. solanacearum* strains (Kawasaki et al., 2007; Yamada et al., 2007), suggesting a unique infection profile for RSS1 and other related phage. While these studies are unique to the *R. solanacearum* strains found in Japan, they pave the way for future phage studies focusing on *R. solanacearum* isolates from other geographic locations. Are these Japanese phages specific for Japanese bacterial isolates, or is the specificity dictated by taxonomy only? Can RSS1 increase the virulence of related but non-Japanese *R. solanacearum* strains?

The genetic 'dissection' of *R. solanacearum* phage is an exciting topic that can now be fully appreciated in light of the genomic revolution. We can now begin to better understand the taxonomic relationships between different phage and the evolutionary history of phage and phage-related sequences found in different genomes. Furthermore, a full understanding of *R. solanacearum* phage biology can assist in our understanding of *R. solanacearum* virulence at the genomic level.

R. solanacearum employs many common virulence themes

Historically, a virulence factor has been defined as a gene product (or products), that is associated with a measurable loss in pathogenicity or virulence when the corresponding gene in question is rendered non-functional. However, a problem arises when the inactivation of a single locus hypothesized to play a role in virulence leads to little or no change in the virulence of a bacterium. Such reductionist thinking loses sight of the forest, so to speak, since we now know (thanks to the age of genomics) that bacteria possess extremely sophisticated genetic networks that function holistically to enable the bacterium to succeed in very specialized niches.

Wassenaar and Gaastra (2001) have summarized several subcategories of virulence genes to address whether a gene can be considered a virulence or merely a 'housekeeping' gene. The authors conclude that it's nearly impossible to draw the line between these two categories, since many housekeeping genes (for example, those involved in LPS biosynthesis) encode well-defined virulence factors even though LPS is also a constituent of non-pathogens. Bacterial pathologists have applied a more subtle definition whereby the entire contents of a bacterium's genome can be compartmentalized into three categories: (1) true virulence genes or genes that are directly causal to pathogenesis, and are not present in non-pathogens; (2) virulence-associated genes which are involved in processing, supporting, and regulating the true virulence genes; and (3) virulence lifestyle genes or those genes that are essential for the pathogenic lifestyle (i.e. multiplication, survival, and competition inside or on the host) (Wassenaar and Gaastra, 2001). By altering the classification system for virulence factors we can more easily consider the contribution each gene makes with regard to the ability of a bacterium to cause disease. This system enables researchers to embrace a more holistic view by which the more subtle nuances of bacterial virulence can be understood.

R. solanacearum employs a large number of conserved pathogenesis mechanisms that are common among both animal and plant pathogens. These mechanisms include secretion of virulence factors (including type III effectors that are targeted to the host cell), enzymatic degradation of host-produced substrates, quorum sensing and signalling, and hierarchical regulation of virulence factors. In the paragraphs that follow, specific virulence mechanisms used by *R. solanacearum* are summarized. See Fig. 9.2 for a schematic of the *R. solanacearum* life cycle.

R. solanacearum possesses an arsenal of secreted and transmembrane proteins

The ability to secrete extracellular proteins is an essential virulence trait for *R. solanacearum* since mutations that disrupt either type II (general secretion pathway) or type III (the so-called *hrp*-dependent secretion system) secretion machinery render the bacteria avirulent (Schell, 2000; see also chapters by Arnold *et al.* and Stavrinides for further details on secretion systems). Emerging data suggest that no single secretion product is solely responsible for *R. solanacearum* pathogenicity, but that all the secreted proteins act synergistically (and redundantly) to promote bacterial entry and colonization of the host plant. To underscore the importance of secreted proteins, it should be pointed out that many genes identified in an *in vivo* expression technology (IVET) screen (Brown and Allen, 2004) encoded for putative transmembrane and secreted proteins. The identity and function of such genes are not known, but the fact that so many are induced in the host plant environment makes them interesting virulence candidates.

The type II secretion system in *R. solanacearum*

The major secreted virulence factor in *R. solanacearum* is an acidic, nitrogen-rich high molecular weight polysaccharide called extracellular polysaccharide (EPS) (Denny and Baek, 1991). EPS is produced in large amounts both *in vitro* and *in planta*. The production of EPS is necessary but not sufficient for the development of bacterial wilt since direct inoculation of EPS-deficient mutants failed to induce wilt symptoms but addition of EPS alone does not have an effect on the plant (Denny and Baek, 1991). Moreover, EPS is not essential for growth in planta since EPS mutants could still attain populations

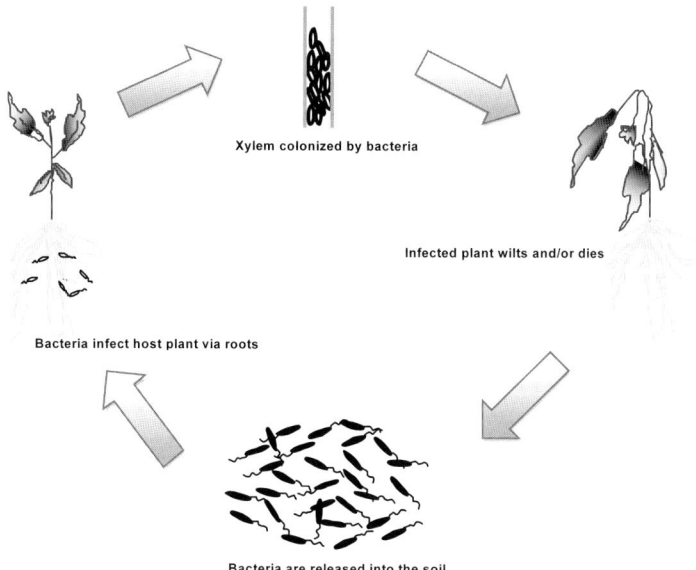

Figure 9.2 The generalized life cycle of *Ralstonia solanacearum*.

similar to wild-type inside the plant, despite the fact that systemic colonization of the plant by the bacterium is hindered in these mutants (Kao and Sequeira, 1992). While the original hypothesis was that EPS functioned to merely block water transport through the xylem (Denny and Baek, 1991), the fact that EPS assists bacterial colonization suggests an additional, yet unknown role in virulence. One hypothesis is that EPS may function to protect the bacteria from plant defences, such as changes in osmolarity or antimicrobial compounds.

In addition to EPS, *R. solanacearum* also produces and secretes three cell wall degrading enzymes termed polygalacturonases (PG) (Roberts et al., 1988; Schell et al., 1988; Allen et al., 1991). PehA (PglA) is an endo-PG that cleaves the pectin polymer internally at random; PehB (PglB) and PehC are exo-PGs that cleave one or two, respectively, terminal galacturonide residues. The polygalacturonic acid (PGA) can be taken up by the bacteria through the PGA transporter, ExuT, and used as a carbon source under in vitro conditions (Gonzalez and Allen, 2003). Interestingly, mutations in PehA and PehB decrease the ability of *R. solanacearum* to cause disease, but a triple PehA/PehB/PehC mutant is actually enhanced for virulence in tomato. This apparent paradox suggests that shortened galacturonate polymers may function as signalling molecules to induce plant defences (Gonzalez and Allen, 2003).

Other type II secreted proteins such as endoglucanases (EG), which degrade cellulose, were also hypothesized to contribute to *R. solanacearum* virulence. However, of the two EGs identified so far, Egl and CbhA, only the Egl-deficient mutant shows a reduction in the speed with which the bacteria cause wilt symptoms (Schell, 2000).

More recently, in an elegant analysis of the type II secretion system (also called the type II Secretin) in the fully sequenced GMI1000 strain (Salanoubat et al., 2002; Liu et al., 2005), it was shown that the six pectin degrading exoenzymes: Egl, PehA, PehB, PehC, CbhA, and Pme, were not the only type II-secreted proteins that contributed to virulence. When all six exoenzymes were deleted from the genome, the resulting mutant strain was still capable of causing wilt symptoms in approximately 80% of the leaves of infected plants, while a strain defective in the type II secretion system (via the inactivation of *sdpD* – secretion-dependent pathway D) resulted in less than a 20% incidence of wilted leaves (Liu et al., 2005). These results indicate that there are likely additional undiscovered virulence determinants secreted through the type II secretion pathway. It seems it will be only a matter of time before the full repertoire of virulence-related proteins secreted via the *R. solanacearum* type II secretin will be identified.

Lastly, the twin arginine translocation (TAT) secretion system was recently identified in several genera of bacteria. The key features of this secretion system are: (1) recognition of a twin arginine motif (S/L-R-R-X-F-L-K) at the N-terminus of the secreted protein and, (2) secretion of pre-folded proteins across the inner membrane (Berks *et al.*, 2003). A mutation in the *R. solanacearum tatC* gene (which encodes the inner-membrane transport protein) is severely attenuated in virulence, exhibits altered EPS production on agar plates, and forms unusual chains of cells (González *et al.*, 2007). The pleiotropic nature of this mutant suggests an important role for the TAT system in *R. solanacearum* virulence and underscores the hypothesis that the sum of all secreted proteins function collectively to promote bacterial wilt disease.

Multidrug efflux pumps are involved in *R. solanacearum* virulence

Two genes identified in an *in vivo* screen for loci that are specifically expressed *in planta* (see below) had high homology to the putative multidrug efflux pumps: *acrA* and *dinF* (Brown *et al.*, 2007). Mutations in these genes decreased resistance of the resulting strains to various plant-derived toxic compounds and also reduced virulence in the susceptible tomato variety, Bonny Best. Furthermore, the regulation of each efflux gene was controlled by a different pathway. For *acrA*, which encodes the periplasmic subunit of the tripartite AcrAB-TolC drug efflux pump, the alternate sigma factor RpoS was required for full expression at high cell densities, while the expression of *dinF* was negatively regulated by HrpB. Together, these results show that *acrAB* and *dinF* encode MDRs in *R. solanacearum* and that they contribute to the overall aggressiveness of this phytopathogen, probably by protecting the bacterium from the toxic effects of host antimicrobial compounds.

The type III secretion system in *R. solanacearum*

Type III secretion systems (TTSS) are ubiquitous among animal and plant pathogenic bacteria (Alfano and Collmer, 2004). TTSSs function to deliver proteins (called effectors) directly into the host cell cytoplasm. Upon release, the effectors modulate the host cell physiology in the pathogen's favour to allow for further bacterial ingress and/or infection. In plant-associated bacteria, the TTSS is encoded by the hypersensitive response and pathogenicity (*hrp*) genes (Buttner and Bonas, 2002). The *R. solanacearum* TTSS gene cluster spans a 23-kb region on the *R. solanacearum* megaplasmid (Van Gijsegem *et al.*, 1995) and comprises over 20 structural and regulatory genes. A mutation in several key *hrp* genes severely affects virulence as well as the induction of the hypersensitive response (HR), a type of programmed cell death elicited in resistant plants exposed to avirulent bacteria (Schell, 2000). In early studies, five proteins have been shown to be secreted through the *R. solanacearum* hrp system: (1) PopA, (2) PopB, (3) PopC, (4) PopP1, and (5) PopP2. PopA is a harpin-like protein (Arlat *et al.*, 1994) and PopC contains characteristic leucine-rich repeats that are associated with TTSS effectors in other bacterial species (Gueneron *et al.*, 2000). PopP1 is responsible for inducing the HR in resistant petunia (Lavie *et al.*, 2002), while PopP2 is recognized by the resistant Nd-R1 and susceptible Col-5 *Arabidopsis thaliana* plant lines via the resistance protein RRS1-R (Deslandes *et al.*, 2003). However, mutants lacking any one of these five proteins exhibit wild-type virulence. The effectors described above were identified in an era before the *R. solanacearum* genome was available. Since 2002, when the GMI1000 genome was published (Salanoubat *et al.*, 2002), much progress has been made in identifying and characterizing other TTSS-associated genes. In a large genome-wide screen, 48 novel TTSS-dependent genes were identified (Cunnac *et al.*, 2004a). Four of these proteins (named RipA, RipB, RipG, RipT for *R*alstonia effector *i*njected into *p*lant cells) are translocated into the host cell cytoplasm, but mutations in the genes encoding these effectors did not affect virulence. In fact, only two genes identified in this study were required for virulence: *brg*8 (an *avrPphD* homologue) and *brg*31. Mutations in these genes showed slightly delayed symptom development (1–2 days later) on tomato plants (Cunnac *et al.*,

2004a). Recently, another set of TTSS effectors have been characterized as a result of careful analysis of the *R. solanacearum* genomes. Using a yeast two-hybrid screen, Angot and coworkers identified a seven-gene cluster of TTSS effectors that harbour leucine-rich repeat (LRR) and F-box domain signatures typical of plant proteins marked for modification or degradation. These proteins, called GALA (for a conserved amino acid sequence in the LRR region) are capable of interacting with *Arabidopsis* (host) proteins that form part of the SCF-type ubiquitin ligase complex. Moreover, when the entire group (but not individually) of GALA effector genes was deleted or inactivated, the result was a lack or reduction of virulence on host plants. The fact that the entire group of GALA effectors had to be inactivated to observe a reduction in virulence underscores the importance of synergistic and overlapping functions for gene products that are required for *R. solanacearum* virulence. Moreover, TTSS effectors are tightly regulated and their function is necessary at a specific time and place. This phenomenon was shown by Kanda and coworkers (Kanda *et al.*, 2003), who constitutively expressed PopA from *R. solanacearum* strain OE-1 in tobacco. Expressing PopA at the inappropriate time (0.5 vs. 3 hours post inoculation) and location (in roots and apoplast) resulted in a loss of virulence. Constitutive expression of this protein did not affect virulence when bacteria were directly inoculated into the xylem, which suggests that PopA functions to facilitate the spread of bacteria from the roots to the stems of the infected plant and is not active once the bacteria are actively colonizing the xylem.

Recently, another type III effector, PopP3, was identified in *R. solanacearum* strain GMI1000 (Lavie *et al.*, 2004). PopP3, along with PopP1 and PopP2, belongs to the AvrRxv/YopJ family of type III effectors present in pathogenic bacteria (Lavie *et al.*, 2002). In GMI1000, however, the *popP3* ORF is inactivated by the insertion of the ISTso13 insertion sequence. A species-wide analysis of over 30 different phylogenetically diverse *R. solanacearum* strains showed that the occurrence and distribution of PopP1, PopP2 and PopP3 is present only in the Asiaticum division, with very few exceptions (for example, GMI1000, which is an Americanum strain).

HrpB regulates the expression of the *R. solanacearum hrp* genes

HrpB is a member of the AraC family of transcriptional activators (Genin *et al.*, 1992). This protein is required for the expression of over 20 structural genes involved in the assembly of the *hrp* TTSS machinery, and a HrpB mutation renders *R. solanacearum* completely avirulent (Genin *et al.*, 1992). Co-culturing *R. solanacearum* with plant cells results in a 20-fold increase in expression of the Hrp system compared to that observed in minimal medium (Marenda *et al.*, 1998). This plant cell-dependent expression relies on the Prh signal transduction system, governed by PrhA, which is thought to directly interact with the host plant cell (Marenda *et al.*, 1998; Brito *et al.*, 1999) (Fig. 9.3). A specific *cis* promoter regulatory motif, TTCGn16TTCG (consensus), called the Hrp$_{II}$box, was recently identified, and a genome-wide analysis of the fully sequenced *R. solanacearum* strain GMI1000 (Salanoubat *et al.*, 2002) revealed 95 putative HrpB-regulated target promoter regions (Cunnac *et al.*, 2004b). Also, microarray analyses have identified 130 genes that constitute the putative HrpB regulon (Ochialini *et al.* unpublished data). However, there is evidence to suggest that other unknown regulatory circuits are involved in the delivery of effector proteins to the host cell since the expression and secretion of the suspected effector proteins PopP1, Rsp0914 and Rsc2137 occurs in the absence of a functional TTSS, even though the genes encoding these proteins are preceded by a consensus Hrp$_{II}$box (Lavie *et al.*, 2002; Cunnac *et al.*, 2004b).

While HrpB is known to positively regulate the expression of TTSS, there are now several lines of evidence that suggest HrpB regulates other virulence associated genes (directly or indirectly) that are not part of the canonical TTSS. The first line of evidence comes from the effect of a *hrpB* deletion on the expression of a multidrug efflux pump, *dinF* (Brown and Allen, 2007). When HrpB is missing, the expression of the *dinF* locus is reduced, which suggests that HrpB promotes the expression of this efflux pump, presumably during plant colonization. However, HrpB can also negatively regulate genes as well

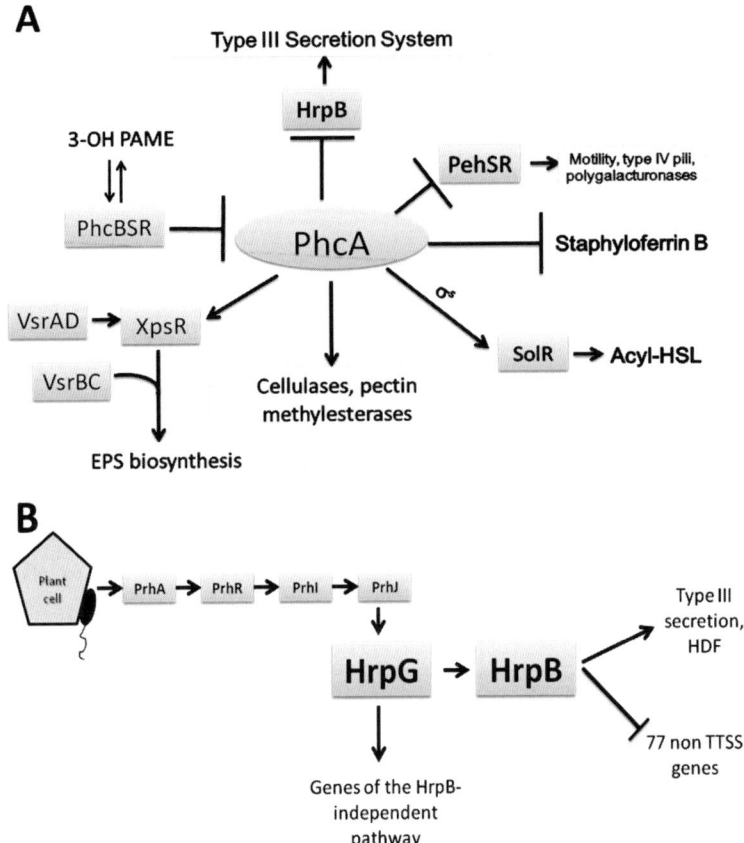

Figure 9.3 The PhcA and HrpB regulatory cascades of *Ralstonia solanacearum* virulence. A represents the virulence factors associated with changes in cell density; regulated by the master regulator, PhcA. B illustrates the plant-cell contact-dependent pathway that ultimately regulates the type III secretion system (adapted from Genin *et al.* 2005).

(Occhialini et al., 2005). Using a microarray, 77 additional loci were identified, many of which are localized on the megaplasmid; of these, *hrpB* was also found to promote the expression of a tryptophan derivative called HDF (for HrpB-dependent factor) that appears to be involved in affecting quorum-sensing pathways in environmental competitors (Delaspre et al., 2007). Thus, it is interesting to learn that the major regulatory protein of the TTSS also seems to function as a master regulator to control the expression of genes involved in making the switch from free living bacterium to plant pathogen.

Tamura et al. (2005) reported the construction and biochemical characterization of a *R. solanacearum* strain that constitutively expresses *hrpB* (*hrpBC*). This strain also constitutively expressed many known effectors. In the future, this and other *hrpBC* strains will be helpful to fully understand the role this transcriptional regulator plays in the early stages of host plant infection.

HrpG controls HrpB expression *and* a HrpB-independent virulence pathway

HrpG (*hrpG*) is a two-component response regulator that was first implicated in positively controlling expression of the TTSS pathway by enhancing *hrpB* transcription. HrpG expression is induced upon host cell contact by a quorum sensing signal, and in minimal medium (Brito et al., 1999; Aldon et al., 2000). Recently, HrpG has gained recognition as a major regulator of both *hrp*-dependent and *hrp*-in-dependent virulence factors (Valls et al., 2006). To define the HrpG regulatory network, Valls and coworkers made use of a GMI1000-based 'pangenomic' microarray to screen for differences in gene expression between three *R. solanacearum* strains: (1) wild-type, (2) ΔHrpG (deletion of the *hrpG* locus) and (3) *hrpG*-constitutive (HrpG overexpressed). The microarray results suggested that HrpG functions as a 'molecular switch' that senses a trio of specific environmental signals (listed above) and acts accordingly. When HrpG is activated, it functions to regulate multiple genetic pathways that are believed to be required for rapid adaptation and growth within the host plant. Many of the genes identified in this screen were already known to play a part in contributing to *R. solanacearum* virulence. Other differentially expressed genes belonged to other functional categories such as: enzymes, transported or membrane-associated, transcription regulators, and those with a hypothetical function. Of particular interest are 12 genes that were shown to be both HrpG-dependent and specifically expressed in the tomato xylem (Valls *et al.*, 2006; Brown and Allen, 2004) (Table 9.4). These genes are interesting because their role in colonization and virulence in the host plant have been identified through two independent and different genomic-based screens, which suggests that they play an interesting role in assisting *R. solanacearum* to successfully colonize the plant.

The phenotype conversion and loss of virulence

In the mid-1950s it was observed that under certain growth conditions, some members of a *R. solanacearum* population undergo a 'phenotype conversion' (PC) by switching from a mucoid to non-mucoid colony morphology (Kelman, 1954). The conditions that favour this phenotype switching include growth in still culture, prolonged growth on agar plates, and growth within tomato plants (Buddenhagen and Kelman, 1964; Brumbley *et al.*, 1993). PC appears to occur spontaneously and PC-mutants are incapable of wilting plants, but are still able to proliferate *in planta* and can induce stunting, stem necrosis, and the formation of adventitious roots along infected tomato stems (Brumbley *et al.*, 1993). PC mutants produce little, if any, EPS and produce less endoglucanase while producing greater quantities of endo-polygalacturonases than wild type. PC mutants are also hyper-motile (Brumbley *et al.*, 1990). Genetic analysis of PC-type mutants revealed that the *phcA* locus was usually responsible for this phenotype. However, there are a couple of exceptions to this finding. For example, the notorious B1 mutant (isolated in the 1950s by Arthur Kelman), which is not a PhcA mutant, exhibits the PC-phenotype. Furthermore, mutations in PhcB, which func-

Table 9.4 Genes dependent on HrpG (but not HrpB) and also identified in an IVET screen adapted from Valls *et al.*, 2006; Brown and Allen, 2004)

ORF	Putative function
Rsp0881	Transmembrane glycosyl hydrolase
Rsp0603	Serine protease, subtilase family
Rsp0536	Transmembrane; sugar-proton transport
Rsp1419	Putative peptide synthase
Rsp1610	Hypothetical protein
Rsp0773	Putative signal peptide protein
Rsc0118	Conserved hypothetical protein
Rsc1651	TAT protein
Rsc3067	Putative transmembrane protein
Rsc2910	Hypothetical signal peptide
Rsc3101	Serine protease, subtilase family

tion to produce 3′ OH-PAME, also give a PC-phenotype.

Brumbley and coworkers (Brumbley et al., 1993) reported that a spontaneous PC-mutant carried a two-bp insertion located near the 5′ end of *phcA*, causing a frameshift mutation that resulted in a truncated C-terminus of PhcA. Furthermore, larger insertions, and several deletions were also found within the *phcA* locus (Brumbley et al., 1993). The addition of *phcA* in trans restored the mutants to the wild-type phenotype. However, it was believed that once formed, PC mutants were incapable of reverting to wild-type. A recent study challenged this long held belief by showing that various *phcA* alleles including eight insertions/transpositions, three tandem duplications, seven deletions and a base substitution reverted to the wild-type (mucoid and virulent) phenotype during growth in planta (Poussier et al., 2003). Wild-type reversion was observed at varying rates with both a 64bp and IS*Rso8* insertions, which reverted at a rate of 10^{-2}. For an IS*Rso1* insertion, reversion occurred at a rate of 10^{-5}. Reversion never occurred under any in vitro conditions with the exception of the 64bp insertion, which reverted in the presence of tomato root exudates (Poussier et al., 2003). Analysis of reverted strains revealed that the insertion sequences were completely absent from the reconstituted *phcA* locus, suggesting that the plant environment can induce genomic rearrangements and transposition events. The implications for these mechanisms are not well understood, but emerging information about pathogenic bacteria suggest that transposition and changes in genomic architecture plays an important role during pathogenesis (Lavie et al., 2004; Canchaya et al., 2004).

PhcA senses population growth via the autoinducer 3-OH-PAME

PhcA is at the centre of a complex regulatory network that regulates virulence in *R. solanacearum* (Fig. 9.3). As described above, inactivation of PhcA through PC results in an inability to wilt plants. PhcA is a LysR-type transcriptional regulator, and like other LysR regulatory proteins, it responds to a chemical autoinducer. In *R. solanacearum*, this autoinducer is the unusual compound 3-hydroxy palmitic acid methyl ester (3′ OH-PAME). Further, the activity of PhcA itself is modulated by the products of the *phcBSR* operon (Schell, 2000). PhcB is believed to synthesize 3-OH-PAME, while PhcSR is a two-component system that functions to repress *phcA* expression. The precise mechanism for this process remains unknown (Schell, 2000).

Fluctuating concentrations of 3-OH-PAME trigger the expression of PhcA. Like a typical auto-inducer, 3-OH-PAME is constitutively produced and freely passes through the cell wall into the extracellular environment. As cell density increases and/or cells grow in a confined space (such as the plant xylem vessels), intracellular levels of 3-OH-PAME increase. When the concentration of 3-OH-PAME reaches an activation threshold of 5nM (at approximately 10^7 cfu/mL), PhcA expression is induced (Schell, 2000).

Recently, it was found that natural bacterial isolates were capable of degrading 3-OH-PAME (Shinohara et al., 2006). In this study, researchers identified an *Ideonella* sp. (beta-proteobacterium) from Japanese soil that was capable of growing on and degrading 3-OH-PAME. The *Ideonella* sp. produced a specific enzyme called, βHPMEH (a β-hydroxypalmitate methylester hydrolase), which could sufficiently hydrolyse 3-OH-PAME to PAME. When the hydrolase gene was introduced to *E. coli*, it was still capable of breaking down the quorum sensing molecule in addition to significantly reducing the production of EPS. These findings indicate that competitive bacteria in the soil environment likely modulate the growth of *R. solanacearum* and that this specific hydrolase could potentially be developed for use as a biocontrol agent to reduce *R. solanacearum* virulence in the field.

Multiple targets of PhcA contribute to virulence

PhcA controls the expression of an unknown number of genes, many of which are involved in causing wilt symptoms on susceptible host plants (Fig. 9.3). The activity of PhcA positively regulates the production of EPS, staphyloferrin B (an iron scavenging siderophore), endoglucanase (EG), pectin-methyl esterase (PME), and the endo-polygalacturonase PehA (PglA). When PhcA is active, the expression of swimming motility, twitching motility, and an exo-PG (PehB) is repressed (Kang et al., 2002). The repression of these phenotypes is due to negative regulation of a two-component regulatory system, PehSR; PhcA represses PehSR expression 12-fold (Allen et al., 1991). PehR is a two-component response regulator of the FixJ subfamily, which has a characteristic N-terminal phosphorylation domain followed by a short variable region, and a C-terminal region that contains a helix-turn-helix domain; another known R. solanacearum regulator, PrhJ (which is involved in regulating the type III secretion system) is also a member of this family (Merkel et al., 1992; Brito et al., 1999). Inactivating PehSR results in a substantial attenuation of virulence; thus, it is likely that PehSR regulates additional, as-yet unknown genes that collectively contribute to the ability of R. solanacearum to cause disease.

PhcA controls the expression of EPS through a complex and indirect mechanism (Fig. 9.3). In its active state, PhcA increases the expression of a regulator called XpsR, which together with the two-component regulatory system, VsrBC, function to increase eps transcription. Another two-component system, VsrAD functions with PhcA to increase XpsR expression; however the inducing signal for VsrAD is unknown (Schell, 2000).

PhcA negatively regulates the production of an iron scavenging siderophore called staphyloferrin B. When *phcA* is inactivated, R. solanacearum exhibits increased iron scavenging activity, but this does not affect virulence (Bhatt and Denny, 2004). This result suggests that siderophore production and iron-scavenging are necessary during the saprophytic stage of the R. solanacearum life cycle and that the plant xylem tissue contains sufficient iron to support rapid bacterial growth.

VsrAD regulates additional genes besides XpsR, since VsrAD mutants that produce wild-type levels of EPS *in trans* from the wild-type promoter are still avirulent (Schell, 2000). Moreover, VsrAD mutants fail to colonize tomato stems, which suggest that the additional gene products regulated by the VsrAD system are specifically required for stem colonization processes. One known target for VsrAD is *cbhA*, which encodes a B-1,4-exocellobiohydrolase, an enzyme involved in cellulose degradation. The other known VsrAD target is a homologue of an outer surface protein, NfrB, with no known function; the expression of NfrB is repressed by VsrAD (Schell, 2000). Moreover, it should be noted that the expression of several IVET-identified genes depends on the presence and/or absence of VsrD, suggesting that this gene plays an important regulatory role for genes that are specifically expressed inside the plant.

There are several reports that the global regulator PhcA negatively regulates the type three secretion system at the level of Hrp gene expression by altering the expression of HrpB. Earlier work from the laboratories of Yasufumi Hikichi and Kougei Onishi has shown conclusively that PhcA represses the Hrp system (K. Ohnishi, unpublished data). Furthermore, the expression of several IVET-identified promoters relied on the presence of both PhcA and HrpB, confirming the hypothesis that the function of these regulators is related (Brown and Allen, 2004). Indeed, Genin et al. (2005) showed direct evidence that PhcA repressed *hrpB* expression and this is dependent on the presence of HrpG. See Fig. 9.3 for the regulatory cascade of R. solanacearum virulence-associated genes.

A second RpoS (σ^S)-dependent quorum sensing system in R. solanacearum

R. solanacearum produces N-octonyl- and N-hexanoyl-homoserine lactones via the SolIR two-component system, which is itself positively regulated by the PhcA global regulatory system (Flavier et al., 1998). The PhcA system is active at 10^7 cfu/mL and the SolIR system is activated when cells reach a threshold population of 10^8 cfu/mL (Schell, 2000). The expression of SolIR is also dependent on a homologue of the sigma factor RpoS, which in E. coli regulates genes involved in stationary phase growth and environmental stress (Hengge-Aronis, 2002). An R. solanacearum *rpoS* mutant produced lower amounts of polygalacturonase, endoglucanase, and EPS and rendered the bacteria less able to survive at pH 4.0 and under starvation conditions. The *rpoS* mutant was also less virulent on tomato when inoculated directly into the tomato xylem vessels (Flavier et al., 1998).

R. solanacearum elicits host defences

Plants can respond to invading bacteria through a basal defence pathway elicited by pathogen-associated molecular patterns (PAMPs) such as the previously characterized flg22 peptide from flagellin, and lipopolysaccharide (LPS) (Nurnberger et al., 2004). Interestingly, Arabidopsis Col-0 seedlings (resistant) and tobacco plants (susceptible) mounted basal defence responses against *R. solanacearum* in the absence of flagellin (Pfund et al., 2004) suggesting that *R. solanacearum* possesses as-yet unknown PAMPS that can trigger resistance gene expression in infected hosts.

Host colonization and disease require motility, chemotaxis and aerotaxis

In order to colonize its host, *R. solanacearum* needs to successfully locate the correct host and it needs an effective means to travel to the host. The ability to identify (via chemical signals) the correct host and travel to it is governed by a process known as chemotaxis. Chemotaxis, which partly involves swimming motility, is required by *R. solanacearum* for virulence (Tans-Kersten et al., 2001; Yao and Allen, 2006; 2007). However, the bacteria are essentially non-motile once inside plant xylem vessels. A mutation in the *fliC* gene that encodes the flagellar subunit significantly reduced the ability of *R. solanacearum* to cause disease in soil-infestation inoculation assays, but the mutant showed wild-type virulence when directly applied to tomato stems (Tans-Kersten et al., 2001). A mutant lacking the master regulator of flagellar biosynthesis, FlhDC, showed the same virulence pattern as the *fliC* mutant, and was even further reduced in virulence on tomato when inoculated through the soil (Tans-Kersten et al., 2004). Thus, it was concluded that FlhDC regulates the motility genes and possibly other genes that assist the bacterium in locating and entering the roots of the host plant.

To assess the specific contribution of chemotaxis to *R. solanacearum* virulence, Yao and Allen (2006) compared the chemotaxis profiles of diverse *R. solanacearum* isolates and observed that these different strains exhibit various chemotactic responses to an array of possible carbon sources. Overall, *R. solanacearum* was attracted to organic acids and amino acids, notably glutamine, proline, and aspartate. All strains tested were attracted to citrate and malate while only two strains (UW551 and UW373) were attracted to sucrose and fructose. Such variability might reflect differences in the metabolic pathways present in the genomes of these isolates. This study also showed that rice root exudate was far less attractive than that of the host plant, tomato. Moreover, two strains were constructed that lacked either CheA or CheW (proteins that are involved in the chemotaxis signalling cascade). Both the CheA and CheW mutant strains were motile, but unable to sense chemotactic signals in the environment. These mutants were assayed for their ability to respond to various compounds present in tomato root exudates and the results showed that chemotaxis (directed motility) was necessary for full virulence. Yao and Allen (2006) also examined the behaviour of GFP-labelled wild-type and CheW mutants along the roots of tomato and observed that bacterial cells were irregularly distributed along the elongation zone and at sites of secondary root emergence, but no bacteria were observed at the root tip. Lastly, competition experiments between the wild type and the non-tactic CheW mutant showed that the wild-type strain out-competed the non-tactic strain in soil-soak experiments as well as direct-inoculation into the xylem.

In another study, Yao and Allen (2007) showed that aerotaxis (a method by which bacteria make use of the electron transport chain to assess internal energy levels) was also important for host location and colonization. Two mutant strains lacking functional *aer1* and/or *aer2* were studied for their ability to colonize susceptible tomatoes. Both the *aer2* and the *aer1/2* double mutant were compromised in their ability to cause disease symptoms early in the infection process. Furthermore, the double aerotactic mutant poorly colonized tomato seedling roots, suggesting that this mutant was not able to sense and respond appropriately to the nutrient-rich root.

Taken together, these results suggest that chemotaxis, aerotaxis and motility are each required for locating, attaching to, and entering the plant and that chemotaxis may also impart a selective advantage inside the host when it comes to obtaining nutrients.

The role of motility in *R. solanacearum*

Using a β-glucuronidase (GUS) reporter gene construct to measure the transcriptional expression of both *fliC* and *flhDC*, it was observed that the expression of these genes was significantly different in the plant compared to expression in culture (Tans-Kersten et al., 2004). In culture, *flhDC* expression depended on PehR, a regulator of early virulence factors; and, in turn, FlhDC was required for *fliC* (flagellin) expression. However, in tomato plants, *flhDC* was expressed in both wild-type and *pehR* mutant backgrounds, despite the fact that PehR is required for motility both in culture and *in planta*. Expression of *flhDC* and *pehSR* were significantly induced in planta relative to culture. Paradoxically, *fliC* was expressed *in planta* at cell densities greater than 10^7 when bacteria are non-motile, as well as in a non-motile *flhDC* mutant. Thus, the uncoupled expression of *flhDC* and *fliC* within the plant environment indicates that additional signals and regulatory circuits repress motility during plant pathogenesis.

In addition to swimming motility, *R. solanacearum* is capable of twitching motility, which is a method of bacterial taxis that is required for movement across semi-solid or solid surfaces (Kang et al., 2002). In *R. solanacearum*, type four pili (Tfp) are composed of a 17-kDa protein, encoded by *pilA*. A *pilA* mutant did not exhibit polar attachment to tobacco suspension culture cells or to tomato roots; further, this mutant was reduced in virulence on tomato plants and in autoaggregation and biofilm formation in broth culture. These results suggest that the ability to attach to and move across root surfaces is requisite for *R. solanacearum* to enter the host plant (Kang et al., 2002).

R. solanacearum thrives in the nutritionally challenging xylem environment by utilizing alternative metabolic pathways and/or coercing the plant to secrete nutrients into the xylem. While the virulence strategies employed by *R. solanacearum* are steadily being revealed at the molecular level through biochemical and genetic analyses, little is known about the myriad factors required for successful colonization of the xylem, a seemingly unsuitable environment for bacterial growth (Pegg, 1985). Most of what we know comes from studies that have focused on 'loss of function' phenotypes. These analyses revealed that many genes identified as virulence factors are major regulatory proteins or a limited set of secreted proteins that, when inactivated, show a large reduction in virulence, which is probably due to the inactivation of a large number of genes. This pleiotropic effect indicates that *R. solanacearum* virulence is dependent upon the sum of many gene products working together to enable this bacterium to locate, attach, invade, and colonize the plant host. Therefore, to more fully understand the nature of bacterial wilt, we need to reassess the definition and concept of a virulence factor, and develop new and improved methods to identify genes that play important roles during the different life stages of *R. solanacearum*.

Which R. solanacearum genes are host specific?

Complex interactions of myriad gene products contribute to the ability of *R. solanacearum* to cause disease, therefore, the hunt for additional virulence factors requires clever experimental design and no doubt a full understanding of the bacterial genome. In recent years, the principle of gene-fusion technology has provided an improved means to measure the expression of host-specific genes. One such technique, *in vivo* expression technology (IVET) has been successfully used to identify bacterial genes involved in growth and/or virulence in many host-associated bacteria and fungi (Rainey and Preston, 2000). In IVET, the host is used to select for genes whose expression is specifically up-regulated in that environment.

To determine which *R. solanacearum* genes were specifically induced in the tomato stem (xylem and surrounding tissue), Brown and Allen (2004) designed an IVET screen based on tryptophan auxotrophy. The strain used was K60, a tomato isolate from North Carolina, USA (Kelman, 1954). The IVET screen was limited to only those genes that were specifically expressed in the tomato stem but not in minimal medium. One-hundred and fifty three stem-specific loci were identified including several known virulence genes (such as *vsrD*, *vsrB*, and *pehR*). Analysis of the putative functions of the host-specific genes reflected a change in the metabolic state of the bacterium once it began growing in the xylem tissue. Further, the IVET screen identified a large number of genes predicted to play a role in regulation, which suggests that the genes required for growth in the xylem are under complex regulatory control. While the IVET screen was performed with strain K60, it was possible to make use of the GMI1000 genomic sequence to assess whether the IVET-identified genes were present in clusters within the genome. This assessment required the assumption that the genomes of the two strains are syntenous (see Figure 4 in Brown and Allen, 2004). Despite the genomic variations between the two strains, some genes appeared in clusters. These kinds of genomic comparisons can be helpful to generate new hypotheses regarding the physical composition and physical evolution of a given genome.

Genome-wide screens are a popular and useful tool for understanding the level to which certain genes are expressed. However, the ultimate power of these studies will be revealed when combined with the better-informed hypotheses that arise from careful analysis of the output data.

Genetic variation between *R. solanacearum* strains; new observations

Genetic and phenotypic diversity among *R. solanacearum* strains in the southern US has been observed over the last 50 years. Specifically, *R. solanacearum* causes severe wilt on tobacco crops grown in both North and South Carolina, but in Georgia and Florida, tomatoes are at a greater risk for succumbing to the disease. In a recent report, Robertson and colleagues (2004) utilized a genomics approach to show how gene diversity at a specific locus impacts the variation of virulence in different geographical regions. The gene studied was *avrA*, which encodes an effector protein that elicits a hypersensitive response (HR) in tobacco and in so doing, restricts the host range of bacteria that harbour this gene (Carney and Denny, 1990). Robertson et al. (2004) amplified the *avrA* locus from 139 *R. solanacearum* isolates. Two classes of *avrA* products were amplified using PCR: a 792 bp allele and a 960 bp allele. Those strains that gave rise to the shorter allele were able to elicit an HR on tobacco while the longer allele was amplified from strains unable to induce an HR. Sequence analysis of the two alleles showed that the longer alleles contained one of two transposable elements that was either 152 bp or 170 bp in length. Interestingly, only the 170bp transposable elements were obtained from *R. solanacearum* isolates found in the Southern US and the Caribbean. Importantly, it was found that the longer *avrA* allele was isolated in over 75% of the isolates from the Carolinas while only 13% of the Georgia isolates showed the longer allele; none of the Florida isolates assayed contained the longer allele. This study leads one to speculate about the possible selective pressures that keep *avrA* intact in the Florida and Georgia ecosystems but allow for insertion of the transposable elements in the Carolina strains. Was there a single event that led to the insertion of the transposable elements and a particular environmental condition specific to the Carolinas that selected for the maintenance of the transposable element? Is this transposable element fixed or can it be induced to exit from the *avrA* gene, and if so, what specific conditions allow the *avrA* gene and its avirulence function to be restored?

Genomic variation within race 1

The genetic variation between *R. solanacearum* strains is addressed below. Brown and Allen (unpublished data) identified genes encoding an integrase and a transposase that are only present in race 1, biovar 1 strains (all in phylotype II, and are sequevar 7 and biotype 7). These genes were not present in strain GMI1000, which is a race 1, biovar 3 strain and is a member of phylotype I (sequevar 12 and biotype 8). One of the strain K60 IVET-isolated genes, *ipx41*, was approximately 14-fold induced inside the tomato xylem, and the strain harbouring the *ipx41* IVET fusion was severely affected in virulence (Brown and Allen, 2004). This gene is predicted to encode a putative transmembrane protein, and the DNA sequence exhibits alternative codon usage. While attempting to clone the full-length *ipx41* open reading frame (ORF) from the K60 genome, it was observed that there were two copies of this gene in K60, but only one in strain GMI1000 (Fig. 9.4). Approximately 30 kb from a cosmid containing a portion of the K60 genome was sequenced. A homologue of the GMI1000 gene, *rsc1706* (which is predicted to encode a hypothetical protein), was present immediately downstream of both copies of *ipx41* in K60. Moreover, the sequence analysis revealed that there are two genes, *tnp*$_{KS}$ and *int*$_{KS}$ (KS = K60 specific), which are predicted to encode a transposase and integrase, respectively. Neither *tnp*$_{KS}$ nor *int*$_{KS}$ are present in strain GM1000. Interestingly, *ipx41A* and the *rsc1706* homologue are immediately upstream of the *tnp*$_{KS}$ gene, while *ipx41B* is immediately downstream of the *int*$_{KS}$ gene (Fig. 9.4).

Moreover, analysis of the K60 chromosomal region showed that the *ipx41* duplication did not occur in tandem but are separated by approximately 25 kb. Within this 25 kb spacer region, the following GMI1000 homologues were found: *rsc1619*, *rsc1620*, *rsc1621*, *rsc2167*, *rsc1700*, *rsc1701*, *rsc1702*, and *rsc1703*. With the exception of *rsc2167*, all loci present in K60 were syntenous with those in GMI1000.

Southern blot analysis using the *tnp*$_{KS}$ and *int*$_{KS}$ as probes against several *R. solanacearum* Race 1 strains isolated from a variety of hosts and geographic regions showed that the *tnp*$_{KS}$ probe hybridized to 12 distinct bands in both K60 and all of the race 1 biovar I (UW134 and UW153) strains, while the same probe appeared to weakly hybridize to a single band in all the other race 1 isolates except for the race 3 strain UW551 (Fig. 9.5A arrow). The *int*$_{KS}$ probe bound two fragments in all the race 1 biovar 1 isolates, and a third band was observed in the K60 lane (Fig. 9.5B arrows). The banding pattern of both the *tnp*$_{KS}$ and the *int*$_{KS}$ probes indicate that these genes are specific to race 1 biovar 1 *R. solanacearum* isolates. Interestingly, K60 is a tomato isolate from the Southern US, while UW134 and UW153 are potato isolates from Kenya and Australia, respectively. These data suggest that these three strains are genetically more related to each other than other race 1 isolates; indeed, two isolates (K60 and UW134) were shown to belong to phylotype II and are both sequevar 7 (Hayward, 1964). Isolate UW153 may also belong to phylotype II, although it differs from K60 and UW134 in the banding pattern observed with the *tnp*$_{KS}$ probe (Fig. 9.5A). Since K60 appears to have a third *int*$_{KS}$-specific band (double arrows), this suggests additional variation among these *R. solanacearum* isolates.

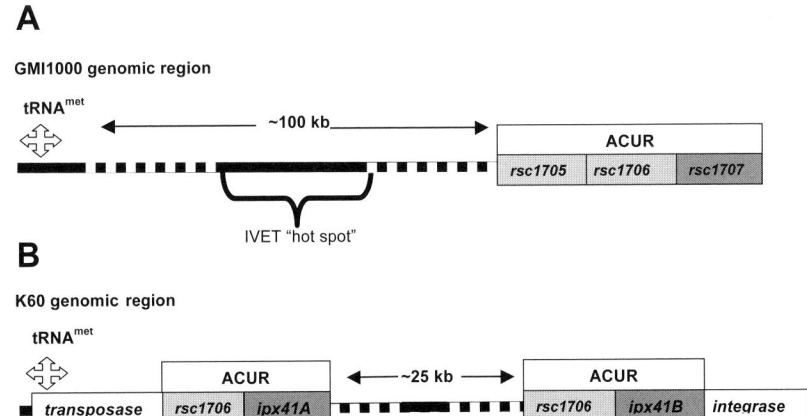

Figure 9.4 Comparison of the *rsc1707/ipx41* genomic regions in *R. solanacearum* strains GMI1000 and K60. In GMI1000 (A) there is only one copy of *rsc1707*, but K60 has two copies of the *rsc1707* homologue *ipx41* (B). The gene *rsc1705* is present in GMI1000, but not K60. The GMI1000 region labelled 'IVET hotspot' is not present in the same genomic location in strain K60, but 12 homologues from this region were identified in a K60 IVET screen. (B) The *ipx41A* locus (LEFT) is flanked by a transposase gene that is specific to K60 and other race 1 biovar 1 strains, the other *ipx41B* locus (right) is flanked by a K60-specific integrase that is also present in the same two other race 1 biovar 1 strains. The tRNA^{met} is the site where the transposase flanking *ipx41* is thought to have inserted. The dashed lines indicate the region of genomic synteny between GMI1000 and K60. Note that figure is not drawn to scale.

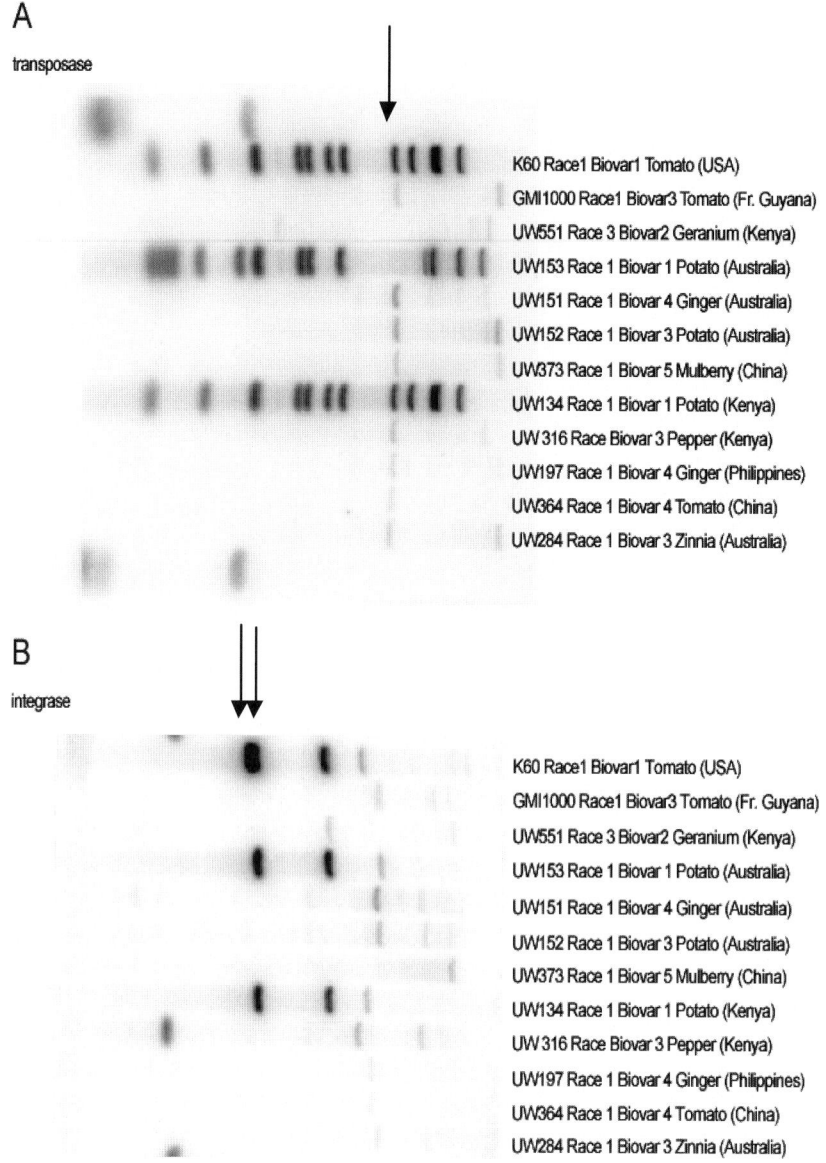

Figure 9.5 Genomic variation within isolates or *Ralstonia solanacearum*. Southern blot of *Eco*RV/*Bgl*II-digested chromosomal DNA from 12 strains of *R. solanacearum* isolates. (A) Probe was a 650 bp fragment of the transposase gene, *tnpKS*. Arrow shows a faint band present in the chromosome of all race 1 strains tested. Note the banding pattern of UW153 is different from K60 and UW134. (B) Probe was an ~800 bp fragment of the integrase gene, *intKS*. Arrows denote a doublet in the K60 lane.

The observation that the tnp_{KS} and the int_{KS} genes were present in only race 1 biovar 1 *R. solanacearum* isolates suggests a practical application for these genes in identifying race 1 biovar 1 strains in the field. Also, the biological role of integrases and transposases in phytopathogenic bacteria is poorly understood. It is possible that the rearrangement observed in K60 may be a transmissible genomic island (GI) since it meets several criteria for classification as a GI (Chen, 2006). Since these genes evolved and have been selected over time, and since there are multiple copies of each, the proteins encoded by these genes play an important role in the biology of a certain group of *R. solanacearum* isolates. Further, since the int_{KS} and tnp_{KS} loci were identified in strains isolated from different hosts on separate continents (US, Australia, and Africa) the presence of these genes might be the remains of an ancestral genetic event. Clearly, these results underscore the importance of comparative genomics when studying the biology and evolution of a phytopathogen as diverse as the R. solanacearum species complex.

Where do we go from here?

Achieving a full understanding of the underlying biology of Ralstonia solanacearum and the pathogen–plant interaction has only just started. In the last decade, R. solanacearum researchers have been fortunate to have taken part in this age of genomics enlightenment, which has facilitated a greater understanding of the underlying genetic mechanisms that drive R. solanacearum pathogenesis, ecological fitness, metabolism and evolution. But we have only started to scratch the surface. We are now primed with more affordable genomics technology and the database infrastructure that will allow us to start testing hypotheses that have heretofore been too technically challenging. What are needed in the immediate future are the genome sequences of a number of diverse R. solanacearum isolates, which will provide even more information about the core genome and its relationship to the so-called 'plastic genome'. With this information, we can start to learn more about the multitude of pathways involved in regulating all the genes required for the adaptive changes that occur when R. solanacearum takes on a role as a plant pathogen versus a soil inhabitant.

Will we ever learn enough to allow us to effectively and sustainably manage R. solanacearum in the field? Perhaps, but it will take all our current technology and understanding in addition to the future insights that will be borne from other emerging technologies such as proteomics and the corresponding databases that can help researchers make sense of the resulting datasets. We are finally on the verge of appreciating the subtle variations between different R. solanacearum isolates which promises to provide us with a greater understanding of the role R. solanacearum plays as a soil inhabitant and how it has evolved into one of the world's most serious plant pathogens.

References

Aldon, D., Brito, B., Boucher, C. and Genin, S. (2000). A bacterial sensor of plant cell contact controls the transcriptional induction of Ralstonia solanacearum pathogenicity genes. EMBO J. 19, 2304–2314.

Alfano, J. and Collmer, A. (2004). Type III secretion system effector proteins: double agents in bacterial disease and plant defense. Annu. Rev. Phytopath. 42, 385–414.

Allen, C., Huang, Y. and Sequeira, L. (1007). Cloning of genes affecting polygalacturonase production in Pseudomonas solanacearum. Mol. Plant-Microbe. Interact. 4, 3389–3407.

Altschul, S. F., Madden, T. L., Scaffer, A. A., Zhang, J., Zhang, Z., Miller, W. and Lipman, D.J. (1997). Gapped BLAST and PSI-BLAST: a new generation of protein database search programs. Nucleic Acid Res. 25, 3389–3407.

Alvarez, B., Lopez, M. M. and Biosca, E. G. (2007). Influence of native microbiota on the survival of Ralstonia solanacearum phylotype II in river water microcosms. Appl. Environ. Microbiol. 73, 7210–7217.

Arlat, M., Van Gijsegem, F., Huet, J. C., Pernollet, J. C. and Boucher, C. A. (1994). PopA1, a protein which induces a hypersensitivity-like response on specific petunia genotypes, is secreted via the Hrp pathway of Pseudomonas solanacearum. EMBO J. 13, 543–553.

Berks, B. C., Palmer, T. and Sargent, F. (2003). The TAT protein translocation pathway and its role in microbial physiology. Adv. Microbial Phys. 47, 187–254.

Bhatt, G. and Denny, T. P. (2004). Ralstonia solanacearum iron scavenging by the siderophore Staphyloferrin B is controlled by PhcA, the global virulence regulator. J. Bacteriol. 186, 7896–7904.

Brito, B., Marenda, M., Barberis, P., Boucher, C. and Genin, S. (1999). prhJ and hrpG: two new components of the plant signal-dependent regulatory cascade controlled by PrhA in Ralstonia solanacearum. Mol. Microbiol. 31, 237–251.

Brown, D. G. and Allen, C. (2004). Ralstonia solanacearum genes induced during growth in tomato: an inside view of bacterial wilt. Mol. Microbiol. 53, 1641–1660.

Brown, D. G., Swanson, J. K. and Allen, C. (2007). Two host-induced Ralstonia solanacearum genes acrA and dinF, encode multidrug efflux pumps and contribute to bacterial wilt virulence. Appl. Environ. Microbiol. 73, 2777–2786.

Brown, D. (2005). Life in the xylem: the secrets of Ralstonia solanacearum pathogenesis revealed by in vivo expression technology. PhD Thesis. University of Wisconsin-Madison.

Brumbley, S. M. and Denny, T. P. (1990). Cloning of wild-type Pseudomonas solanacearum phcA, a gene that when mutated alters expression of multiple traits that contribute to virulence. J. Bacteriol. 172, 5677–5685.

Brumbley, S. M., Carney, B. F. and Denny, T. P. (1993). Phenotype conversion in Pseudomonas solanacearum due to spontaneous inactivation of PhcA, a putative LysR transcriptional regulator. J. Bacteriol. 175, 5477–5487.

Buddenhagen, I. W. and Kelman, A. (1964). Biological and physiological aspects of bacterial wilt caused by *Pseudomonas solanacearum*. Annu. Rev. Phytopathol. 2, 203–230.

Buddenhagen, I. W., Sequeira, L. and Kelman, A. (1962). Designation of races of *Pseudomonas solanacearum*. Phytopathology 52, 726–732.

Buttner, D. and Bonas, U. (2002). Getting across – bacterial type III effector porteins on their way to the plant cell. EMBO J. 21, 5313–5322.

Canchaya, C., Fournouse, G. and Brussow, H. (2004). The impact of prophages on bacterial chromosomes. Mol. Microbiol. 53, 9–18.

Carney, B. F. and Denny, T. P. (1990). A cloned avirulence gene from *Pseudomonas solanacearum* determines incompatibility on *Nicotiana tabacum* at the host species level. J. Bacteriol. 172, 4836–4843.

Castillo, J. A. and Greenburg, J. T. (2007). Evolutionary dynamics of *Ralstonia solanacearum*. Appl. Environ. Microbiol. 73, 1225–1238.

Chen, L.-L. (2006). Identification of genomic islands in six plant pathogens. Gene 374, 134–141.

Coenye, T. and Vandamme, P. (2003). Simple sequence repeats and compositional bias in the bipartite *Ralstonia solanacearum* GMI1000 genome. BMC Genomics 4, 10–13.

Cunnac, S. (2004). Inventory and functional analysis of the large Hrp regulon in *Ralstonia solanacearum*: identification of novel effector proteins translocated to plant host cells through the type III secretion system. Mol. Microbiol. 53, 115–128.

Cunnac, S., Boucher, C. and Genin, S. (2004). Characterization of the cis-acting regulatory element controlling HrpB-mediated activation of the type III secretion syste and effector genes in *Ralstonia solanacearum*. J. Bacteriol. 186, 2309–2318.

Delaspre, F., Nieto-Penalver, C. G., Saurel, O., Kiefer, P., Gras, E., Milon, A., Boucher, C., Genin, S. and Vorholt, J.A. (2007). The *Ralstonia solanacearum* pathogenicity regulator HrpB induces 3-hydroxyoxindole synthesis. Proc. Natl. Acad. Sci. USA 104, 15870–15875.

Denny, T. P. and Baek, S. R. (1991). Genetic evidence that extracellular polysaccharide is a virulence factor of *Pseudomonas solanacearum*. Mol. Plant–Microbe Interact. 4, 198–206.

Deslandes, L., Olivier, J., Peeters, N., Feng, D. X., Khounlotham, M., Boucher, C., Somssich, I., Genin, S. and Marco, Y. (2003). Physical interaction between RRS1-R, a protein conferring resistance to bacterial wilt, and PopP2, a type III effector targeted to the plant nucleus. Proc. Natl. Acad. Sci. USA 100, 8024–8029.

Deslandes, L., Olivier, J., Theulieres, F., Hirsch, J., Feng, D. X., Bittner-Eddy, P., Beynon, J. and Marco, Y. (2002). Resistance to *Ralstonia solanacearum* in *Arabidopsis thaliana* is conferred by the recessive RRS1-R gene, a member of a novel family of resistance genes. Proc. Natl. Acad. Sci. USA 99, 2404–2409.

Deslandes, L., Pileur, L. and Liaubet, S. (1998). Genetics characterization of RRS1, a recessive locus in *Arabidopsis thaliana* that confers resistance to the bacterial soil borne pathogen *Ralstonia solanacearum*. Mol. Plant-Microbe. Interact. 11, 659–667.

Fegan, M. and Prior, P. (2005). How complex is the *Ralstonia solanacearum* species complex? In Bacterial wilt: the disease and the Ralstonia solanacearum species complex, C. Allen, P. Prior and J. Elphinstone, eds. (St. Paul, USA: APS Press), pp. 449–461.

Feng, J. (2005). A novel antimicrobial protein isolated from porato (*Solanum tuberosum*) shares homology with an acid phosphatase. Biochem. J. 376, 481–487.

Flavier, A. B., Schell, M. A. and Denny, T. P. (1998). An RpoS homologue regulates acylhomoserine lactondependent autoinduction in *Ralstonia solanacearum*. Mol. Microbiol. 28, 475–486.

Gabriel, D., Allen, C. and Schell, M. (2006). Identification of reading frames unique to a select agent: *Ralstonia solanacearum* Race 3 biovar 2. Mol. Plant–Microbe Interact. 19, 69–79.

Genin, S. (1992). Evidence that the hrpB gene encodes a positive regulator of pathogenicity genes from *Pseudomonas solanacearum*. Mol. Plant–Microbe Interact. 6, 107–134.

Genin, S. and Boucher, C. (2004). Lessons learned from the genome analysis of *Ralstonia solanacearum*. Annu. Rev. Phytopathol. 42, 107–134.

Genin, S., Brito, B., Denny, T. and Boucher, C. (2005). Control of the *Ralstonia solanacearum* type III secretion system (Hrp) genes by the global virulence regulator PhcA. FEBS Lett. 579, 2077–2081.

Gonzalez, E. and Allen, C. (2003). Characterization of a *Ralstonia solanacearum* operon required for polygalacturonate degradation and uptake of galacturonic acid. Mol. Plant–Microbe Interact. 16, 536–544.

Gonzalez, E., Brown, D., Swanson, J. and Allen, C. (2007). Using the *Ralstonia solanacearum* TAT secretome to identify bacterial wilt virulence factors. Appl. Envron. Microbiol. 73, 3779–3786.

Grimault, V., Prior, P. and Anais, G. (1995). A monogenic dominant resistance gne of tomato to bacterial wilt in Hawaii7996 is associated with plant colonization by *Pseudomonas solanacearum*. J. Phytopathol. 143, 349–352.

Grover, A., Azmi, W., Gadewar, A., Pattanayak, D., Naik, P., Shekhawat, G. and Chakrabarti, S.K. (2006). Genotypic diversity in a localized population of *Ralstonia solanacearum* as revealed by random amplified polymorphic DNA markers. J. Appl. Microbiol. 101, 798–806.

Gueneron, M. (2000). Two novel proteins: PopB, which has functional nuclear localization signals, and PopC, which has a large leucine-rich repeat domain, are secreted through the hrp-secretion apparatus of *Ralstonia solanacearum*. Mol. Microbiol. 36, 261–277.

Guidot, A., Prior, P., Schoenfeld, S., Carrere, S., Genin, S. and Boucher, C. (2007). Genomic structure and phylogeny of the plant pathogen *Ralstonia solanacearum* inferred from gene distrubution analysis. J. Bacteriol. 189, 377–387.

Hauck, P., Thilmony, R. and He, S. (2003). A Pseudomonas syringae type III effector suppresses cell wall-based extracellular defense in susceptible Arabidopsis plants. Proc. Natl. Acad. Sci. USA 100, 8577–8782.

Hayward, A. (1991). Biology and epidemiology of bacterial wilt caused by *Pseudomonas solanacearum*. Annu. Rev. Phytopathol. *29*, 65–87.

Hayward, A. (1964). Characteristics of *Pseudomonas solanacearum*. J. Appl. Bacteriol. *27*, 265–277.

Hengge-Aronis, R. (2002). Signal transduction and regulatory mechanisms involved in control of the sigma (RpoS) subunit of RNA polymerase. Microbiology *66*, 373–395.

Jackson, M. and Gonzalez, L. (1981). Persistence of *Pseudomonas solanacearum* (race 1) in a naturally infested soil in Costa Rica. Phytopathology *71*, 690–693.

Janse, J., Beld, D. and van Den, H. (2004). Introduction to Europe of *Ralstonia solanacearum* biovar 2, race 3 in Pelargonium zonale cuttings. J. Plant Pathol. *86*, 147–155.

Jenkins, S., Morton, D. and Dukes, P. (1966). Comparison of techniques for detection of *Pseudomonas solanacearum* in artificially infested soils. Phytopathology *57*, 25–27.

Kanda, A. (2003). Ectopic expression of *Ralstonia solanacearum* effector protein PopA early in invasion reslts in loss of virulence. Mol. Plant-Microbe. Interact. *16*, 447–455.

Kang, Y. (2002). *Ralstonia solanacearum* requries type 4 pili to adhere to multiple surfaces and for natural transformation and virulence. Mol. Microbiol. *2*, 427–437.

Kao, C. and Sequeira, L. (1992). Extracellular polysaccharide is required for wild-type virulence of *Pseudomonas solanacearum*. J. Bacteriol. *174*, 1068–1071.

Kawasaki, T., Nagata, S., Fujiwara, A., Satsuma, H., Fujie, M., Usami, S. And Yamada, T. (2007). Genomic characterization of the filamentous integrative bacteriophages pRSS1 and pRSM1, which infect *Ralstonia solanacearum*. J. Bacteriol. *189*, 5792–5802.

Kelman, A. (1954). The relationship of pathogenicity of *Pseudomonas solanacearum* to colony appearance in a tetrazolium medium. Phytopathology *44*, 693–695.

Lamessa, F. and Zeller, W. (2007). Isolation and characterization of *Ralstonia solanacearum* strains from Solanaceae crops in Ethiopia. J. Basic Microbiol. *47*, 40–49.

Lavie, M. (2004). Distribution and sequence analysis of a family of type III-dependent effectors correlate with the phylogeny of *Ralstonia solanacearum* strains. Mol. Plant–Microbe Interact. *17*, 931–940.

Lavie, M. (2002). PopP1, a new member of the YopJ/AvrRxv family of type III effector proteins, acts as a host specificity factor and modulates aggressiveness of *Ralstonia solanacearum*. Mol. Plant–Microbe Interact. *15*, 1058–1068.

Liu, H., Zhang, S., Schell, M. and Denny, T. (2005). Pyramiding unmarked deletions in *Ralstonia solanacearum* shows that secreted proteins in addition to plant cell-wall degrading enzymes contribute to virulence. Mol. Plant–Microbe Interact. *18*, 1286–1305.

Mangin, B. (1999). Temporal and multiple quantitative trait loci analyses of resistance to bacterial wilt in tomato permit the resolution of linked loci. Genetics *151*, 1165–1172.

Manoil, C. and Beckwith, J. (1985). TnphoA: a transposon probe for protein export signals. Proc. Natl. Acad. Sci. USA *82*, 8129–8133.

Marenda, M. (1998). PrhA controls a novel regulatory pathway required for the specific induction of *Ralstonia solanacearum* hrp genes in the presence of plant cells. Mol. Microbiol. *27*, 437–453.

McGarvey, J., Denny, T. and Schell, M. (1999). Spatial, temporal, and quantitative analysis of growth and EPS I production by R. solanacearum in resistant and susceptible tomato cultivars. Phytopathology *89*, 1233–1239.

Merkel, T. (1992). Promoter elements required for positive control of the *Escherichia coli uhpT* gene. J. Bacteriol. *174*, 2761–2770.

Nakaho, K. (2004). Distribution and multiplication of *Ralstonia solanacearum* in tomato plants with resistance derived from different origins. J. Gen. Plant Pathol. *70*, 115–119.

Nakaho, K., Hibino, H. and Miyagawa, H. (2000). Possible mechanisms limiting movement of *Ralstonia solanacearum* in tomato resistant tissues. Phytopathology *148*, 181–190.

Nurnberger, T. (2004). Innate immunity in plants and animals: striking similarities and obvious differences. Immunol. Rev. *198*, 249–266.

Occhialini, A., Cunnac, S., Reymond, N., Genin, S. and Boucher, C. (2005). Genome-wide analysis of gene expression in *Ralstonia solanacearum* reveals that the *hrpB* gene acts as a regulatoryswitch controlling multiple virulence pathways. Mol. Plant–Microbe Interact. *18*, 938–949.

Pegg, G. (1985). Presidential address: life in a black hole – the micro-environment of the vascular pathogen. Trans. Brit. Mycolog. Soc. *85*, 1–20.

Pfund, C., Tans-Kersten, J. and Allen, C. (2004). Flagellin is not a major defense elicitor in *Ralstonia solanacearum* cells or extracts applied to *Arabidopsis thaliana*. Mol. Plant–Microbe Interact. *17*, 696–706.

Poussier, S. (2003). Host plant-dependent phenotypic reversion of *Ralstonia solanacearum* from non-pathogenic to pathogenic forms via alterations in the *phcA* gene. Mol. Microbiol. *49*, 991–1003.

Pradhanang, P., Elphinstone, J. and Rox, R. (2000). Identification of crop and weed hosts of *Ralstonia solanacearum* biovar 2 in the hills of Nepal. Plant Pathol. *49*, 403–413.

Prior, P. and Fegan, M. (2005). Recent developments in the phylogeny and classification of *Ralstonia solanacearum*. ACTA Hort. *695*, 127–136.

Rainey, P.B and Preston, G.M. (2000). *In vivo* expression technology stategies: valuable tools for biotechnology. Curr. Opin. Biotechnol. *11*, 440–444.

Roberts, D., Denny, T. and Schell, M. (1988). Cloning of the *egl* gene of *Pseudomonas solanacearum* and analysis of its role in phytopathogenicity. J. Bacteriol. 1445 1451.

Robertson, A., Wechter, W., Denny, T., Fortnum, B. and Kluepful, D. (2004). Relationship between avirulence gene (*avrA*) diversity in *Ralstonia solanacearum* and bacterial wilt incidence. Mol. Plant–Microbe Interact. *17*, 1376–1384.

Salanoubat, M. (2002). Genome sequence of the pant pathogen *Ralstonia solanacearum*. Nature *415*, 497–502.

Schell, M. (2000). Control of virulence and pathogenicity genes of *Ralstonia solanacearum* by an elaborate sensory network. Ann. Rev. Phytopathol. *38*, 263–292.

Schell, M., Roberts, D. and Denny, T. (1988). Analysis of the *Pseudomonas solanacearum* polygalacturonase encoded by *pglA*, and its involvement in phytopathogenicity. J. Bacteriol. *170*, 4501–4508.

Shinohara, M., Nakajima, N. and Uehara, Y. (2006). Purification and characterization of a novel esterase (beta-hydroxypalmitate methyl ester hydrolase) and prevention of the expression of virulence by *Ralstonia solanacearum*. J. Appl. Microbiol. *103*, 152–162.

Smith, E. (1896). A bacterial disease of tomato, eggplant, and Irish potato (*Bacillus solanacearum* sp. nov.). US Dep. Agric. Div. Veg. Physiol. Path. Bull. *12*, 1–28.

Swanson, J. (2005). Behavior of *Ralstonia solanacearum* race 3 biovar 2 during latent and active infection of geranium. Phytopathology *95*, 136–143.

Swanson, J., Montes, L., L., M. and Allen, C. (2007). Detection of latent infections of *Ralstonia solanacearum* race 3 biovar 2 in geranium. Phytopathology *91*, 828–834.

Tamura, N., Murata, Y. and Mukaihara, T. (2005). Isolation of *Ralstonia solanacearum hrpB* constitutive mutants and secretion analysis of *hrpB*-regulated gene products that share homology with known type III effectors and enzymes. Microbiology *151*, 2873–2884.

Tans-Kersten, J., Brown, D. and Allen, C. (2004). Swimming motility, a virulence trait of *Ralstonia solanacearum*, is regulated by FlhDC and the plant host environment. Mol. Plant–Microbe Interact. *17*, 686–695.

Tans-Kersten, J., Huang, H. and Allen, C. (2001). *Ralstonia solanacearum* needs motilty for invasive virulence on tomato. J. Bacteriol. *183*, 3597–3605.

Tusiime, G., Adipala, E., Opio, F. and Bhagsari, A. (1998). Weeds as latent hosts of *Ralstonia solanacearum* in highland Uganda: implications to development of an integrated control package for bacterial wilt. In Bacterial Wilt Disease: Molecular and Ecological Aspects, P. Prior, C. Allen and J. Elphinstone, eds. (Berlin, Germany: Springer), pp. 413–419.

Vailleau, F., Sartorel, E., Jardinaud, M.F., Chardon, F., Genin, S., Huguet, T., Gentzbittel, L. and Petitprez, M. (2007). Characterization of the interaction between the bacterial wilt pathogen *Ralstonia solanacearum* and the model legume plant *Medicago truncatula*. Mol. Plant–Microbe Interact. *20*, 159–167.

Valls, M., Genin, S. and Boucher, C. (2006). Integrated regulation of the type III secretion system and other virulence determinants in *Ralstonia solanacearum*. PloS Pathogens *2*, 82–89.

Van Gijsegem, F. (1995). The hrp gene locus of *Pseudomonas solanacearum*, which controls the production of a type III secretion system, encodes eight proteins related to components of the bacterial flagellar biogenesis complex. Mol. Microbiol. *15*, 1095–1114.

Van Sluys, M. (2002). Comparative genomic analysis of plant-associated bacteria. Annu. Rev. Phytopathol. *40*, 169–189.

Vasse, J., Danoun, S. and Trigalet, A. (2005). Microscopic studies of root infection in resistant tomato cv. Hawaii7996. In Bacterial Wilt: The disease and the *Ralstonia solanacearum* species complex, C. Allen, P. Prior and A. Hayward, eds. (St. Paul, USA: APS Press), pp. 275–291.

Vasse, J., Frey, P. and Trigalet, A. (1995). Microscopic studies of intercellular infection and protoxylem invasion of tomato roots by *Pseudomonas solanacearum*. Mol. Plant–Microbe Interact. *8*, 241–251.

Wassenar, T. and Gaastra, W. (2001). Bacterial virulence: can we draw the line? FEMS Microbiol. Lett. *201*, 1–7.

Wenneker, M., Verdel, M., Groeneveld, R., Kempenaar, C., van Beuningen, A. and Janse, J. (1999). *Ralstonia* (*Pseudomonas*) *solanacearum* race 3 (biovar 2) in surface water and natural weed hosts: first report on stinging nettle (*Urtica dioica*). Eur. J. Plant Pathol. *105*, 307–315.

Wicker, E., Grassart, L., Caranson-Beaudu, R., Mian, D., Guilbaud, C., Fegan, M. and Prior, P. (2007). *Ralstonia solanacearum* strains from Martinique (French West Indies) exhibiting a new pathogenic potential. Appl. Environ. Microbiol. *73*, 6790–6801.

Yabuuchi, E., Kosako, Y., Yano, I., Hotta, H. and Nishiuchi, Y. (1995). Transfer of two Burkholderia and an Alcaligenes species to Ralstonia gen. nov.: proposal of Ralstonia pickettii (Ralston, Palleroni, and Doudoroff 1973) comb. nov., *Ralstonia solanacearum* (Smith 1896) comb nov. and *Ralstonia eutropha* (Davis 1969) comb. nov. Microbiol. Immunol. *39*, 897–904.

Yamada, T., Kawasaki, T., Nagata, S., Fujiwara, A., Usami, S. and Fujie, M. (2007). New bacteriophages that infect the phytopathogen *Ralstonia solanacearum*. Microbiology *153*, 2630–2639.

Yang, C. and Ho, G. (1998). Resistance and susceptibility of *Arabidopsis thaliana* to bacterial wilt caused by *Ralstonia solanacearum*. Phytopathology *88*, 330–334.

Yao, J. and Allen, C. (2006). Chemotaxis is required for virulence and competitve fitness of the bacterial wilt pathogen *Ralstonia solanacearum*. J. Bacteriol. *188*, 3687–3708.

Yao, J. and Allen, C. (2007). The plant pathogen *Ralstonia solanacearum* needs aerotaxis for normal biofilm formation and itneractions with its tomato host. J. Bacteriol. *189*, 6415–6424.

Zolobowska, L., and van Gijsegem, F. (2006). Induction of lateral root structure formation on petunia roots: a novel effect of GMI1000 *Ralstonia solanacearum* infection in Hrp mutants. Mol. Plant–Microbe Interact. *19*, 597–606.

Pseudomonas syringae Genomics Provides Important Insights to Secretion Systems, Effector Genes and the Evolution of Virulence

Dawn L. Arnold, Scott. A. C. Godfrey and Robert. W. Jackson

Abstract

The start of the twenty-first century was a watershed for *Pseudomonas syringae* genomics, with the completion of three genome sequences for different *P. syringae* pathovars. The release of these sequences permitted a series of investigations designed at unravelling the biology of this group of plant pathogens. One area that has benefited has been the identification of different secretion systems and their substrate effectors; some of these secretion systems can deliver proteins into the plant environment to subvert the plants' resistance machinery and to gain access to plant nutrients. The most investigated secretion system in *P. syringae* is the type III system which delivers effector proteins into the plant cytoplasm to disrupt the plants' cellular pathways, including signalling mechanisms that would otherwise trigger defence mechanisms. The genome sequences have been analysed to predict the number of effector genes present in the different strains, gain insights into the function of these proteins and examine how the bacteria can evolve and change its effector repertoire in order to overcome plant resistance. There is still much scope for further analysis, particularly of poorly understood secretion systems. For example, very little is known about the role of non-type III secretion systems in *P. syringae*. Indeed, the genome sequences have allowed us to identify putative orthologues of the *vas-vgr* type VI system in *P. syringae* and intriguingly there appears to be some variation in gene content and synteny between the pathovars.

Introduction

Pseudomonas syringae is a Gram negative plant pathogenic bacterium that is responsible for a range of disease on plants with symptoms such as blights, spots and galls (Fig. 10.1). Strains of *P. syringae* are assigned to one of over 50 pathovars depending on their host range. Typically, *P. syringae* infects and causes disease in only a limited range of plant hosts. In non-host plants, *P. syringae* strains usually trigger plant resistance mechanisms, often including a type of programmed cell death known as the hypersensitive response (HR). The HR typically consists of rapid plant cell death at foci of infections that limits the availability of nutrients to the potential pathogen.

Over the last few years the genome of three *P. syringae* strains, which belong to three different pathogenic varieties (pathovars), have been sequenced, annotated and published. The first of these was *P. syringae* pv. *tomato* (*Pto*) strain DC3000 which is pathogenic on tomato and *Arabidopsis thaliana* (Buell et al., 2003). The publication of this sequence was quickly followed by *P. syringae* pv. *syringae* (*Psy*) strain B728a, the cause of brown spot on bean (Feil et al., 2005) and *P. syringae* pv. *phaseolicola* (*Pph*) strain 1448A, the cause of halo blight of bean (Joardar et al., 2005) (Fig. 10.1). The complete genome sizes range from 6.1–6.5 megabases with *Pto* DC3000 and *Pph* 1448A containing a chromosome and two plasmids and *Psy* B728a just a chromosome. Further details of the genome projects are presented in Table 10.1. The wealth

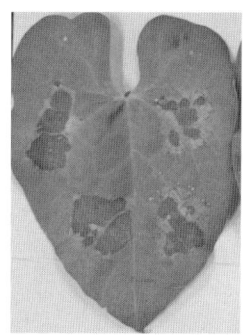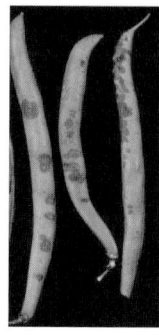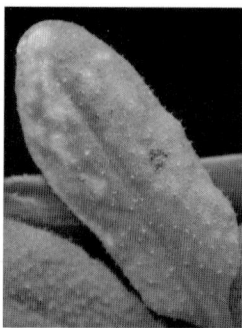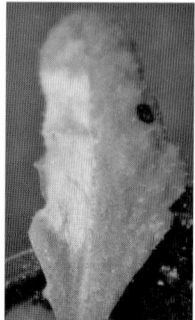

Figure 10.1 Disease symptoms produced by *Pseudomonas syringae* in host plants. From left to right: artificial inoculations of *P. syringae* pv. *phaseolicola* into a *Phaseolus vulgaris* leaf shows the watersoaked disease lesions surrounded by a chlorotic halo, characteristic of halo blight (photo courtesy of Helen Lovell); naturally occurring disease lesions in *P. vulgaris* pods; artificial inoculation of *P. syringae* pv. *tomato* into leaves of *Arabidopsis thaliana* shows early (left) and late (right) disease symptoms defined by coronatine-induced chlorosis followed by tissue collapse (photos courtesy of John Mansfield).

of data arising from these projects has allowed a number of comparative studies to be undertaken that has produced new insights into this group of plant pathogenic bacteria. This chapter will focus on the different secretion systems identified and/or predicted to be used by *P. syringae* with emphasis on the type III secretion system and the effector proteins it delivers.

Secretion systems of *P. syringae* pathovars

As with all pathogenic bacteria, an important part of the plant pathogens arsenal is their ability to secrete proteins. These bacterial proteins have various functions including allowing them to adhere to and degrade plant cell walls, to suppress plant defence responses, and to deliver DNA and proteins into the cytoplasm of plant cells (Preston *et al.*, 2005). There are several different bacterial secretion systems (Gerlach and Hensel, 2007) and the *P. syringae* sequencing projects have enabled the identification of a number of these (Fig. 10.2; see also chapter by Stavrinides).

Type I secretion system

All three sequenced genomes encode homologues of ATP-binding cassette (ABC) transporters that secrete proteins and are known as the type I secretion system. The type I secretion system consists of three components; an ABC ATP-binding protein, a membrane fusion protein that forms a bridge between the outer and inner membrane, and an outer membrane pore channel protein (Schmitt and Tampé, 2002). The type I secretion system allows single step secretion of a wide range of substrates from the cytoplasm to the

Table 10.1 Summary of the general features of the *Pseudomonas syringae* genome sequencing projects

Strain	Sequenced by	Chromosome		Plasmids		Reference
		bp	Number of ORFs	bp	Number of ORFs	
Pto DC3000	TIGR	6 397 12	5615	A. 73 661 B. 67 473	71 77	Buell *et al.* (2003)
Psy B278a	DOE-JGI	6 093 698	5217	NA	NA	Feil *et al.* (2005)
Pph 1448A	TIGR	5 928 785	5144	A. 131 950 B. 51 711	149 60	Joardar *et al.* (2005)

Pto, *P.s.* pv. *tomato*; *Psy*, *P.s.* pv. *syringae*; *Pph*, *P.s.* pv. *phaseolicola*; TIGR, The Institute for Genomics Research; DOE-JGI, The Department of Energy Joint Genome Institute. NA, not applicable. [For more detail go to http://www.pseudomonas-syringae.org/home.html.]

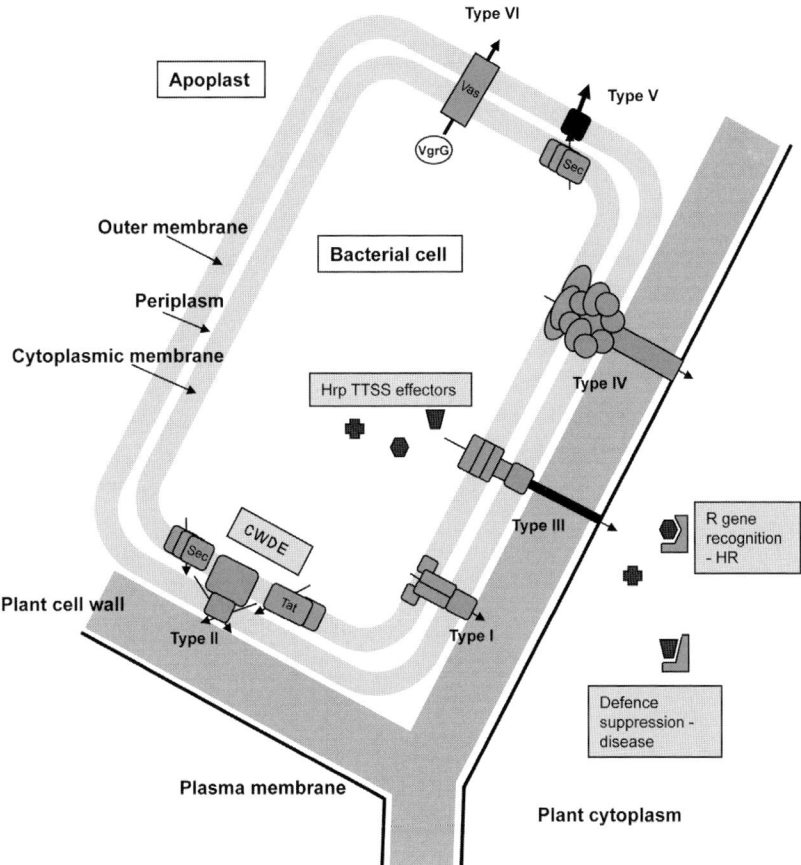

Figure 10.2 A schematic diagram showing the various bacterial secretion systems expressed by *Pseudomonas syringae*. The type I, II and III secretion systems are known to be used for secreting factors that promote virulence and enhanced fitness. Hrp TTSS effectors, hypersensitive response and pathogenicity proteins delivered by the type III secretion system; CWDE, cell wall degrading enzymes.

extracellular space without a stable periplasmic intermediate (Gerlach and Hensal, 2007). The proteins known to be exported by ABC systems in plant pathogenic bacteria are predominantly proteases, lipases and haemolysins (Preston et al., 2005). *Pto* DC3000 has 15 ABC transporter systems with predicted specificities for arabinose, xylose, ribose, and other plant-derived sugars (Buell et al., 2003). In *Psy* B728A it is suggested that one function of the type I secretion system is the secretion of an antifreeze protein that moderates the ice nucleation activity of *P. syringae* strains (Feil et al., 2005). Many strains of *Pph* produce and secrete a toxin called phaseolotoxin, which is secreted via an oligopeptide permease (Opp) type I transporter system (Staskawicz and Panopoulos, 1980). A second toxin, syringomycin, is produced by several pathovars of *P. syringae*. Quigley et al. (1993) examined the *syrD* gene and found that a *syrD* mutant did not produce four large proteins of the syringomycin complex and had reduced virulence on plants. SyrD is highly similar to ABC transporters and is therefore predicted to be involved in secretion of syringomycin.

More recently, Chen and Beattie (2007) identified four putative ABC transporter systems predicted to function in osmoprotection in *P. syringae* pv. *tomato* strain *Pto* DC3000. One system, encoded by PSPTO_4575-4578, was found to be important for osmoprotection by transporting betaine osmoprotectant. The ATPase encoded by PSPTO_4575 is highly similar to the OpuC transporter found in various bacteria. Further characterization showed that OpuC is a high-affinity betaine transporter and low-affinity choline transporter. Inactivation of OpuC did not completely eliminate choline uptake suggesting the presence of a further uptake system. *In silico* analysis identified PSPTO_5629 as a putative choline transporter (Chen and Beattie, 2008). This gene is similar to the betaine/carnitine/choline (BCCT) transporter BetT from *Escherichia coli*. Unusually, BetT was found to have a low affinity for choline, suggesting that *P. syringae* has adapted to a choline-rich environment.

Type II secretion system

The type II secretion system is used to secrete enzymes and toxins by a wide variety of Gram-negative bacteria. The system is a two stage process where proteins to be secreted are first translocated (or exported) across the cytoplasmic membrane by a translocase (either the Sec or Tat exporter – see below), and then transported across the outer membrane via the type II secretion system (Russel, 1998). The system consists of a multiple component secretion apparatus which spans both the inner and outer membranes. Between 12 and 15 genes (*gsp*A to O and S) are essential for a functional type II secretion system.

Fourteen *gsp* genes have been predicted as components of a single type II secretion system in *Pto* DC3000 (Buell *et al.*, 2003). Buell *et al.* (2003) were unable to identify genes similar to *gspC* and *gspO*, but they did observe a homologue of the prepilin peptidase PilD present in the genome that can replace the function of GspO. Also *gspC* is not always present in all systems suggesting it may be functionally redundant.

The *Pph* 1448A genome contains genes for two distinct type II secretion systems for which candidate substrates include two cellulases, two pectate lyases, a pectin lyase, and a polugalacturonase (Joardar *et al.*, 2005). In a recent review, Jha *et al.* (2005) surveyed the type II secretion system of a number of Gram-negative plant pathogens and found examples in all the major groups of bacterial plant pathogens. This is an interesting review that suggests that a major theme emerging in type II secretion systems is that they are controlled by various quorum-sensing systems, such that protein secretion is maximal at high cell densities. Jha *et al.* (2005) suggest that by adopting such a cell-density-dependent strategy, bacteria may be ensuring that the host defence responses, which can be induced by type II secreted proteins, are elicited only when the bacterial population is sufficient to cope with the host responses.

In order for the type II secretion system to transport proteins across the outer membrane, they are first delivered into the periplasm by one of two additional systems, Sec or Tat. Whereas the Sec system transports unfolded proteins into the periplasm using a multi-subunit translocon (Gerlach and Hensal, 2007), the Twin-Arginine Translocation (Tat) system translocates pre-folded proteins across the cytoplasmic membrane using the transmembrane proton gradient as the driving force (Palmer and Berks, 2003). Bronstein *et al.* (2005) created and characterized a *Pto* DC3000 strain containing a mutation in *tatC* and showed this caused inactivation of the Tat system which resulted in multiple complex phenotypes including loss of motility on soft agar plates, deficiency in siderophore synthesis and thus iron acquisition, sensitivity to copper, loss of extracellular phospholipase activity, and attenuated virulence in host plant leaves. Interestingly, for this final point, they demonstrated that inactivation of the Tat system reduced the efficiency of type III secretion translocation by >1.5-fold. They suggest that one reason for this might be due to a compromised outer membrane or a defect in the assembly of the type III secretion machinery. In the wild-type they also confirmed that several of the Tat substrates were translocated across the outer membrane by the type II system indicating that the Tat and type II systems operate in synergy to form a two-step process for the secretion of *P. syringae* virulence factors such as phospholipases. An analogous study of *tat* deletion mutants of *Pto* DC3000 and *P. s.* pv. *maculicola* strain ES4326 showed the mutants displayed pleiotropic phenotypic changes as described above, but also including decreased resistance to sodium dodecyl sulphate (Caldelari *et al.*, 2006). They also identified a number of putative Tat substrates for secretion and showed that some were necessary for full virulence. A similar system has previously been identified in *P. aeruginosa* where it was demonstrated that the Tat system is essential for secreting two extracellular virulence determinants, both phospholipases, via the type II machinery (Voulhoux et al., 2001).

Type III secretion system
This system is described in detail below.

Type IV secretion system
Type IV secretion systems are believed to have evolved from bacterial conjugation machineries and are characterized by an ability to extracellularly translocate proteins (or complexes) and single-stranded DNA (Gerlach and Hensel, 2007). The best-studied example of type IV secretion is the VirB system of *Agrobacterium tumefaciens*, which transports a nucleoprotein complex from bacterium into a plant cell (Christie, 2004; see chapter by Setubal et al.). Type IV secretion systems are divided into two subclasses based on sequence comparisons; type IVA (conjugation system) where machines are assembled from VirB homologues and type IVB (transfer system) where machines are assembled from IncI Tra homologues (Christie and Vogel, 2000).

A functional type IV secretion system has not been identified in *Pto* DC3000 (Buell et al., 2003) or in *Psy* B728a. However, *Pph* 1448A does have a large number of genes with high similarity to the type IV secretion genes of *A. tumefaciens* (Joardar et al., 2005). Observations that these *Pph* 1448A genes are plasmid borne, contain a *virB5* homologue (systems transferring DNA) and have a small protein between *virB5* and *virB6* (found only in conjugal systems), all led Joardar et al. (2005) to suggest the *Pph* 1448A type IV homologues are involved in conjugal transfer of DNA rather than virulence-related protein translocation.

Further studies have assessed type IV secretion systems in other *P. syringae* pathovars. For example, the complete nucleotide sequence of the five-plasmid complement of *P. syringae* pv. *maculicola* ES4326 (a pathogen of radish and *Arabidopsis*) was obtained by Stavrinides and Guttman (2004). A complete type IV secretion system was found on the 46 697 bp plasmid, pPMA4326A, which suggested it was a self-transmissible plasmid. Another study used a macroarray containing 161 genes to estimate and compare the gene contents of 31 plasmids from 12 *P. syringae* pathovars (Zhao et al., 2005). The results revealed that the plasmids could be distinguished by the type IV secretion system they encoded and separated into four groups. Fifteen plasmids encoded type IVA (VirB-VirD4 conjugative system), whereas 12 encoded a type IVB (*tra* system), 2 plasmids encoded both type IV systems and 6 plasmids carried none or only a few genes of either the type IVA or IVB. These results showed that in addition to having type IV secretion systems potentially involved in conjugation, some *P. syringae* bacteria have type IV secretion systems that are potentially capable of delivering effector proteins (or toxins) into host cells during infection. Such delivery of virulence factors is likely as it has been demonstrated previously in other bacterial species (Christie and Vogel, 2000; Burns, 2003; Cambronne and Roy, 2006).

Type V secretion system
The type V secretion system is referred to as the autotransporter system and is dependent on the Sec system (Gerlach and Hensal, 2007). One type V mechanism is where proteins are secreted through the inner membrane via the Sec system. Here the C-terminal of the protein forms a translocation unit, a beta-barrel structure, in the outer membrane that enables secretion of the passenger domain. The passenger domain can then be cleaved from the translocation unit and released extracellularly. Preston et al. (2005) conducted a bioinformatics analysis of *Pto* DC3000 and eight plant pathogenic bacterial genomes and found that *Pto* DC3000 contained the highest number of autotransporter-like proteins, with nine candidate proteins. Some of these had recognizable domains such as serine peptidase, pertactin, acid phosphatase and lipase. However, they conclude that there is relatively little experimental evidence for the biological roles of type V-secreted proteins in plant pathogenic bacteria.

Type VI secretion system
A prototypic type VI secretion system has been described for the extracellular translocation of proteins lacking N-terminal hydrophobic leader sequences (Pukatzki et al., 2006). Virulence-associated secretion (*vas*) genes are shown to be responsible for *Vibrio cholerae* cytotoxicity

towards *Dictyostelium* amoebae and mammalian J774 macrophages. As this study describes a number of Gram-negative bacterial pathogens carrying orthologues to *vas* genes, it was suggested that such type VI secretion systems may be important in microbial pathogenesis (Pukatzki et al., 2006). As such, we searched for *V. cholerae vas* orthologues in the genomes of *Pto* DC3000, *Pph* 1448A and *Psy* B728A. In each of the three sequenced strains, there are clusters of genes with similarity to the *V. cholerae vas* gene cluster (Fig. 10.3). This suggests type VI systems may contribute to *P. syringae* pathogenesis and such putative systems warrant biological investigation to assess functionality.

Type III (TTSS) section system of *P. syringae*

The type III secretion system (TTSS) has been found in all of the genome-sequenced *P. syringae* strains and has the role of secreting proteins from the bacterium into the plant cytoplasm. The TTSS is known to play a key role in the pathogenicity of *P. syringae* where strains carrying mutations in the TTSS are unable to cause disease. Intriguingly, TTSS mutants do not trigger the HR in non-host plants. The TTSS is a complex, supramolecular structure which spans the inner membrane, the periplasmic space, the outer membrane, the extracellular space (or plant cell wall) and the host cellular membrane (Gerlach and Hensal, 2007). This structure is evolutionarily related to flagella systems (He, 1997; Sawada et al., 1999; Nguyen et al., 2000).

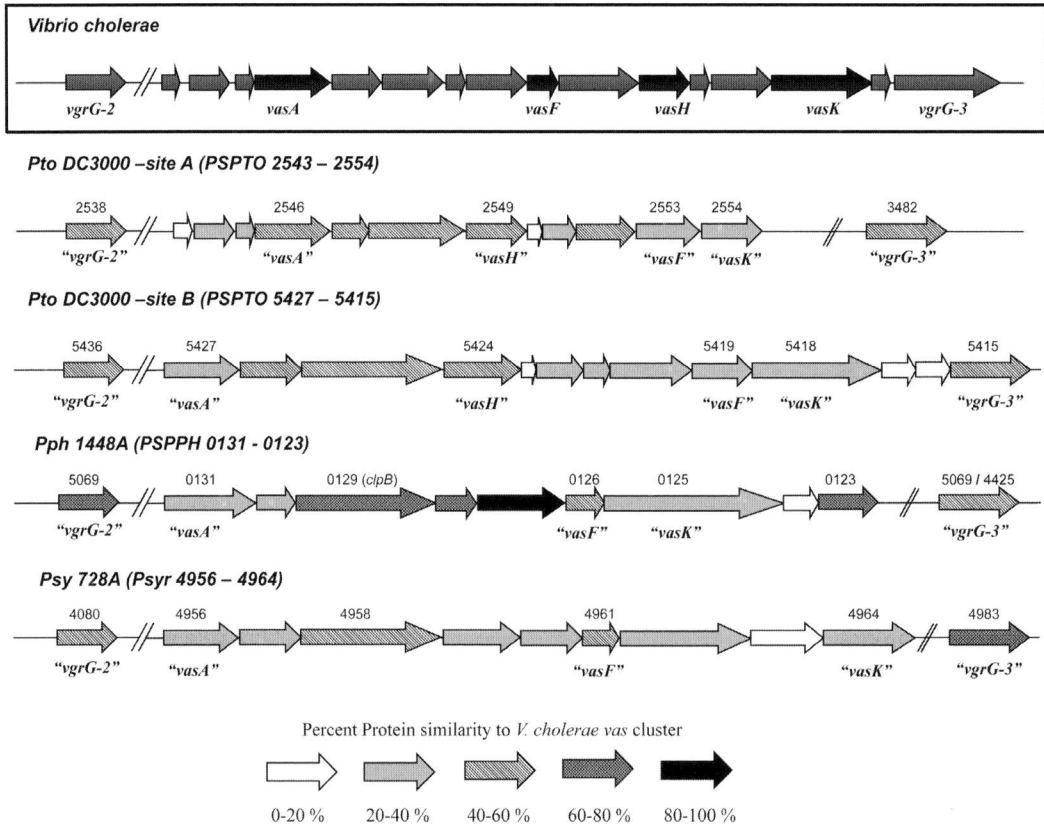

Figure 10.3 Type VI secretion system *vas* gene orthologues in *Pto* DC3000, *Pph* 1448A and *Psy* B728A. Comparative TBLASTX analysis of the *vas* gene cluster from *Vibrio cholerae* (Pukatzki et al., 2006) with DC3000, 1448A and B728A. Continuous regions of similarity are presented showing average amino-acid percent-identity within respective genes. Gene designations are those provided at http://www.pseudomonas-syringae.org/home.html. DC3000 encodes two regions of continuous *vas* similarity that are presented as site A and site B.

The TTSS is encoded by *hrp/hrc* (reflecting the their key role in <u>h</u>ypersensitive <u>r</u>esponse and <u>p</u>athogenicity/<u>c</u>onserved) genes (Alfano and Collmer, 1997) that form the pore and pilus complex for the secretion of proteins out of the bacterial cell into the plant cell (Casper-Lindley *et al.*, 2002; Szurek *et al.*, 2002)(Fig. 10.4). In all examined strains of *P. syringae*, the *hrp* locus is chromosomal borne, consisting of 27 open reading frames (ORFs) that are organized into six operons (Jin *et al.*, 2003). Eight *hrc* genes show a high degree of similarity to flagellum assembly genes and are found in most bacteria that carry a TTSS. The structure of the TTSS is similar, but not identical, to the flagellum system (Alfano and Collmer, 1997; Brown *et al.*, 2001), but only a few of the structural genes have been studied in detail. Two of the first to be studied were membrane proteins *hrpH* and *hrcC* (Huang *et al.*, 1992; Deng *et al.*, 1998). *hrpH* encodes an envelope protein similar to PulD outer membrane proteins (Huang *et al.*, 1992). HrpH is found only in the membrane and not in the bacterial cytoplasm or periplasm. HrpH has now been discovered to be the primary lytic transglycosylase (LT) for the TTSS (Fereirra *et al.*, 2006; Oh *et al.*, 2007) involved in breakdown of bacterial peptidoglycan in the periplasm to allow the secretion pore to be constructed. Two other LTs, HopP1 and HopAJ1, were also discovered and have overlapping functions (Oh *et al.*, 2007). The ATPase powerhouse of the pore, HrcN, has now been studied in greater detail (Pozidis *et al.*, 2003; Müller *et al.*, 2006). HrcN is a peripheral protein that assembles in clusters at the cell inner membrane. Dodecamerization of HrcN as two hexameric rings stimulates its ATP hydrolytic activity to catalyse protein secretion. The HrpA pilin protein is secreted via the pore and is the major subunit that makes the Hrp pilus (Roine *et al.*, 1997a, b). The pilus acts as a conduit for translocation of proteins from bacterial cell to plant cell (Brown *et al.*, 2001).

A recent study of the *Psy* 61 TTSS demonstrated the structural proteins HrpB, HrpD, HrpF, HrpJ and HrpP could all be translocated into plant cells, were important for triggering a full strength HR in *Nicotiana tabacum* and were essential for the secretion of AvrPto1 into *N. benthamiana* cells (Ramos *et al.*, 2007).

Translocation of effector proteins from the bacterial cell to the plant cell is thought to be aided by translocator proteins HrpK1 and harpins (Petnicki-Ocwieja *et al.*, 2005; Kvitko *et al.*, 2007) acting at the plant cell membrane. *hrpK1* is located in the exchangeable effector locus (EEL) adjacent to the TTSS structural genes and the protein shares similarity to the HopF translocator of *Xanthomonas campestris* pv. *vesicatoria*. Harpins are small glycine-rich, cysteine-free proteins regulated by the TTSS system and capable of eliciting a HR in host and non-host plants when they are injected into the intercellular spaces of plant leaves (Jin

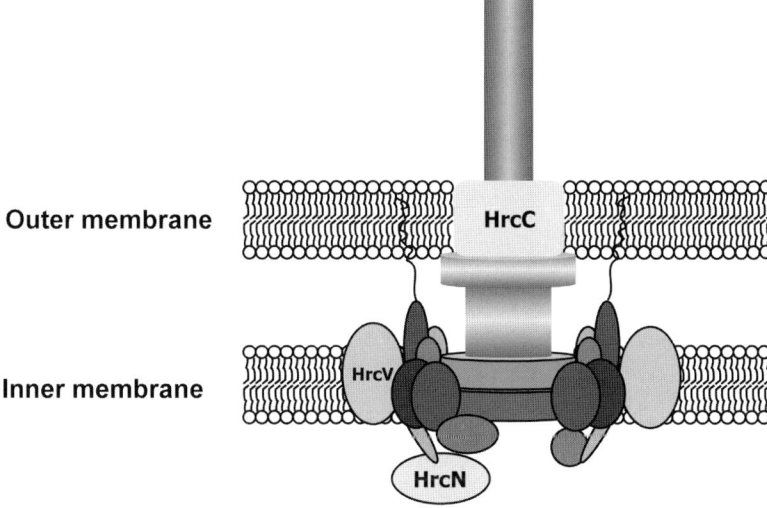

Figure 10.4 The type III secretion system pore spans the bacterial inner and outer membrane to form a pilus that penetrates the plant cell. Adapted from the Kyoto Encyclopedia of Genes and Genomes (KEGG).

et al., 2003). Pto DC3000 encodes at least 4 harpins, HrpZ1, HrpW1, HopAK1 and HopP1 (Kvitko et al., 2007). *hrpZ1* is encoded within the TTSS cluster next to *hrpA*, whereas *hrpW1* is within the conserved effector locus (CEL) adjacent to the TTSS structural genes; *hopAK1* and *hopP1* are located elsewhere in the chromosome. HrpZ1 is able to bind to lipids and form ion-conducting pores in lipid bilayers (Lee et al., 2001). Interestingly, the harpins have similar N-terminal domains (HrpZ has only a single domain), but HrpW1 and HopAK1 have pectate lyase C-terminal domains and HopP1 has a soluble lytic transglycosylase C-terminal domain. Kvitko et al. (2007) hypothesized that three classes of functionally overlapping translocator exist, comprising (i) HrpK1, the HrpF/NopX class, (ii) HrpZ1, the one-domain harpin class and (iii) HrpW1/HopAK1/HopP1, the two-domain harpin class. Two models were proposed for the function of the translocators. One model suggests that the HrpA pilus relies on the translocation protein to breach the plant cell wall with a two-step process consisting of the harpins aiding the crossing of the plant cell wall and the HrpF/NopX translocator (HrpK1) aiding the crossing of the plant plasma membrane. An alternative model is that HrpA pilus can breach the plant cell wall and that the translocators act as a consortium at the tip of the pilus to facilitate effector exit into the plant cell – this is analogous to the strategy used by the animal pathogen, *Yersinia*, and is the favoured model.

As with the flagellum system, there is strict regulation of the TTSS (Dasgupta et al., 2003; Tang et al., 2006). Experimental evidence has shown that the expression of the Hrp TTSS is induced only in plant tissues, apoplastic fluid or in an *hrp*-inducing medium, which mimics the conditions encountered when bacteria colonize the intercellular space (apoplast) of the plant during infection (Rahme et al., 1992; Xiao et al., 1992; Rico and Preston, 2008). Three intracellular positive regulators (HrpR, HrpS and HrpL) and one negative regulator (HrpV) are involved in the transcriptional regulation of the TTSS in *P. syringae*. The expression of the *hrpRS* operon is elevated in *hrp*-inducing medium (Jin et al., 2003). *hrpRS* has been shown to be regulated by a number of mechanisms, including the GacS/GacA two-component system (where GacS is a sensory histidine kinase and GacA is a cognate response regulator), the ATP-dependent Lon protease, which negatively regulates the levels of the HrpR protein and the HrpG protein which negatively regulates HrpV that suppresses HrpS (Preston et al., 1998; Bretz et al., 2002; Chatterjee et al., 2003; Wei et al., 2005). HrpR and HrpS are required for maximal expression of the *hrpL* gene. HrpL belongs to a family of sigma factors with extracytoplasmic function and directs the transcription of all TTSS-associated genes in *P. syringae* by recognizing a consensus sequence motif, known as the '*hrp* box', present upstream of these genes (Jin et al., 2003). Interestingly, the HrpRS and HrpL regulators also control expression of a number of non-TTSS-related genes indicating a much broader, global role in regulation (Ferreira et al., 2006; Lan et al., 2006). HrpV is a negative regulator of TTSS genes.

Sequence analysis of the TTSS from a number of plant pathogenic bacteria has lead to the conclusion that the TTSS gene cluster is a pathogenicity island (PAI) which has a tripartite mosaic structure composed of an EEL, the *hrp/hrc* gene cluster, and a CEL (Alfano et al., 2000). The Hrp PAI of *P. syringae* has several properties of PAIs found in animal pathogens including the presence of many virulence-associated genes (several with relatively low G + C content, which indicates potential horizontal DNA transfer) in a large chromosomal region linked to a tRNA locus and absent from corresponding regions in closely related *Pseudomonas* species (Hacker et al., 1997; Alfano et al., 2000). The EEL harbours exchangeable effector genes and makes only a quantitative contribution to parasitic fitness in host plants whereas the CEL contains a number of conserved ORFs, shown to be essential for *Pto* pathogenicity in tomato.

Effectors are the substrates for the TTSS

The TTSS functions to translocate effector proteins from the bacterial cell into the plant cytoplasm (see also chapter by Boch). Some effectors may trigger a defensive reaction in the plant that leads to a resistance response, termed the hypersensitive reaction (HR). Such a protein has historically been termed an avirulence (*avr*) gene/protein (Keen, 1990). Alternatively, secreted proteins that lead to the development of disease symptoms in the plant were designated virulence (*vir*) genes/proteins. The broader term 'effector' has been adopted to encompass genes whose protein products contribute to the plant–microbe interaction, irrespective of function as avirulence or virulence factors (van Dijk *et al.*, 1999). Many effectors have been identified by their ability to transverse the TTSS pathway and these are designated Hop (Hrp outer protein) (Alfano and Collmer, 1997). Lindeberg *et al.* (2005) proposed nomenclature guidelines for type III Hop effector proteins in *P. syringae* to try and overcome the lack of an established systematic naming procedure which, with the advent of the genome sequencing projects, was causing problems such as one Hop being assigned multiple names.

Several research groups have identified potential secretion signals that target effectors to the TTSS apparatus. The exact nature of the secretion signal remains contentious; one theory is that the signal is encoded in the first 15 codons of the effector mRNA (Anderson *et al.*, 1999; Hienonen *et al.*, 2002) while another theory suggests that the secretion signal is encoded within the amino acid (aa) sequence (Lloyd *et al.*, 2001). Bioinformatics combined with genomics has allowed the analysis of all predicted TTSS effectors and the discovery that secreted proteins have a particular amino acid profile in the first 50 amino acids. The rules for secretion are that the first 50 amino acids of the effector N-terminal region should contain >10% serines, an aliphatic aa (isoleucine, leucine, valine, alanine or proline) in the 3rd or 4th position and no acidic aa (aspartic acid or glutamic acid) in the first 12 aa (Guttman *et al.*, 2002; Petnicki-Ociewa *et al.*, 2002; Schechter *et al.*, 2004). Several effectors, for example HopPtoF (AvrPphF), HopPsyA and HopPtoV, also require type III chaperones (TTC) to aid stability and secretion (van Dijk *et al.*, 2002; Badel *et al.*, 2003; Shan *et al.*, 2004; Wehling *et al.*, 2004; Guo *et al.*, 2005). Specific Hop chaperone (Shc) proteins have been shown to interact with their cognate effectors (e.g. ShcF$_{Pto}$ with HopF$_{Pto}$, the co-transcribed genes that form the HopPtoF locus) via yeast two-hybrid and pull down assays; the N-terminus appears to be important for this interaction.

Identification of *P. syringae* effectors by genetic and genomic screens

In 1984, the first effector was identified from *P. syringae* pv. *glycinea* by Staskawicz *et al.* (1984) and was designated *avrA*. In 2000, a review on bacterial effector genes by Vivian and Arnold (2000) estimated that 30 effector genes had been identified from a range of *P. syrinage* pathovars. A number of experimental approaches were used to identify these 30 TTSS effectors, one common method being 'gain-of-function' assays. For example, the effector gene *avrPpiB* (*hopAM1*) was first identified from race 3 strain 870A of *P. syringae* pv. *pisi* (*Ppi*) by using a cosmid library to restore avirulence activity to race 1 strain 299A that had lost an endogenous plasmid responsible for an avirulence phenotype (Bavage *et al.*, 1991). The cosmid conferring avirulence was then used for subcloning and was partially sequenced and a gene identified that could restore avirulence activity (Cournoyer *et al.*, 1995). Other approaches to identify TTSS effectors involved sequence characterization of the genomic location of known effector genes to identify others. For example, DNA sequences flanking two effector genes from *Ppi* showed a high degree of similarity and specific primers were designed from these conserved regions and used in PCR amplifications of all *Ppi* races. In addition to the amplification of expected genes, two further fragments were isolated. Both fragments encoded novel effector genes, *avrPpiC* on one and *avrPpiG* on the other (Arnold *et al.*, 2001).

Further studies have looked at specific genomic areas to try and identify effectors. The EEL was a good target for such analysis having previously been shown to contain a number of effector genes (Alfano *et al.*, 2000). Twelve EELs were sequenced from a range of *P. syringae* pathovars and from these, homologues of known effector genes were identified (Deng *et al.*, 2003). In a similar study, 86 *P. syringae* strains were screened and 33 EELs were amplified to provide six new alleles of known effectors that differed in sequence and/or size (Charity *et al.*, 2003). This study showed EEL evolution was independent of the central conserved region of the *hrp* PAI, most likely through integron-like assembly of transposed gene cassettes.

By 2002 there was a draft of the *Pto* DC3000 genome sequence publicly available which led to a number of investigations into predicting the number of effectors in this strain. Several approaches were used by a number of groups and are summarized in Table 10.2. One of the first approaches used was a modified *in vivo* expression technology (IVET) system (Boch et al. 2002). The aim of this investigation was to identify genes from *Pto* DC3000 that were induced upon infection of the model organism, *Arabidopsis thaliana*. Over 500 *in planta*-expressed (*ipx*) promoter fusions were identified. Sequence analysis of 79 of these fusions revealed several known virulence genes (including *hrp/hrc*, *avr*, and coronatine biosynthetic genes) as well as a number of newly described genes. Of the 79 *ipx* genes, 22 contained a well-conserved *hrp* box motif (10 of which were potentially novel effector genes). The *hrp* box motif is a consensus promoter sequence found upstream of effectors that are known to be expressed by *hrpL*. The motif consists of the sequence 5′-GGAACCNA-$N_{13\text{-}14}$CCACNNA-3′ (Innes et al., 1993; see also MacLean and Studholme chapter in this book). This sequence has been instrumental in scanning genome sequences to predict putative effector genes in plant pathogens.

Fouts et al. (2002) began identifying effector genes when only eight effector genes had been identified in *Pto* DC3000; three of these were well characterized *avr* (effector) genes and the remaining five poorly characterized. Fouts et al. (2002) initially used a reporter transposon screen for HrpL-activated genes and found 18 mutants downstream of 18 different *hrp* boxes, thus revealing several novel genes. This increased the number and diversity of *hrp* boxes that could be used as a training set to produce a Hidden Markov Model (HMM) to find additional Hrp promoters in the *Pto* DC3000 genome. Following this HMM analysis, 48 Hrp promoter sequences were predicted from the *Pto* DC3000 draft sequence, which was a higher number than expected at the time. Fouts et al. (2002) concluded that there are limitations in this type of promoter-based search and that there was a need for additional criteria to refine this bioinformatic approach.

In an extension of the Fouts et al. (2002) study, Petnicki-Ocwieja et al. (2002) examined the N-terminal regions of a set of Hrp-secreted proteins and found a number of conserved patterns and properties. These patterns included a serine content of at least 10%, certain aliphatic amino acids in positions three and four and a lack of acidic amino acids in the first 12 residues – these observations formed the basis of a set of rules for defining effector proteins. These patterns were used to construct a training set of 28 non-redundant proteins thought to be secreted by the *P. syringae* TTSS (Petnicki-Ocwieja et al., 2002). Interestingly they noted that four representative proteins that are expressed by the Hrp regulon, but are not secreted by the TTSS (CorS, IaaL, HrcC and HrpL), failed multiple rules for effector features. The training set was applied to the draft *Pto* DC3000 genome and 400 hits were found. This list was narrowed down to 32 effector candidates based on manually checking for features such as G+C content skewed from the *Pto* DC3000 genome, ORFs near Hrp promoters and ORFs with sequence identity to non-secreted proteins. Two of these candidate proteins were tested and shown to travel the Hrp pathway. They concluded that by various methods they had increased the inventory of *Pto* DC3000 Hrp-secreted proteins to 22, but the remaining candidate effectors needed to be tested for secretion.

Other approaches to effector identification have used functional methods such as identifying proteins that are secreted via the TTSS. *P. syringae* pv. *maculicola* strain ES4326 was used to develop an *in vivo* screen using the COOH-terminus of AvrRpt2 effector and its cognate R protein Rps2 (Guttman et al 2002). A transposon was devised containing the DNA coding for the COOH-terminus of AvrRpt2 effector to capture type III secretion signals from unknown effector genes. Insertions of the transposon that created translational fusions with effector genes generated an ES4326 strain that induced the HR (in an Rps2-dependent manner) upon infection of *Arabidopsis thaliana*. Using this approach 13 different effector genes were identified in the ES4326 genome. These investigators then

Table 10.2 Summary of functional genomic analyses for the prediction of effector proteins in *Pseudomonas syringae*

Reference	Strain	Initial Investigation	Genome screen used	Confirmation method(s)	Predicted number of effector candidates
Boch et al. (2002)	DC3000	NA	*In vivo* expression technology (IVET)	Sequencing, *In planta* GUS expression analysis, selected deletion analysis	>500 *in planta* expressed fusion, 79 known and potential virulence gene fusions
Fouts et al. (2002)	DC3000	Identification of *hrp* boxes by mini-Tn5*gus* mutagenesis	Hidden Markov model (HMM) screen based on *hrp* boxes	Selected microarray and RNA blot analyses to confirm HrpL-dependency, selected gene expression	48 (draft seq)
Petnicki-Ocwieja et al., (2002)	DC3000	HMM Hrp box screen (Fouts et al., 2002)	Algorithms based on N-terminal amino acid biases and manual confirmation	Secretion assays	22 confirmed
Guttman et al. (2002)	DC3000	AvrRpt2 *in vivo* screen (translocation assay)	N-terminal amino acid biases and conserved *hrp* box promoter elements	HR assays, Western blot expression assays, induced in MM, not rich media	38 (draft seq)
Zwiesler-Vollick et al. (2002)	DC3000	Assembly of conserved published *hrp* motifs	*hrp* box screening algorithms	Microarray, Northern blot, expression analysis, translocation assays	73 putative (draft seq), 24 confirmed experimentally
Chang et al. (2005)	DC3000, 1448A	Assembly of previously identified effector protein and helper motifs	FACS based screen, protein fusions to HR reporter	AvrRpt2 translocation assay	29 DC3000, 19 1448A
Greenberg and Vinatzer (2003)	B728a	NA	N-terminal amino acid biases and conserved *hrp* box promoter elements	NA	29 (draft seq)
Vinatzer et al. (2006)	B728a	Predicted in Greenberg and Vinatzer (2003)	AvrRpt2 transduction assay	NA	22 secreted
Vencato et al. (2006)	1448A	NA	N-terminal amino acid biases and conserved *hrp* box promoter elements	RT-PCR and translocation assays	27 effectors and related TTSS substrates
Lindeberg et al. (2006)	B728a	Previous literature	N-terminal amino acid biases and conserved *hrp* box promoter elements	NA	18
Lindeberg et al. (2006)	DC3000, 1448A	Previous literature	NA	NA	48 DC300, 29 1448A

NA, not applicable.

went on to analyse the NH$_2$-terminal region of a number of effector genes and screened the draft *Pto* DC3000 genome based upon characteristic NH$_2$-terminal amino acid biases (high Ser and low Asp) and the conserved *hrp* box promoter element. They estimated from these analyses there were 38 effectors in *Pto* DC3000 (Guttman *et al.*, 2002).

Another report published in the same year also used the available *Pto* DC3000 draft sequence to search for genes downstream from the *hrp* box-like promoter sequence as an indication of HrpL regulation (Zwiesler-Vollick *et al.* 2002). This study identified 73 hits in total and then used microarrays and Northern blotting experiments to identify 24 genes/operons that had higher expression levels in *hrp*-inducing medium (a medium that activates HrpL) than in LB medium. Interestingly, during their screen they failed to detect three effector genes previously identified in *Pto* DC3000. It was noted however that two of these genes, *avrPphE* (*hopX1*) (Mansfield *et al.*, 1994) and *orf4* of the CEL (Alfano *et al.*, 2000), contained a G, instead of a C, at the second position of the second motif in the *hrp* box consensus sequence which was not accommodated in their original search parameters. Indeed, when they repeated their search using a revised *hrp* box consensus, they obtained twice as many hits (151 compared with 73). So although these two genes represented a small minority of *hrp*-regulated genes, twice as many candidate effectors would have to be screened to find them.

Further developments in effector detection have included a high-throughput, near-saturating screen for type III effector genes from *P. syringae* developed by Chang *et al.* (2005). This screen relied on a modification of the technique termed differential fluorescence induction (DFI). Genomic libraries of *P. syringae* were created in DFI vectors and screened for clones expressing Green Fluorescent Protein (GFP) in a HrpL-dependent manner using fluorescence-activated cell sorting (FACS). Fluorescent clones were then sequenced and the resulting contiguous DNA sequences were examined for characteristics of type III effector genes. Forty three such genes were identified in *Pto* DC3000 and 41 in *Pph* 1448A. These screens were followed by cloning all of the predicted effector genes and testing their ability to be secreted via the *avrRpt2* translocation assay. After this was completed, the number of predicted effector genes in *Pto* DC3000 was cut to at least 24 (possibly 28) and in *Pph* 1448A, at least 17 (possibly 19). The authors suggested that predicted numbers based on previous informatics approaches (see above) had been significantly over-estimated.

Other investigators made use of the other *P. syringae* genome sequences as they became available. Greenberg and Vinatzer (2003) predicted 29 effector genes in *Psy* B728a based on the presence of the conserved *hrp*-promoter element upstream of effector-encoding genes and biased amino acid content in the first 50 N-terminal amino acids. They also compared these predictions with those of *Pto* DC3000 and found that there were 25 effectors common to both *Pto* DC3000 and *Psy* B728A, 33 only present in *Pto* DC3000 and 4 only present in *Psy* B728a. These authors then went on to experimentally confirm 22 of the *Psy* B728a predicted effectors and cloned most of the *Psy* B728a effector repertoire (Vinatzer *et al.*, 2006).

It was now evident that *P. syringae* strains contain a number of effector genes that are functional in one strain but not in another. These non-functional genes may act as a reservoir of new effectors that could become functional if the conditions of their hosts change. As such, many studies have aimed to complete the *Pto* DC3000 effector inventory to include effectors that are silent, non-functional, or disrupted by transposable elements. One such approach involved scanning the *Pto* DC3000 genome to predict candidate effector genes and then eliminated those previously confirmed as *hop* genes (Schechter *et al.*, 2006). This left 44 potentially new *hop* genes that were then tested by using a *Bordetella pertussis* calmodulin-dependent adenylate cyclase (Cya) translocation reporter assay (Schechter *et al.* 2004). The translocation of type III effectors carrying C-terminal fusions with Cya can be monitored by the calmodulin-dependent production of cyclic AMP (cAMP) within eukaryote cells. Ten of the 44 high-probability candidate effectors were translocated into plant cells. This work concluded that *Pto* DC3000 harbours a total of 53 *hop/avr* genes and pseudogenes (encoding both injected effectors and TTSS substrates likely to be released into the apoplast); 33 of these genes are likely functional in *Pto* DC3000, 12 are non-functional *Pto* DC3000 members of valid Hop families, and the remaining 8 may or may not be produced at functional levels. However, it was also shown that growth of *Pto* DC3000 in tomato and *Arabidopsis* Col-0 was not impaired by constitutive expression of repaired versions of two *hops* that were disrupted naturally by transposable elements or *hop* genes that are naturally cryptic (Schechter et al., 2006).

A further approach to identify effector genes was taken by Ferreira et al. (2006). Here, a whole-genome microarray of *Pto* DC3000 was constructed to comprehensively identify genes differentially expressed in wild-type and Δ*hrpL* strains. These experiments revealed the wild-type had 119 up-regulated and 76 down-regulated genes compared with the Δ*hrpL* strain. Interestingly, in addition to HrpL controlling the expected effector genes, this analysis also revealed HrpL involvement in regulation of twin-arginine transport (Tat) substrates, regulatory proteins, and proteins involved in the synthesis or metabolism of phytohormones, phytotoxins, and myoinositol.

An excellent review of effector identification in all three *P. syringae* strains was published in 2006 by Lindeberg et al. (2006). This review draws information from all previously published work and draws many conclusions including that *Pto* DC3000 has the largest effector inventory of the three published strains and *Psy* B728a has the smallest. They also reiterate that each strain has several inactive effector genes and, interestingly, only five of the 46 known effector families has an active member represented in each of the three sequenced strains. The authors also recommend the *Pseudomonas*-Plant interaction web site (http://www.pseudomonas-syringae.org/home.html) as a resource that not only provides detailed coverage of the three sequenced strains, but also harbours the main Hop database that contains over 20 fields of information for 300 Hop/effector/helper proteins from many *P. syringae* strains.

Effector functions in *P. syringae*

Avirulence gene products effectively signal to plants that they are being attacked and allow them to mount a resistant response. However, this cannot be the true intended bacterial function of these gene products (that we now refer to as effectors). The beneficial bacterial roles of effectors are predicted to include the suppression of host defences, release of nutrients from host tissue and bacterial transmission (Greenburg and Vintazer, 2003). For years it was very difficult to identify the functions of individual effectors as knocking out a single effector rarely lead to a noticeable change in phenotype due to the functional redundancy of many effectors. Several insights have now been proposed as to the function of effector genes in *P. syringae* through analysis of genome sequence and expression profiling data. A number of noteworthy examples of these are given below. Further detailed reviews of this area include: Abramovitch et al. (2006a), Desveaux et al. (2006), Grant et al. (2006), chapter by Boch in this book.

Blocking effector recognition

One function of some effectors is suppression of host genes responsible for detection and initiation of a resistance response. An example of this was observed with the identification of the virulence gene *virPphA* (*hopAB1*) (Jackson et al., 1999). In this work, a 154-kb plasmid was cured from race 7 strain 1449B of *Pph* resulting in the cured strain losing virulence towards bean and causing an HR in previously susceptible cultivars. Restoration of virulence was achieved by trans-complementation using cosmid clones, which when sequenced, were shown to contain a number of previously identified effector genes [including *avrD*, *avrPphC* (*avrB2*) and *avrPphF* (*hopF1*)]. Partial virulence was also restored by the introduction of *virPphA* alone. The proximity of several *avr* and *vir* genes lead the authors to conclude that they had identified a plasmid-borne PAI. One particularly interesting aspect of this work was that the loss of the plasmid led to the uncovering of a previously suppressed HR. This suggested that gene(s) able to trigger the HR are present elsewhere in the *Pph* chromosome, but the presence of the plasmid-borne *virPphA* dominates the observable phenotype.

This phenomenon has been examined in detail for the effector gene *avrPtoB* from *P. syringae* pv. *tomato* (Abramovitch et al., 2003). AvrPtoB and VirPphA are orthologues that share approximately 53% amino acid sequence identity. The regions of identity are dispersed throughout the proteins, suggesting AvrPtoB and VirPphA are related, but divergent proteins (Jackson et al., 2002; Abramovitch et al., 2003). Abramovitch et al. (2003) showed that AvrPtoB acts inside the plant cell to inhibit programmed cell death (PCD) initiated by Pto and Cf9 disease resistance proteins and the pro-apoptotic mouse protein Bax. AvrPtoB also was shown to suppress PCD in yeast, thus demonstrating that AvrPtoB functions as a cell death inhibitor across a variety of organisms (AvrPtoB activity is discussed in greater detail below). This ability of effectors to suppress different types of PCD in different organisms suggests that they act on conserved pathways (reviewed in Alfano and Collmer, 2004).

In a related study, Jamir et al. (2004) expressed a cloned type III secretion system, and effector gene *hopPsyA*, in *P. fluorescens* to elicit an HR in tobacco. They then systematically tested 19 effectors from *Pto* DC3000 for their ability to suppress the HR elicited by HopPsyA and found five effectors could suppress the HR. They concluded that the high proportion of effectors that suppress PCD suggests that suppressing plant immunity is one of the primary roles for *Pto* DC3000 effectors and a central requirement for *P. syringae* pathogenesis. A further example of plant immunity suppression is AvrPphC suppressing the bean HR that is triggered by the recognition of AvrPphF from *Pph* (Tsiamis et al., 2000).

Enzymatic activities

Within the effector repertoire of *Pto* DC3000 are three genes, *hopO1-1*, *hopU1* and *hopO1-2*, that were predicted to encode proteins containing potential active sites of mono-ADP ribosyltransferases (ADP-RTs) (Fu et al., 2007). Fu et al. (2007) characterized *hopU1* showing firstly that it is injected into plants via the TTSS and then when deleted from *Pto* DC3000, the Δ*hopU1* mutant exhibited a six-fold reduction in growth in plant tissue and had reduced ability to cause disease symptoms in *A. thaliana* Col-0. The Δ*hopU1* mutant also elicited the HR on tobacco at cell densities below the threshold needed for wild-type *Pto* DC3000 which suggested that *hopU1* acts to suppress the HR in non-hosts. Fu et al. (2007) also demonstrated that HopU1 functions as an ADP-RT and that the HopU1 substrates in *A. thaliana* extracts were RNA-binding proteins possessing RNA-recognition motifs. The authors concluded that the ADP-ribosylation of RNA-binding proteins quells host immunity by affecting RNA metabolism and this may reduce the amount of immunity-related mRNAs available in the plant, thus favouring establishment of the pathogen.

Other enzymatic activities of effectors that have been implicated in suppression of plant innate immunity include cysteine protease, tyrosine phosphatase and E3 ubiquitin ligase (Abramovitch et al., 2006b; Grant et al., 2006). Ubiquitination is the specific and covalent addition of ubiquitin to proteins that, depending on the polyubiquitin chain topology, can either act as a signal for the 26S proteasome to destroy such earmarked proteins or its addition can result in new protein properties. Both of these functions have been shown to be essential for the host's immune response to pathogens (Angot et al., 2007). It has been suggested that altering the host's ubiquitination system is a common strategy among pathogenic bacteria to either control the timing of their effectors' action by programming them for degradation, to block specific intermediates in mammalian or plant innate immunity, or to target host proteins for degradation by mimicking specific ubiquitin/proteasome system components (Angot et al., 2007). For example, the effector protein AvrPtoB has a carboxy-terminal domain that encodes a E3 ubiquitin ligase (Abramovitch et al., 2006; Janjusevic et al., 2006). It has been demonstrated that a host kinase, Fen, physically interacts with AvrPtoB$_{1-387}$ and is responsible for activating the plant immune response. However, the intact AvrPtoB specifically ubiquitinates Fen and promotes its degradation in a proteasome-dependent manner, which leads to disease susceptibility in Fen-expressing tomato lines (Rosebrock et al., 2007).

Recently, Nimchuck et al. (2007) have reported on the enzymatic activity of the *hopX* family of effector genes. HopX (previously AvrPphE) is an effector originally identified in *Pph* (Stevens et al., 1998). Genomic studies have uncovered *hopX* family members in diverse *P. syringae* strains as well as in unrelated TTSS-harbouring phytopathogenic bacteria including *Ralstonia solanacearum* and *Xanthomonas campestris* (see Nimchuck et al., 2007 for references). Nimchuck et al. (2007) demonstrated that the HopX family members are modular proteins that contain a putative cysteine-based catalytic triad and a potential protein-cofactor interaction platform. Although they were unable to demonstrate *in vitro* enzyme activity, bioinformatics and genetic analysis strongly suggested that the HopX family encode enzymes of the transglutaminase (TGase) catalytic triad superfamily. They concluded that the HopX family members are therefore putative enzymes whose activity is required for both resistance gene-mediated recognition and for potential virulence functions in susceptible hosts. They further observed that the *Legionella pneumophila* genome contained a protein similar to HopX and suggested such proteins may also contribute to animal pathogenesis.

Altering host pathways

Phytohormones such as ethylene, are known to be involved in plant defence responses and are also reported to play a role in disease progression. Using a tomato cDNA microarray (consisting of 8600 genes) to assess plant responses to both *Pto* DC3000 and the *hrp/hrc* deletion mutant strain, Cohn and Martin (2005) showed over 300 genes to be differentially expressed within 24 h post inoculation in a susceptible tomato line. Of these, many were genes encoding proteins associated with hormone response or hormone biosynthesis pathways. They also used isogenic mutant strains of *Pto* DC3000 to monitor the host transcription changes in response to two specific effector proteins. They found that during a compatible interaction, AvrPto and AvrPtoB enhance necrosis in tomato leaves by regulating the expression of two genes encoding the ethylene-forming enzyme, ACC oxidase and thus promoting ethylene production. In a similar study, de Torres-Zabala et al. (2007) showed that *Pto* hijacks the *Arabidopsis* abscisic acid (ABA) signalling pathway to cause disease. Global transcription profiling revealed a prominent group of effector-induced genes that were associated with ABA biosynthesis and response machinery. They also showed that expression of AvrPtoB *in planta* could enhance bacterial growth and elevate plant ABA levels. It was suggested that ABA biosynthesis may be regulated by bacterial effectors through induction of *NCED3* (a key enzyme of ABA biosynthesis).

AvrPto and AvrPtoB have also been found to modify basal defence phenotypes. Ultrastructural studies of *Pseudomonas* bacteria initiating a defence response on a resistant host have shown the plant cell wall to thicken, forming a papilla [cell wall appositions composed of callose, phenolics, hydroxyproline-rich glycoproteins and other materials (Bestwick et al., 1995)]. This type of reaction is not seen in susceptible plants and therefore suggests the function of some effectors may be suppressing this cell wall-based plant defence response. This hypothesis was investigated using a combination of large-scale gene expression profiling, *in planta* expression of the *Pto* DC3000 effector AvrPto, and cytological examination (Hauck et al., 2003). These investigations confirmed the effector AvrPto to be a suppressor of the papillae-associated cell wall defence in *Arabidopsis*. They also showed that the TTSS of *Pto* DC3000 down-regulated the expression of a set of *Arabidopsis* genes that encode putative secreted cell wall and defence proteins in a salicylic acid-independent manner. A similar response of suppression of plant cell wall alterations was observed by *in planta* expression of AvrPtoB (de Torres et al., 2006).

Structural mimicry

The primary amino acid sequence for most effector genes provides no insight into their biochemical function. However, recent developments in elucidating the crystal structure of effector proteins has lead to advances in functional understanding (Desveaux et al., 2006). Microbial pathogens of animals have been shown to use structural mimicry as an important strategy to mimic host proteins so as to manipulate host physiology and cellular functions for their benefit (Stebbins and Galán, 2001). Determination of three-dimensional structures of *P. syringae* type III effectors has also revealed structural mimicry of host factors as a virulence strategy for plant pathogens (Desveaux et al., 2006).

For example, the effector protein AvrPphB from *Pph* is a member of a novel class of cysteine proteases that includes the *Yersinia* type III effector YopT (Shao et al., 2002). The AvrPphB crystal structure was the first reported for a type III effector from a plant pathogen (Zhu et al., 2004) and it showed similarity in its predicted active site to papain-like cysteine proteases, thus suggesting a similar molecular mechanism for proteolysis. Another example is AvrPphF (originally isolated from *Pph*) which consists of two ORFs that are both required for virulence/avirulence function (Tsiamis et al., 2000). Shan et al. (2004) demonstrated that the *Pto* DC3000 AvrPphF allele (HopPtoF) contains one ORF encoding a type III chaperone, ShcF, which interacts with (and stabilizes) the effector protein encoded by the second ORF, HopF1. The crystal structures of the HopPtoF ORFs revealed that ShcF had high structural similarity to type III chaperones from mammalian pathogens. HopF1 adopts a novel mushroom-like three-dimensional structure that can be subdivided into 'head' and 'stalk' subdomains. The head subdomain has limited structural similarity to the catalytic domain of ADP-ribosyltransferase (ADPRT) diphtheria toxin (Singer et al., 2004). Although it could not be demonstrated that HopF1 displayed any *in vitro* ADPRT-associated activity or NAD-binding activity, it was shown by mutagenesis that the amino acids Arg-72 and Asp-174 might represent catalytic residues that are important for enzymatic activity or, alternatively, might be required for interaction with specific host proteins or cofactors (Desveaux et al., 2006).

Another example of the use of crystal structure to elucidate function is that of AvrPtoB, which has revealed its enzymatic function as an E3 ubiquitin ligase (as discussed above) (Janjusevic et al., 2006). Mutation of conserved residues involved in the binding of E2 ubiquitin-conjugating enzymes abolishes this activity *in vitro*, as well as anti-PCD activity in tomato leaves, thus dramatically reducing virulence. The authors concluded that these results show that *P. syringae* uses a mimic of host E3 ubiquitin ligases to inactivate plant defences.

AvrB is recognized by the products of the *Arabidopsis RPM1* and soybean *RPG1* resistance genes. The RPM1-interacting protein 4 (RIN4) is required for RPM1 function during recognition of AvrB (Mackey et al., 2002). The crystal structure of AvrB revealed a novel bilobal fold, with a large lobe that contains a deep cleft and a distinct small lobe attached to the back and side of the large lobe (Lee et al., 2004) although its biochemical function remained elusive. In order to investigate this further, Desveaux et al. (2007) determined the structure of AvrB complexed with RIN4 and identified interacting residues in the upper lobe of AvrB that are required for both RIN4 binding and activation of RPM1. A model was postulated for the function of both AvrB virulence and avirulence (triggering of disease resistance). In essence they suggest that AvrB binds to a nucleotide or another small molecule of similar shape inside the host cell. This leads to the recruitment of a plant cofactor that transforms AvrB into a kinase capable of autophosphorylation. The phosphorylated AvrB then becomes myristoylated and directed to the plasma membrane where it interacts with RIN4 and its conformation is altered. RIN4 is phosphorylated and triggers RPM1-mediated activation of disease resistance. Alternatively, if RPM1 is absent, the interaction of AvrB with RIN4 and other proteins leads to a suppression of basal defence responses and the effector contributes to disease. For a fuller explanation of this model, see Desveaux et al. (2007). Interestingly, RIN4 is proposed to interact with another effector protein, AvrRpt2, but via a different mechanism. In this case AvrRpt2, which is a cysteine protease, is autoprocessed inside the host cell where it activates the resistance protein RPS2 by causing removal of RIN4 by protease activity (Kim et al., 2005). A broader review of effectors and the plant targets is detailed in the chapter by Jens Boch.

How do pathogens evolve to avoid triggering plant resistance?

We have described in this chapter how the function of one effector gene can be suppressed by another, but the establishment of durable plant resistance can also be difficult due to mutations in effector genes, which results in evasion of host recognition. The antimicrobial environment provided by plants undergoing the HR presents a strongly selective environment for mutants that can avoid triggering resistance. Changes in effector genes may either be by mutation of the gene itself or by loss of larger pieces of DNA from the genome carrying the gene(s) (Fig. 10.5; see also chapter by Stavrinides).

Mutation and loss of avirulence function

One way in which functional avirulence of an effector gene is lost, is by changes to the gene via mutational insertions, deletions or rearrangements. Even single base pair changes in effector genes can cause non-functionality. There are a number of examples of this in plant pathogenic bacteria. For example, homologues of the *avr* gene *avrPphE* are present in all nine races (subdivisions of the pathovar based on interaction phenotypes with host cultivars) of the bean pathogen *Pph*, but only races 2, 4, 5 and 7 are avirulent on cultivars of bean with the corresponding *R2* resistance gene (Stevens *et al.*, 1998). The other homologues have been inactivated either via single basepair changes (conferring amino acid substitutions) in races 1, 3, 6 and 9 or by insertion of 104 bp in the allele of race 8. These mutations in the *avrPphE* alleles do not affect the ability of *Pph* to colonize susceptible bean hosts.

Seven races of *P. syringae* pv. *glycinea* contain non-functional *avrD* alleles (Keith *et al.*, 1997). These alleles all contain mutations collectively altering only nine amino acid positions. It was suggested that the gene had mutated to escape defence surveillance in soybean plants containing resistance gene *Rpg4*.

Insertion of mobile genetic elements

Inactivation of gene function can also be caused by gene disruption by mobile genetic elements such as insertion sequences and transposons. The recent increase in bacterial pathogen genomes available for comparison has led to the identification of a number of genes that are disrupted in one strain and not another. For example, analysis of the completed *Pto* DC3000 genome revealed

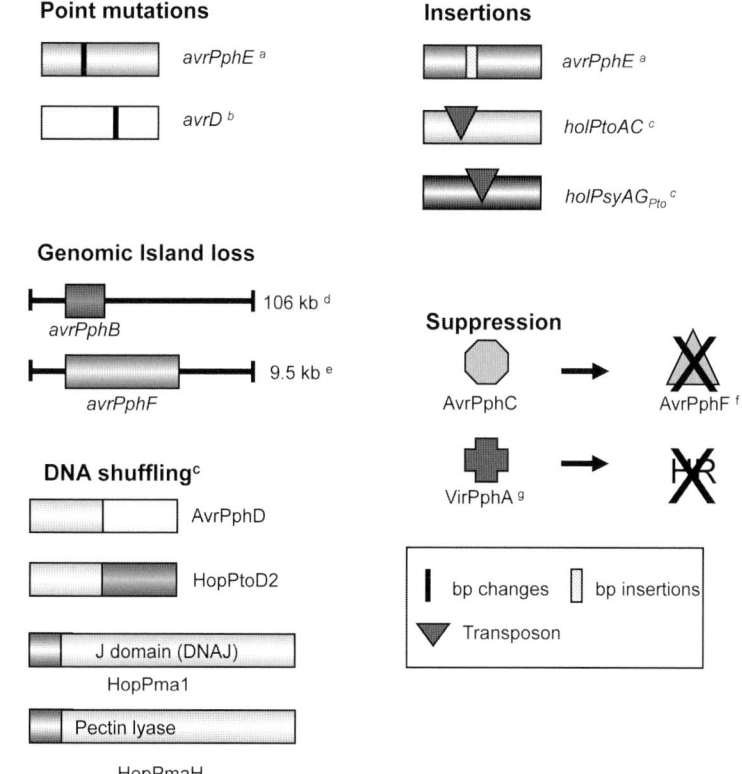

Figure 10.5 Mechanisms by which the function of *Pseudomonas syringae* effector genes can be lost or suppressed. A number of representative examples are shown. bp, base pair; HR, hypersensitive response. [a]Stevens *et al.* (1998). [b]Keith *et al.* (1997). [c]Greenberg and Vinatzer (2003). [d]Pitman *et al.* (2005). [e]Rivas *et al.* (2005). [f]Tsiamis *et al.* (2000). [g]Jackson *et al.* (1999).

31 effectors and seven additional proteins secreted by the type III system (Buell et al., 2003) that when compared to orthologues in the draft genome of *Psy* B728A, showed at least four of these effectors to be disrupted by mobile genetic elements.

Greenberg and Vinatzer (2003) reviewed methods used to identify effector genes and then used comparative genomics to predict novel effectors potentially disrupted by strain-specific mutations in the genomes of *Pto* DC3000 and *Psy* B728A. They reported transposon-disrupted genes such as *holPtoAC* and *holPsyAGPto* from *Pto* DC3000, whose orthologues are intact in *Psy* B728A. In fact, Greenberg and Vinatzer (2003) suggested that differences in effector repertoires of highly related phytopathogenic strains could be responsible for the different host range and/or disease characteristics induced. Candidate effector genes that carry transposon insertions using translocation assays have also been identified by Schechter et al. (2004).

DNA shuffling

Pto DC3000 contains many more transposon copies than Psy B728A and this may be driving a high level of DNA shuffling within the genome giving rise to novel effectors and effector operons. Greenberg and Vinatzer (2003) illustrated a number of effector chimeras that could have arisen (or become inactivated) by DNA shuffling, including three examples of effectors similar to other bacterial proteins [pectin lyase, transglycosylase, and J domain (DNAJ)] that have acquired type III secretion signals from unknown origins. This plasticity of the genome may be beneficial by allowing bacteria to evade host defence mechanisms.

Stavrinides et al. (2006) used computational and evolutionary approaches to identify numerous mosaic and truncated TTSS effectors among animal and plant pathogens. They proposed that these genes have evolved through a shuffling process that they termed 'terminal reassortment'. In this process, the termini of existing effectors are mobilized within the genome, creating random genetic fusions that result in chimeric genes that are subsequently selected for due to improved host colonization. For example, *P. syringae* effectors HopD1 and HopAO1 are N-terminal chimeras that have different biological functions due to their unique C termini. They concluded that terminal reassortment may permit the near instantaneous evolution of new effectors and appears to be responsible for major evolutionary modifications to effector activity and function.

DNA loss

The entire coding region of effector genes may be lost from the cell if the gene is carried on mobile DNA elements that can be lost during cell replication or through rare spontaneous deletions (Arnold et al., 2007). For example, the loss of the genomic island PPHGI-1 from the chromosome of *Pph* race 4 strain 1302A leads to loss of the effector gene *avrPphB* (Pitman et al., 2005). Loss of this island causes the bacteria to become pathogenic on a bean cultivar carrying the *R3* resistance gene. The island was sequenced (106-kb) and shown to share features with integrative and conjugative elements (ICElands) and also PAIs in diverse bacteria (Hacker and Carniel, 2001, Van der Meer and Sentchilo, 2003). Loss of PPHGI-1 was detected after passaging 1302A through leaves of a R3 resistant cultivar undergoing the HR. It was postulated that the antimicrobial environment of the AvrPphB-induced HR imposed strong selective pressure for bacteria to lose the *avrPphB* gene and it appeared that the causal mechanism was loss of the PPHGI-1 genomic island. Indeed, the island was found to be excised from chromosomal *att* loci (mediated by a *xerC*-like integrase coded for on the island) to form an episome that is rapidly lost from bacteria in leaves undergoing the HR. Eventually, virulent forms of the bacteria, lacking PPHGI-1 (and therefore *avrPphB*) emerged as the dominant *Pph* genotype in bean tissue.

Loss of gene function can occur not just by loss of an entire PAI but by loss of a section within it. For example, virulence towards bean and soybean is determined by effector genes in a plasmid-borne PAI in race 7 strain 1449B of *Pph* (Jackson et al., 1999). Rivas et al. (2005) carried out a comparative analysis of the PAI in a number of *Pph* strains. They found that the PAI is generally well conserved in strains of *Pph*, but that the PAI encoded gene, *avrPphF*, was absent from the genome of strain *Pph* 1448A due to deletion of a continuous 9.5 kb fragment. The left junction of the deleted region consisted of a chimeric transposable element generated from the fusion of homologues of IS*1492* from *P. putida* and IS*1090* from *Ralstonia eutropha*. The conservation of this border over a number of isolates suggests that the *avrPphF* deletions were mediated by activity of a chimeric mobile element. The authors also found evidence that most of the strains belonging to races lacking *avrPphF* were derived from an common *Pph* ancestor after a unique deletion event in the PAI.

Conclusion

The analysis of the published *P. syringae* genome sequences has lead to a number of insights into their biology. Bioinformatic approaches have predicted the presence of all six major secretion systems in some or all of the strains analysed. A number of these have been shown to be functional in *P. syringae*. Although the TTSS has been the main focus of work it is likely that some or all of the other secretion systems play a role in the life cycle of *P. syringae*, whether during its interaction with plants or in other niches it may inhabit (Morris *et al.*, 2007). As far as the TTSS is concerned it is interesting to compare the methods that have been used for effector prediction and the numbers that have been predicted. In general the conclusions have been in broad agreement and it can be seen that a combination of *in silico* and *in vitro* approaches is appropriate. Elucidating the effector content of these strains has lead to a number of insights into effector function and evolution. Hopefully the advances in sequencing technologies will result in more stains of *P. syringae* having their genome sequenced allowing more detailed comparisons to be made to elucidate the key effectors present in these strains. Comparisons should also allow identification of further functional groups of effectors and advances in protein structure prediction will allow a great understanding of their mechanisms. Genome resequencing technology (see chapter by MacLean and Studholme) will also allow us to follow the evolution of *Pseudomonas syringae* and their virulence systems. We would also suggest that there is a need to examine the other secretion systems in greater detail to gain a more holistic insight to disease processes. These developments will ultimately lead to a better understanding of the 'enemy' and allow the development of targeted disease management strategies.

References

Abramovitch, R.B., Kim, Y.J., Chen, S., Dickman, M.B., and Martin, G.B. (2003). *Pseudomonas* type III effector AvrPtoB induces plant disease susceptibility by inhibition of host programmed cell death. EMBO J. 22, 60–9.

Abramovitch, R.B., Anderson, J.C., and Martin, G.B. (2006a). Bacterial elicitation and evasion of plant innate immunity. Nat. Rev. Mol. Cell Biol. 7, 601–11.

Abramovitch, R.B., Janjusevic, R., Stebbins, C.E., and Martin, G.B. (2006b). Type III effector AvrPtoB requires intrinsic E3 ubiquitin ligase activity to suppress plant cell death and immunity. Proc. Natl. Acad. Sci. USA 103, 2851–6.

Alfano, J.R., and Collmer, A. (1997). The type III (Hrp) secretion pathway of plant pathogenic bacteria: trafficking harpins, Avr proteins, and death. J. Bacteriol. 179, 5655–62.

Alfano, J.R., and Collmer, A. (2004). Type III secretion system effector proteins: double agents in bacterial disease and plant defense. Annu. Rev. Phytopathol. 42, 385–414.

Alfano, J.R., Charkowski, A.O., Deng, W., Badel, J.L., Pentnicki-Ocwieja, T., van Dijk, K., and Collmer, A. (2000). The *Pseudomonas syringae* Hrp pathogenicity island has a tripartite mosaic structure composed of a cluster of type III secretion genes bounded by exchangeable effector and conserved effector loci that contribute to parasitic fitness and pathogenicity in plants. Proc. Natl. Acad. Sci. USA 97, 4856–4861.

Anderson, D.M., Fouts, D.E., Collmer, A., and Schneewind, O. (1999). Reciprocal secretion of proteins by the bacterial type III machines of plant and animal pathogens suggests universal recognition of mRNA targeting signals. Proc. Natl. Acad. Sci. USA, 96, 12839–43.

Angot, A., Vergunst, A., Genin, S., and Peeters, N. (2007). Exploitation of eukaryotic ubiquitin signaling pathways by effectors translocated by bacterial type III and type IV secretion systems. PLoS Pathog. 3, e3.

Arnold, D.L., Jackson, R.W., Fillingham, A.J., Goss, S.C., Taylor, J.D., Mansfield, J.W., and Vivian, A. (2001). Highly conserved sequences flank avirulence genes: isolation of novel avirulence genes from *Pseudomonas syringae* pv. *pisi*. Microbiology 147, 1171–82.

Arnold, D.L., Jackson, R.W., Waterfield, N.R., and Mansfield, J.W. (2007). Evolution of microbial virulence: the benefits of stress. Trends Genet. 23, 293–300.

Badel, J.L., Nomura, K., Bandyopadhyay, S., Shimizu, R., Collmer, A., and He, S.Y. (2003). *Pseudomonas syringae* pv. *tomato* DC3000 HopPtoM (CEL ORF3) is important for lesion formation but not growth in tomato and is secreted and translocated by the Hrp type III secretion system in a chaperone-dependent manner. Mol. Microbiol. 49, 1239–51.

Bavage, A.D., Vivian. A., Atherton. G.T., Taylor. J.D., and Malik, A.N. (1991). Molecular genetics of *Pseudomonas syringae* pathovar *pisi*: plasmid involvement in cultivar-specific incompatibility. J. Gen. Microbiol. *137*, 2231–9.

Bestwick, C.S., Bennett, M.H., and Mansfield, J.W. (1995). Hrp mutant of *Pseudomonas syringae* pv *phaseolicola* induces cell wall alterations but not membrane damage leading to the hypersensitive reaction in lettuce. Plant Physiol. *108*, 503–516.

Boch, J., Joardar, V., Gao, L., Robertson, T.L., Lim, M., and Kunkel, B.N. (2002). Identification of *Pseudomonas syringae* pv. *tomato* genes induced during infection of Arabidopsis thaliana. Mol. Microbiol. *44*, 73–88.

Bretz, J., Losada, L., Lisboa, K., and Hutcheson, S.W. (2002). Lon protease functions as a negative regulator of type III protein secretion in *Pseudomonas syringae*. Mol. Microbiol. *45*, 397–409.

Bronstein, P.A., Marrichi, M., Cartinhour, S., Schneider, D.J., and DeLisa, M.P. (2005). Identification of a twin-arginine translocation system in *Pseudomonas syringae* pv. *tomato* DC3000 and its contribution to pathogenicity and fitness. J. Bacteriol. *187*, 8450–61.

Brown, I.R., Mansfield, J.W., Taira, S., Roine, E., and Romantschuk, M. (2001). Immunocytochemical localization of HrpA and HrpZ supports a role for the Hrp pilus in the transfer of effector proteins from *Pseudomonas syringae* pv. *tomato* across the host plant cell wall. Mol. Plant–Microbe Interact. *14*, 394–404.

Buell, C.R., Joardar, V., Lindeberg, M., Selengut, J., Paulsen, I.T., Gwinn, M.L., Dodson, R.J., Deboy, R.T., Durkin, A.S., Kolonay, J.F., Madupu, R., Daugherty, S., Brinkac, L., Beanan, M.J., Haft, D.H., Nelson, W.C., Davidsen, T., Zafar, N., Zhou, L., Liu, J., Yuan, Q., Khouri, H., Fedorova, N., Tran, B., Russell, D., Berry, K., Utterback, T., Van Aken, S.E., Feldblyum, T.V., D'Ascenzo, M., Deng, W.L., Ramos, A.R., Alfano, J.R., Cartinhour, S., Chatterjee, A.K., Delaney, T.P., Lazarowitz, S.G., Martin, G.B., Schneider, D.J., Tang, X., Bender, C.L., White, O., Fraser, C.M., and Collmer, A. (2003). The complete genome sequence of the Arabidopsis and tomato pathogen *Pseudomonas syringae* pv. *tomato* DC3000. Proc. Natl. Acad. Sci. USA *100*, 10181–6.

Burns, D.L. (2003). Type IV transporters of pathogenic bacteria. Curr. Opin. Microbiol. *6*, 29–34.

Caldelari, I., Mann, S., Crooks, C., and Palmer, T. (2006) The Tat pathway of the plant pathogen *Pseudomonas syringae* is required for optimal virulence. Mol. Plant–Microbe Interact. *19*, 200–12.

Cambronne, E.D., and Roy, C.R. (2006) Recognition and delivery of effector proteins into eukaryotic cells by bacterial secretion systems. Traffic *7*, 929–39.

Casper-Lindley, C., Dahlbeck, D., Clark, E.T., and Staskawicz, B.J. (2002). Direct biochemical evidence for type III secretion-dependent translocation of the AvrBs2 effector protein into plant cells. Proc. Natl. Acad. Sci. USA *11*, 8336–41.

Chang, J.H., Urbach, J.M., Law, T.F., Arnold, L.W., Hu, A., Gombar, S., Grant, S.R., Ausubel, F.M., and Dangl, J.L. (2005). A high-throughput, near-saturating screen for type III effector genes from *Pseudomonas syringae*. Proc. Natl. Acad. Sci. USA *102*, 2549–54.

Charity, J.C., Pak, K., Delwiche, C.F., and Hutcheson, S.W. (2003). Novel exchangeable effector loci associated with the *Pseudomonas syringae* hrp pathogenicity island: evidence for integron-like assembly from transposed gene cassettes. Mol. Plant–Microbe Interact. *16*, 495–507.

Chatterjee, A., Cui, Y., Yang, H., Collmer, A., Alfano, J.R., and Chatterjee, A.K. (2003). GacA, the response regulator of a two-component system, acts as a master regulator in *Pseudomonas syringae* pv. *tomato* DC3000 by controlling regulatory RNA, transcriptional activators, and alternate sigma factors. Mol. Plant–Microbe Interact. *16*, 1106–17.

Chen, C., and Beattie, G.A. (2007). Characterization of the osmoprotectant transporter OpuC from *Pseudomonas syringae* and demonstration that cystathionine-beta-synthase domains are required for its osmoregulatory function. J. Bacteriol. *189*, 6901–12.

Chen, C., and Beattie, G.A. (2008). *Pseudomonas syringae* BetT is a low affinity choline transporter that is responsible for superior osmoprotection by choline over glycine betaine. J. Bacteriol. *190*, 2717–2725.

Christie, P.J. (2004). Type IV secretion: the *Agrobacterium* VirB/D4 and related conjugation systems. Biochim. Biophys. Acta *1694*, 219–34.

Christie, P.J., and Vogel, J.P. (2000) Bacterial type IV secretion: conjugation systems adapted to deliver effector molecules to host cells. Trends Microbiol. *8*, 354–60.

Cohn, J.R., and Martin, G.B. (2005). *Pseudomonas syringae* pv. *tomato* type III effectors AvrPto and AvrPtoB promote ethylene-dependent cell death in tomato. Plant J. *44*, 139–154.

Cournoyer, B., Sharp, J.D., Astuto, A., Gibbon, M.J., Taylor, J.D., and Vivian, A. (1995). Molecular characterization of the *Pseudomonas syringae* pv. *pisi* plasmid-borne avirulence gene avrPpiB which matches the R3 resistance locus in pea. Mol. Plant–Microbe Interact. *8*, 700–8.

Dasgupta, N., Wolfgang, M.C., Goodman, A.L., Arora, S.K., Jyot, J., Lory, S., and Ramphal, R. (2003). A four-tiered transcriptional regulatory circuit controls flagellar biogenesis in *Pseudomonas aeruginosa*. Mol. Microbiol. *50*, 809–24.

de Torres, M., Mansfield, J.W., Grabov, N., Brown, I.R., Ammouneh, H., Tsiamis, G., Forsyth, A., Robatzek, S., Grant, M., and Boch, J. (2006). *Pseudomonas syringae* effector AvrPtoB suppresses basal defence in Arabidopsis. Plant J. *47*, 368–82.

de Torres-Zabala, M., Truman, W., Bennett, M.H., Lafforgue, G., Mansfield, J.W., Rodriguez Egea, P., Bögre, L., and Grant, M. (2007). *Pseudomonas syringae* pv. *tomato* hijacks the Arabidopsis abscisic acid signalling pathway to cause disease. EMBO J. *7*, 1434–43.

Deng, W.L., Preston, G., Collmer, A., Chang, C.J., and Huang, H.C. (1998). Characterization of the *hrpC* and *hrpRS* operons of *Pseudomonas syringae* pathovars *syringae*, *tomato*, and *glycinea* and analysis of the ability of hrpF, hrpG, hrcC, hrpT, and hrpV mutants to elicit the hypersensitive response and disease in plants. J. Bacteriol. *180*, 4523–3.

Deng, W.L., Rehm, A.H., Charkowski, A.O., Rojas, C.M., and Collmer, A. (2003). *Pseudomonas syringae* exchangeable effector loci: sequence diversity in representative pathovars and virulence function in *P. syringae* pv. *syringae* B728a. J. Bacteriol. *185*, 2592–602.

Desveaux, D., Singer, A.U., and Dangl, J.L. (2006). Type III effector proteins: doppelgangers of bacterial virulence. Curr. Opin. Plant Biol. *9*, 376–82.

Desveaux, D., Singer, A.U., Wu, A.J., McNulty, B.C., Musselwhite, L., Nimchuk, Z., Sondek, J., and Dangl, J.L. (2007). Type III effector activation via nucleotide binding, phosphorylation, and host target interaction. PLoS Pathog. *3*, e48

Ding, Z., Atmakuri, K., and Christie, P.J. (2003). The outs and ins of bacterial type IV secretion substrates. Trends Microbiol. *11*, 527–35.

Feil, H., Feil, W.S., Chain, P., Larimer, F., DiBartolo, G., Copeland, A., Lykidis, A., Trong, S., Nolan, M., Goltsman, E., Thiel, J., Malfatti, S., Loper, J.E., Lapidus, A., Detter, J.C., Land, M., Richardson, P.M., Kyrpides, N.C., Ivanova, N., and Lindow, S.E. (2005). Comparison of the complete genome sequences of *Pseudomonas syringae* pv. *syringae* B728a and pv. *tomato* DC3000. Proc. Natl. Acad. Sci. USA. *102*, 11064–9.

Ferreira, A.O., Myers, C.R., Gordon, J.S., Martin, G.B., Vencato, M., Collmer, A., Wehling, M.D., Alfano, J.R., Moreno-Hagelsieb, G., Lamboy, W.F., DeClerck, G., Schneider, D.J., and Cartinhour, S.W. (2006). Whole-genome expression profiling defines the HrpL regulon of *Pseudomonas syringae* pv. *tomato* DC3000, allows de novo reconstruction of the Hrp cis clement, and identifies novel coregulated genes. Mol. Plant–Microbe Interact. *19*, 1167–79.

Fouts, D.E., Abramovitch, R.B., Alfano, J.R., Baldo, A.M., Buell, C.R., Cartinhour, S., Chatterjee, A.K., D'Ascenzo, M., Gwinn, M.L., Lazarowitz, S.G., Lin, N.C., Martin, G.B., Rehm, A.H., Schneider, D.J., van Dijk, K., Tang, X., and Collmer, A. (2002). Genomewide identification of *Pseudomonas syringae* pv. *tomato* DC3000 promoters controlled by the HrpL alternative sigma factor. Proc. Natl. Acad. Sci. USA *19*, 2275–80.

Fu, Z.Q., Guo, M., Jeong, B., Tin, F., Elthon, T.E., Cerny, R.L., Staiger, D., and Alfano, J.R. (2007). A type III effector ADP-ribosylates RNA-binding proteins and quells plant immunity. Nature *447*, 284–289.

Gerlach, R.G, and Hensel, M. (2007). Protein secretion systems and adhesins: the molecular armory of Gram-negative pathogens. Int. J. Med. Microbiol. *297*, 401–15.

Grant, S.R., Fisher, E.J., Chang, J.H., and Dangl, J.L. (2006). Subterfuge and manipulation: type III effector proteins of phytopathogenic bacteria. Annu. Rev. Microbiol. *60*, 425–49.

Greenburg, J.T., and Vintazer, B.A. (2003). Identifying type III effectors of plant pathogens and analyzing their interaction with plant cells. Curr. Opin. Microbiol. *6*, 20–28.

Guo, M., Chancey, S.T., Tian, F., Ge, Z., Jamir, Y., and Alfano, J.R. (2005). *Pseudomonas syringae* type III chaperones ShcO1, ShcS1, and ShcS2 facilitate translocation of their cognate effectors and can substitute for each other in the secretion of HopO1–1. J. Bacteriol. *187*, 4257–69.

Guttman, D.S., Vinatzer, B.A., Sarkar, S.F., Ranall, M.V., Kettler, G., Greenberg, J.T. (2002). A functional screen for the type III (Hrp) secretome of the plant pathogen *Pseudomonas syringae*. Science *295*, 1722–6.

Hacker, J., and Carniel, E. (2001). Eological fitness, genomic islands and bacterial pathogenicity. A Darwinian view of the evolution of microbes. EMBO Rep. *2*, 376–81.

Hacker, J., Blum-Oehler, G., Muhldorfer, I., and Tschape. H. (1997). Pathogenicity islands of virulent bacteria: structure, function and impact on microbial evolution. Mol. Microbiol. *23*, 1089–97.

Hauck, P., Thilmony, R., and He, S.Y. (2003). A *Pseudomonas syringae* type III effector suppresses cell wall-based extracellular defense in susceptible *Arabidopsis* plants. Proc. Natl. Acad. Sci. USA. *100*, 8577–8582.

He, S.Y. (1997). Hrp-controlled interkingdom protein transport: learning from flagellar assembly? Trends Microbiol. *5*, 489–95.

Hienonen, E., Roine, E., Romantschuk, M., and Taira, S. (2002). mRNA stability and the secretion signal of HrpA, a pilin secreted by the type III system in *Pseudomonas syringae*. Mol. Genet. Genomics. *266*, 973–8.

Huang, H.C., He, S.Y., Bauer, D.W., and Collmer, A. (1992). The *Pseudomonas syringae* pv. *syringae* 61 hrpH product, an envelope protein required for elicitation of the hypersensitive response in plants. J. Bacteriol. *174*, 6878–85.

Innes, R.W., Bent, A.F., Kunkel, B.N., Bisgrove, S.R., and Staskawicz, B.J. (1993). Molecular analysis of avirulence gene *avrRpt2* and identification of a putative regulatory sequence common to all known *Pseudomonas* syringae avirulence genes. J. Bacteriol. *175*, 4859–69.

Jackson, R.W., Athanassopoulos, E., Tsiamis, G., Mansfield, J.W., Sesma, A., Arnold, D.L., Gibbon, M.J., Murillo, J., Taylor, J.D., and Vivian, A. (1999). Identification of a pathogenicity island, which contains genes for virulence and avirulence, on a large native plasmid in the bean pathogen *Pseudomonas syringae* pathovar *phaseolicola*. Proc. Natl. Acad. Sci. USA. *96*, 10875–80.

Jackson, R.W., Mansfield, J.W., Ammouneh, H., Dutton, L.C., Wharton, B., Ortiz-Barredo, A., Arnold, D.L., Tsiamis, G., Sesma, A., Butcher, D., Boch, J., Kim, Y.J., Martin, G.B., Tegli, S., Murillo, J. and Vivian, A. (2002). Location and activity of members of a family of *virPphA* homologues in pathovars of *Pseudomonas syringae* and P. *savastanoi*. Mol. Plant Pathol. *3*, 205–216.

Jamir, Y., Guo, M., Oh, H.S., Petnicki-Ocwieja, T., Chen, S., Tang, X., Dickman, M.B., Collmer, A., and Alfano, J.R. (2004). Identification of *Pseudomonas syringae* type III effectors that can suppress programmed cell death in plants and yeast. Plant J. 37, 554–65.

Janjusevic, R., Abramovitch, R.B., Martin, G.B., and Stebbins, C.E. (2006). A bacterial inhibitor of host programmed cell death defenses is an E3 ubiquitin ligase. Science 311, 222–6.

Jha, G., Rajeshwari, R., Sonti, R.V. (2005). Bacterial type two secretion system secreted proteins: double-edged swords for plant pathogens. Mol. Plant–Microbe Interact. 18, 891–8.

Jin, Q., Thilmony, R., Zwiesler-Vollick, J., and He, S.Y. (2003). Type III protein secretion in *Pseudomonas syringae*. Microbes Infect. 5, 301–10.

Joardar, V., Lindeberg, M., Jackson, R.W., Selengut, J., Dodson, R., Brinkac, L.M., Daugherty, S.C., Deboy, R., Durkin, A.S., Giglio, M.G., Madupu, R., Nelson, W.C., Rosovitz, M.J., Sullivan, S., Crabtree, J., Creasy, T., Davidsen, T., Haft, D.H., Zafar, N., Zhou, L., Halpin, R., Holley, T., Khouri, H., Feldblyum, T., White, O., Fraser, C.M., Chatterjee, A.K., Cartinhour, S., Schneider, D.J., Mansfield, J., Collmer, A., Buell, and Keen, C.R. (2005). Whole-genome sequence analysis of *Pseudomonas syringae* pv. *phaseolicola* 1448A reveals divergence among pathovars in genes involved in virulence and transposition. J. Bacteriol. 187, 6488–98.

Keith, L.W., Boyd, C., Keen, N.T., and Partridge, J.E. (1997). Comparison of avrD alleles from *Pseudomonas syringae* pv. *glycinea*. Mol. Plant–Microbe Interact. 10, 416–22.

Keen, N.T. (1990). Gene-for-gene complementarity in plant–pathogen interactions. Annu. Rev. Genet. 24, 447–63.

Kim, M.G., da Cunha, L., McFall, A.J., Belkhadir, Y., DebRoy, S., Dangl, J.L., and Mackey, D. (2005). Two *Pseudomonas syringae* type III effectors inhibit RIN4-regulated basal defense in Arabidopsis. Cell 121, 749–59.

Kvitko, B.H., Ramos, A.R., Morello, J.E., Oh, H.S., and Collmer, A.J. (2007). Identification of harpins in *Pseudomonas syringae* pv. *tomato* DC3000, which are functionally similar to HrpK1 in promoting translocation of type III secretion system effectors. J. Bacteriol. 189, 8059–72.

Lan, L., Deng, X., Zhou, J., and Tang, X. (2006). Genome-wide gene expression analysis of *Pseudomonas syringae* pv. *tomato* DC3000 reveals overlapping and distinct pathways regulated by hrpL and hrpRS. Mol. Plant–Microbe Interact. 19, 976–87.

Lee, J., Klusener, B., Tsiamis, G., Stevens, C., Neyt, C., Tampakaki, A.P., Panopoulos, N.J., Nöller, J., Weiler, E.W., Cornelis, G.R., Mansfield, J.W., and Nürnberger, T. (2001). HrpZ(Psph) from the plant pathogen *Pseudomonas syringae* pv. *phaseolicola* binds to lipid bilayers and forms an ion-conducting pore *in vitro*. Proc. Natl. Acad. Sci. USA 98, 289–94.

Lee, C.C., Wood, M.D., Ng, K., Andersen, C.B., Liu, Y., Luginbühl, P., Spraggon, G., and Katagiri, Fumiaki. (2004). Crystal Structure of the Type III Effector AvrB from *Pseudomonas syringae*. Structure 12, 487–494.

Lindeberg, M., Stavrinides, J., Chang, J.H., Alfano, J.R., Collmer, A., Dangl, J.L., Greenberg, J.T., Mansfield, J.W., and Guttman, D.S. (2005). Proposed guidelines for a unified nomenclature and phylogenetic analysis of type III Hop effector proteins in the plant pathogen *Pseudomonas syringae*. Mol. Plant–Microbe Interact. 18, 275–82.

Lindeberg, M., Cartinhour, S., Myers, C.R., Schechter, L.M., Schneider, D.J., and Collmer, A. (2006). Closing the circle on the discovery of genes encoding Hrp regulon members and type III secretion system effectors in the genomes of three model *Pseudomonas syringae* strains. Mol. Plant–Microbe Interact. 19, 1151–8.

Lloyd, S.A., Norman, M., Rosqvist, R., and Wolf-Watz, H. (2001). *Yersinia* YopE is targeted for type III secretion by N-terminal, not mRNA, signals. Mol. Microbiol. 39, 520–31.

Mackey, D., Holt, B.F. 3rd., Wiig, A., and Dangl, J.L. (2002). RIN4 interacts with *Pseudomonas syringae* type III effector molecules and is required for RPM1-mediated resistance in Arabidopsis. Cell 108, 743–54.

Mansfield, J., Jenner, C., Hockenhull, R., Bennett, M.A., and Stewart, R. (1994). Characterization of avrPphE, a gene for cultivar-specific avirulence from *Pseudomonas syringae* pv. *phaseolicola* which is physically linked to hrpY, a new hrp gene identified in the halo-blight bacterium. Mol. Plant–Microbe Interact. 7, 726–39.

Morris, C.E., Kinkel, L.L., Xiao, K., Prior, P., and Sands, D.C. (2007). Surprising niche for the plant pathogen *Pseudomonas syringae*. Infect. Genet. Evol. 7, 84–92.

Müller, S.A., Pozidis, C., Stone, R., Meesters, C., Chami, M., Engel, A., Economou, A., and Stahlberg, H. (2006). Double hexameric ring assembly of the type III protein translocase ATPase HrcN. Mol. Microbiol. 61, 119–25.

Nguyen, L., Paulsen, I.T., Tchieu, J., Hueck, C.J., and Saier, M.H. Jr. (2000). Phylogenetic analyses of the constituents of Type III protein secretion systems. J. Mol. Microbiol. Biotechnol. 2, 125–44.

Nimchuck, Z.L., Fisher, E.J., Desveaux, D., Chang, J.H., and Dangl, J.L. (2007). The HopX (AvrPphE) family of *Pseudomonas syringae* type III effectors require a catalytic traid and a novel N-terminal Domain for function. Mol. Plant–Microbe Interact. 20, 346–357.

Oh, H.S., Kvitko, B.H., Morello, J.E., and Collmer, A. (2007). *Pseudomonas syringae* lytic transglycosylases coregulated with the type III secretion system contribute to the translocation of effector proteins into plant cells. J. Bacteriol. 189, 8277–89.

Palmer, T., and Berks, B.C. (2003). Moving folded proteins across the bacterial cell membrane. Microbiology. 149, 547–56.

Petnicki-Ocwieja, T., Schneider, D.J., Tam, V.C., Chancey, S.T., Shan, L., Jamir, Y., Schechter, L.M., Janes, M.D., Buell, C.R., Tang, X., Collmer, A., and Alfano, J.R. (2002). Genomewide identification of proteins secreted by the Hrp type III protein secretion system of *Pseudomonas syringae* pv. *tomato* DC3000. Proc. Natl. Acad. Sci. USA 99, 7652–7.

Petnicki-Ocwieja, T., van Dijk, K., and Alfano, J.R. (2005). The hrpK operon of *Pseudomonas syringae* pv. *tomato* DC3000 encodes two proteins secreted by the

Pitman, A.R., Jackson, R.W., Mansfield, J.W., Kaitell, V., Thwaites, R., and Arnold, D.L. (2005). Exposure to host resistance mechanisms drives evolution of bacterial virulence in plants. Curr. Biol. 15, 2230–5.

Pozidis, C., Chalkiadaki, A., Gomez-Serrano, A., Stahlberg, H., Brown, I., Tampakaki, A.P., Lustig, A., Sianidis, G., Politou, A.S., Engel, A., Panopoulos, N.J., Mansfield, J., Pugsley, A.P., Karamanou, S., and Economou, A. (2003). Type III protein translocase: HrcN is a peripheral ATPase that is activated by oligomerization. J. Biol. Chem. 278, 25816–24.

Preston, G., Deng, W.L., Huang, H.C., and Collmer, A. (1998). Negative regulation of hrp genes in Pseudomonas syringae by HrpV. J. Bacteriol. 180, 4532–7.

Preston, G.M., Studholme, D.J., and Caldelari, I. (2005). Profiling the secretomes of plant pathogenic Proteobacteria. FEMS Microbiol. Rev. 29, 331–60.

Pukatzki, S., Ma, A.T., Sturtevant, D., Krastins, B., Sarracino, D., Nelson, W.C., Heidelberg, J.F., and Mekalanos, J.J. (2006). Identification of a conserved bacterial protein secretion system in Vibrio cholerae using the Dictyostelium host model system. Proc. Natl. Acad. Sci. USA. 103, 1528–1533.

Quigley, N.B., Mo, Y.Y., and Gross, D.C. (1993). SyrD is required for syringomycin production by Pseudomonas syringae pathovar syringae and is related to a family of ATP-binding secretion proteins. Mol. Microbiol. 9, 787–801.

Rahme, L.G., Mindrinos, M.N., and Panopoulos, N.J. (1992). Plant and environmental sensory signals control the expression of hrp genes in Pseudomonas syringae pv. phaseolicola. J. Bacteriol. 174. 3499–507.

Ramos, A.R., Morello, J.E., Ravindran, S., Deng, W.L., Huang, H.C., and Collmer, A. (2007). Identification of Pseudomonas syringae pv. syringae 61 type III secretion system Hrp proteins that can travel the type III pathway and contribute to the translocation of effector proteins into plant cells. J. Bacteriol. 189, 5773–8.

Rico, A., and Preston, G.M. (2008). Pseudomonas syringae pv. tomato DC3000 Uses Constitutive and Apoplast-Induced Nutrient Assimilation Pathways to Catabolize Nutrients That Are Abundant in the Tomato Apoplast. Mol. Plant–Microbe Interact. 21, 269–82.

Rivas, L.A., Mansfield, J., Tsiamis, G., Jackson R.W., and Murillo, J. (2005). Changes in race-specific virulence in Pseudomonas syringae pv. phaseolicola are associated with a chimeric transposable element and rare deletion events in a plasmid-borne pathogenicity island. Appl. Environ. Microbiol. 71, 3778–85.

Roine, E., Saarinen, J., Kalkkinen, N., and Romantschuk, M. (1997a). Purified HrpA of Pseudomonas syringae pv. tomato DC3000 reassembles into pili. FEBS Lett. 417, 168–72.

Roine, E., Wei, W., Yuan, J., Nurmiaho-Lassila, E.L., Kalkkinen, N., Romantschuk, M., and He, S.Y. (1997b). Hrp pilus: an hrp-dependent bacterial surface appendage produced by Pseudomonas syringae pv. tomato DC3000. Proc. Natl. Acad. Sci. USA. 94, 3459–64.

Rosebrock, T.R., Zeng, L., Brady, J.J., Abramovitch, R.B., Xiao, F., and Martin, G.B. (2007). A bacterial E3 ubiquitin ligase targets a host protein kinase to disrupt plant immunity. Nature 448, 370–375.

Russel, M. (1998). Macromolecular assembly and secretion across the bacterial cell envelope: type II protein secretion systems. J. Mol. Biol. 279, 485–99.

Sawada, H., Suzuki, F., Matsuda, I., and Saitou, N. (1999). Phylogenetic analysis of Pseudomonas syringae pathovars suggests the horizontal gene transfer of argK and the evolutionary stability of hrp gene cluster. J. Mol. Evol. 49, 627–44.

Schechter, L.M., Roberts, K.A., Jamir, Y., Alfano, J.R., and Collmer, A. (2004). Pseudomonas syringae type III secretion system targeting signals and novel effectors studied with a Cya translocation reporter. J. Bacteriol. 186, 543–55.

Schechter, L.M., Vencato, M., Jordan, K.L., Schneider, S.E., Schneider, D.J., and Collmer, A. (2006). Multiple approaches to a complete inventory of Pseudomonas syringae pv. tomato DC3000 type III secretion system effector proteins. Mol. Plant–Microbe Interact. 19, 1180–92.

Schmitt, L., and Tampé, R. (2002). Structure and mechanism of ABC transporters. Curr. Opin. Struct. Biol. 6, 754–60.

Shan, L., Oh, H.S., Chen, J., Guo, M., Zhou, J., Alfano, J.R., Collmer, A., Jia, X., and Tang, X. (2004). The HopPtoF locus of Pseudomonas syringae pv. tomato DC3000 encodes a type III chaperone and a cognate effector. Mol. Plant–Microbe Interact. 17, 447–55.

Shao, F., Merritt, P.M., Bao, Z., Innes, R.W., and Dixon, J.E. (2002). A Yersinia effector and a Pseudomonas avirulence protein define a family of cysteine proteases functioning in bacterial pathogenesis. Cell 109, 575–88.

Singer, A.U., Desveaux, D., Betts, L., Chang, J.H., Nimchuk, Z., Grant, S.R., Dangl, J.L., and Sondek, J. (2004). Crystal structures of the type III effector protein AvrPphF and its chaperone reveal residues required for plant pathogenesis. Structure 12, 1669–81.

Staskawicz, B.J., and Panopoulos, N.J. (1980). Phaseolotoxin transport in Escherichia coli and Salmonella typhimurium via the oligopeptide permease. J. Bacteriol. 142, 474–9.

Staskawicz, B.J., Dahlbeck, D., and Keen, N.T. (1984). Cloned avirulence gene of Pseudomonas syringae pv. glycinea determines race-specific incompatibility on Glycine max (L.) Merr. Proc. Natl. Acad. Sci. USA 81, 6024–6028.

Stavrinides, J., and Guttman, D.S. (2004). Nucleotide sequence and evolution of the five-plasmid complement of the phytopathogen Pseudomonas syringae pv. maculicola ES4326. J. Bacteriol. 186, 5101–15.

Stavrinides, J., and Ma, W., and Guttman, D.S. (2006). Terminal reassortment drives the quantum evolution of type III effectors in bacterial pathogens. PLoS Pathog. 2, e104.

Stebbins, C.E., and Galán, J.E. (2001). Structural mimicry in bacterial virulence. Nature 412, 701–705.

Stevens, C., Bennett, M.A., Athanassopoulos, E., Tsiamis, G., Taylor, J.D., and Mansfield, J.W. (1998). Sequence variations in alleles of the avirulence gene *avrPphE.R2* from *Pseudomonas syringae* pv. *phaseolicola* lead to loss of recognition of the AvrPphE protein within bean cells and a gain in cultivar-specific virulence. Mol. Microbiol. *29*, 165–77.

Szurek, B., Rossier, O., Hause, G., and Bonas, U. (2002). Type III dependent translocation of the *Xanthomonas* AvrBs3 protein into the plant cell. Mol Microbiol. *46*, 13–23.

Tang, X., Xiao, Y., and Zhou, J.M. (2006). Regulation of the type III secretion system in phytopathogenic bacteria. Mol. Plant–Microbe Interact. *19*, 1159–66.

Tsiamis, G., Mansfield, J.W., Hockenhull, R., Jackson, R.W., Sesma, A., Athanassopoulos, E., Bennett, M.A., Stevens, C., Vivian, A., Taylor, J.D., and Murillo, J. (2000). Cultivar-specific avirulence and virulence functions assigned to *avrPphF* in *Pseudomonas syringae* pv. *phaseolicola*, the cause of bean halo-blight disease. EMBO J. *19*, 3204–14.

van der Meer., J.R., and Sentchilo, V. (2003). Genomic islands and the evolution of catabolic pathways in bacteria. Curr. Opin. Biotechnol. *14*, 248–54.

van Dijk, K., Fouts, D.E., Rehm, A.H., Hill, A.R., Collmer, A., and Alfano, J.R. (1999). The Avr (effector) proteins HrmA (HopPsyA) and AvrPto are secreted in culture from *Pseudomonas syringae* pathovars via the Hrp (type III) protein secretion system in a temperature- and pH-sensitive manner. J. Bacteriol. *181*, 4790–7.

van Dijk, K., Tam, V.C., Records, A.R., Petnicki-Ocwieja, T., and Alfano, J.R. (2002). The ShcA protein is a molecular chaperone that assists in the secretion of the HopPsyA effector from the type III (Hrp) protein secretion system of *Pseudomonas syringae*. Mol. Microbiol. *44*, 1469–81.

Vinatzer, B.A., Teitzel, G.M., Lee, M.W., Jelenska, J., Hotton, S., Fairfax, K., Jenrette, J., and Greenberg, J.T. (2006). The type III effector repertoire of *Pseudomonas syringae* pv. *syringae* B728a and its role in survival and disease on host and non-host plants. Mol. Microbiol. *62*, 26–44.

Vivian, A. and Arnold, D.L. (2000). Bacterial effector genes and their role in host–pathogen interactions. J. Plant Pathol. *82*, 163–178.

Voulhoux, R., Ball, G., Ize, B., Vasil, M.L., Lazdunski, A., Wu, L.F., and Filloux, A. (2001). Involvement of the twin-arginine translocation system in protein secretion via the type II pathway. EMBO J. *20*, 6735–41.

Wehling, M.D., Guo, M., Fu, Z.Q., and Alfano, J.R. (2004). The *Pseudomonas syringae* HopPtoV protein is secreted in culture and translocated into plant cells via the type III protein secretion system in a manner dependent on the ShcV type III chaperone. J. Bacteriol. *186*, 3621–30.

Wei, C.F., Deng, W.L., and Huang, H.C. (2005). A chaperone-like HrpG protein acts as a suppressor of HrpV in regulation of the *Pseudomonas syringae* pv. *syringae* type III secretion system. Mol. Microbiol. *57*, 520–36.

Xiao, Y., Lu, Y., Heu, S., Hutcheson, S.W. (1992). Organization and environmental regulation of the *Pseudomonas syringae* pv. *syringae* 61 hrp cluster. J. Bacteriol. *174*, 1734–41.

Zhao, Y., Ma, Z., and Sundin, G.W. (2005). Comparative genomic analysis of the pPT23A plasmid family of *Pseudomonas syringae*. J. Bacteriol. *187*, 2113–26.

Zhu, M., Shao, F., Innes, R.W., Dixon, J.E., and Xu, Z. (2004). The crystal structure of Pseudomonas avirulence protein AvrPphB: a papain-like fold with a distinct substrate-binding site. Proc. Natl. Acad. Sci. USA. *101*, 302–7.

Zwiesler-Vollick, J., Plovanich-Jones, A.E., Nomura, K., Bandyopadhyay, S., Joardar, V., Kunkel, B.N., and He, S.Y. (2002). Identification of novel hrp-regulated genes through functional genomic analysis of the *Pseudomonas syringae* pv. *tomato* DC3000 genome. Mol. Microbiol. *45*, 1207–18.

MAMPs/PAMPs – Elicitors of Innate Immunity in Plants

Gitte Erbs and Mari-Anne Newman

Abstract

Plants perceive several general elicitors from both host and non-host pathogens. These elicitors are essential structures for pathogen survival and are for that reason conserved among pathogens. These conserved microbe-specific molecules, also referred to as microbe- or pathogen-associated molecular patterns (MAMPs or PAMPs), are recognized by the plant innate immune systems pattern recognition receptors (PRRs). General bacterial elicitors, like lipopolysaccharides (LPS), flagellin (Flg), elongation factor Tu (EF-Tu), cold shock protein (CSP), peptidoglycan (PGN) and the enzyme superoxide dismutase (SodM) are known to act as MAMPs and induce immune responses in plants or plant cells (Gómez-Gómez and Boller, 2000; Erbs and Newman, 2003; Felix and Boller, 2003; Kunze et al., 2004; Watt et al., 2006, Gust et al., 2007; Erbs et al., 2008). The corresponding PRRs for some of these bacterial elicitors have, in recent years, been identified. Here, the current knowledge regarding bacterial elicitors of innate immunity in plants is presented.

Introduction

In an environment that is rich in potentially pathogenic microorganisms, the survival of higher eukaryotic organisms depends on efficient pathogen sensing and rapidly mounted defence responses. Such protective mechanisms are found in all multicellular organisms and are collectively referred to as innate immunity. Innate immunity is the first line of defence against invading microorganisms in vertebrates and the only line of defence in invertebrates and plants (van Baarlen et al., 2007). Plants interact with a variety of microorganisms, and like insects and mammals, they respond to a range of microbial molecules. The recognition of non-self induces plant defence responses such as the oxidative burst, nitric oxide generation, extracellular pH increase, cell-wall strengthening and pathogenesis-related (PR) protein accumulation, leading to basal resistance or innate immunity. Recognition of non-self, i.e. an invading pathogen, is crucial for an effective defence response.

Microbe-associated molecular patterns (MAMPs)

As a first line of defence, the plant innate immune system recognizes conserved microbe specific molecules, also referred to as microbe- or pathogen-associated molecular patterns (MAMPs or PAMPs), by pattern recognition receptors (PRRs). Hereafter, the term MAMP will be used, as MAMPs refer to molecules from both pathogenic and non-pathogenic microbes. MAMPs have been defined as evolutionarily conserved microbe-derived molecules that distinguish hosts from pathogens (Janeway, 1992; Ausubel, 2005).

As the plant senses the presence of a bacterium it activates defence mechanisms, regardless of whether the microbe is able to colonize the plant or not. Innate defence mechanisms in plants have been shown to have similarity to the innate defence system known in mammals and insects (Nürnberger and Brunner, 2002). The innate immunity system in mammals perceives invading pathogens through Toll-like receptors (TLRs), an interleukin 1 receptor (IL-1R) (Medzhitov et al., 1997), that resembles the Toll receptor found in Drosophila (Hashimoto et al., 1988; Lemaitre et al., 1996). The TLRs, one class of PRRs, comprise a family of transmembrane receptors that have an extracellular leucine-rich repeat (LRR) domain, by which pathogens are recognized, and a cytoplasmic Toll/IL-1R (TIR) domain, through which the MAMP signal is transduced (Fig. 11.1). Exactly how this cascade is initiated is not completely understood. Once the TLRs are activated by MAMP recognition, adaptor molecules are recruited to initiate downstream signalling, that is, activation of transcription factors and MAP kinases (reviewed by Carpenter and O'Neill, 2007). Several mammalian TLRs are known: for instance, TLR4 recognizes lipopolysaccharides (LPS), TLR5 recognizes flagellin (reviewed by Akira and Takeda, 2004) and TLR3 has been found to mediate the recognition of double-stranded viral RNA (Alexopoulou et al., 2001). Another class of PRRs is the nucleotide-binding oligomerization domain (NOD)-like receptors (NLRs) that recognize microbial components present in the host cytosol (Strober et al., 2006). These intracellular receptors, NLRs, have a pyrin domain (PYD) or a caspase-recruitment domain (CARD), a nucleotide binding site (NBS) and a LRR domain (reviewed by Creagh and O'Neill, 2006) (Fig. 11.1). In contrast to the mammalian and insect systems, the recognition of MAMPs and the triggering of innate immunity are far less studied in plants.

Plants perceive various MAMPs, from a range of pathogens including oomycetes, fungi, viruses and bacteria. Examples include the necro-

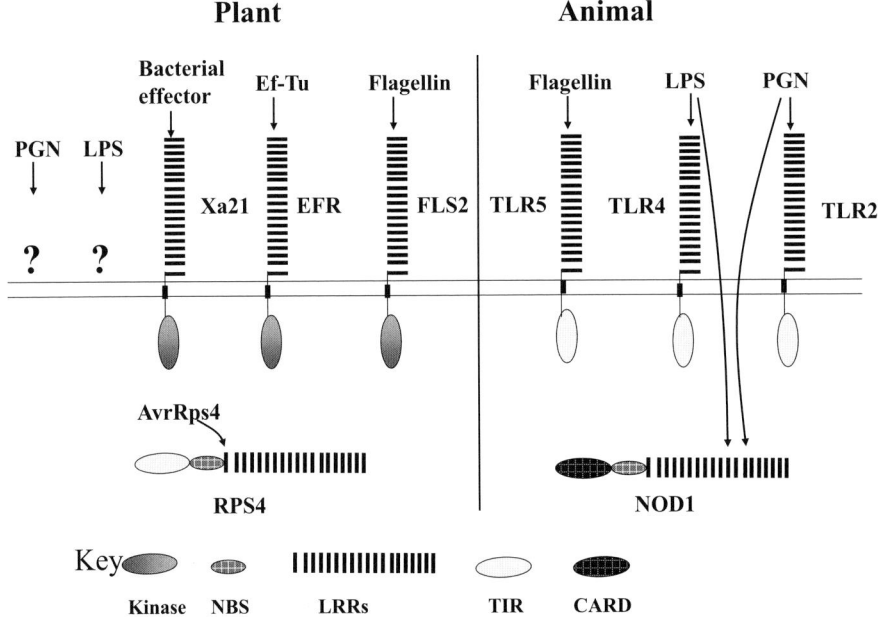

Figure 11.1 Schematic representation of pattern recognition receptors (PRRs) involved in microbe sensing in plant and animals. In animals (right-hand side), the receptor kinases (RKs) TLR5 and TLR4, with TIR-LRR domains, recognize flagellin and lipopolysaccharide (LPS), respectively, and the RK TLR2 is thought to recognize peptidoglycan (PGN). In addition, NOD factors, with CARD-NBS-LRRs domains, have been shown to recognize both LPS and PGN (reviewed by Akira and Takeda, 2004). In plants (left-hand side), the *Arabidopsis* LRR-RLK FLS2 recognizes flagellin (Gómez-Gómez and Boller, 2000), and the *Arabidopsis* LRR-RLK EFR recognizes elongation factor EF-Tu (Zipfel et al., 2006). No PRRs for LPS and PGN have been characterized in plants yet. An uncharacterized effector molecule from *Xanthomonas oryzae* pv. *oryzae* is recognized through the rice LRR-RLK Xa21 (Song et al., 1995), and the intracellular resistance protein RPS4, with TIR-NBS-LRR domains, recognizes the type III secreted bacterial effector AvrRPS4 (Gassmann et al., 1999). Abbreviations: Kinase, serine-threonine kinase; NBS, nucleotide binding/apoptotic ATPase domain; LRRs, leucine-rich repeats; TIR, Toll-interleukin-1 receptor domain; CARD, caspase-recruitment domain. The figure is adapted and modified after Newman et al. (2007).

sis-inducing *Phytophthora* protein 1 (NPP1) from *P. parasitica* (Fellbrich *et al.*, 2002), fungal chitin, β-glucan from *Phytophthora megasperma*, and from bacteria lipopolysaccharides (LPS), flagellin, the elongation factor Tu (EF-Tu), the cold shock protein (CSP), peptidoglycan (PGN) and the enzyme superoxide dismutase (SodM) (Gómez-Gómez and Boller, 2000; Erbs and Newman, 2003; Felix and Boller, 2003; Kunze *et al.*, 2004; Watt *et al.*, 2006; Erbs *et al.*, 2008). However, corresponding PRRs have only been identified in a few cases: the soybean β-glucan binding protein (GBP) recognizes *P. megasperma* β-glucan (Umemoto *et al.*, 1997); the tomato EIX1 and EIX2 proteins bind *Trichoderma viride* ethylene-inducing xylanase (EIX) (Ron and Avni, 2004); the chitin oligosaccharide elicitor binding protein (CEBiP) recognizes fungal chitin (Kaku *et al.*, 2006); the *Arabidopsis thaliana* Elongation factor Tu (EF-Tu) receptor (EFR) recognizes EF-Tu (Zipfel *et al.*, 2006); and FLS2 and LeFLS2 (Gómez-Gómez and Boller, 2000; Robatzek *et al.*, 2007), which recognize flagellin in *Arabidopsis* and tomato, respectively. No receptor for LPS or PGN has, so far, been found in plants. Pathogens carrying specific effector molecules can suppress the host's basal defence responses or immune responses, leading to disease. These effector proteins, originally referred to as avirulence (Avr) proteins (Jones and Dangl, 2006), are injected into the host plant by the bacterial type III secretion system (TTSS) (reviewed by Alfano and Collmer, 2004) and modulate plant physiology to promote the growth and dissemination of the pathogen. Effectors are recognized by specialized resistance (*R*) genes in the plant (gene-for-gene resistance), activating plant defences. With time, pathogens evolve to alter their recognizable effector molecules to regain the lost virulence on the host plant (reviewed by Bent and Mackey, 2007; see also chapters by Stavrinides, and Boch).

Plant perception of MAMPs

Lipopolysaccharides (LPS)
Lipopolysaccharides (LPS), the major component of the outer membrane of Gram-negative bacteria, have been shown to have multiple roles in plant–microbe interactions (reviewed by Newman *et al.*, 2007). LPS is thought to contribute to the restrictive Gram-negative membrane permeability, allowing bacterial growth in unfavourable environments. LPS and its derivatives act as MAMPs and induce innate immune responses in plants (Newman *et al.*, 1995; Dow *et al.*, 2000; Bedini *et al.*, 2005; Silipo *et al.*, 2005). Earlier studies in plants have shown that LPS can prevent the hypersensitive response (HR) induced by avirulent bacteria. The HR, which is characterized by localized cell necrosis at the infection site, is the triggering of gene-for-gene resistance in plants caused by invading pathogens carrying avirulence (*avr*) genes that are recognized by corresponding *R* genes in the plant (reviewed by Greenberg, 1997). The mechanisms behind HR prevention are still unknown, but the effects of LPS pre-treatment are considered to be associated with enhanced resistance of the plant tissue to pathogenic bacteria, which is thought to occur through a LOS-dependent potentiation of plant defence responses (Newman *et al.*, 2002).

LPS consists of a lipid, a core oligosaccharide and an O-polysaccharide part. The lipid, referred to as lipid A, is embedded in the outer part of the phospholipid bilayer. Lipid A and the core oligosaccharide are linked, usually by the sugar 3-deoxy-D-manno-2-octulosonate (KDO). The core oligosaccharide consists of a short series of sugars and ends in the O-antigen, which is composed of repeating oligosaccharide units (Raetz and Whitfield, 2002). The O-antigen of the LPS from many phytopathogenic bacteria has shown to consist of oligorhamnans (Bedini *et al.*, 2002). In order to know more about the structures within LPS that trigger immune responses in plants, synthetic O-antigen polysaccharides, oligorhamnans of increasing chain lengths, were tested in *Arabidopsis*. Tri-, hexa- and nonasaccharides were synthesized and found to suppress the HR, as well as act as MAMPs and elicit the induction of the pathogenesis-related genes *PR1* and *PR2* in *Arabidopsis*. The efficiency of HR suppression and *PR* gene induction improved with increasing chain lengths of sugars in the synthetic O-antigen. In addition, a coiled structure was observed with the increasing chain length, indicating a role for this structure as a MAMP and by correlation a role for the O-antigen from phytopathogenic bacteria in plant innate immunity (Bedini et al., 2005).

Studies in mammalian cells have shown that LPS is recognized through their lipid A moiety and this recognition was shown to govern the interactions with the innate immune system (Loppnow et al., 1989). In addition to this, the molecular shape of lipid A was found to directly correlate with its activity as a conical shape of lipid A was associated with endotoxicity and a cylindrical shape with antagonistic activity. A net negative charge of lipid A was found to influence its molecular conformation, and with that, its biological activity (Schromm et al., 1998; Schromm et al., 2000). To study if the innate immune system from the mammalian system has parallels in the plant system, the role and mechanisms of action of LPS and its derivatives, the core oligosaccharide and the lipid A moiety, in plant-bacteria interactions were investigated in *Arabidopsis*. Initially, the complete structure of purified *Xanthomonas campestris* pathovar (pv.) *campestris* (*Xcc*) lipo-oligosaccharides (LOS), LPS without the O-chain, was determined. *Xcc* LOS was found to be a unique molecule with a high negative charge density and a phosphoramide group never found in such molecules before (Silipo et al., 2005). *Xcc* LOS and derivatives have been shown to elicit induction of the pathogenesis-related *PR* genes in *Arabidopsis*. LOS was found to induce the defence-related *PR1* and *PR2* genes in two temporal phases: the core oligosaccharide induced only the early phase and the lipid A moiety only the later phase, which suggests that both the core oligosaccharide and the lipid A are recognized by plant cells, e.g. both act as elicitors. These findings support the role of *Xcc* lipid A and the *Xcc* core oligosaccharide as MAMPs of innate immunity in plants. Silipo and co-workers (2005) speculated that the different LPS fragments are recognized by different plant receptors. The fact that *Xcc* lipid A has been shown to act as an elicitor of innate immunity in plants correlates very well with earlier findings where lipid A from various bacteria was found to induce a rapid burst of nitric oxide (NO) production and with that the induction of defence-relates genes in *Arabidopsis* (Zeidler et al., 2004). Contrary to this, studies in tobacco cells, have shown that neither the lipid A nor the O-chain of the *Xcc* LPS molecule could induce the oxidative burst alone, but rather it was the inner core part of the LPS molecule that was responsible (Braun et al., 2005). The conflict in results could reflect the different defence responses measured after treatment with LPS and its derivatives in different plants. Pre-treatment of *Arabidopsis* leaves with LOS and its derivatives was found to prevent the HR caused by avirulent strains of *Pseudomonas syringae* pv. *tomato* (*Pst*) carrying the *avrRpm1* or the *avrRps4* genes, a phenomenon referred to as localized induced response (LIR) (Newman et al., 2002; Silipo et al., 2005). In correlation to studies in the mammalian system, where it is well established that the phosphorylation pattern of lipid A affects its biological activity (reviewed by Gutsmann et al., 2007), it was tested whether de-phosphorylated *Xcc* LOS could be recognized in plants. After de-phosphorylation of *Xcc* LOS the molecule maintained only the negative charge of the KDO residue, and rendered the molecule unable to induce LIR, suggesting that the charged groups present in LOS play a key role in inducing defence responses in *Arabidopsis* (Silipo et al., 2005). Furthermore, from these experiments it could be concluded that the electrostatic interactions involving the phosphate groups seem to have a crucial function in binding not only lipid A, but also the core oligosaccharide, to putative receptors in plants (Silipo et al., 2005). The core oligosaccharide from *Escherichia coli* and *Ralstonia solanacearum* have previously been shown not to prevent the HR or induce defence-related genes (Newman et al., 1997), indicating that the effect of the *Xcc* core oligosaccharide could be due to its unique phosphoramide group (Silipo et al., 2005). LPS has been found, not only to induce defence responses, but also to prime expression of plant defence responses upon subsequent bacterial inoculation, e.g. promote an early triggering of the synthesis of the antimicrobial compounds feruloyl tyramine (FT) and p-coumaroyl tyramine (CT) (Newman et al., 2001; Newman et al., 2002; Prime-A-Plant Group, 2006).

The O-antigen part of the LPS molecule is thought to be responsible for ISR (induced systemic resistance) in *Arabidopsis*. Early studies showed that LPS from the rhizobacteria *Pseudomonas fluorescens*, as well as the live bacteria, induced ISR in carnation and radish, whereas mutant bacteria, lacking the O-antigen side chain could not induce ISR (Leeman et al., 1995; van Loon et al., 1998). In contrast to the rhizobacteria-mediated ISR, systemic activation of defence-related responses in plants upon local necrotizing pathogen infection is referred to as systemic acquired resistance (SAR). SAR is accompanied by a systemic increase in salicylic acid (SA), and SA is required for SAR signalling (Ryals et al., 1996; Schneider et al., 1996). However, recent studies suggest that recognition of the MAMPs, LPS or flagellin, and not necrotic lesion formations contribute to the bacterial induction of SAR in *Arabidopsis*. Treatment of *Arabidopsis* with *Pseudomonas aeruginosa* LPS, flagellin or non-host bacteria were shown to be associated with accumulation of SA, expression of the *PR* genes and expression of the SAR marker gene *Flavin-dependent monooxygenase 1* in treated as well as in distant leaves (Mishina and Zeier, 2006; Mishina and Zeier, 2007).

Until recently, the activity of LPS in plants has only been described in dicots, but current studies in rice cells have revealed that LPS, from various pathogenic and non-pathogenic bacteria, induce a generation of reactive oxygen species and defence-related gene expression in monocots, indicating that the machinery recognizing LPS is evolutionary conserved in monocots and dicots (Desaki et al., 2006). Furthermore, the two MAMPs, LPS and chitin oligosaccharide, induced a close correlation of genes in rice cells, indicating a convergence in signalling cascades downstream of recognition. In addition, the effect of LPS from various bacteria was shown to be associated with a programmed cell death (PCD) in rice cells. In contrast, LPS has never been shown to elicit PCD in dicots (Desaki et al., 2006).

The mechanism by which LPS is perceived by plants is still not understood. Recent studies with fluorescein-labelled *Xcc* LPS in cultured *Nicotiana tabacum* cells revealed that LPS was rapidly bound to the cell wall and then internalized into the cell, and eventually, LPS was found exclusively inside the vacuole. These findings suggests endocytosis, comparable to the mammalian system, of *Xcc* LPS in tobacco cells (Gross et al., 2005). However, no PRRs for LPS and its derivatives have been characterized in plants. In the mammalian immune system, LPS form complexes with LPS-binding proteins (LBP), and this LPS–LBP complex is recognized by the membrane-bound CD14 receptor, a glycosylphosphatidyl inositol (GPI)-anchored glycoprotein (Wright et al., 1990), which again is thought to associate with TLR4-MD2 to participate in LPS-induced signalling (Jiang et al., 2000; Miyake, 2004). The mammalian system is able to sense very low amounts of LPS (Miyake, 2004); however, a higher level of LPS is required to trigger defence responses in plants, so this may indicate another perception system for LPS and its derivatives in plants than the TLRs.

Peptidoglycan (PGN)

Peptidoglycan (PGN) provides rigidity and structure to the bacterial cell wall in both Gram-negative and Gram-positive bacteria. PGN is found as a thick layer in the cell wall of Gram-positive bacteria, whereas only a thin layer is present in the cell wall of Gram-negative bacteria. The carbohydrate backbone of PGN is conserved in all bacteria, while the peptide displays diversity. Several types of PGN, classified by the nature of the third residue of the stem peptide are commonly found. Typically, this is m-diaminopimelic acid (mDAP) PGN in Gram-negative bacteria and in some Gram-positive bacilli (genus *Bacillus* and *Clostridium*), whereas most other Gram-positive bacteria have L-lysine (LYS) PGN. PGN, a molecule never found in eukaryotes, is an essential and unique cell wall component of all bacteria, making it an excellent target for the eukaryotic innate immune system (reviewed by Dziarski and Gupta, 2005a; McDonald et al., 2005).

Early experiments with plant cells showed that the Gram-positive human pathogen *Staphylococcus aureus* PGN was active as an elicitor in inducing extracellular alkalization of cultured tobacco cells, while no response was observed in cultured tomato cells, indicating a different perception system for PGN within the Solanaceae (Felix and Boller, 2003). Surprisingly, a very recent plant study examining *Arabidopsis* challenged with *S. aureus*, a human pathogen, PGN showed that the PGN sugar backbone was responsible for triggering immune responses

(Gust et al., 2007), and not the breakdown product of PGN, the muramyl dipeptide or the muropeptide dimer, which is known to be the minimal chemical structure required for triggering the innate immune system in vertebrates and insects (reviewed by Traub et al., 2006). In fact, the vast majority of plant pathogenic bacteria are Gram-negatives. Using two Gram-negative pathogens we have shown that PGN from *Xcc* and *Agrobacterium tumefaciens* (*At*) act as elicitors and trigger the innate immune response in *Arabidopsis*. Muropeptides were significantly more effective than the intact PGN molecule. PGN and fragments from *Xcc* were more potent defence elicitors than from *At*, possibly reflecting the biotrophic mode of parasitism of the latter (Erbs et al., 2008).

The apparent difference in the minimal chemical structure of PGN required to induce innate immunity in *Arabidopsis* is intriguing. Further research should distinguish if plants have a perception system for PGN which distinguishes between the PGN from Gram-positive and Gram-negative pathogens, or different perception systems for plant pathogens and non-plant pathogens (i.e. PGN from the human pathogen *Staphylococcus aureus*). For instance, the ability of *Drosophila* to distinguish Gram-positive from Gram-negative bacteria is based on recognition of specific forms of PGN, and not as earlier thought, on the detection of LPS (Leulier et al., 2003).

Thus far, no PGN recognition receptors have been identified in plants. In animals, several PGN recognition molecules are known, including CD14 (Dziarski et al., 1998) which is also known to bind the LPS-LBP complex, the NOD containing proteins (Franchi et al., 2006), the TLR2, even though this is controversial (Dziarski et al., 2005b; Travassos et al., 2004), and the PGN recognition proteins (PGRPs) (reviewed by Guan and Mariuzza, 2007).

Flagellin

Flagella are essential structures for the pathogenic bacteria as they provide motility and often increase adhesion of the bacteria to its host. Flagellin, the main building block of bacterial flagella, is well established as a major activator of innate immunity in animals (reviewed by Ramos et al., 2004). Some of the first MAMP recognition studies in plants were done using flagellin. Studies in mammals have shown that at least one of the conserved domains in the N-terminal and C-terminal part of the bacterial flagellin, found to be involved in bacterial motility as well, is recognized by Toll-like receptor 5 (TLR5) (Hayashi et al., 2001; Smith et al., 2003). Studies in *Arabidopsis*, tomato and other plants, revealed that plants respond to a highly conserved domain in the N terminal part of the bacterial flagellin, a 22 amino acid peptide, flg22 (Felix et al., 1999). In order to identify the gene involved in recognition and transduction of the flg22 elicitor signal, Gómez-Gómez and Boller (2000), used a genetic approach to screen *Arabidopsis* mutants after flg22 treatment and isolated several *flagellin sensing 2* (*FLS2*) mutants, which mapped to the *FLS2* locus on chromosome 5. The *Arabidopsis* PRR FLS2 was identified.

FLS2 belongs to the receptor-like kinase (RLK) family and has an extracellular domain with 28 LRRs, a transmembrane domain and an intracellular serine/threonine kinase domain. No high-affinity binding site was found, after treatment with a radiolabelled derivative of flg22, in the flagellin insensitive *Arabidopsis* ecotype Ws-0 and in plants carrying mutations in the LRR domain of the *FLS2* gene, indicating a role for LRR in flagellin binding (Gómez-Gómez and Boller, 2000; Bauer et al., 2001). Later work revealed that both an extracellular LRR domain and kinase activity of FLS2 were necessary for high affinity binding and binding specificity for flagellin (Gómez-Gómez et al., 2001). Based on immunoprecipitation and chemical cross-linking, Chinchilla et al. (2006) showed the specific interaction of flg22 with FLS2 in *Arabidopsis*. The perception of flagellin in tomato cells differs from that of *Arabidopsis*. Flg15, an N-terminally shortened version of flg22, was shown to be highly active in Solanaceae (tomato), while it only elicits immune responses at higher concentrations in Brassicaceae (*Arabidopsis*). The functionality of the FLS2 receptor was tested by heterologous expression of the *Arabidopsis* FLS2 receptor in tomato cells. In these expression studies, tomato cells gained the flagellin perception system characteristic for *Arabidopsis*, demonstrating that FLS2 represents the PRR that determines the specificity of flagellin perception (Meindl et al., 2000; Bauer et al., 2001; Chinchilla et al., 2006). The difference in recognition of the flagellin epitope is not restricted to different plant families; variations have also been found between species in the same family. A 15 amino acid peptide derived from *E. coli* flagellin was shown only to be highly active in *Lycopersicum esculentum* (tomato), but not in *Nicotiana benthamiana* (tobacco). Furthermore, the tomato flagellin receptor, *Lycopersicum esculentum* FLS2 (LeFLS2), an orthologue of the *Arabidopsis* FLS2 receptor, has now been identified and used in expression studies with *N. benthamiana*, where *N. benthamiana* expressing LeFLS2 gained the flagellin perception system specific for tomato (Robatzek et al., 2007). In addition to this, studies focusing on host recognition of *Xcc* flagellin have revealed within-species and within-pathovar variations for defence-eliciting activity of flagellins among *Xcc* strains (Sun et al., 2006).

Confirmation of FLS2 as a surface receptor came with studies using transgenic Ws-0, expressing FLS2 fused to the green fluorescent protein (GFP), which revealed a cell membrane localization of FLS2. Additionally, FLS2 was found to undergo ligand-induced endocytosis; it is thought that this subcellular redistribution of FLS2, or any other surface receptor, from the plasma membrane to cytoplasmic vesicles may be a central point in signalling during immune responses (McCoy et al., 2004; Robatzek et al., 2006).

Flagellin induced activation of FLS2 in *Arabidopsis*, involves a complex formation with the Brassinosteroid-Insensitive 1 (BRI1)-associated receptor kinase 1 (BAK1) (Chinchilla et al., 2007). BAK1 has also been reported to be involved in BRI1 endocytosis (Russinova et al., 2004). *bak1* mutants showed a delayed and reduced oxidative burst in response to treatment with flg22 or elf26, a fully active EF-Tu derivative (see text below), compared to wild type plants. Furthermore, the activities of MAP kinases (MAPK) were delayed and reduced or even absent in response to flg22 or elf26 treatment in *bak1* mutants, indicating that BAK1 could act as a positive regulator of MAMP signalling in *Arabidopsis*. In addition, it was revealed that FLS2 interacts with BAK1 in a ligand-dependent manner (Chinchilla et al., 2007). This is one of the first steps in trying to uncover the link between receptor action and intracellular signal transduction in plants. This elicitor-induced complex, which is known to initiate defence responses, is also known from the mammalian LPS recognition system.

Elongation factor Tu (EF-Tu)

In protein biosynthesis, the ribosome translates the sequence of nucleotides in mRNA into the sequence of amino acids in a protein. During the phase of elongation the ribosome is associated by elongation factors. One such elongation factor is the translation elongation factor, EF-Tu, the most abundant protein in the bacterial cell (Jeppesen et al., 2005). So far, EF-Tu has only been found to elicit innate immunity in members of the Brassicaceae family (Zipfel et al., 2006). EF-TU is not freely exposed to the cell surface, so how and why EF-Tu is released from the bacterial cell is still not known. It is speculated that the MAMP could be released by a bacterial export system during the infection process or by leakiness of the infecting bacteria caused by lytic enzymes of the host (Kunze et al., 2004).

The elicitor activity is attributed to a highly conserved part of the N-terminus of EF-Tu, a 26 amino acid peptide, elf26. The perception of EF-Tu is independent of flagellin perception, as EF-Tu is active in plants carrying mutations in FLS2 (Kunze et al., 2004). Studies using cross-linking assays in *Arabidopsis* cells, confirmed that elf26, and flg22 bind to different high-affinity binding receptors. Nevertheless, elf26 and flg22 were found by microarray analysis to induce the same pool of genes, and also a common set of responses in *Arabidopsis*. In addition to this, a combined treatment with both MAMPs, elf26 and flg22, was shown to induce the same kinases without an additive effect (Zipfel et al., 2006). Using a growth inhibition assay, an *efr-1* mutant was identified. This mutant was insensitive to EF-Tu, and did not respond with an oxidative burst, increased ethylene biosynthesis or induced resistance to *Pst* DC3000 in response to EF-Tu-derived elicitors, whereas *Arabidopsis* Col-0 and the *fls2* mutant did respond to EF-Tu elicitors. The EF-Tu receptor, EFR, was identified. Heterologous expression studies of EFR in *N. benthamiana*, a plant lacking a perception system for EF-Tu (Kunze *et al.*, 2004), resulted in *N. benthamiana* with a perception system for EF-Tu, confirming the role for EFR as a functional receptor for EF-Tu (Zipfel *et al.*, 2006). In addition to this, *efr* mutants were found to be more susceptible to *Agrobacterium tumefaciens*-mediated transformation than wild-type plants, indicating that EF-Tu recognition and the subsequent defence responses reduce *Agrobacterium*-mediated plant transformation (Zipfel *et al.*, 2006).

Similar to FLS2, EFR belongs to the RLK family and has an extracellular domain with 24 LRRs, a single transmembrane domain and an intracellular serine/threonine kinase domain (Zipfel *et al.*, 2006). Both FLS2 and EFR are members of the subfamily LRR-XII of RLKs. Besides FLS2 and EFR from *Arabidopsis*, the rice pathogen recognition receptor, XA21, which confers resistance to *Xanthomonas oryzae* pv. *oryzae* strains is also a member of the LRR-XII subfamily (Song *et al.*, 1995; Shiu *et al.*, 2004; Lee *et al.*, 2006).

Cold shock protein (CSP)

Prokaryotic low-temperature adaptation has been studied in *E. coli*, and temperature drop has been shown to induce cold shock proteins (CSPs). These CSPs have been found to be essential for the cells to resume growth at low temperatures (reviewed by Thieringer *et al.*, 1998). The CSP from *Micrococcus lysodeikticus* (*Staphylococcus aureus*) has been identified as a MAMP and has been shown to induce defence responses in tobacco cells (Felix and Boller, 2003). A 22 amino acid peptide of the N terminus of the cold shock domain, csp22, was synthesized based on the consensus sequence of bacterial CSP and proven to be the active domain. Studies using various solanaceous plants revealed that treatment of leaf tissues with csp15, the most conserved part of csp22 which includes an RNA-binding motif (RNP-1), induced oxidative bursts as well as an increase in ethylene biosynthesis, indicating a perception system for CSP in these plants. A CSP receptor still remains to be found, even though studies have shown that CSP- and flagellin-derived elicitors induce the same set of responses, suggesting a converging signal transduction pathway and possibly a receptor-mediated process for CSP similar to flagellin. A perception system for CSP-related elicitors has not been found outside the Solanaceae (Felix and Boller, 2003).

Superoxide dismutase (SodM)

One of the most abundant proteins of the extracellular proteome of *Xcc*, the cytocylic superoxide dismutase (SodM) protein, has been isolated from two-dimensional gels loaded with the extracellular proteome extracted from the culture supernatants of an *Xcc* culture. SodM was able to induce oxidative bursts in tobacco cells (Watt *et al.*, 2005). Superoxide dismutase (SodM) is involved in catalyzing the reduction of superoxide anions to hydrogen peroxide. Furthermore, *SodM* sequences are highly conserved among plant pathogenic bacteria and the protein is therefore a good candidate as a MAMP. How plants recognize SodM remains to be revealed, but a model has been proposed: as plant PRRs recognize MAMPs, reactive oxygen species (ROS) are produced and this rise in ROS activates the transcription of the *sodM* gene. The rise in SodM is then thought to induce plant defence responses via a SodM-specific receptor (Watt *et al.*, 2006).

MAMP-induced immune responses

Thus far, perception of MAMPs in plants is known to occur at the cell surface and the only known plant PRRs are plasma membrane-resident proteins RLKs (receptor-like kinases; Gómez-Gómez and Boller, 2000; Zipfel *et al.*, 2006; Robatzek *et al.*, 2007). The intracellular signal transduction cascades that link the recognition of MAMPs to defence responses in plants show similarities to that of the animal innate immune system (reviewed by Nürnberger *et al.*, 2004). Signalling events acting downstream of the recognition of MAMPs are thought to include changes in cytoplasmic calcium levels, production of ROS and NO as well as activation of mitogen-activated protein kinases (MAPKs) which leads to gene induction through the WRKY transcription factors (Asai *et al.*, 2002; Kunze *et al.*, 2004; Zipfel *et al.*, 2004; Mészáros *et al.*, 2006).

The recognition of MAMPs in plants may not necessary induce a MAMP specific defence response as in mammals, but rather a basal defence response (Zipfel et al., 2006). The innate immune signalling components, the activation of MAPK cascades and WRKY transcription factors, have been found to be encoded by functionally redundant genes in Arabidopsis (Zhang and Klessig, 2001; Asai et al., 2002). The fact that these redundant components exist within the cascade may reflect a strong need to back up fundamental regulatory functions.

In tobacco cells, *Burkholderia cepacia* LPS has been found to induce several genes, of which a receptor-like protein kinase, a binding protein for the type III effector protein harpin as well as a virus resistance gene has been identified (Sanabria and Dubery, 2006). In addition to this, a gene expression search has revealed that FLS2 and EFR are induced by (besides their respective MAMPs) bacterial LPS, fungal chitin and oomycete-derived NPP1, indicating that plants might not distinguish between bacteria, fungi and oomycetes, but rather respond with a basal defence when they are exposed to MAMPs (Zipfel et al., 2006). Recently, it was shown that not only pathogen-derived elicitors, but also plant-derived peptides can induce components of innate immunity in *Arabidopsis*. AtPep1, an endogenous 23 amino acid peptide elicitor isolated from *Arabidopsis* leaves has been shown to activate defence genes associated with innate immune responses in *Arabidopsis*. AtPep1 was isolated from the precursor protein, PROPEP1, whose gene is expressed in response to methyl jasmonate (MeJA) and ethylene, among other things, and thought to regulate the expression of the defensin protein PDF1.2 (Huffaker et al., 2006; Huffaker and Ryan, 2007). An AtPep1-binding protein PEPR1 was isolated from the cell surface of cultured *Arabidopsis* cells and kinetic analyses as well as gain-of-function experiments, revealed that PEPR1 was the AtPep1 receptor. PEPR1 was found to belong to the LRR-XI RLK subfamily, having an extracellular domain with 26 LRRs, a transmembrane domain and an intracellular protein kinase domain (Yamaguchi et al., 2006).

It has been proposed that these plant-derived compounds are referred to as microbe-induced molecular patterns (MIMPs) (Mackey and McFall, 2006).

Suppression of innate immunity

The perception of MAMPs by plant surface PRRs triggering plant innate immunity is the first line of defence, constituting the basis of non-host resistance of the host plant (Thordal-Christensen, 2003). MAMP triggered immunity can be undermined by microbial effector proteins. Many phytopathogenic bacteria inject their virulence effector proteins directly into the host cell through a TTSS (reviewed by Alfano and Collmer, 2004; see also chapter by Boch). These effector molecules have been demonstrated to suppress innate immunity (Jamir et al., 2004; He et al., 2006; Nomura et al., 2006), resulting in effector triggered susceptibility. However, recognition of a given effector through a set of *R* genes results in effector triggered immunity, that is, disease resistance or HR (reviewed by Jones and Dangl, 2006). In addition, bacterial extracellular polysaccharides (EPS) have recently also been shown to suppress MAMP triggered immunity. This appears to be via the sequestration of apoplastic Ca^{2+} ions, influx of which is a prerequisite for defence signalling (Aslam et al., 2008).

Receptors or proteins that have been found to recognize MAMPs and effector molecules are proteins with a NBS and LRRs, RLKs and receptor-like proteins (RLPs) (Shiu and Bleecker, 2003; Belkhadir et al., 2004).

Concluding remarks

From plant science to drug discovery?
Plant receptor-like kinases (RLK) are related to animal receptor kinases (RK). In plants, the RLKs known to date have serine/threonine kinase specificity, whereas animal RKs often are tyrosine kinases. The majority of the plant RLKs are LRR-RLKs (Shiu and Bleeker, 2001). The LRR domain, found in a number of defence-related proteins, can bind pathogen produced ligands (reviewed by Diévart and Clark, 2003). For a schematic overview of some known PPRs in plants and animals, see Fig. 11.1.

The unravelling of the components involved in plant innate immunity is important for understanding the complex interactions between plants and bacterial pathogens, and perhaps, between bacterial pathogens and mammals. The similarities of the innate immune system in plants and mammals, and the microbial capacity for cross-kingdom pathogenicity of human pathogens, has led a few researchers to use the *Arabidopsis* model plant to study human microbial pathogenicity factors (Rahme et al., 1997). Human pathogens, like *Pseudomonas aeruginosa* and *Staphylococcus aureus* have been shown to infect both plants and animals (reviewed by Prithiviraj et al., 2005b). *P. aeruginosa* genes, not earlier known to be involved in pathogenesis-related functions, were identified by studying pathogenesis of *P. aeruginosa* in *Arabidopsis* and *Caenorhabditis elegans* (nematodes), indicating that there could exist general virulence mechanisms, used by *P. aeruginosa*, that are conserved across phylogeny (Rahme et al., 2000). In addition, studies with *Arabidopsis* as a model host for *S. aureus* revealed that the virulence factors essential for animal pathogenesis were also important for plant pathogenesis, and in both cases, mediated by SA (Prithiviraj et al., 2005a). The results obtained using *Arabidopsis* and nematodes as hosts for human pathogens, support the use of disparate hosts as models for mammals.

The advantages of using the plant system compared to the animal system is that the plant is easily used in molecular studies, it has a short generation time and there are no ethical concerns. The *Arabidopsis* genome is completely sequenced, making it easy to study insertion mutations for almost every gene in the genome (Alonso et al., 2003). This provides unprecedented possibilities for studying microbial pathogenicity factors and their cognate host targets using thousands of genetically similar host plants. In the near future, perhaps, the drug discovery industry will benefit from plant science.

References

Akira, S., and Takeda, K. (2004). Toll-like receptor signalling. Nat. Immunol. *4*, 499–511.

Alfano, J.R., and Collmer, A. (2004). Type III secretion system effector proteins: Double agents in bacterial disease and plant defense. Annu. Rev. Phytopathol. *42*, 385–414.

Alexopoulou, L., Holt, A.C., Medzhitov, R., and Flavell, R.A. (2001). Recognition of double-stranded RNA and activation of NF-kappa B by Toll-like receptor 3. Nature *413*, 732–738.

Alonso, J.M., Stepanova, A.N., Leisse, T.J., Kim, C.J., Chen, H., Shinn, P., Stevenson, D.K., Zimmerman, J., Barajas, P., Cheuk, R., Gadrinab, C., Heller, C., Jeske, A., Koesema, E., Meyers, C.C., Parker, H., Prednis, L., Ansari, Y., Choy, N., Deen, H., Geralt, M., Hazari, N., Hom, E., Karnes, M., Mulholland, C., Ndubaku, R., Schmidt, I., Guzman, P., Aguilar-Henonin, L., Schmid, M., Weigel, D., Carter, D.E., Marchand, T., Risseeuw, E., Brogden, D., Zeko, A., Crosby, W.L., Berry, C.C., and Ecker, J.R. (2003). Genome-wide insertional mutagenesis of *Arabidopsis thaliana*. Science *301*, 653–657.

Asai, T., Tena, G., Plotnikova, J., Willmann, M.R., Chiu, W.L., Gomez-Gomez, L., Boller, T., Ausubel, F.M., and Sheen, J. (2002). MAP kinase signalling in *Arabidopsis* innate immunity. Nature *415*, 977–983.

Aslam, S., Jackson, R.W., Morrissey, K., Knight, M.R., Chinchilla, D., Boller, T., Erbs, G., Tandrup Jensen, T., Newman, M.A. and Cooper, R.M. (2008). Bacterial extracellular polysaccharides promote pathogenicity by suppressing plant basal defences. In: Biology of Plant–Microbe interactions, M. Lorito, S. L. Woo, F. Scala, ed. (St. Paul, Minnesota, USA: International Society for Molecular Plant–Microbe Interactions), Vol 6 (*In Press*).

Ausubel, F. (2005). Are innate immune signalling pathways in plants and animals conserved? Nat. Immunol. *6*, 973–979.

Bauer, Z., Gómez-Gómez, L., Boller, T., and Felix, G. (2001). Sensitivity of different ecotypes and mutants of *Arabidopsis* thaliana toward the bacterial elicitor flagellin correlates with the presence of receptor-binding sites. J. Biol. Chem. *276*, 45669–45676.

Bedini, E., De Castro, C., Erbs, G., Mangoni, L., Dow, J.M., Newman, M.-A., Parrilli, M., and Unverzagt, C. (2005). Structure-dependent modulation of a pathogen response in plants by synthetic O-antigen polysaccharides. J. Am. Chem. Soc. *127*, 2414–2416.

Bedini, E., Parrilli, M., and Unverzagt, C. (2002). Oligomerization of a rhamnanic trisaccharide repeating unit of O-chain polysaccharides from phytopathogenic bacteria. Tetrahedron Lett. *43*, 8879–8882.

Belkhadir, Y., Subramaniam, R., and Dangl, J.L. (2004). Plant disease resistance protein signalling: NBS-LRR proteins and their partners. Curr. Opin. Plant Biol. *7*, 391–399.

Bent, A.F., and Mackey, D. (2007). Elicitors, effectors, and R genes: The new paradigm and a lifetime supply of questions. Annu. Rev. Phytopathol. *45*, 399–436.

Braun, S.G., Meyer, A., Holst, O., Pühler, A., and Niehaus, K. (2005). Characterization of the *Xanthomonas campestris* pv. *campestris* lipopolysaccharide substructures essential for elicitation of an oxidative burst in tobacco cells. Mol. Plant–Microbe Interact. *18*, 674–681.

Carpenter, S., and O'Neill, L.A.J. (2007). How important are Toll-like receptors for antimicrobial responses? Cell. Microbiol. *9*, 1891–1901.

Chinchilla, D., Bauer, Z., Regenass, M., Boller, T., and Felix, G. (2006). The *Arabidopsis* receptor kinase FLS2 binds flg22 and determines the specificity of flagellin perception. Plant Cell *18*, 465–476.

Chinchilla, D., Zipfel, C., Robatzek, S., Kemmerling, B., Nürnberger, T., Jones, D.G.J., Felix, G., and Boller, T. (2007). A flagellin-induced complex of the receptor FLS2 and BAK1 initiates plant defence. Nature 448, 497–500.

Creagh, E.M., and O'Neill, L.A.J. (2006). TLRs, NLRs and RLRs: a trinity of pathogen sensors that co-operate in innate immunity. Trends in Immunol. 27, 352–357.

Desaki, Y., Miya, A., Venkatesh, B., Tsuyumu, S., Yamane, H., Kaku, H., Minami, E., and Shibuya, N. (2006). Bacterial lipopolysaccharides induce defense responses associated with programmed cell death in rice cells. Plant Cell Physiol. 47, 1530–1540.

Diévart, A., and Clark, S.E. (2003). Using mutant alleles to determine the structure and function of leucine-rich repeat receptor-like kinases. Curr. Opin. Plant Biol. 6, 507–516.

Dziarski, R., and Gupta, D. (2005a). Peptidoglycan recognition in innate immunity. J. Endotoxin Res. 11, 304–310.

Dziarski, R., and Gupta, D. (2005b). *Staphylococcus aureus* peptidoglycan is a Toll-like receptor 2 activator: a reevaluation. Infect. Immun. 73, 5212–5216.

Dziarski, R., Tapping, R.I., and Tobias, P.S. (1998). Binding of bacterial peptidoglycan to CD14. J. Biol. Chem. 273, 8680–8690.

Dow, M., Newman, M.-A., and von Roepenack, E. (2000). The induction and modulation of plant defence responses by bacterial lipopolysaccharides. Annu. Rev. Phytopathol. 38, 241–261.

Erbs, G., and Newman, M.-A. (2003). The role of lipopolysaccharides in induction of plant defence responses. Mol. Plant Pathol. 4, 421–425.

Erbs, G., Silipo, A., Aslam, S., De Castro, C., Liparoti, V., Flagiello, A., Pucci, P., Lanzetta, R., Parrilli, M., Molinaro, A., Newman, M.-A. And Cooper, R.M. (2008). Peptidoglycan and muropeptides from pathogens Agrobacterium and Xanthomanas elicit innate immunity: structure and activity. Chemistry and Biology, doi: 10.1016/j. chembiol. 2008.03.017.

Felix, G., Duran, J.D., Volko, S., and Boller, T. (1999). Plants have a sensitive perception system for the most conserved domain of bacterial flagellin. Plant J. 18, 265–276.

Felix, G. and Boller, T., (2003). Molecular sensing of bacteria in plants. J. Biol. Chem. 278, 6201–6208.

Fellbrich, G., Romanski, A., Varet, A., Blume, B., Brunner, F., Engelhardt, S., Felix, G., Kemmerling, B., Krzymowska, M., and Nürnberger, T. (2002). NPP1, a *Phytophthora*-associated trigger of plant defense in parsley and *Arabidopsis*. Plant J. 32, 375–390.

Franchi, L., McDonald, C., Kanneganti, T.D., Amer, A., and Nunez, G. (2006). Nucleotide-binding oligomerization domain-like receptors: Intracellular pattern recognition molecules for pathogen detection and host defense. J. Immunol. 177, 3507–3513.

Gassmann, W., Hinsch, M.E., and Staskawicz, B.J. (1999). The *Arabidopsis* RPS4 bacterial-resistance gene is a member of the TIR-NBS-LRR family of disease-resistance genes. Plant J. 20, 265–277.

Gómez-Gómez, L., Bauer, Z., and Boller, T. (2001). Both the extracellular leucine-rich repeat domain and the kinase activity of FLS2 are required for flagellin binding and signalling in *Arabidopsis*. Plant Cell 13, 1155–1163.

Gómez-Gómez, L., and Boller, T. (2000). FLS2: An LRR receptor-like kinase involved in the perception of the bacterial elicitor flagellin in *Arabidopsis*. Mol. Cell 5, 1003–1011.

Greenberg, J.T. (1997). Programmed cell death in plant–pathogen interactions. Annu. Rev. Plant Physiol. Plant Mol. Biol. 48, 525–545.

Gross, A., Kapp, D., Nielsen, T., and Niehaus, K. (2005). Endocytosis of *Xathomonas campestris* pathovar *campestris* lipopolysaccharides in non-host plant cells of *N. benthamiana*. New Phytol. 165, 215–226.

Guan, R., and Mariuzza, R.A. (2007). Peptidoglycan recognition proteins of the innate immune system. Trends in Microbiol. 15, 127–134.

Gust, A., Biswas, R., Lenz, H.D., Rauhut, T., Ranf, S., Kemmerling, B., Götz, F., Glawischnig, E., Lee, J., Felix, G., and Nürnberger, T. (2007). Bacteria-derived peptidoglycans constitute pathogen-associated molecular patterns triggering innate immunity in *Arabidopsis*. J. Biol. Chem. 282, 32338–32348.

Gutsmann, T., Schromm, A.B., and Brandenburg, K. (2007). The physiochemistry of endotoxins in relation to bioactivity. J. Med. Microbiol. 297, 341–352.

Hashimoto, C., Hudson, K.L., and Anderson, K.V. (1988). The Toll gene of *Drosophila*, required for dorsal-ventral embryonic polarity, appears to encode a transmembrane protein. Cell 52, 269–279.

Hayashi, F., Smith, K.D., Ozinsky, A., Hawn, T.R., Yi, E.C., Goodlett, D.R., Eng, J.K., Akira, S., Underhill, D.M., and Aderem, A. (2001). The innate immune response to bacterial flagellin is mediated by Toll-like receptor 5. Nature 410, 1099–1103.

He, P., Shan, L., Lin, N.-C., Martin, G.B., Kemmerling, B., Nürnberger, T., and Sheen, J. (2006). Specific bacterial suppressors of MAMP signalling upstream of MAPKKK in *Arabidopsis* innate immunity. Cell 125, 563–575.

Huffaker, A., Pearce, G., and Ryan, C.A. (2006). An endogenous peptide signal in *Arabidopsis* activates components of the innate immune response. Proc. Natl. Acad. Sci. USA 103, 10098–10103.

Huffaker, A., and Ryan, C.A. (2007). Endogenous peptide defense signals in *Arabidopsis* differentially amplify signalling for the innate immune response. Proc. Natl. Acad. Sci. USA 104, 10732–10736.

Jamir, Y., Guo, M., Oh, H.-S., Petnicki-Ocwieja, T., Chen, S., Tang, X., Dickman, M.B., Collmer, A., and Alfano, J.R. (2004). Identification of *Pseudomonas syringae* type III effectors that can suppress programmed cell death in plants and yeast. Plant J. 37, 554–565.

Janeway, C.A. (1992). The immune-system evolved to discriminate infectious nonself from non-infectious self. Immunol. Today 13, 11–16.

Jeppesen, M.G., Navratil, T., Spremulli, L.L., and Nyborg, J. (2005). Crystal structure of the bovine mitochondrial elongation factor Tu.Ts complex. J. Biol. Chem. 280, 5071–5081.

Jiang, Q., Akashi, S., Miyake, K., and Petty, H.R. (2000). Cutting edge: Lipopolysaccharide induces physical proximity between CD14 and Toll-like receptor 4

(TLR4) prior to nuclear translocation of NF-κB. J. Immunol. *165*, 3541–3544.

Jones, J.D.G., and Dangl, J.L. (2006). The plant immune system. Nature *444*, 323–329.

Kaku, H., Nishizawa, Y., Ishii-Minami, N., Akimoto-Tomiyama, C., Dohmae, N., Takio, K., Minami, E., and Shibuya, N. (2006). Plant cells recognize chitin fragments for defense signaling through a plasma membrane receptor. Proc. Natl. Acad. Sci. USA *103*, 11086–11091.

Kunze, G., Zipfel, C., Robatzek, S., Niehaus, K., Boller, T., and Felix, G. (2004). The N terminus of bacterial elongation factor Tu elicits innate immunity in *Arabidopsis* plants. Plant Cell *16*, 3496–3507.

Lemaitre, B., Nicolas, E., Michaut, L., Reichhart, J.-M., and Hoffmann, J.A. (1996). The dorsoventral regulatory gene cassette spätzle/Toll/cactus controls the potent antifungal response in *Drosophila* adults. Cell *86*, 973–983.

Lee, S.-W., Han, S.-W., Bartley, L.E., and Ronald, P.C. (2006). Unique characteristics of *Xanthomonas oryzae* pv. *oryzae* AvrXa21 and implications for plant innate immunity. Proc. Natl. Acad. Sci. USA *103*, 18395–18400.

Leeman, M., Vanpelt, J.A., Denouden, F.M., Heinsbroek, M., Pahm, B., and Schippers, B. (1995). Induction of systemic resistance against fusarium-wilt of radish by lipopolysaccharides of *Pseudomonas fluorescens*. Phytopathol. *85*, 1021–1027.

Leulier, F., Parquet, C., Pili-Floury, S., Ryu, J.-H., Caroff, M., Lee, W.-J., Mengin-Lecreulx, D., and Lemaitre, B. (2003). The *Drosophila* immune system detects bacteria through specific peptidoglycan recognition. Nat. Immunol. *4*, 478–484.

Loppnow, H., Brade, H., Durrbaum, I., Dinarello, C.A., Kusumoto, S., Rietschel, E.T., and Flad, H.D. (1989). IL-1 induction-capacity of defined lipopolysaccharide partial structures. J. Immunol. *142*, 3229–3238.

Mackey, D., and McFall, A.J. (2006). MAMPs and MIMPs: proposed classifications for inducers of innate immunity. Mol. Microbiol. *61*, 1365–1371.

McCoy, S.L., Kurtz, S.E., Hausman, F.A., Trune, D.R., Bennett, R.M., and Hefeneider, S.H. (2004). Activation of RAW264.7 macrophages by bacterial DNA and lipopolysaccharides increases cell surface DNA binding and internalization. J. Biol. Chem. *279*, 17217–17223.

McDonald, C., Inohara, N., and Nuñez, G. (2005). Peptidoglycan signalling in innate immunity and inflammatory disease. J. Biol. Chem. *280*, 20177–20180.

Medzhitov, R., Preston-Hurlburt, P., and Janeway Jr., C.A. (1997). A human homologue of the Drosophila Toll protein signals activation of adaptive immunity. Nature *388*, 394–397.

Meindl, T., Boller, T., and Felix, G. (2000). The bacterial elicitor flagellin activates its receptor in tomato cells according to the address-message concept. Plant Cell *12*, 1783–1794.

Mészáros, T., Helfer, A., Hatzimasoura, E., Magyar, Z., Serazetdinova, L., Rios, G., Bardóczy, V., Teige, M., Koncz, C., Peck, S., and Bögre, L. (2006). The *Arabidopsis* MAP kinase kinase MKK1 participates in defence responses to the bacterial elicitor flagellin. Plant J. *48*, 485–498.

Mishina, T.E., and Zeier, J. (2006). The *Arabidopsis* flavin-dependent monooxygenase FMO1 is an essential component of biologically induced systemic acquired resistance. Plant Physiol. *141*, 1666–1675.

Mishina, T.E., and Zeier, J. (2007). Pathogen-associated molecular pattern recognition rather than development of tissue necrosis contributes to bacterial induction of systemic acquired resistance in *Arabidopsis*. Plant J. *50*, 500–513.

Miyake, K. (2004). Innate recognition of lipopolysaccharide by Toll-like receptor 4-MD-2. Trends in Microbiol. *12*, 186–192.

Newman, M.-A., Daniels, M.J., and Dow, J.M. (1995). Lipopolysaccharide from *Xanthomonas campestris* induces defence-related gene expression in *Brassica campestris*. Mol. Plant Microbe Interact. *8*, 778–780.

Newman, M.-A., Daniels, M.J., and Dow, J.M. (1997). The activity of lipid A and core components of bacterial lipopolysaccharides in the prevention of the hypersensitive response in pepper. Mol. Plant–Microbe Interact. *10*, 926–928.

Newman, M.-A., Dow, J.M., Molinaro, A., and Parrilli, M. (2007). Priming, induction and modulation of plant defence responses by bacterial lipopolysaccharides. J. Endotoxin Res. *13*, 69–84.

Newman, M.-A., von Roepenack-Lahaye, E., Parr, A., Daniels, M.J., and Dow, J.M. (2001). Induction of hydroxycinnamoyl-tyramine conjugates in pepper by *Xanthomonas campestris*, a plant defense response activated by *hrp* gene-dependent and *hrp* gene-independent mechanisms. Mol. Plant Microbe Interact. *14*, 785–792.

Newman, M.-A., von Roepenack-Lahaye, E., Parr, A., Daniels, M.J., and Dow, J.M. (2002). Prior exposure to lipopolysaccharide potentiates expression of plant defenses in response to bacteria. Plant J. *29*, 485–497.

Nomura, K., DebRoy, S., Lee, Y.H., Pumplin, N., Jones, J., and He, S. Y. (2006). A bacterial virulence protein suppresses host innate immunity to cause plant disease. Science *313*, 220–223.

Nürnberger, T., and Brunner, F. (2002). Innate immunity in plants and animals: emerging parallels between the recognition of general elicitors and pathogen-associated molecular patterns. Curr. Opin. Plant Biol. *5*, 1–7.

Nürnberger, T., Brunner, F., Kemmerling, B., and Piater, L. (2004). Innate immunity in plants and animals: striking similarities and obvious differences. Immunol. Rev. *198*, 249–266.

Prime-A-Plant Group: Conrath, U., Beckers, G.J.M., Flors, V., García-Agustín, P., Jakab G., Mauch, F., Newman, M.-A., Pieterse, C.M.J., Poinssot, B., Pozo, M.J., Pugin, A., Schaffrath, U., Ton, J., Wendelhenne, D., Zimmerli, L., and Mauch-Mani, B. (2006). Priming: Getting ready, for battle. Mol. Plant Microbe Interact. *19*, 1062–1071.

Prithiviraj, B., Bais, H.P., Jha, A.K., and Vivanco, J.M. (2005a). *Staphylococcus aureus* pathogenicity on *Arabidopsis thaliana* is mediated either by a direct effect of salicylic acid on the pathogen or by SA-dependent, NPR1-independent host responses. Plant J. 42, 417–432.

Prithiviraj, B., Weir, T., Bais, H.P., Schweizer, H.P., and Vivanco, J.M. (2005b). Plant models for animal pathogenesis. Cell. Microbiol. 7, 315–324.

Raetz, C.R.H., and Whitfield, C. (2002). Lipopolysaccharide endotoxins. Annu. Rev. Biochem. 71, 635–700.

Rahme, L.G., Ausubel, F.M., Cao, H., Drenkard, E., Goumnerov, B.C., Lau, G.W., Mahajan-Miklos, S., Plotnikova, J., Tan, M.-W., Tsongalis, J., Walendziewicz, C.L., and Tompkins, R.G. (2000). Plants and animals share functionally common bacterial virulence factors. Proc. Natl. Acad. Sci. USA 97, 8815–8821.

Rahme, L.G., Tan, M.-W., Le, L., Wong, S.M., Tompkins, R.G., Calderwood, S.B., and Ausubel, F.M. (1997). Use of model plant hosts to identify *Pseudomonas aeruginosa* virulence factors. Proc. Natl. Acad. Sci. USA 94, 13245–13250.

Ramos, H.C., Rumbo, M., and Sirard, J.-C. (2004). Bacterial flagellins: mediators of pathogenicity and host immune responses in mucosa. Trends Microbiol. 12, 509–517.

Robatzek, S., Bittel, P., Chinchilla, D., Köchner, P., Felix, G., Shiu, S.H., and Boller, T. (2007). Molecular identification and characterization of the tomato flagellin LeFLS2, an orthologue of *Arabidopsis* FLS2 exhibiting characteristically different perception specificities. Plant Mol. Biol. 64, 539–547.

Robatzek, S., Chinchilla, D., and Boller, T. (2006). Ligand-induced endocytosis of the pattern recognition receptor FLS2 in *Arabidopsis*. Genes Dev. 20, 537–542.

Ron, M., and Avni, A. (2004). The receptor for the fungal elicitor ethylene-inducing xylanase is a member of a resistance-like gene family in tomato. Plant Cell 16, 1604–1615.

Russinova, E., Borst, J.-W., Kwaaitaal, M., Caño-Delgado, A., Yin, Y., Chory, J., and de Vries, S.C. (2004). Heterodimerization and endocytosis of *Arabidopsis* brassinosteroid receptors BRI1 and AtSERK3 (BAK1). Plant Cell 16, 3216–3229.

Ryals, J.A., Neuenschwander, U.H., Willits, M.G., Molina, A., Steiner, H.-Y., and Hunt, M.D. (1996). Systemic acquired resistance. Plant Cell 8, 1809–1819.

Sanabria, N.M., and Dubery, I.A. (2006). Diifferential display profiling of the *Nicotiana* response to LPS reveals elements of plant basal resistance. Biochem. Bioph. Res. Co. 344, 1001–1007.

Schneider, M., Schweizer, P., Meuwly, P., and Métraux, J.P. (1996). Systemic acquired resistance in plants. Int. Rev. Cytol. 168, 303–340.

Schromm, A.B., Brandenburg, K., Loppnow, H., Moran, A.P., Koch, M.H.J., Rietschel, E.Th., and Seydel, U. (2000). Biological activities of lipopolysaccharides are determined by the shape of their lipid A portion. Eur. J. Biochem. 267, 2008–2013.

Schromm, A.B., Brandenburg, K., Loppnow, H., Zahringer, U., Rietschel, E.T., Carroll, S.F., Koch, M.H.J., Kusumoto, S., and Seydel, U. (1998). The charge of endotoxin molecules influences their conformation and IL-6-inducing capacity. J. Immunol. 161, 5464–547.

Shiu, S.-H., and Bleecker, A.B. (2001). Receptor-like kinases from *Arabidopsis* form a monophyletic gene family related to animal receptor kinases. Proc. Natl. Acad. Sci. USA 98, 10763–10768.

Shiu, S.-H., and Bleecker, A.B. (2003). Expansion of the receptor-like kinase/pelle gene family and receptor-like proteins in *Arabidopsis*. Plant Physiol. 132, 530–543.

Shiu, S.-H., Karlowski, W.M., Pan, R., Tzeng, Y.-H., Mayer, K.F.X., and Li, W.-H. (2004). Comparative analysis of the receptor-like kinase family in *Arabidopsis* and rice. Plant Cell 16, 1220–1234.

Silipo, A., Molinaro, A., Sturiale, L., Dow, J.M., Erbs, G., Lanzetta, R., Newman, M.-A., and Parrilli, M. (2005). The elicitation of plant innate immunity by lipooligosaccharide of *Xanthomonas campestris*. J. Biol. Chem. 280, 33660–33668.

Smith, K.D., Andersen-Nissen, E., Hayashi, F., Strobe, K., Bergman, M.A., Rassoulian Barrett, S.L., Cookson, B.T., and Aderem, A. (2003). Toll-like receptor 5 recognizes a conserved site on flagellin required for protofilament formation and bacterial mobility. Nat. Immunol. 4, 1247–1253.

Song, W.Y., Wang, G.L., Chen, L.L., Kim, H.S., Pi, L.Y., Holsten, T., Gadner, J., Wang, B., Zhai, W.X., Zhu, L.H., Fauquet, C., and Ronald, P.C. (1995). A receptor kinase-like protein encoded by the rice disease resistance gene Xa21. Science 270, 1804–1806.

Strober, W., Murray, P.J., Kitani, A., and Watanabe, T. (2006). Signalling pathways and molecular interactions of NOD1 and NOD2. Nat. Rev. Immunol. 6, 9–20.

Sun, W., Dunning, F.M., Pfund, C., Weingarten, R., and Bent, A.F. (2006). Within-species flagellin polymorphism in *Xanthomonas campestris* pv *campestris* and its impact on elicitation of *Arabidopsis* FLAGELLIN SENSING2-dependent defenses. Plant Cell 18, 764–779.

Thieringer, H.A., Jones, P.G., and Inouye, M. (1998). Cold shock and adaptation. Bioessays 20, 49–57.

Thordal-Christensen, H. (2003). Fresh insights into processes of nonhost resistance. Curr. Opin. Plant Biol. 6, 351–357.

Traub, S., von Aulock, S., Hartung, T., and Herman, C. (2006). MDP and other muropeptides – direct and synergistic effects on the immune system. J. Endotoxin Res. 12, 69–85.

Travassos, L.H., Girardin, S.E., Philpott, D.J., Blanot, D., Nahori, M.-A., Werts, C., and Boneca, I.G. (2004). Toll-like receptor 2-dependent bacterial sensing does not occur via peptidoglycan recognition. EMBO J. 5, 1000–1006.

Umemoto, N., Kakitani, M., Iwamatsu, A., Yoshikawa, M., Yamaoka, N., and Ishida, I. (1997). The structure and function of a soybean β-glucan-elicitor-binding-protein. Proc. Natl. Acad. Sci. USA 94, 1029–1034.

van Baarlen, P., van Belkum, A. and Thomma, P.H.J. (2007). Disease induction by human microbial pathogens in plant-model systems: potential, problems and prospects. Drug Discovery Today 12, 167–173.

van Loon, L.C., Bakker, P.A., and Pieterse, C.M. (1998). Systemic resistance induced by rhizosphere bacteria. Annu. Rev. Phytopathol. *36*, 453–483.

Watt, S.A., Tellström, V., Patschkowski, T., and Niehaus, K. (2006). Identification of the bacterial superoxide dismutase (SodM) as plant-inducible elicitor of an oxidative burst reaction in tobacco cell suspension cultures. Journal of Biotech. *126*, 78–86.

Watt, S.A., Wilke, A., Patschkowski, T., and Niehaus, K. (2005). Comprehensive analysis of the extracellular proteins from *Xanthomonas campestris* pv. *campestris* B100. Proteomics 5, 153–167.

Wright, S.D., Ramos, R.A., Tobias, P.S., Ulevitch, R.J., and Mathison, J.C. (1990). CD14 serves as the cellular receptor for complexes of lipopolysaccharides with lipopolysaccharide binding protein. Science *249*, 1431–1433.

Yamaguchi, Y., Pearce, G., and Ryan, C.A. (2006). The cell surface leucine-rich repeat receptor for AtPep1, an endogenous peptide elicitor in *Arabidopsis*, is functional in transgenic tobacco cells. Proc. Natl. Acad. Sci. USA *103*, 10104–10109.

Zhang, S., and Klessig, D.F. (2001). MAPK cascades in plant defence signalling. Trends Plant Sci. *6*, 520–527.

Zeidler, D., Zahringer, U., Gerber, I., Dubery, I., Hertung, T., Bors, W., Hutzler, P., and Durner, J. (2004). Proc. Natl. Acad. Sci. USA *101*, 15811–15816.

Zipfel, C., Kunze, G., Chinchilla, D., Caniard, A., Jones, J.D.G., Boller, T., and Felix, G. (2006). Perception of the bacterial PAMP EF-Tu by the receptor EFR restricts *Agrobacterium*-mediated transformation. Cell *125*, 749–760.

Zipfel, C., Robatzek, S., Navarro, L., Oakeley, E.J., Jones, J.D.G., Felix, G., and Boller, T. (2004). Bacterial disease resistance in *Arabidopsis* through flagellin perception. Nature *428*, 764–767.

The Art of Manipulation: Bacterial Type III Effectors and their Plant Targets

Jens Boch

Abstract

A successful plant pathogen has to accomplish several tasks during infection of a plant host. It has to gain entry into the tissue, acquire nutrients, multiply, and spread to uninfected tissues or neighbouring plants. Pathogens have evolved different virulence factors to accomplish this. Key to this are bacterial effector proteins that are directly translocated into plant cells via a type III protein secretion system. These effectors are potent devices to manipulate the eukaryotic cell from within. Bacterial infections are antagonized by the plant which carries a sophisticated surveillance system to detect invading microbes and respond with defence reactions to prevent pathogen proliferation. The conflicting interests have spawned a complex pathogen–plant interaction network between effectors in pathogenic bacteria and protective plant defence systems. The net outcome is of grave importance for both interaction partners and the evolutionary pressure has led to the development of a large set of effectors in plant pathogenic bacteria which accomplish diverse virulence activities.

Introduction

Although plant biomass resembles a huge nutrient resource for microbes, most plants in nature appear healthy. Indeed, plants are very capable to prevent microbes from exploiting this habitat. First, plant tissue is protected by several passive layers. The epidermis with its hydrophobic cuticle and gated entry sites (stomata) prohibits direct access to more exposed plant cells and nutrient flow. Every plant cell is surrounded by a cell wall which is an additional barrier that blocks direct contact between pathogen and host cell. Second, plants have evolved several active layers of defence to prevent microbial access to these nutrients.

The first layer is the basal defence that recognizes the presence of potentially detrimental microbes through key molecular components. The recognized molecules are most often parts of typical microbial proteins or compounds and called PAMPs (pathogen-associated molecular patterns) or more accurately MAMPs (microbe-associated molecular patterns) or DAMPs (damage-associated molecular patterns) (Fig. 12.1; Nürnberger et al., 2004; Mackey and McFall, 2006; Bittel and Robatzek, 2007; Lotze et al., 2007; see also chapter by Erbs and Newman). Known MAMPs from plant pathogenic bacteria are for example derivatives of (1) flagellin, the subunit of the extracellular bacterial propulsion apparatus, (2) bacterial elongation factor EF-Tu, (3) lipopolysaccharides (LPS) from the bacterial envelope, and (4) harpin (HrpZ), a bacterial virulence protein (Nürnberger et al., 2004; Bittel and Robatzek, 2007). Also, plant components can serve as MAMPs, when they occur basically only in the presence of microbes. Attacking microbes which secrete hydrolyzing enzymes can liberalize cell wall components which are then detected by the plant (Bittel and Robatzek, 2007).

Plant cells carry cell-surface receptors to detect many MAMP molecules probably via direct binding to the extracellular leucine-rich repeat (LRR)-domain of these receptors. The receptors are grouped into receptor-like kinases (RLKs)

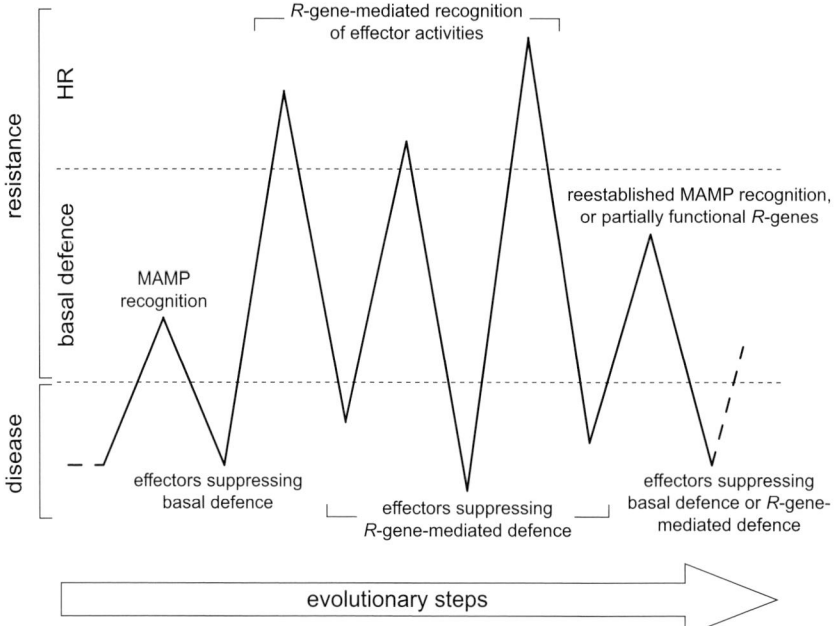

Figure 12.1 Development of plant–pathogen interactions. Recognition of microbe-associated microbial patterns (MAMPs) enabled plant basal defence and restricted bacterial growth. The development of a type III secretion system (T3SS) lead to suppression of basal defence by bacterial effectors. Recognition of effector activities by plant resistance (R) proteins lead to a hypersensitive response (HR) and established plant resistance again. Novel effector activities then repressed R-gene-mediated defence. Alternating novel detection specificities of plant R-proteins and novel bacterial effector activities lead to resistance and disease, accordingly. The HR reactions and the diseases can be of different speed or strength as indicated by peak size. Development of novel MAMP recognition specificities, only partially functional effectors, or partially suppressed R-gene activities can establish basal defence again. The model is adapted from Jones and Dangl (2006).

if they contain an intracellular serine/threonine kinase-signalling domain and receptor-like proteins if the kinase is absent (Bittel and Robatzek 2007). The best studied MAMP receptor is FLS2 from *A. thaliana* (Gomez-Gomez and Boller, 2000) which detects bacterial flagellin and flg22, a 22 amino acid fragment from the highly conserved N-terminus of flagellin (Felix et al., 1999; Chinchilla et al., 2006). After binding of flg22, FLS2 forms a complex with the LRR receptor-like kinase BAK1 which not only regulates the brassinosteroid receptor BRI1, but is also needed for signalling via FLS2 and possibly other cell-surface receptors (Chinchilla et al., 2007; Heese et al., 2007).

After MAMP recognition, the signal is transduced inside the plant cell and elicits a diverse set of defence reactions. Receptor-interaction leads to a rapid influx of Ca^{2+} to the cytoplasm, activation of mitogen-activated protein kinase (MAPK) signalling cascades, and activation of WRKY transcription factors (Fig. 12.2; Asai et al., 2002; Nürnberger et al., 2004; Eulgem and Somssich, 2007). In response, defence-related genes are induced, production of secreted proteins up-regulated (de Torres et al., 2003; Tao et al., 2003), and vesicle trafficking increased that is aimed to secrete antimicrobial compounds and strengthen the cell wall at the site of attack (Brown et al., 1995). Reactive oxygen species and nitric oxide are produced and cell wall-associated callose synthase is activated (Nürnberger and Scheel, 2001; Torres et al., 2006). These responses lead amongst others to the formation of papillae composed of callose and phenolic compounds which protect the plant cell from microbial access.

A successful pathogen has evolved means to overcome this line of defence and specific virulence factors play a key role in this. Most Gram-negative plant and animal pathogens have employed a device called the type III secretion

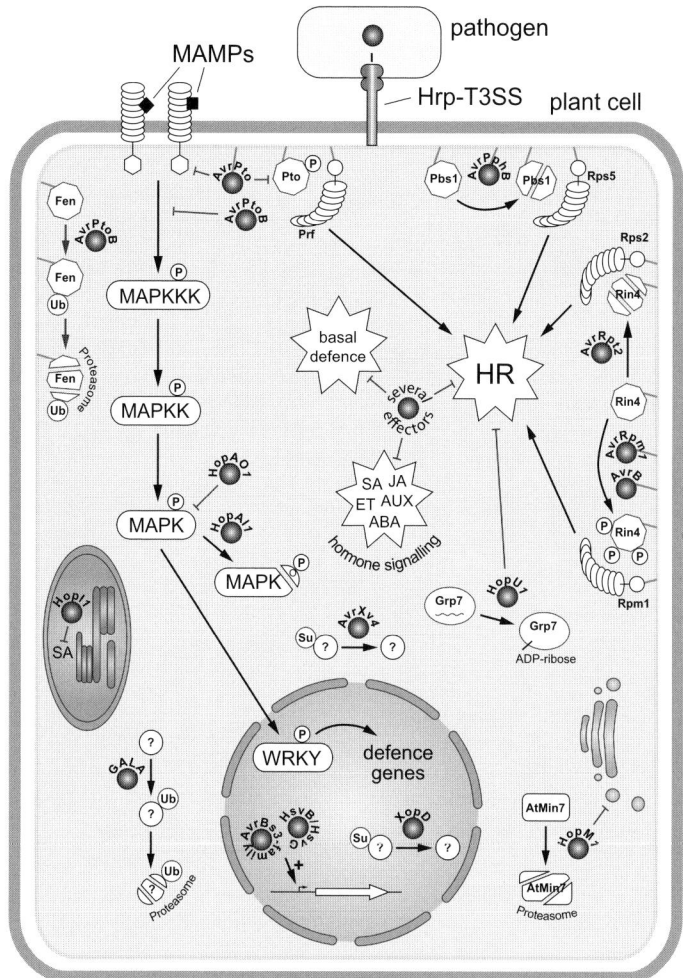

Figure 12.2 Overview of bacterial effector activities and targets within the plant cell. Bacterial pathogens inject effector proteins via a type III secretion system (T3SS) into plant cells. These effectors interfere with plant cell processes to promote disease. Some effector activities are detected by plant resistance proteins leading to a hypersensitive response (HR), a programmed cell death that limits bacterial proliferation. MAMPs (microbe-associated molecular patterns) are pathogen-derived molecules that are detected by the plant and elicit the plant basal defence. For a detailed description of effector activities see the main text. Phytohormones: SA, salicylic acid; JA, jasmonic acid; ET, ethylene; AUX, auxin; ABA, abscisic acid.

system (T3SS) that has become a key virulence factor essential for infection of eukaryotic hosts (He et al., 2004b; Tampakaki et al., 2004; see also chapter by Arnold et al.). This unique secretion system is designed to translocate bacterial virulence proteins across the bacterial and the eukaryotic cell membranes into the cytoplasm of host cells. The transport occurs through the basal T3SS machinery inserted into both bacterial membranes and through the lumen of an extracellular appendage that spans the distance to the surface of the eukaryotic cell (He et al., 2004b; Tampakaki et al., 2004). In Hrp (hypersensitive response and pathogenicity)-T3SSs of plant pathogens this appendage extends into a several μM long pilus that can traverse the width of the plant cell wall (He and Jin, 2003). The final step across the host membrane is facilitated by a translocon that is inserted into the eukaryotic membrane (Büttner and Bonas, 2002). The translocated effectors manipulate the host cell from within to the benefit of the pathogen. Several effectors contribute to the suppression of the basal plant defence.

Nevertheless, plants are still no easy prey for microbes, because plant cells have evolved an additional layer of defence. This second layer is based on resistance proteins which detect the activity of specific effectors (Jones and Dangl, 2006). Although a ligand–receptor type of interaction between effector and resistance protein could be shown in a few cases, often resistance proteins indirectly sense the presence of effectors. This function is best described by the so-called guard hypothesis (Van der Biezen and Jones, 1998). Following this hypothesis, plant resistance proteins guard key plant proteins and signal the disturbances of these components by specific effector activities. The activation of resistance proteins typically induces a programmed cell death called the hypersensitive response (HR) (Greenberg and Yao, 2004). Although the HR itself is not sufficient to stop pathogens, together with concurrent reactions the plant effectively limits microbial multiplication and entraps bacteria to a limited site to prevent spreading of the infection. Structurally, many resistance proteins contain a nucleotide-binding site (NB) and an LRR. Resistance proteins can further be classified according to their N-terminus which either contains a Toll-interleukin-receptor-like (TIR) domain or a different domain (often coiled-coil). This N-terminal domain and the NB-site determine the requirements for specific downstream signalling pathways (DeYoung and Innes, 2006). Whereas some resistance proteins detecting fungal or oomycete infections are integral membrane proteins with extracellular LRR domains, resistance proteins detecting bacterial infections are typically localized within the plant cell (Chisholm et al., 2006; Jones and Dangl, 2006).

The development of plant resistance proteins and the strong cell death response they trigger is an evolutionary masterpiece that successfully prevents infection by pathogens that rely on certain effectors for their virulence. This forced the bacteria to upgrade their virulence inventory and they have done so by developing novel effectors. The possibility to transport virulence proteins via the T3SS into plant cells offered the pathogen diverse means of manipulation that were not functionally limited to suppression of basal resistance. Accordingly, several bacterial effectors have now been described that effectively interfere with resistance protein signalling and defence responses including the HR (Fig. 12.2). This again exerted strong selective pressure on the plant to develop novel recognition specificities for the new effectors and this struggle is believed to have resulted in alternating evolutionary steps in bacteria and plants to gain virulence or resistance, respectively (Fig. 12.1; Jones and Dangl, 2006). The zig-zag of virulence and resistance results in a complex interaction network of effectors and plant resistance components.

Keys to the world – identification of type III effectors

The first Hrp-T3SS-substrates, now called effectors, were identified because their presence lead to an HR in resistant plants (Staskawicz et al., 1984; Staskawicz et al., 1987). The outcome of this interaction was unfavourable for the infecting bacteria and these effectors were therefore initially called avirulence factors (abbreviated as Avr). The HR turned out to be dependent on the T3SS, which indicated that the avirulence factors were substrates of this secretion system (Lindgren et al., 1988). It soon became evident that not all T3SS-secreted substrates elicit a plant defence and the term 'avirulence' is counterintuitive to the functions of these factors to promote bacterial virulence. Therefore, T3SS-secreted substrates that are translocated into host cells are nowadays generally called effectors (abbreviated: Hop (*Pseudomonas syringae*); Xop (*Xanthomonas* spp.); Pop (*Ralstonia solanacearum*)). Because T3SSs play a key role in virulence of the bacteria, the substrates of T3SSs have gained major attention and have initiated a hunt which is still ongoing to identify new effectors (Grant et al., 2006; Gürlebeck et al., 2006; Lindeberg et al., 2006). Several genetic screens were performed to identify effector genes in plant pathogenic bacteria and together with genome sequencing of plant pathogenic bacteria a large collection of effectors has been assembled to date (Cunnac et al., 2004; Gürlebeck et al., 2006; Lindeberg et al., 2006). These effectors group into more than 50 families based on similarities while up to 38 functional effectors can be found in one strain of *P. syringae* (Lindeberg et al., 2005; Lindeberg et al., 2006).

Destination anywhere – subcellular localization guides type III effectors to their host targets

The cytoplasm

Most bacterial effectors are believed to display their activity within the cytoplasm of the host cell. Translocated effectors are transported through the T3SS in a partially unfolded state and probably refold in the host cytoplasm. Following translocation, some bacterial effectors are activated within the eukaryotic cell by cytoplasmic host components (Coaker et al., 2005). Effectors with virulence targets in the host cytoplasm are then ready to perform their specific activity. The cytoplasm is the cellular location for signalling cascades, mRNAs and the generation of secretory vesicles. All these processes have been identified as virulence targets of effectors. In contrast, some effectors have to localize to specific subcellular structures and compartments within the host cell to perform their specific task (Table 12.1).

The nucleus

The first subcellular compartment that was identified to be targeted by effectors was the host nucleus (Van den Ackerveken et al., 1996). The largest class of effectors that localize to the host nucleus are members of the diverse AvrBs3-family of effectors from *Xanthomonas* spp. and *Ralstonia* spp. (Gürlebeck et al., 2006). These effectors carry as distinct structural features 2–3 nuclear localization sequences (NLS) in their C-terminal domain that interact with plant importin-α which mediates its nuclear import (Table 12.1; Szurek et al., 2001). In addition, XopD from *X. campestris* pv. *vesicatoria* is a SUMO-isopeptidase that contains an NLS in the C-terminus of the protein. XopD localizes to subnuclear foci within the nucleus, but its targets in this sub-compartment are not known (Hotson et al., 2003). PopP2 from *R. solanacearum* also localizes to the plant nucleus, but here a bipartite NLS in the N-terminal domain of the protein is used (Deslandes et al., 2003). Finally, HsvB and HsvG, two effectors from *Pantoea agglomerans* localize to the plant nucleus, but NLSs were not identified so far (Nissan et al., 2006). These examples show that bacterial effectors can successfully employ the nuclear import machinery of the eukaryotic host cell to reach their cellular targets.

The plasma membrane

Signal perception is a key event in plant–pathogen interactions and is initiated at the plant plasma membrane. To inhibit these processes it may be advantageous for the pathogen to localize effectors to the cytoplasmic face of the plasma membrane. Pioneering work showed that the effectors AvrB, AvrRpm1, and AvrPphB (HopAR1) from *P. syringae* are targeted to the plant plasma membrane by a modification applied by the host cell itself (Nimchuk et al., 2000). This modification is an N-terminally attached fatty acid (myristic acid) that functions as a hydrophobic anchor to hook the effector up to the membrane (Nimchuk et al., 2000). The modification requires an N-terminal recognition signal containing a terminal glycine residue that is N-myristoylated. AvrPphB autoproteolytically processes its own N-terminus and thus reveals the N-myristoylation site (Nimchuk et al., 2000; Shao et al., 2002; Tampakaki et al., 2002). A N-myristoylation-signal has since been identified in several effectors from *P. syringae*, *Xanthomonas* spp., *R. solanacearum* and *P. agglomerans* (Table 12.1; Nimchuk et al., 2000; Shan et al., 2000; Maurer-Stroh and Eisenhaber, 2004; Robert-Seilaniantz et al., 2006; Thieme et al., 2007). In contrast, the effector HopM1 localizes to endomembranes of the plant cell, but the molecular basis of this is not known (Nomura et al., 2006).

The chloroplast

A few effectors from *P. syringae* contain N-terminal chloroplast-targeting signals and have therefore been speculated to localize to these plant cell organelles (Guttman et al., 2002). HopI1 (HopPmaI) from *P. syringae* was shown to localize to chloroplasts (Jelenska et al., 2007). This effector represses chloroplast-mediated defences and induces remodelling of the chloroplast thylakoids (Jelenska et al., 2007). Many eukaryotic chloroplast-targeted proteins contain an N-terminal signal sequence that is cleaved upon

Table 12.1 Effectors containing N-myristoylation, chloroplast or nuclear localization signals

Effector[a]	Organism[a]	Localization[b,c]	Sequence motif[d]	Reference[e]
Nuclear localization				
AvrBs3-family	*Xanthomonas* spp.	Nucleus[b]	-KRVKP-; -RKRSRS-; -RVKRPR-	Van den Ackerveken *et al.* (1996), Zhu *et al.* (1998), Szurek *et al.* (2001)
XopD	*X.c.* pv. *vesicatoria*	Nucleus[b]	-KKKKSK-	Hotson *et al.* (2003)
PopP2	*R. solanacearum*	Nucleus[b]	-RRRR-X$_{11}$-RRQRQ-	Deslandes *et al.* (2003)
HsvB, HsvG	*P. agglomerans*	Nucleus[b]	n.d.	Nissan *et al.* (2006)
Plasma membrane localization				
AvrB (AvrB1)	*P.s.* pv. *glycinea*	Plasma membrane[b]	-GCVSSKSTTVLSPQT-	Nimchuk *et al.* (2000)
AvrRpm1	*P.s.* pv. *maculicola*	Plasma membrane[b]	-GCVSSTSRSTGYYSG-	Nimchuk *et al.* (2000)
AvrPphB (HopAR1)	*P.s.* pv. *phaseolicola*	Plasma membrane[b]	-GCASSSGVSLEDDSHT	Nimchuk *et al.* (2000)
AvrC (AvrB2)	*P.s.* pv. *glycinea*	Plasma membrane[c]	-GNVCFRPSRSHVSQE-	AAA88428
AvrPto (AvrPto1)	*P.s.* pv. *tomato*	Plasma membrane[b]	-GNICVGGSRMAHQVN-	Shan *et al.* (2000)
AvrPphF (HopF1)	*P.s.* pv. *phaseolicola*	Plasma membrane[c]	-GNICNSGGVSRTYSP-	CAM12736
HopF2 (HopPtoF)	*P.s.* pv. *tomato*	Plasma membrane[b]	-GNICGTSGSRHVYSP-	Robert-Seilaniantz *et al.* (2006)
HopO1-1 (HopPtoO)	*P.s.* pv. *tomato*	Plasma membrane[c]	-GNICGTSGSNHVYSP-	NP_808677
HopZ1 (HopPmaD)	*P.s.* pv. *glycinea*	Plasma membrane[c]	-GNICIGGPRMSQQVY-	ABK13729
HopZ2 (AvrPpiG1)	*P.s.* pv. *pisi*	Plasma membrane[c]	-GICVSKPSVRHDYNE-	CAC16700
HopX2 (HopPmaB)	*P.s.* pv. *maculicola*	Plasma membrane[c]	-GLCVSKGSTASSPQH-	AAL84240
HopAF1 (HopPtoJ)	*P.s.* pv. *tomato*	Plasma membrane[c]	-GLCISKHSGSSYSYS-	NP_791393
XopE1	*X.c.* pv. *vesicatoria*	Plasma membrane[b]	-GLCISKPAMSGSSVA-	Thieme *et al.* (2007)

XopE2	X.c. pv. vesicatoria	Plasma membrane[b]	-GLCSSKPSVVGSPVA-	Thieme et al. (2007)
XopJ	X.c. pv. vesicatoria	Plasma membrane[b]	-GLCVSKPSVAGSPEH-	Thieme et al. (2007)
AvrXacE1	X.a. pv. citri	Plasma membrane[c]	-GLCVSRPATSGSSVA-	AAM35178
AvrXacE3	X.a. pv. citri	Plasma membrane[c]	-GLCSSKPSVAGSPVA-	AAM39257
XAC3230	X.a. pv. citri	Plasma membrane[c]	-GLCTSKPSVVGSPVA-	AAM38074
AvrXccB	X.c. pv. campestris	Plasma membrane[c]	-GLCNSKSVAGSVVGS-	AAM42989
AvrXccC	X.c. pv. campestris	Plasma membrane[c]	-GLCASKPSVAGSPAR-	Wang et al. (2007b)
AvrXccE1	X.c. pv. campestris	Plasma membrane[c]	-GLCVSKPSVAGSPDH-	AAM40923
PthG	P.a. pv. gypsophilae	Plasma membrane[c]	-GCFNVTGASGRANNY-	AAC24862
RSp0213	R. solanacearum	Plasma membrane[c]	-GCFNVTGTSGTASNY-	NP_521774
RSp0304	R. solanacearum	Plasma membrane[c]	-GNLQIKASSAPYALL-	NP_521865
Chloroplast localization				
HopI1 (HopPmaI)	P.s. pv. tomato	Chloroplast[b]	n.d.	Jelenska et al. (2007)

[a]Only representative family members are listed. Alternative effector names are in brackets. P.s., Pseudomonas syringae; X.c., Xanthomonas campestris; X.a., Xanthomonas axonopodis; P.a., Pantoea agglomerans; R. solanacearum, Ralstonia solanacearum.
[b]Subcellular localization in host cells was experimentally shown.
[c]Subcellular localization in host cells is predicted.
[d]Sequences directing subcellular localization are predicted or experimentally verified. N-terminal methionines are removed before myristoylation and omitted here. Potentially myristoylated glycines and potentially palmitoylated cysteines are bold. n.d., not determined
[e]Reference or accession number.

entry into the chloroplast (Jarvis and Robinson, 2004). Surprisingly, a N-terminal consensus sequence is not required for uptake of HopI1 into the stroma of the chloroplast. The effector protein is not processed after uptake and instead seems to employ a non-canonical mechanism for chloroplast membrane transit (Jarvis and Robinson, 2004; Jelenska et al., 2007).

Comfortably numb – host targets of type III effectors

Suppression of basal plant defence

P. syringae and Xanthomonas pathogens with mutations in their T3SS elicit plant cell wall depositions and induce defence-related plant genes (Hauck et al., 2003; Keshavarzi et al., 2004; Truman et al., 2006). Both responses are indicative of an activated basal plant defence. In contrast, wild-type strains only transiently and slightly induce expression of corresponding genes and no callose deposition occurs (Hauck et al., 2003; Keshavarzi et al., 2004). These results indicate that bacterial pathogens can effectively suppress the basal defence responses of plant cells through the virulence activities of effectors (Fig. 12.2 and Table 12.2).

Effectors blocking perception and signalling
The first candidate for a suppressor of plant defence was the effector AvrPto from P. syringae pv. tomato. Expression of avrPto in transgenic Arabidopsis plants suppresses callose deposition and prohibits gene induction elicited by a T3SS mutant of P. syringae (Hauck et al., 2003). The AvrPto activity is even sufficient to allow significant growth of a P. syringae pv. tomato T3SS-mutant in avrPto-transgenic plants. This demonstrates that suppression of the basal defence alleviates to some extend the need for a T3SS of a pathogen and suggests that suppression of basal plant defence is a major function of the T3SS and its effectors.

Two studies have monitored the induction of the defence-related gene *FRK1* (flg22-induced receptor-like kinase 1) from *Arabidopsis* encoding an RLK (Asai et al., 2002; He et al., 2006). *FRK1* expression is induced at 2h after challenge with flg22, a T3SS mutant or wild-type *P. syringae* pv. *tomato*, but only wild-type bacteria are able to suppress *FRK1* expression at subsequent time points. This again reveals the suppression of MAMP signalling by effectors secreted from virulent bacteria. Detailed analysis showed that only AvrPto and AvrPtoB (HopAB2) out of eleven tested *P. syringae* effectors (AvrPto, AvrPtoB, HopAO1, HopE1, HopK1, AvrRpm1, AvrB, AvrRpt2, AvrBsT, VirPphA, HopAI1) can suppress *FRK1* induction (He et al., 2006), demonstrating that both can suppress MAMP signalling. AvrPto and AvrPtoB blocked MAMP perception very early between receptor binding and MAPKKK activation (He et al., 2006). AvrPto functions as an inhibitor of activated kinases (Xing et al., 2007) and is able to bind and inhibit directly the kinase domain of the MAMP receptors FLS2, EFR, and possibly others (Xiang et al., 2008).

Basal defence responses can also be induced by a pathogen that carries a functional T3SS, but is unable to successfully mount an infection on a non-host plant. *P. syringae* pv. *phaseolicola* is a pathogen of bean (Lindgren et al., 1986) that can not multiply or cause disease in *Arabidopsis*, but does not elicit an HR either. Infiltration of *P. syringae* pv. *phaseolicola* strains into the *Arabidopsis* leaf apoplast lead to deposition of callose and the induction of defence-related genes indicative of an activated plant basal defence (Li et al., 2005; de Torres et al., 2006; Ham et al., 2007). Although *P. syringae* pv. *phaseolicola* carries a functional T3SS it cannot suppress this basal defence in *Arabidopsis* possibly because its effector repertoire is not suitable for this plant host. *P. syringae* pv. *phaseolicola* therefore can be used to deliver additional effectors and test whether they enhance virulence of these bacteria on *Arabidopsis*.

The basal defences acting against *P. syringae* pv. *phaseolicola* can be suppressed by AvrPto and AvrPtoB (de Torres et al., 2006; He et al., 2006). Delivery of AvrPtoB by *P. syringae* pv. *phaseolicola* enables the bacteria to multiply to some extent and to develop disease symptoms on *Arabidopsis* Ws-3 (de Torres et al., 2006). In comparison, transgenic expression of *avrPtoB* before challenge by pathogens has a more thorough effect on basal defence suppression than delivery by

Table 12.2 Effector activities and plant targets

Effector[a]	Organism	Feature	Plant target[b]	Phenotype	Selected references[c]
AvrB (AvrB1)	P. syringae	Possible protokinase; phosphorylated *in planta*	RIN4	RIN4 phosphorylation; manipulation of JA signalling	Grant et al. (2006), Desveaux et al. (2007)
AvrB2 (AvrPphC)	P. syringae		n.d.	Suppression of HopF1 cell death in bean cultivar Canadian Wonder	Tsiamis et al. (2000)
AvrE (AvrE1)	P. syringae		n.d.	Suppression of SA-dependent basal defence responses; enhances disease symptoms and *in planta* growth	Badel et al. (2006), Grant et al. (2006), Nomura et al. (2006), Ham et al. (2007)
AvrPphB (HopAR1)	P. syringae	Papain-like cysteine protease; CA clan, C58 peptidase	PBS1, serine-threonine kinase	PBS1 cleavage; manipulation of JA signalling	Grant et al. (2006)
AvrPto (AvrPto1)	P. syringae	Kinase inhibitor	Pto, kinase	Suppression of SA-independent basal defence signalling, inhibition of Pto,FLS2 and EFR; manipulation of ethylene signalling	Grant et al., (2006), He et al. (2007), Xing et al. (2007), Xiang et al. (2008)
VirPphA (HopAB1)	P. syringae	Ubiquitin E3 ligase	Pto, kinase; FLS2, EFR, receptor-like kinases	Suppression of HR in bean	Jackson et al. (1999), Jackson et al. (2002), Kim et al. (2002)
AvrPtoB (HopAB2)	P. syringae	Ubiquitin E3 ligase; phosphorylated *in planta*	Pto, Fen, kinases	Ubiquitination of Fen and related kinases; suppression of cell death; suppression of basal defence signalling; manipulation of ABA and ethylene signalling	de Torres et al. (2006), Grant et al. (2006), He et al. (2006), de Torres-Zabala et al. (2007), Rosebrock et al. (2007), Xiao et al. (2007a,b)
AvrRpm1	P. syringae		RIN4	RIN4 phosphorylation; suppression of basal defence	Grant et al., (2006), Ham et al. (2007)
AvrRpt2	P. syringae	Staphopain-like cysteine protease; CA clan, C47 peptidase; activation *in planta* by cyclophilin	RIN4	RIN4 cleavage; suppression of RPM1 HR; SA-independent suppression of basal defence; manipulation of JA signalling; increase of cellular auxin signalling	Coaker et al. (2006), Grant et al. (2006), Chen et al. (2007)
HopC1 (HopPtoC)	P. syringae		n.d.	Suppression of basal defence	Li et al. (2005)
HopE1 (HopPtoE)	P. syringae		n.d.	Suppression of HopPsyA cell death	Jamir et al. (2004)

Table 12.2 continued

Effector[a]	Organism	Feature	Plant target[b]	Phenotype	Selected references[c]
HopF1 (AvrPphF)	P. syringae		n.d.	Suppression of cell death in bean cultivar Tendergreen	Tsiamis et al. (2000)
HopF2 (HopPtoF)	P. syringae		n.d.	Suppression of HopPsyA cell death; suppression of basal defence	Jamir et al. (2004); Li et al. (2005)
HopG1 (HopPtoG)	P. syringae		n.d.	Suppression of cell death	Jamir et al. (2004);
HopI1 (HopPmaI)	P. syringae	J-domain; repeats; phosphorylated in planta	n.d.	Blocks SA accumulation and basal defence signalling; alters thylakoid ultrastructure	Jelenska et al. (2007)
HopK1	P. syringae		n.d.	Manipulation of JA signalling	He et al. (2004a)
HopM1 (HopPtoM)	P. syringae		AtMIN7, ARF-GEF	Suppression of SA-dependent basal defence responses; targets AtMIN7 for proteasome degradation; enhances disease symptoms and in planta growth	Badel et al. (2006), Grant et al. (2006), Nomura et al. (2006), Ham et al. (2007)
HopN1 (HopPtoN)	P. syringae	Papain-like cysteine protease; CA clan, C58 peptidase	n.d.	Suppression of cell death in HR and disease	Lopez-Solanilla et al. (2004)
HopS1	P. syringae		n.d.	Suppression of basal defence	Li et al. (2005)
HopT1 (HopPtoT1)	P. syringae		n.d.	Suppression of basal defence	Li et al. (2005)
HopU1 (HopPtoS2)	P. syringae	ADP-ribosyltransferase	GRP7, RNA-binding protein	Suppression of basal defences; delay of HR; blocks RNA binding by GRP7	Fu et al. (2007)
HopX1 (AvrPphE)	P. syringae	Putative cysteine-based catalytic triad	n.d.	Manipulation of JA signalling; suppression of HopPsyA cell death	Nimchuk et al. (2007)
HopZ-family	P. syringae	Protease activity; CE clan, C55 peptidase	n.d.	n.d.	Grant et al. (2006); Ma et al. (2006)
HopAA1 (CEL ORF5)	P. syringae		n.d.	Suppression of basal defence	Li et al. (2005)

Effector	Organism	Domain/Activity	Host target	Function	Reference
HopAF1 (HopPtoJ)	P. syringae		n.d.	Suppression of basal defence	Li et al. (2005)
HopAI1	P. syringae	Phosphothreonine lyase	MPK3, MPK6	blocks MAPK signalling; suppression of basal defences	Zhang et al. (2007)
HopAM1 (AvrPpiB)	P. syringae		n.d.	Suppression of HopPsyA cell death	Jamir et al. (2004)
HopAO1 (HopPtoD2)	P. syringae	Tyrosine phosphatase domain	Possibly MAPK	Blocks MAPK signalling and cell death in N. tabacum; suppression of basal defence responses	Bretz et al. (2003), Espinosa et al. (2003), Underwood et al. (2007)
HsvB, HsvG	P. agglomerans	Transcriptional activator	n.d.	Promotes gall formation	Nissan et al. (2006)
AvrBs2	X. campestris	Similarity to glycerophosphoryl diester phosphodiesterase	n.d.	Enhances virulence	Grant et al. (2006)
AvrBs3 (PthA)-family	Xanthomonas and R. solanacearum	Transcriptional activator	upa20 (AvrBs3); Os8N3 (PthXo1)	Induces expression of target host genes; suppression of basal defence	Gürlebeck et al. (2006), Yang et al. (2006), Kay et al. (2007), Römer et al. (2007)
AvrXv4	X. campestris	CE clan, C55 peptidase	n.d.	Reduction of SUMOylated proteins	Roden et al. (2004)
XopD	X. campestris	SUMO isopeptidase; CE clan, C48 peptidase	n.d.	deSUMOylation of nuclear host proteins	(Hotson et al., 2003)
GALA1-7	R. solanacearum	F-box, leucine-rich repeat	n.d.	supports bacterial virulence; possibly forms complex with ubiquitinase system to degrade plant proteins	(Angot et al., 2006)

[a] Alternative effector names are in brackets

[b] n.d., not determined

[c] Effector activities are reviewed in Grant et al. (2006). Additional results are cited separately.

the T3SS of *P. syringae* pv. *phaseolicola* (de Torres et al., 2006). This indicates that the timing and quantity of effector activities are crucial for a successful suppression of host defences.

AvrPto and AvrPtoB are not the only effectors that block MAMP signalling. A screen was performed based on the induction of the defence-related gene *NHO1* (non-host 1) from *Arabidopsis* whose expression is induced upon flg22 perception (Kang et al., 2003; Li et al., 2005). *NHO1* encodes a glycerol kinase that is involved in resistance to bacterial and fungal pathogens (Lu et al., 2001; Kang et al., 2003). Nine out of 19 tested *P. syringae* pv. *tomato* DC3000 effectors (HopS1, HopAI1, HopAF1, HopT1-1, HopT1-2, HopAA1-1, HopF2, HopC1 and AvrPto) block flg22-mediated induction of *NHO1* when the effector genes are expressed in transfected protoplasts (Li et al., 2005). It is clear that several effectors are able to disrupt MAMP signalling. Interestingly, HopAI1 suppresses *NHO1* but not *FRK1* induction which demonstrates that MAMP signalling encompasses separate signalling pathways (Li et al., 2005).

Not all effectors intercept MAMP signalling at very early stages like AvrPto and AvrPtoB, but some interfere with the subsequent signalling cascade. The effector HopAI1 specifically blocks activation of the MAP kinase cascade by a novel mechanism demonstrating the variability of effector activities (Zhang et al., 2007). HopAI1 interacts *in vivo* with the *Arabidopsis* MAP kinases MPK3 and MPK6 (Zhang et al., 2007), which are both specifically involved in flg22-mediated FLS2 signalling (Asai et al., 2002). The effector displays phosphothreonine lyase activity to remove the phosphate group from MPK3, MPK6 and even human MAP kinase p-Erk2 (Li et al., 2007; Zhang et al., 2007). Phosphothreonine lyase is a novel enzymatic activity that leaves the threonine residue without a free hydroxyl group. This prevents its rephosphorylation and HopAI1 thereby irreversibly inactivates MAP kinases (Li et al., 2007). The activity of HopAI1 efficiently blocks flg22-induced defence-gene induction and cell wall modifications which are triggered through the MAP kinase signalling cascade (Zhang et al., 2007).

Effectors blocking the establishment of cellular basal defence responses

The interference of effectors with basal plant defence can occur at different steps. Whereas the examples given so far showed a block of MAMP signal transduction before the induction of marker genes, some effectors seem to interfere with the establishment of basal plant defence at a later time point.

The effector HopAO1 (HopPtoD2) is a multi-domain protein with an N-terminal similarity to the effector HopD1 (AvrPphD) and a C-terminal protein tyrosine phosphatase domain (Bretz et al., 2003; Espinosa et al., 2003). Transgenic expression of *hopAO1* in *Arabidopsis* suppresses several basal plant defence responses and enables growth of a *P. syringae* T3SS mutant (Underwood et al., 2007). This activity is dependent on the enzymatic phosphatase activity of this effector. Although HopAO1 blocks flg22-induced responses (Underwood et al., 2007), it does not interfere with *FRK1* induction (He et al., 2006). HopAO1 does not seem to interfere with MAMP signalling itself, but rather with the later induction of defence-related genes (e.g. several pathogenesis-related (PR)-genes) and the establishment of defence reactions. *Arabidopsis* MAP kinases are not targets for the HopAO1 phosphatase because MPK3 or MPK6 signalling is not down-regulated by this effector, but rather show an enhanced activation (Underwood et al., 2007). At which step HopAO1 interferes between enhanced activation of the MAMP-signalling pathway and the establishment of defence is not clear yet.

At least two other effectors interfere with a separate branch of MAMP signalling and seem to also suppress the establishment of basal defence rather than early signalling. AvrE (AvrE1) and HopM1 (HopPtoM) are able to suppress MAMP-triggered callose deposition at the plant cell wall, similar to several of the effectors described above. Nevertheless, both inhibit only salicylic acid (SA)-dependent callose deposition (DebRoy et al., 2004) whereas AvrPto for example suppresses SA-independent responses (Hauck et al., 2003). SA is a plant hormone involved in defence signalling. This indicates that AvrE and HopM1 interfere with a defence signalling pathway that involves SA.

In addition, AvrE and HopM1 promote disease lesion formation and *in planta* growth (Badel *et al.*, 2006). AvrE and HopM1 have partially redundant activity although they don't share amino acid sequence similarities. The *avrE* and *hopM1* genes are localized in the conserved effector locus neighbouring the *hrp* gene cluster (Alfano *et al.*, 2000) and members of these effector families are widely conserved in different pathosystems. Mutations in both genes severely compromise pathogen virulence which indicates that these effectors fulfil an important virulence function (Gaudriault *et al.*, 1997; Bogdanove *et al.*, 1998; Alfano *et al.*, 2000; DebRoy *et al.*, 2004).

One of the molecular host targets of HopM1 was identified (Nomura *et al.*, 2006). HopM1 interacts with the immunity-associated protein AtMIN7 from *A. thaliana* and mediates its degradation via the host proteasome (Nomura *et al.*, 2006). AtMIN7 is one of eight adenosine diphosphate (ADP) ribosylation factor (ARF) guanine nucleotide exchange factors (GEF) in *Arabidopsis* which are key components of the vesicle trafficking system. Vesicle trafficking plays an important role in plant innate immunity and is the prerequisite for cell wall-associated defences which are blocked by HopM1. Accordingly, an AtMIN7 mutant doesn't show callose deposition when challenged with strains deficient in *avrE* and *hopM1* whereas a wild-type *Arabidopsis* does (Nomura *et al.*, 2006). HopM1 also suppresses induction of the defence-related gene PR-1 (Ham *et al.*, 2007), but it is unclear how this is achieved. Although HopM1 and AvrE are partially functionally redundant (DebRoy *et al.*, 2004), AtMIN proteins are not destabilized by AvrE indicating that both effectors eventually target different host components (Nomura *et al.*, 2006).

MAMP perception leads to the transcription of defence-related genes in plant cells which supports the establishment of an antimicrobial environment. At least one effector seems to interfere with late defence by manipulating mRNA-related processes. Transgenic expression of the effector *hopU1* (*hopPtoS2*) in *Arabidopsis* reduces the callose depositions after flg22 challenge indicating a suppression of MAMP responses (Fu *et al.*, 2007). HopU1 displays mono-ADP-ribosyltransferase activity towards plant cell proteins. ADP-ribosyltransferases covalently modify target proteins with an ADP-ribose moiety and thereby often inactivate their targets. Plant cells don't contain enzymes with mono-ADP-ribosyltrasferase activity themselves, but this activity is well known from toxins (e.g. diphtheria toxin, cholera toxin) of mammalian pathogens. HopU1 modifies a plant glycine-rich RNA-binding protein (GRP7) and prevents its binding to RNA (Fu *et al.*, 2007). It is not known so far, which RNA molecules are normally bound by GRP7, but the inactivation of GRP7 and the following impact on the RNA household of the plant cell seem to compromise plant defence. An *Arabidopsis grp7* mutant allows higher growth of wild-type *P. syringae* pv. *tomato* DC3000 and a T3SS mutant strain (Fu *et al.*, 2007). These results demonstrate that plant RNA pathways can be virulence targets of bacterial effectors to block the establishment of basal defence.

Basal resistance causes reduced vascular flow into minor veins of plant leaves. The vascular flow was visualized using a neutral red solution and leaf staining was reduced by T3SS mutant strains of *P. syringae* pv. *tomato* or flg22 (Oh and Collmer, 2005). This reduced vascular staining is suppressed by either wild-type bacteria or non-pathogenic *Pseudomonas fluorescens* expressing a heterologous T3SS and one of a set of effectors (AvrPto, AvrE, HopM1, HopF2, or HopG1; Oh and Collmer, 2005). Although this screen demonstrates the influence of several effectors on basal plant defence, it does not distinguish at which point the effectors block this establishment. Some of the effectors have been described above, but it is unclear so far, where HopF2 and HopG1 interfere with basal resistance.

Although the molecular targets are not known, some effectors exhibit a distinct phenotype associated with disease development. Mutations in *hopAA1-1* and its paralogue *hopAA1-2* from *P. syringae* pv. *tomato* DC3000 have no influence on bacterial growth *in planta* or disease symptom formation, but they contribute to the successful formation of microcolonies in the apoplast of infected tomato leaves (Badel *et al.*, 2002).

The effector XopX from *Xanthomonas* targets basal plant defence and displays a particular phenotype (Metz *et al.*, 2005). *xopX* was cloned as a typical avirulence gene from *X. campestris* pv. *vesicatoria* by conferring enhanced necrosis to an *X. campestris* pv. *campestris* strain that elicits only mild symptoms on *N. benthamiana*. Surprisingly, the enhanced necrosis mediated by *xopX* is not associated with less bacterial growth typical of a plant HR, but rather with increased bacterial titers (Metz et al., 2005). In addition, the necrosis-promoting activity of XopX is not seen when the gene is expressed in *N. benthamiana*, but only if pathogenic bacteria are simultaneously infected. *xopX* confers enhanced virulence not only to *Xanthomonas*, but also to *P. syringae* strains emphasizing a more general effect that was interpreted as suppression of basal defence, but the molecular targets and the mechanism of XopX effector activity are unknown so far (Metz et al., 2005).

Multiple layers of defence
P. syringae pv. *phaseolicola* not only elicits basal defences in its non-host plant *Arabidopsis* comparable to T3SS mutant bacteria, but in addition triggers plant defence responses through its T3SS (Ham et al., 2007). The responses include an additional subset of large callose deposits and a strong induction of the defence-related *PR-1* gene. These responses can be differentially suppressed by heterologous effectors. AvrRpm1 and HopM1 both suppress callose deposition, but only HopM1 also efficiently suppresses PR-1 gene induction. On the other hand, transgenically expressed *avrRpm1* supported some *P. syringae* pv. *phaseolicola* growth whereas HopM1 delivered from bacteria does not (Ham et al., 2007). High-level growth of *P. syringae* pv. *phaseolicola* is only achieved when multiple layers of defence are compromised either by the combined activity of several appropriate effectors or plant mutants in defence signalling (Ham et al., 2007). This demonstrates that non-host bacteria elicit separate plant defence responses via MAMPs and their T3SS. Their lack of pathogenicity is based on their inability to suppress these responses. It also supports the model that basal plant defence is not a single pathway leading to basal resistance. Instead, basal defence seems to rely on multiple signalling pathways and different layers of cellular responses that together prevent microbial growth. Plant pathogens have to overcome all of these layers to mount a successful infection (Thordal-Christensen, 2003).

Suppression of resistance gene-based plant defence

Plant resistance genes signal effector-mediated disturbances of plant cellular components often eliciting an HR. This strategy successfully prevents growth of pathogens expressing specific effectors. A growing body of evidence supports the observation that certain effectors can suppress this additional layer of defence and block resistance gene-based plant defence (Table 12.2). Alternating new effector activities and new resistance specificities seem to have evolved into a complicated network of interdependencies (Fig. 12.1). In addition, plants seem to have evolved molecular mousetraps that attract pathogen effectors to resistance complexes to inadvertently elicit resistance signalling.

Effectors suppressing the HR
The first example of an effector that can suppress plant HR reactions was VirPphA (HopAB1) from *P. syringae* pv. *phaseolicola* 1449B (Jackson et al., 1999). Deletion of a virulence plasmid in the bean pathogen *P. syringae* pv. *phaseolicola* causes the strain to elicit an HR on normally susceptible bean cultivars (Jackson et al., 1999). Reintroduction of *virPphA* renders the bacteria virulent again, indicating that this effector can suppress a plant defence reaction against bacterial factors that would otherwise trigger an HR. Subsequent studies have shown that the VirPphA-related effector AvrPtoB which has already been described above to block basal plant defence can also efficiently suppress cell death when transiently expressed in *N. benthamiana* (Abramovitch et al., 2003) or delivered by non-pathogenic *P. fluorescens* (Jamir et al., 2004). AvrPtoB can even prevent programmed cell death triggered by the proapoptotic Bax protein from mouse and stress-induced programmed cell death in yeast (Abramovitch et al., 2003). Nevertheless, AvrPtoB is not a general cell death suppressor, because specific HRs against several effectors are not influenced (de Torres et al., 2006; He et al., 2006).

AvrPtoB itself can also elicit an HR in resistant tomato plants. This occurs through interaction with a protein complex consisting of the tomato serine-threonine protein kinase Pto and the NB-LRR protein Prf (Kim et al., 2002; Mucyn et al., 2006). The recognition depends on direct interaction of Pto and AvrPtoB (Kim et al., 2002). The sequence-unrelated effector AvrPto, which has also already been mentioned above to block basal defences in a similar way like AvrPtoB, is detected by the same Pto/Prf complex (Scofield et al., 1996; Tang et al., 1996). The interaction is based on conserved structural motifs present in both AvrPto and AvrPtoB (Kim et al., 2002; Wu et al., 2004; Abramovitch and Martin, 2005; Xing et al., 2007). Analysis of the crystal structure of an AvrPto-Pto complex confirms that AvrPto only binds to the active, phosphorylated state of Pto (Tang et al., 1996; Sessa et al., 2000; Xing et al., 2007). Importantly, the binding resembles a kinase-inhibitor interaction and indeed AvrPto binding strongly decreases Pto kinase activity. While the interaction of AvrPto and Pto itself triggers Prf activation, the original function of AvrPto likely is the inhibition of kinases, e.g. receptor-like kinases involved in basal defence or resistance gene-mediated defence signalling (He et al., 2006; Xing et al., 2007; Xiang et al., 2008).

The interaction between these effectors and Pto reveals an interesting view to the evolution behind effectors and their host targets. *Prf*, *Pto*, and four additional *Pto*-homologues, are clustered at a single genomic locus in tomato, but only two *Pto* homologues, Pto and Fen, are functional kinases (Chang et al., 2002; Pedley and Martin, 2003). The *Pto* resistance locus and AvrPtoB seem to have coevolved and have established a network of alternating recognition and suppression activities. Recent results have begun to untangle this interaction network and show that AvrPtoB is a multi-domain protein. The N-terminal region of AvrPtoB is sufficient to contribute to bacterial virulence by suppressing basal defences (de Torres et al., 2006; He et al., 2006; Xiao et al., 2007b), but triggers an HR in tomato plants by interaction with Fen and Prf (Rosebrock et al., 2007). In contrast, full length AvrPtoB contains a C-terminal domain with E3 ubiquitin ligase activity that mediates ubiquitination of Fen and causes its degradation via the host proteasome (Abramovitch et al., 2006; Janjusevic et al., 2006; Rosebrock et al., 2007). Nevertheless, the Pto kinase has evolved to mediate recognition of AvrPtoB without being susceptible to ubiquitination by the C-terminal E3 ubiquitin ligase domain of AvrPtoB (Rosebrock et al., 2007). HopPmaL (HopAB3-1) is a short allele of AvrPtoB that doesn't contain the C-terminal E3 ubiquitin ligase domain. Accordingly, HopPmaL is recognized by Pto and Prf, but does not confer cell death suppression (Lin et al., 2006). Together, these results indicate that AvrPtoB has acquired different modular activities.

AvrPtoB presumably interacts with other kinases in addition to Pto and Fen, because it is phosphorylated independent of *Pto* and *Prf* within the plant cell and this phosphorylation is required for its virulence activity (Xiao et al., 2007a). It is still unclear whether the Fen and Pto kinases have solely evolved to function as mousetraps for AvrPto and AvrPtoB and mimic virulence targets or whether they confer another function for the plant that makes them valid virulence targets.

The effector AvrPphB (HopAR1) is a papain-like cysteine protease (Shao et al., 2002; Zhu et al., 2004) that also targets a host kinase. The target of AvrPphB in *Arabidopsis* is PBS1, a serine-threonine kinase which is specifically cleaved by AvrPphB (Shao et al., 2003). PBS1 forms a complex with the NB-LRR resistance protein RPS5 and cleavage of PBS1 is detected by RPS5 and subsequently triggers an HR in resistant plants (Shao et al., 2003; Ade et al., 2007). PBS1 therefore acts as a mousetrap protein for AvrPphB activity and it is not known whether PBS1 has any additional functions in the plant. *Arabidopsis* kinases similar to PBS1, which carry a motif that is required for cleavage of PBS1, are not substrates of AvrPphB and it is unclear whether any alternative virulence targets of AvrPphB exist (Shao et al., 2003).

HopN1 (HopPtoN) from *P. syringae* pv. *tomato* is a papain-like cysteine protease that suppresses host cell death. Interestingly, both cell death associated with defence responses, as well as cell death of necrotic speck lesion typical for *P. syringae* pv. *tomato* disease is reduced (Lopez-Solanilla et al., 2004). Cell death symptoms occurring in later stages of infections and cell death associated with the HR are thought to be basically similar cellular processes involving common physiological changes, but their development occurs at different rates (Klement, 1982). It is therefore possible that effector activities influence both processes. In addition, although the *hopN1* mutant strain produces more necrotic disease symptoms than wild-type *P. syringae* pv. *tomato*, both strains grow to the same level in infected leaves (Lopez-Solanilla *et al.*, 2004), indicating that bacterial proliferation is not immediately associated with disease symptoms.

Effector interdependencies
The early reports on HR suppression activity of effectors were followed by analyses that showed that this activity is widespread among effectors and a network of specific effector interdependencies has been exposed. Deletion of the virulence plasmid in *P. syringae* pv. *phaseolicola* which rendered the bacteria avirulent on former susceptible bean cultivars can not only be restored by addition of *virPphA*, but also by reintroducing the effector *hopF1* (*avrPphF*) (Jackson *et al.*, 1999; Tsiamis *et al.*, 2000). In contrast to VirPphA, the activity of HopF1 is more limited though, because it enables virulence on bean cultivar Tendergreen, but the strains now elicit an even faster HR on cultivar Canadian Wonder. The additional introduction of AvrB2 (AvrPphC) blocks the HopF1 HR on cultivar Canadian Wonder and enables virulence again (Tsiamis *et al.*, 2000).

These reports introduced the idea that some effectors can support the activity of other specific effectors and suppress their recognition by resistance proteins. In some cases this might be achieved because different effectors can target the same host component. It is believed that especially critical elements of plant cell integrity or plant defence might be virulence targets of several effectors. In parallel, these critical elements can be guarded by several plant resistance proteins (Jones and Dangl, 2006).

The best example for this is RIN4 from *Arabidopsis* (Mackey *et al.*, 2002). RIN4 is protected by the resistance proteins RPM1 and RPS2 which are designed to detect the activities of the bacterial effectors AvrRpm1 and AvrB (both by RPM1) and AvrRpt2 (by RPS2), respectively (Mackey *et al.*, 2002; Mackey *et al.*, 2003). The function of RIN4 for the plant cell is still not fully understood, but overexpression of RIN4 suppresses callose deposition in response to flg22 and enables T3SS mutant bacteria to grow (Kim *et al.*, 2005b). In contrast, plants lacking RIN4 display an enhanced MAMP-response and allow less bacterial growth in comparison to wild-type *Arabidopsis* (Kim *et al.*, 2005b). These results indicate that RIN4 may function as a negative regulator of plant basal defence, but it is also possible that it functions basically as mousetrap for effectors. AvrRpm1 and AvrB induce phosphorylation of RIN4 (Mackey *et al.*, 2002). Although no kinase activity of AvrB could so far be measured, analysis of the AvrB crystal structure shows certain structural similarities to kinases (Desveaux *et al.*, 2007). The authors believe that AvrB resembles a novel form of protokinase which is activated by a host factor inside the plant cell and subsequently phosphorylates itself, RIN4, and probably other targets (Desveaux *et al.*, 2007; Shang *et al.*, 2006). This effector activity possibly aims to inactivate host targets, but manipulation of RIN4 is detected by the resistance protein RPM1 which triggers an HR, accordingly (Mackey *et al.*, 2002).

In 1996 it was observed that AvrRpt2 can mask recognition of AvrRpm1 by RPM1 and therefore suppress this HR (Ritter and Dangl, 1996). It is now known that this suppression activity is mediated through proteolytic cleavage of the RIN4 protein by AvrRpt2 (Axtell and Staskawicz, 2003; Mackey *et al.*, 2003). AvrRpt2 is a staphopain-class cysteine protease that is activated inside the host cell by the plant peptidyl-prolyl cis/trans isomerase cyclophilin (Axtell *et al.*, 2003; Jin *et al.*, 2003; Coaker *et al.*, 2005; Coaker *et al.*, 2006). Upon activation, AvrRpt2 self-processes its N-terminus and cleaves RIN4 at two positions thus effectively quenching the RPM1 surveillance of RIN4 (Mudgett and Staskawicz, 1999; Chisholm *et al.*, 2005; Kim *et al.*, 2005a). However, this novel type of RIN4 manipulation can also be recognized by the plant. In the presence of the resistance protein RPS2 the cleavage of RIN4 is detected and again an HR occurs. Two resistance proteins together efficiently protect plant RIN4 from manipulation by the bacteria (Axtell and Staskawicz, 2003; Mackey *et al.*, 2003; Day *et al.*, 2005). Most likely RIN4 is not the only virulence target of AvrRpt2 and AvrRpm1, because both effectors enhance pathogen growth in an *Arabidopsis rin4* mutant (Belkhadir et al., 2004; Chen et al., 2004; Lim and Kunkel, 2004). Accordingly, the recognition motif of the AvrRpt2 cleavage site can be found in several *Arabidopsis* proteins (Chisholm et al., 2005; Kim et al., 2005a), but it is unclear so far whether any of these proteins are important for the virulence function of AvrRpt2.

Jamir et al. (2004) executed a systematic screen for *P. syringae* effectors that could suppress the HR elicited by another effector. The screen was based on HopPsyA (HopA1) which induces cell death in *N. benthamiana*. Five out of 19 effectors (HopE1, HopX1, HopAM1, AvrPtoB, and HopF2) from *P. syringae* pv. *tomato* suppress this HR (Jamir et al., 2004). Accordingly, strains carrying mutations in genes encoding these effectors elicit a stronger HR on non-host tobacco plants (Jamir et al., 2004) indicating that the HR phenotype of wild-type bacteria is the sum of HR suppressing and eliciting activities. Deletion of HR suppressors reveals the underlying HR elicitation activity of other effectors. While four of the effectors also suppress Bax-induced cell death, HopAM1 does not, demonstrating that the effectors may target different signalling pathways (Jamir et al., 2004). This study emphasizes that cell death suppression and effector interdependency is a far more widespread activity among effectors than previously thought. Nevertheless it is not clear which molecular targets within the plant cell are manipulated to prevent HR elicitation.

Targeting both layers of the plant defence
Some effectors were shown to influence the basal defence as well as the HR of the plant. AvrPtoB suppresses cell death reactions (Abramovitch and Martin, 2005), and in addition, suppresses basal defence components (Shan et al., 2000; Lin et al., 2006). In this case, different domains of the effector are responsible for these activities. The N-terminal region of AvrPtoB is sufficient for basal defence suppression and the C-terminus facilitates cell death suppression (Abramovitch et al., 2006; de Torres et al., 2006).

In contrast, both activities have been assigned to one domain in other effectors. HopAO1 inhibits basal defence signalling, as described above (Underwood et al., 2007), but in addition efficiently blocks the HR (Espinosa et al., 2003). HopAO1 contains a tyrosine phosphatase domain and this is required for both activities (Espinosa et al., 2003; Underwood et al., 2007). HopAO1 suppresses cell death elicited not only by *P. syringae* pv. *phaseolicola* in *N. benthamiana*, but also by an autoactive NtMEK2^{DD} derivative (Espinosa et al., 2003). NtMEK2 is the MAPK kinase upstream of SIPK and WIPK, two tobacco MAP kinases known to activate plant defences including the HR (Nürnberger and Scheel, 2001). MAP kinase pathways are known to be involved in activating the HR as well as basal defence signalling (Nakagami et al., 2005; Pedley and Martin, 2005). HopAO1 seems to interrupt a cell-death signalling pathway in tobacco, but does not interfere with MAMP-triggered MAPK activation in *Arabidopsis* (He et al., 2006; Underwood et al., 2007). In addition, HopAO1 does not inhibit HopPsyA-triggered cell death in *N. benthamiana* (Jamir et al., 2004). The identification of virulence targets of HopAO1 will reveal how this effector can suppress both basal resistance and cell death.

The effector HopU1 is an ADP-ribosyltransferase that also interferes with both layers of the plant defence. A *Pto* DC3000 derivative carrying a mutation in *hopU1* elicits a stronger HR on non-host tobacco plants than the wild-type strain (Fu et al., 2007). In addition, transgenic expression of *hopU1* in *Arabidopsis* reduces callose deposition after flg22 challenge and delays RPS2-mediated HR. Together these results demonstrate that HopU1 influences both layers of plant defence probably via common RNA-containing pathways (Fu et al., 2007).

Manipulation of post-translational modifications

Bacterial effectors employ sophisticated means to manipulate host target proteins. Several examples show that effectors even apply or remove post-translational protein modifications to control the activity and the stability of host factors.

Several effectors from plant and mammalian pathogens are predicted to be cysteine proteases based on structural similarities and conserved residues that might form a catalytic core (Hotson and Mudgett, 2004; da Cunha et al., 2007). Mutations exchanging these conserved amino acids typically render the effector inactive (Hotson and Mudgett, 2004) attesting the importance of these residues for effector activity. These predicted cysteine proteases are classified into different clans and further into families according to their structural similarities (Hotson and Mudgett, 2004). Whereas some of them (e.g. AvrRpt2, AvrPphB, YopT) directly cleave target proteins, a few seem to manipulate host targets differently.

The effector YopJ from the mammalian pathogen *Yersinia* is the best studied example of this latter group. YopJ binds several members of the MAPK kinase superfamily and blocks their phosphorylation thus quenching the inflammatory host response (Orth et al., 1999; Viboud and Bliska, 2005). Due to the similarities of YopJ with Ulp1, a SUMO protease from yeast, it was thought that YopJ might possess similar activity. SUMO is a small protein that is linked to proteins by a conjugation mechanism similar to ubiquitination and controls the activity of its protein substrates (Hotson and Mudgett, 2004; Yang and Sharrocks, 2004). Indeed, expression of *yopJ* decreases the amount of free SUMO and SUMOylated host proteins, but no direct protease activity is observed *in vitro* (Orth et al., 2000). Still, YopJ was considered to be a SUMO protease (Orth et al., 2000). It was shown that YopJ in addition has deubiquitinase activity (Zhou et al., 2005), but both activities do not well explain the specific inhibition of MAP kinase pathways by YopJ. Recently, this mystery was solved by showing that YopJ is not a cysteine protease after all, but rather an acetyltransferase (Mittal et al., 2006; Mukherjee et al., 2006; Mukherjee et al., 2007). YopJ specifically acetylates MAPK kinases at critical serine and threonine residues thereby competing with their phosphorylation (Mittal et al., 2006; Mukherjee et al., 2006). This acetylation efficiently suppresses the MAP kinase pathway. The misconception that YopJ might be a SUMO protease stemmed from the fact that acetyltransferases and cysteine proteases actually carry out similar enzymatic reactions and thus can contain a similar catalytic core although they use different substrates (Mukherjee et al., 2007). Whereas acetyltransferases use acetyl-CoA and the hydroxyl group of serines or threonines in target proteins, cysteine proteases use a peptide and water. Additional protein domains are believed to regulate substrate choice and target protein interaction thereby clearly differentiating between both enzymatic reactions under natural conditions *in vivo*.

The effector XopD from *Xanthomonas* is also considered to be a SUMO protease. XopD drastically decreases the amount of SUMOylated plant proteins *in vitro* and *in vivo* (Hotson et al., 2003). By structural similarity, XopD belongs to the C48 peptidase family and is therefore more closely related to the SUMO protease Ulp1 from yeast (also a C48 peptidase) than the acetyltransferase YopJ (a C55 peptidase). In addition, XopD displays enzymatic activity *in vitro* which could not be shown for a member of the C55 family. This strongly suggests that XopD is a SUMO protease and not an acetyltransferase. XopD is plant specific and only cleaves SUMO from plant and not from mammalian origin which is in contrast to the promiscuous Ulp1 SUMO protease from yeast (Hotson et al., 2003). Crystal structural analysis and site directed mutagenesis of XopD reveal residues responsible for the substrate specificity of XopD (Chosed et al., 2007). The subcellular localization of XopD to nuclear foci indicates that host proteins in this compartment are its preferential virulence targets. SUMO-modifications are involved in repression of transcription (Verger et al., 2003; Yang and Sharrocks, 2004) and therefore XopD might function to manipulate the plant transcriptome.

Several additional effectors with predicted cysteine protease/acetyltransferase activity are found in *Xanthomonas*, *P. syringae*, and *R. solanacearum*, but for most of them a virulence activity is not known. A subset of these effectors including AvrXv4, AvrBsT, the HopZ-family, and PopP2 group into the YopJ-family of C55 peptidases (Hotson and Mudgett, 2004) suggesting that they might act as acetyltransferases. Nevertheless, expression of *avrXv4 in planta* leads to a reduction of SUMO-modified proteins similar to the activity of XopD. In contrast, no SUMO protease activity could be measured *in vitro* which does not permit the distinction between a direct and an indirect influence of AvrXv4 on the plant cell SUMOylation status (Roden et al., 2004). AvrXv4 localized to the plant cytoplasm and host targets of AvrXv4 have not been identified so far (Roden et al., 2004). HopZ-family effectors from *P. syringae* exhibit protease activity in an *in vitro* assay (Ma et al., 2006) indicating that possibly not all C55 peptidases are acetyltransferases. Additional analyses are needed to determine if AvrXv4 or HopZ also exhibit acetyltransferase activity. Interestingly, the effector AvrBsT, another member of the C55 peptidase family, elicits an HR in resistant *Arabidopsis* Pi-0 plants which is dependent on the recessive *sober1* allele encoding a carboxylesterase (Cunnac et al., 2007). The inactivation of the carboxylesterase in the Pi-0 *Arabidopsis* ecotype reveals the activity of AvrBsT to the plant defence resulting in an HR. In contrast, the active carboxylesterase from ecotype Col-0 quenches recognition. Carboxylesterases of this class are known to hydrolyse lysophospholipids and acylated proteins. It is therefore tempting to speculate that AvrBsT and SOBER1 may act on the same host target to alter its acylation or SUMOylation status (Cunnac et al., 2007).

Several effectors from plant and mammalian pathogens exploit the eukaryotic ubiquitination system to manipulate the host cell (Angot et al., 2007). Ubiquitination is the post-translational addition of one, or a chain, of ubiquitin (Ub) moieties to target proteins thereby modifying their localization, activity, or stability (Kerscher et al., 2006). Polyubiquitination is particularly well known to constitute a signal for subsequent degradation by the 26S proteasome (Pickart and Cohen, 2004). The ubiquitination machinery involves several steps: An Ub-activating enzyme (E1) binds Ub and transfers it to the Ub-conjugating enzyme (E2). An Ub-ligase complex (E3) then recruits target proteins for ubiquitination (Angot et al., 2007). The E3-Ub-ligase complex contains F-box proteins that mediate the recruitment of target proteins. Two examples of T3SS effectors from plant pathogenic bacteria were already mentioned that used the host ubiquitination machinery to manipulate the plant cellular machinery. First, AvrPtoB, which carries a E3-ligase domain to ubiquitinate plant targets and thereby suppresses host cell death (Abramovitch et al., 2006; Janjusevic et al., 2006). Second, HopM1 mediates proteasome-dependent degradation of AtMIN7, but features no E3 similarity (Nomura et al., 2006). Instead, HopM1 may function as an adaptor protein between AtMIN7 and the ubiquitin proteasome system (Angot et al., 2007). In addition, a family of effectors from *R. solanacearum* usurps the plant ubiquitination system in a particular fashion. *R. solanacearum* expresses seven effectors which exhibit an N-terminal F-box motif and interact with the SKP1 component of the E3 Ub-ligase complex (Angot et al., 2006). In addition, these effectors contain a C-terminal leucine-rich repeat with conserved amino acid residues that spawned the name 'GALA' for this family of effectors (Angot et al., 2006). The GALA effectors possibly function by forming composite E3-Ub-ligase complexes with components of host cells and subsequently mediate ubiquitination and degradation of plant target proteins (Angot et al., 2006; Angot et al., 2007). Although the virulence targets of these GALA effectors are not known, their contribution to virulence of *R. solanacearum* suggests that they play an important role (Angot et al., 2007).

Manipulation of host gene expression

Effectors have evolved an ingenious way to directly control the expression of target host genes. This is accomplished by the large AvrBs3 family of effectors from *Xanthomonas* and *R. solanacearum* (see also chapters by Meyer and Bogdanove, and Brown). They localize to the host cell nucleus and employ a C-terminal acidic activation sequence to direct expression of target genes. The most prominent feature of these effectors is a central domain of several repeats of 34 or 35 amino acids in length. These near identical repeats vary in number (1.5 to 28.5) between different AvrBs3-family members and contain certain variable regions (Gürlebeck et al., 2006). The central repeat domain encodes the information for the target specificity of these effectors and accordingly, AvrBs3-like effectors differ predominantly in the order and nature of single repeats in this region. Recently it was shown that AvrBs3 can bind specifically to promoter regions of target genes and the repeat region is essential for this (Kay et al., 2007; Romer et al., 2007).

AvrBs3 from the pepper and tomato pathogen *X. campestris* pv. *vesicatoria* induces several pepper genes and elicits hypertrophy of host tissue (Marois et al., 2002). One of these induced genes (*upa20*) encodes a cell-size regulator that facilitates the hypertrophy of plant cells (Kay et al., 2007). Hypertrophy (cell enlargement) and hyperplasia (cell division) lead to typical *Xanthomonas* disease symptoms like cankers and ruptures of the plant leaf epidermis (Swarup et al., 1991; Gürlebeck et al., 2006). Hereby, bacterial release from the infected tissue in later infection stages is enhanced. Other members of this effector family stimulate pathogen virulence in different ways. AvrXa7 from *X. oryzae* pv. *oryzae* enhances lesion length in rice (Bai et al., 2000) and Avrb6 from *X. campestris* pv. *malvacearum* promotes watersoaking in cotton (Yang et al., 1996).

A few additional target genes of AvrBs3-like effectors have been identified. The AvrBs3-like effector PthXo1 from *Xanthomonas oryzae* pv. *oryzae* specifically induces the *Os8N3* gene in susceptible rice cultivars (Yang et al., 2006). Expression of this gene is essential for infection of rice by this pathogen strain and *Os8N3* was therefore termed a susceptibility gene. The differences between resistant and susceptible cultivars are located to deletions in the promoter region (Yang et al., 2006). These differences might be responsible for differential binding of the effector and subsequent differential gene induction in both rice cultivars (Yang et al., 2006). In addition, the effector AvrXa27 induces expression of the *Xa27* resistance gene in resistant rice cultivars (Gu et al., 2005) and AvrBs3 directs expression of the *Bs3* resistance gene in resistant pepper lines (Römer et al., 2007). These processes are not advantageous for the pathogen and might resemble an inadvertent interaction. In these cases the host has successfully installed a transcriptional mousetrap to lure the AvrBs3-like effectors to express resistance genes and elicit an HR. Again, the promoter regions of *Bs3* and *Xa27* in resistant and susceptible cultivars differ indicating differential interaction with the effector (Gu et al., 2005; Römer et al., 2007). In *P. agglomerans*, two effectors (HsvG, HsvB) have been identified which act as transcriptional activators in the plant nucleus (Nissan et al., 2006). They contain five helix-turn-helix motifs and one-to-two direct repeats but show no sequence similarity to the AvrBs3-family (Nissan et al., 2006). This indicates that it is advantageous for the pathogen to manipulate the host transcriptome.

The chemistry between us – altering the plant hormone homeostasis

The plant hormones salicylic acid (SA), jasmonic acid (JA), and ethylene (ET) play important roles in plant defence signalling. The levels of SA increase in plant tissues after infection and mediate signalling events that limit pathogen growth locally as well as systemically in a process called systemic acquired resistance (SAR; Grant and Lamb, 2006). *Arabidopsis* mutants which are unable to accumulate SA show a higher susceptibility to different pathogens, emphasizing the importance of SA for defence (Kunkel and Brooks, 2002; Durrant and Dong, 2004). In contrast, JA controls multiple aspects of plant development and often acts antagonistically to the SA signalling pathway (Feys and Parker, 2000; Kunkel and Brooks, 2002). It is believed that SA signalling is mainly involved in responses to biotrophic pathogens, whereas JA signalling promotes defence against necrotrophic pathogens. Activation of one pathway often inhibits the other (Kunkel and Brooks, 2002). Although SA plays a major role in alerting distal tissues in the SAR after local infections, the translocated signal most probably is JA (Durrant and Dong, 2004; Truman et al., 2007). ET is involved in resistance as well as susceptibility to pathogens, but is generally seen as supporting the JA pathway. These three plant hormones have long been the focus of attention in plant–pathogen interactions, but recent data reveal a network that also includes at least auxin, gibberellic acid (GA) and abscisic acid (ABA) (Robert-Seilaniantz et al., 2007). This network includes hormone interdependencies, e.g. SA causes a global repression of auxin-related genes and this inhibition is part of the induced disease resistance mechanism (Wang et al., 2007a). Pathogen activities as well as plant responses cause changes in several points of the phytohormone network. This demonstrates that phytohormones on the one hand are an essential part of plant defence whereas microbes on the other hand have adapted to manipulate these responses.

Salicylic acid and jasmonic acid

It has long been known that plant pathogenic bacteria as well as many pathogenic fungi can themselves produce plant hormones like ethylene, auxin, and cytokinins (Robert-Seilaniantz et al., 2007). In addition, several P. syringae strains produce the phytotoxin coronatine, a structural mimic of jasmonoyl-isoleucine (JA-Ile) (Staswick 2008). JA-Ile induces several JA-responsive plant reactions and blocks SA-dependent plant defence to promote bacterial growth (Feys and Parker, 2000). Coronatine has a second function for P. syringae virulence. It induces stomata opening by suppression of ABA signalling and thus enables leaf surface-localized bacteria to gain entry to the apoplast (Melotto et al., 2006).

Besides the production of phytohormones, bacterial pathogens also influence hormone signalling pathways within the plant cell through effector activities. Infection with virulent pathogens increases the levels of SA, JA, ET, auxin, and ABA (O'Donnell et al., 2003; Schmelz et al., 2003). Expression of the *RAP2.6* gene encoding an ethylene response factor (ERF)-transcription factor is induced upon infection with virulent P. syringae (Chen et al., 2002; He et al., 2004a). This induction is dependent on coronatine and several T3SS effectors (AvrB, AvrRpt2, AvrPphB, HopK1, HopX1). AvrB is the most potent activator of *RAP2.6* gene induction and requires the JA and ET pathways for its full activity (He et al., 2004a). In the absence of its cognate resistance protein RPM1, AvrB induces chlorosis in *Arabidopsis* and promotes growth of a non-virulent P. syringae strain (Shang et al., 2006). This effect depends on interaction of AvrB with a host complex containing RAR1. RAR1 controls the stability of NB-LRR resistance proteins and acts as a negative regulator of basal defence (Shen and Schulze-Lefert, 2007). The AvrB activity depends on a second host factor, the F-box protein COI1, a component of SCF-^{COI1} ubiquitin-ligase complexes required for JA signalling (Xu et al., 2002; Shang et al., 2006). Together these data imply that AvrB influences plant defences possibly through manipulation of JA signalling components.

The plant hormone SA and precursors of JA are produced in chloroplasts and the effector HopI1 targets this organelle to alter phytohormone signalling. HopI1 localizes to the stroma of plant chloroplasts, alters thylakoid ultrastructure and blocks SA accumulation and thus also SA-dependent defence signalling. HopI1 carries P/Q-rich repeats which may facilitate protein-protein interactions and a functional J-domain which mediates interaction with HSP70 (Jelenska et al., 2007). Which plant factors are manipulated by HopI1 is unknown so far. Interestingly, HopI1 is phosphorylated both within the bacterium and within the plant cell which may contribute to activation of this effector inside the host cell. The effector AvrRpt2 is a cysteine protease degrading RIN4 and potentially other target proteins in plant cells. AvrRpt2 reduces induction of PR genes during a bacterial infection, which are common indicators of SA signalling. Nevertheless, AvrRpt2 acts downstream or independent of SA itself (Chen et al., 2004).

Auxin

Auxin is involved in many aspects of plant cell development including cell wall loosening, membrane permeability (Buchanan et al., 2002), stomatal opening (Dietrich et al., 2001), and abiotic as well as biotic stress adaptation (Park et al., 2007). These effects are potentially important for pathogen proliferation and accordingly auxin signalling is targeted by plant pathogens. For a long time it has been known that different plant pathogenic bacteria produce auxin directly, possibly to manipulate plant signalling (Glickmann et al., 1998). On the other hand, the plant responds to an infection by suppressing the auxin-signalling pathway and the control of plant auxin receptors seems to be crucial for this (Navarro et al., 2006; Wang et al., 2007a). Challenge of *Arabidopsis* tissue with the bacterial MAMP flg22 or a P. syringae T3SS mutant strain induces production of a plant micro RNA (miR393) which mediates silencing of F-box proteins that function as auxin receptors (Navarro et al., 2006; Fahlgren et al., 2007). The defence-related plant hormone SA also causes repression of auxin-related genes, including auxin receptors (Wang et al., 2007) thereby successfully quenching the auxin-responsiveness of the plant cell, but this activity is independent of micro RNAs. Decreased auxin signalling supports plant resistance while exogenously applied auxin analogues enhance disease symptoms (Navarro et al., 2006; Chen et al., 2007). This indicates that auxin homeostasis is a key control point for pathogen and host, and suppression of the auxin response is involved in plant basal defence.

The auxin response of the plant is also manipulated by T3SS effectors. AvrRpt2 alters the auxin physiology of plant cells leading to increased sensitivity to exogenous auxin as well as slightly increased auxin levels within *Arabidopsis* seedlings (Chen et al., 2007). The molecular targets of AvrRpt2 that cause this phenotype are not known and it is possible that the impact of AvrRpt2 on cellular integrity indirectly causes aberrant auxin signalling. In contrast, the effector AvrBs3 from *X. campestris* pv. *vesicatoria* influences auxin signalling at a late step in the cascade. AvrBs3 indirectly induces expression of auxin-responsive genes independent of auxin synthesis (Marois et al., 2002). This gene induction might efficiently bypass the plants attempt to block auxin-signalling.

Abscisic acid and ethylene

The phytohormone ABA is involved in adaptation to drought, cold, osmotic changes, and important during seed germination (Nambara and Marion-Poll, 2005). In addition to this classic role in abiotic stress control, ABA is also involved in plant responses to pathogens. Exogenous application of ABA enhances disease development and attenuates basal defences, whereas plant mutants in ABA biosynthesis are more resistant to infections. Therefore, it is believed that ABA acts as a negative regulator of plant defence (Thaler and Bostock, 2004; de Torres-Zabala et al., 2007).

Accordingly, the ABA signalling pathway seems to be a key target of *P. syringae* and its effectors. Microarray analysis revealed that ABA biosynthesis and ABA-responsive genes are induced by virulent *P. syringae* pv. *tomato* and ABA levels increase accordingly (de Torres-Zabala et al., 2007). ABA levels are not changed after infection with a T3SS mutant which implies that effectors of virulent *P. syringae* pv. *tomato* manipulate ABA levels within the plant. One of these effectors is AvrPtoB which has been implicated in inhibition of basal resistance and cell death suppression already. Transgenic expression of AvrPtoB in *Arabidopsis* is sufficient to induce ABA signalling responses and to raise plant ABA levels which promotes bacterial growth (de Torres-Zabala et al., 2007).

The effector AvrPtoB was also shown to manipulate the ethylene response of the plant (Cohn and Martin, 2005; Xiao et al., 2007b). cDNA microarray analysis was performed using tomato plants infected with wild-type *P. syringae* pv. *tomato* and a T3SS mutant. Many differentially expressed genes are involved in hormone response or hormone biosynthesis, indicating that the effectors of virulent bacteria modify the hormone homeostasis. Comparative analyses of infections with *avrPto* and *avrPtoB* mutant strains reveals that both effectors induce tomato genes involved in ethylene biosynthesis and signalling. In addition, on susceptible plants, AvrPto and AvrPtoB promote necrotic disease symptom development in infected tomato leaves by up-regulating ethylene production (Cohn and Martin, 2005; Lin and Martin, 2005). The ethylene response is probably important for symptom development in later stages of disease development. How the ethylene and abscisic acid response is manipulated by AvrPtoB is unclear so far, but the concomitant manipulation by the same effector indicates that shared or similar components of the ethylene and the ABA pathway are targeted by AvrPtoB.

Together these reports show that manipulation of plant hormone homeostasis is a key element of bacterial pathogenesis. The bacterial benefit of this seems to stem at least in part from suppression of plant defence responses, but it is possible that the hormonal changes cause additional cellular responses advantageous for the bacteria. It is important to consider that manipulation of one phytohormone level likely influences others in the hormone network. Thus the activity of single effectors might have effects on several phytohormone signalling pathways.

Master of puppets – how to control eukaryotic cells

An evolving effector variety

Effectors are extremely useful tools for pathogens to manipulate eukaryotic host cells. For this, individual bacterial strains carry a number of different effectors with different activities. Possibly several steps are needed to fully control a eukaryotic host cell. Nevertheless, plant pathogenic bacteria seem to contain an exceptionally large repertoire of effectors and this may also partly reflect their lifestyle. In nature these pathogens might encounter different plant hosts in consecutive infections. Some effectors function in different hosts (Jackson et al., 2002), but in addition, the pathogen might carry effectors that function in specific hosts. Still, the pathogen has to make sure that its effectors do not elicit resistance gene-mediated defence responses and this chance increases with a larger individual effector repertoire. The presence or absence of single effectors can influence the host range of a strain towards different plant species (Wei et al., 2007).

The pathogen constantly incorporates changes in its virulence repertoire, because host plants evolve novel detection specificities possibly leading to plant resistance. Plant pathogenic bacteria have achieved this by enabling an exchange of effectors via horizontal gene transfer (see chapters by Arnold et al. and Stavrinides). Frequently, effector genes are located adjacent to or even on mobile genetic elements like transposons, plasmids, and bacteriophages (Kim et al., 1998; Arnold et al., 2000; Jackson et al., 2000; Vivian et al., 2001; Landgraf et al., 2006; see also chapter by Sundin and Murillo). In addition, genuinely novel effectors can be generated through a process called 'terminal reassortment'. This process combines promoter elements and N-termini of effectors containing the T3SS secretion signal with C-terminal domains exerting the actual effector activity (Stavrinides et al., 2006). Hereby novel proteins can acquire the ability to be transported via the T3SS into eukaryotic host cells and novel effectors can arise. A second mechanism is the sequential combination of different activities within one effector protein which transforms this effector into a versatile tool with superior functionality. This process is favoured by the typically modular structure of effectors which employ distinct and exchangeable domains for T3SS secretion, translocation, subcellular localization, and different virulence functions. Together, horizontal transfer and rearrangement of virulence genes result in an extraordinary genome flexibility and effector variability of plant pathogenic bacteria.

Assignment of multifarious tasks

Many experimental approaches have revealed that the majority of effectors suppress plant defence reactions. This is either achieved by direct interference with host components that control the onset of plant resistance or indirectly via modulation of cellular processes which interfere with defence establishment, for example hormone imbalances. The manipulation by bacterial effectors occurs at different steps in the plant defence pathway. Recognition, signal transduction, gene expression, and response reactions are all targets of bacterial effectors. This emphasizes how diverse effector activities can support a similar aim of the pathogen. The effector repertoires from different pathogens can be substantially different, but the pathogens are nevertheless able to infect the same host plants. This indicates that sequence-unrelated effectors can have redundant activity and perform similar virulence tasks. It might reflect an evolutionary way to gain complexity and lower the chance to be detected by plant resistance mechanisms.

The basal defence is the first and probably also the more ancient layer of the plant defence. It is tempting to speculate that effectors which target the basal defence are also more ancient (Fig. 12.1). Comparison of effectors from different bacterial strains and pathogens revealed that some effectors are more widespread than others and may resemble these ancient and key effectors (Jones and Dangl, 2006). Plant resistance genes are believed to have been developed later and accordingly, effectors displaying an activity to block this part of the plant defence have probably also been more recently acquired (Fig. 12.1). Interestingly, some host targets of effectors seem to represent mousetraps and their only role is to lure effectors into an interaction which is then detected by a resistance protein.

It is not clear whether effectors have additional tasks besides suppression of host defences. The bacteria have to acquire nutrients and water for their propagation. Apoplastic sugar and mineral estimates indicate that enough nutrients are already present in this compartment (Hancock and Huisman, 1981), but an additional release from plant cells might benefit bacterial growth. It is conceivable that some effectors act to redirect the flow of nutrients and water from the plant cell to the bacteria, but no unambiguous candidates have been identified so far. Alternatively, bacteria might create an artificial sink in leaf tissue that misleads the plant to supply an increased concentration of nutrients without the need for T3SS effectors.

Effectors – a prevalent invention

The unique transport ability of the T3SS renders it a very suitable tool to manipulate eukaryotic cells. Accordingly, T3SS can not only be found in plant- and animal pathogenic bacteria, but also in several symbionts and non-pathogenic plant-associated bacteria (Pallen et al., 2005; Troisfontaines and Cornelis, 2005). Although the T3SSs of these eukaryote-associated bacteria are functionally analogous to T3SS of pathogens, the tasks for associated effectors are not focused on virulence. Conveniently, similar T3SSs can deploy different effector repertoires. Although some effectors of symbiotic bacteria have been identified, their activity within the host cell is largely not known (Pühler et al., 2004; Süß et al., 2006). Not all symbiotic *Rhizobia* carry a T3SS, but several T3SS-less strains contain a type IV secretion system instead that could potentially also facilitate the transport of proteins into eukaryotic host cells (Hubber et al., 2004).

The definition of an 'effector' has been extended since it became known that pathogenic oomycetes and fungi can also translocate proteins into the cytoplasm of host cells (Kamoun, 2007). The role of these fungal and oomycete effectors for virulence is often not understood, but the suppression of basal plant defence similar to bacterial effectors has been observed (Sohn et al., 2007). Predictions using conserved translocation signals revealed a high number of potential effectors (hundreds) in the genomes of pathogenic oomycetes, indicating that effectors also play a key role for these pathogens. In addition, nematodes also deploy proteins into the cytoplasm of host cells to enable their pathogenic lifestyle (Vanholme et al., 2004). It seems that the transport of virulence proteins into the cells of target eukaryotic cells is a predominant feature of bacteria–eukaryote and eukaryote–eukaryote interactions. This method allows an extensive manipulation of host cells and effectors are the sophisticated tools to perform this task.

Future perspectives

The identification of effector targets will be a major task for future studies on plant pathogens. It will be especially important to identify plant components that are targeted by several effectors and possibly resemble key control points for pathogenesis. Here, it will be important to distinguish between primary targets and secondary cellular effects of effector activities. The identification of key virulence targets will eventually reveal the strategy of how pathogens aim to control host cells. In this regard, the cross-kingdom analysis of effector functions from bacterial, oomycete, and fungal pathogens will shed light on different pathogen strategies and possibly common exploits of plant cells. This knowledge might finally help to breed resistant crops and minimize threats to food productions that are thousands of years old.

Acknowledgements

I apologize to those colleagues whose work was not cited due to space restrictions. I am grateful to Sabine Kay, Frank Thieme and Sebastian Schornack for suggestions on the manuscript and Ulla Bonas for continued support.

References

Abramovitch, R.B., Janjusevic, R., Stebbins, C.E., and Martin, G.B. (2006). Type III effector AvrPtoB requires intrinsic E3 ubiquitin ligase activity to suppress plant cell death and immunity. Proc. Natl. Acad. Sci. USA *103*, 2851–2856.

Abramovitch, R.B., Kim, Y.-J., Chen, S., Dickman, M.B., and Martin, G.B. (2003). *Pseudomonas* type III effector AvrPtoB induces plant disease susceptibility by inhibition of host programmed cell death. EMBO J. *22*, 60–69.

Abramovitch, R.B., and Martin, G.B. (2005). AvrPtoB: a bacterial type III effector that both elicits and suppresses programmed cell death associated with plant immunity. FEMS Microbiol. Lett. *245*, 1–8.

Ade, J., DeYoung, B.J., Golstein, C., and Innes, R.W. (2007). Indirect activation of a plant nucleotide binding site-leucine-rich repeat protein by a bacterial protease. Proc. Natl. Acad. Sci. USA *104*, 2531–2536.

Alfano, J.R., Charkowski, A.O., Deng, W.-L., Badel, J.L., Petnicki-Ocwieja, T., van Dijk, K., and Collmer, A. (2000). The *Pseudomonas syringae* Hrp pathogenicity island has a tripartite mosaic structure composed of a cluster of type III secretion genes bounded by exchangeable effector and conserved effector loci that contribute to parasitic fitness and pathogenicity in plants. Proc. Natl. Acad. Sci. USA *97*, 4856–4861.

Angot, A., Peeters, N., Lechner, E., Vailleau, F., Baud, C., Gentzbittel, L., Sartorel, E., Genschik, P., Boucher, C., and Genin, S. (2006). *Ralstonia solanacearum* requires F-box-like domain-containing type III effectors to promote disease on several host plants. Proc. Natl. Acad. Sci. USA *103*, 14620–14625.

Angot, A., Vergunst, A., Genin, S., and Peeters, N. (2007). Exploitation of eukaryotic ubiquitin signaling pathways by effectors translocated by bacterial type III and type IV secretion systems. PLoS Pathog. *3*, e3.

Arnold, D.L., Jackson, R.W., and Vivian, A. (2000). Evidence for the mobility of an avirulence gene, *avrPpiA1*, between the chromosome and plasmids of races of *Pseudomonas syringae* pv. *pisi*. Mol. Plant Pathol. *1*, 195–199.

Asai, T., Tena, G., Plotnikova, J., Willmann, M.R., Chiu, W.L., Gomez-Gomez, L., Boller, T., Ausubel, F.M., and Sheen, J. (2002). MAP kinase signalling cascade in *Arabidopsis* innate immunity. Nature *415*, 977–983.

Axtell, M.J., Chisholm, S.T., Dahlbeck, D., and Staskawicz, B.J. (2003). Genetic and molecular evidence that the *Pseudomonas syringae* type III effector protein AvrRpt2 is a cysteine protease. Mol. Microbiol. *49*, 1537–1546.

Axtell, M.J., and Staskawicz, B.J. (2003). Initiation of RPS2-specified disease resistance in *Arabidopsis* is coupled to the AvrRpt2-directed elimination of RIN4. Cell *112*, 369–377.

Badel, J.L., Charkowski, A.O., Deng, W.L., and Collmer, A. (2002). A gene in the *Pseudomonas syringae* pv. *tomato* Hrp pathogenicity island conserved effector locus, *hopPtoA1*, contributes to efficient formation of bacterial colonies in planta and is duplicated elsewhere in the genome. Mol. Plant–Microbe Interact. *15*, 1014–1024.

Badel, J.L., Shimizu, R., Oh, H.S., and Collmer, A. (2006). A *Pseudomonas syringae* pv. *tomato avrE1/hopM1* mutant is severely reduced in growth and lesion formation in tomato. Mol. Plant–Microbe Interact. *19*, 99–111.

Bai, J., Choi, S.-H., Ponciano, G., Leung, H., and Leach, J.E. (2000). *Xanthomonas oryzae* pv. *oryzae* avirulence genes contribute differently and specifically to pathogen aggressiveness. Mol. Plant–Microbe Interact. *13*, 1322–1329.

Belkhadir, Y., Nimchuk, Z., Hubert, D.A., Mackey, D., and Dangl, J.L. (2004). *Arabidopsis* RIN4 negatively regulates disease resistance mediated by RPS2 and RPM1 downstream or independent of the NDR1 signal modulator and is not required for the virulence functions of bacterial type III effectors AvrRpt2 or AvrRpm1. Plant Cell *16*, 2822–2835.

Bittel, P., and Robatzek, S. (2007). Microbe-associated molecular patterns (MAMPs) probe plant immunity. Curr. Opin. Plant Biol. *10*, 335–341.

Bogdanove, A.J., Bauer, D.W., and Beer, S.V. (1998). *Erwinia amylovora* secretes DspE, a pathogenicity factor and functional AvrE homolog, through the Hrp (type III secretion) pathway. J. Bacteriol. *180*, 2244–2247.

Bretz, J.R., Mock, N.M., Charity, J.C., Zeyad, S., Baker, C.J., and Hutcheson, S.W. (2003). A translocated protein tyrosine phosphatase of *Pseudomonas syringae* pv. *tomato* DC3000 modulates plant defence response to infection. Mol. Microbiol. *49*, 389–400.

Brown, I., Mansfield, J., and Bonas, U. (1995). *hrp* genes in *Xanthomonas campestris* pv. *vesicatoria* determine ability to suppress papilla deposition in pepper mesophyll cells. Mol. Plant–Microbe Interact. *8*, 825–836.

Büttner, D., and Bonas, U. (2002). Port of entry-the type III secretion translocon. Trends Microbiol *10*, 186–192.

Buchanan, B.B., Gruissem, W., and Jones, R.L. (2000). Biochemistry and molecular biology of plants (Rockville, USA: American Society of Plant Physiologists).

Chang, J.H., Tai, Y.S., Bernal, A.J., Lavelle, D.T., Staskawicz, B.J., and Michelmore, R.W. (2002). Functional analyses of the *Pto* resistance gene family in tomato and the identification of a minor resistance determinant in a susceptible haplotype. Mol. Plant–Microbe Interact. *15*, 281–291.

Chen, W., Provart, N.J., Glazebrook, J., Katagiri, F., Chang, H.S., Eulgem, T., Mauch, F., Luan, S., Zou, G., Whitham, S.A., *et al.* (2002). Expression profile matrix of Arabidopsis transcription factor genes suggests their putative functions in response to environmental stresses. Plant Cell *14*, 559–574.

Chen, Z., Kloek, A.P., Cuzick, A., Moeder, W., Tang, D., Innes, R.W., Klessig, D.F., McDowell, J.M., and Kunkel, B.N. (2004). The *Pseudomonas syringae* type III effector AvrRpt2 functions downstream or independently of SA to promote virulence on *Arabidopsis thaliana*. Plant J. *37*, 494–504.

Chen, Z., Agnew, J.L., Cohen, J.D., He, P., Shan, L., Sheen, J., and Kunkel, B.N. (2007). *Pseudomonas syringae* type III effector AvrRpt2 alters *Arabidopsis thaliana* auxin physiology. Proc. Natl. Acad. Sci. USA *104*, 20131–20136.

Chinchilla, D., Bauer, Z., Regenass, M., Boller, T., and Felix, G. (2006). The *Arabidopsis* receptor kinase FLS2 binds *flg22* and determines the specificity of flagellin perception. Plant Cell *18*, 465–476.

Chinchilla, D., Zipfel, C., Robatzek, S., Kemmerling, B., Nürnberger, T., Jones, J.D., Felix, G., and Boller, T. (2007). A flagellin-induced complex of the receptor FLS2 and BAK1 initiates plant defence. Nature *448*, 497–500.

Chisholm, S.T., Coaker, G., Day, B., and Staskawicz, B.J. (2006). Host–microbe interactions: shaping the evolution of the plant immune response. Cell *124*, 803–814.

Chisholm, S.T., Dahlbeck, D., Krishnamurthy, N., Day, B., Sjolander, K., and Staskawicz, B.J. (2005). Molecular characterization of proteolytic cleavage sites of the *Pseudomonas syringae* effector AvrRpt2. Proc. Natl. Acad. Sci. USA *102*, 2087–2092.

Chosed, R., Tomchick, D.R., Brautigam, C.A., Mukherjee, S., Negi, V.S., Machius, M., and Orth, K. (2007). Structural analysis of *Xanthomonas* XopD provides insights into substrate specificity of ubiquitin-like protein proteases. J. Biol. Chem. *282*, 6773–6782.

Coaker, G., Falick, A., and Staskawicz, B. (2005). Activation of a phytopathogenic bacterial effector protein by a eukaryotic cyclophilin. Science *308*, 548–550.

Coaker, G., Zhu, G., Ding, Z., Van Doren, S.R., and Staskawicz, B. (2006). Eukaryotic cyclophilin as a molecular switch for effector activation. Mol. Microbiol. *61*, 1485–1496.

Cohn, J.R., and Martin, G.B. (2005). *Pseudomonas syringae* pv. *tomato* type III effectors AvrPto and AvrPtoB promote ethylene-dependent cell death in tomato. Plant J. *44*, 139–154.

Cunnac, S., Occhialini, A., Barberis, P., Boucher, C., and Genin, S. (2004). Inventory and functional analysis of the large Hrp regulon in *Ralstonia solanacearum*: identification of novel effector proteins translocated to plant host cells through the type III secretion system. Mol. Microbiol. *53*, 115–128.

Cunnac, S., Wilson, A., Nuwer, J., Kirik, A., Baranage, G., and Mudgett, M.B. (2007). A conserved carboxylesterase is a SUPPRESSOR OF AVRBST-ELICITED RESISTANCE in *Arabidopsis*. Plant Cell *19*, 688–705.

da Cunha, L., Sreerekha, M.V., and Mackey, D. (2007). Defense suppression by virulence effectors of bacterial phytopathogens. Curr. Opin. Plant Biol. *10*, 349–357.

Day, B., Dahlbeck, D., Huang, J., Chisholm, S.T., Li, D., and Staskawicz, B.J. (2005). Molecular basis for the RIN4 negative regulation of RPS2 disease resistance. Plant Cell *17*, 1292–1305.

de Torres, M., Mansfield, J.W., Grabov, N., Brown, I.R., Ammouneh, H., Tsiamis, G., Forsyth, A., Robatzek, S., Grant, M., and Boch, J. (2006). *Pseudomonas syringae* effector AvrPtoB suppresses basal defence in *Arabidopsis*. Plant J. *47*, 368–382.

de Torres, M., Sanchez, P., Fernandez-Delmond, I., and Grant, M. (2003). Expression profiling of the host response to bacterial infection: the transition from basal to induced defence responses in RPM1-mediated resistance. Plant J. *33*, 665–676.

de Torres-Zabala, M., Truman, W., Bennett, M.H., Lafforgue, G., Mansfield, J.W., Rodriguez Egea, P., Bogre, L., and Grant, M. (2007). *Pseudomonas syringae* pv. *tomato* hijacks the Arabidopsis abscisic acid signalling pathway to cause disease. EMBO J. *26*, 1434–1443.

DebRoy, S., Thilmony, R., Kwack, Y.B., Nomura, K., and He, S.Y. (2004). A family of conserved bacterial effectors inhibits salicylic acid-mediated basal immunity and promotes disease necrosis in plants. Proc. Natl. Acad. Sci. USA *101*, 9927–9932.

Deslandes, L., Olivier, J., Peeters, N., Feng, D.X., Khounlotham, M., Boucher, C., Somssich, I., Genin, S., and Marco, Y. (2003). Physical interaction between RRS1-R, a protein conferring resistance to bacterial wilt, and PopP2, a type III effector targeted to the plant nucleus. Proc. Natl. Acad. Sci. USA *100*, 8024–8029.

Desveaux, D., Singer, A.U., Wu, A.J., McNulty, B.C., Musselwhite, L., Nimchuk, Z., Sondek, J., and Dangl, J.L. (2007). Type III effector activation via nucleotide binding, phosphorylation, and host target interaction. PLoS Pathog. *3*, e48.

DeYoung, B.J., and Innes, R.W. (2006). Plant NBS-LRR proteins in pathogen sensing and host defense. Nature Immunol. *7*, 1243–1249.

Dietrich, P., Sanders, D., and Hedrich, R. (2001). The role of ion channels in light-dependent stomatal opening. J. Exp. Bot. *52*, 1959–1967.

Durrant, W.E., and Dong, X. (2004). Systemic acquired resistance. Annu. Rev. Phytopathol. *42*, 185–209.

Espinosa, A., Guo, M., Tam, V.C., Fu, Z.Q., and Alfano, J.R. (2003). The *Pseudomonas syringae* type III-secreted protein HopPtoD2 possesses protein tyrosine phosphatase activity and suppresses programmed cell death in plants. Mol. Microbiol. *49*, 377–387.

Eulgem, T., and Somssich, I.E. (2007). Networks of WRKY transcription factors in defense signaling. Curr. Opin. Plant Biol. *10*, 366–371.

Fahlgren, N., Howell, M.D., Kasschau, K.D., Chapman, E.J., Sullivan, C.M., Cumbie, J.S., Givan, S.A., Law, T.F., Grant, S.R., Dangl, J.L., and Carrington, J. C. (2007). High-throughput sequencing of *Arabidopsis* microRNAs: evidence for frequent birth and death of MIRNA genes. PLoS ONE *2*, e219.

Felix, G., Duran, J.D., Volko, S., and Boller, T. (1999). Plants have a sensitive perception system for the most conserved domain of bacterial flagellin. Plant J. *18*, 265–276.

Feys, B.J., and Parker, J.E. (2000). Interplay of signaling pathways in plant disease resistance. Trends Genet. *16*, 449–455.

Fu, Z.Q., Guo, M., Jeong, B.R., Tian, F., Elthon, T.E., Cerny, R.L., Staiger, D., and Alfano, J.R. (2007). A type III effector ADP-ribosylates RNA-binding proteins and quells plant immunity. Nature *447*, 284–288.

Gaudriault, S., Malandrin, L., Paulin, J.P., and Barny, M.A. (1997). DspA, an essential pathogenicity factor of *Erwinia amylovora* showing homology with AvrE of *Pseudomonas syringae*, is secreted via the Hrp secretion pathway in a DspB-dependent way. Mol. Microbiol. *26*, 1057–1069.

Glickmann, E., Gardan, L., Jacquet, S., Hussain, S., Elasri, M., Petit, A., and Dessaux, Y. (1998). Auxin production is a common feature of most pathovars of *Pseudomonas syringae*. Mol. Plant–Microbe Interact. *11*, 156–162.

Gomez-Gomez, L., and Boller, T. (2000). FLS2: an LRR receptor-like kinase involved in the perception of the bacterial elicitor flagellin in Arabidopsis. Mol. Cell *5*, 1003–1011.

Grant, M., and Lamb, C. (2006). Systemic immunity. Curr. Opin. Plant Biol. *9*, 414–420.

Grant, S.R., Fisher, E.J., Chang, J.H., Mole, B.M., and Dangl, J.L. (2006). Subterfuge and manipulation: type III effector proteins of phytopathogenic bacteria. Annu. Rev. Microbiol. *60*, 425–449.

Greenberg, J.T., and Yao, N. (2004). The role and regulation of programmed cell death in plant–pathogen interactions. Cell Microbiol. 6, 201–211.

Gu, K., Yang, B., Tian, D., Wu, L., Wang, D., Sreekala, C., Yang, F., Chu, Z., Wang, G. L., White, F.F., and Yin, Z. (2005). R gene expression induced by a type-III effector triggers disease resistance in rice. Nature 435, 1122–1125.

Gürlebeck, D., Thieme, F., and Bonas, U. (2006). Type III effector proteins from the plant pathogen *Xanthomonas* and their role in the interaction with the host plant. J. Plant Physiol. 163, 233–255.

Guttman, D.S., Vinatzer, B.A., Sarkar, S.F., Ranall, M.V., Kettler, G., and Greenberg, J.T. (2002). A functional screen for the type III (Hrp) secretome of the plant pathogen *Pseudomonas syringae*. Science 295, 1722–1726.

Ham, J.H., Kim, M.G., Lee, S.Y., and Mackey, D. (2007). Layered basal defenses underlie non-host resistance of *Arabidopsis* to *Pseudomonas syringae* pv. *phaseolicola*. Plant J. 51, 604–616.

Hancock, J.G., and Huisman, O.C. (1981). Nutrient movement in host–pathogen systems. Annu. Rev. Phytopathol. 19, 309–331.

Hauck, P., Thilmony, R., and He, S.Y. (2003). A *Pseudomonas syringae* type III effector suppresses cell wall-based extracellular defense in susceptible *Arabidopsis* plants. Proc. Natl. Acad. Sci. USA 100, 8577–8582.

He, P., Chintamanani, S., Chen, Z., Zhu, L., Kunkel, B.N., Alfano, J.R., Tang, X., and Zhou, J.M. (2004a). Activation of a COI1-dependent pathway in Arabidopsis by *Pseudomonas syringae* type III effectors and coronatine. Plant J. 37, 589–602.

He, P., Shan, L., Lin, N.C., Martin, G.B., Kemmerling, B., Nürnberger, T., and Sheen, J. (2006). Specific bacterial suppressors of MAMP signaling upstream of MAPKKK in *Arabidopsis* innate immunity. Cell 125, 563–575.

He, P., Shan, L., and Sheen, J. (2007). Elicitation and suppression of microbe-associated molecular pattern-triggered immunity in Plant–Microbe interactions. Cell Microbiol. 9, 1385–1396.

He, S.Y., and Jin, Q. (2003). The Hrp pilus: learning from flagella. Curr. Opin. Microbiol. 6, 15–19.

He, S.Y., Nomura, K., and Whittam, T.S. (2004b). Type III protein secretion mechanism in mammalian and plant pathogens. Biochim. Biophys. Acta 1694, 181–206.

Heese, A., Hann, D.R., Gimenez-Ibanez, S., Jones, A.M., He, K., Li, J., Schroeder, J.I., Peck, S.C., and Rathjen, J.P. (2007). The receptor-like kinase SERK3/BAK1 is a central regulator of innate immunity in plants. Proc. Natl. Acad. Sci. USA 104, 12217–12222.

Hotson, A., Chosed, R., Shu, H., Orth, K., and Mudgett, M.B. (2003). *Xanthomonas* type III effector XopD targets SUMO-conjugated proteins in planta. Mol. Microbiol. 50, 377–389.

Hotson, A., and Mudgett, M.B. (2004). Cysteine proteases in phytopathogenic bacteria: identification of plant targets and activation of innate immunity. Curr. Opin. Plant Biol. 7, 384–390.

Hubber, A., Vergunst, A.C., Sullivan, J.T., Hooykaas, P.J., and Ronson, C.W. (2004). Symbiotic phenotypes and translocated effector proteins of the *Mesorhizobium loti* strain R7A VirB/D4 type IV secretion system. Mol. Microbiol. 54, 561–574.

Jackson, R.W., Athanassopoulos, E., Tsiamis, G., Mansfield, J.W., Sesma, A., Arnold, D.L., Gibbon, M.J., Taylor, J.D., and Vivian, A. (1999). Identification of a pathogenicity island, which contains genes for virulence and avirulence, on a large native plasmid in the bean pathogen *Pseudomonas syringae* pathovar phaseolicola. Proc. Natl. Acad. Sci. USA 96, 10875–10880.

Jackson, R.W., Mansfield, J.W., Ammouneh, H., Dutton, L.C., Wharton, B., Ortiz-Barredo, A., Arnold, D.L., Tsiamis, G., Sesma, A., Butcher, D., et al. (2002). Location and activity of members of a family of *virPphA* homologues in pathovars of *Pseudomonas syringae* and *P. savastanoi*. Mol. Plant Pathol. 3, 205–216.

Jackson, R.W., Mansfield, J.W., Arnold, D.L., Sesma, A., Paynter, C.D., Murillo, J., Taylor, J.D., and Vivian, A. (2000). Excision from tRNA genes of a large chromosomal region, carrying *avrPphB*, associated with race change in the bean pathogen, *Pseudomonas syringae* pv. *phaseolicola*. Mol. Microbiol. 38, 186–197.

Jamir, Y., Guo, M., Oh, H.S., Petnicki-Ocwieja, T., Chen, S., Tang, X., Dickman, M.B., Collmer, A., and Alfano, J.R. (2004). Identification of *Pseudomonas syringae* type III effectors that can suppress programmed cell death in plants and yeast. Plant J. 37, 554–565.

Janjusevic, R., Abramovitch, R.B., Martin, G.B., and Stebbins, C.E. (2006). A bacterial inhibitor of host programmed cell death defenses is an E3 ubiquitin ligase. Science 311, 222–226.

Jarvis, P., and Robinson, C. (2004). Mechanisms of protein import and routing in chloroplasts. Curr. Biol. 14, R1064–1077.

Jelenska, J., Yao, N., Vinatzer, B.A., Wright, C.M., Brodsky, J.L., and Greenberg, J.T. (2007). A J domain virulence effector of *Pseudomonas syringae* remodels host chloroplasts and suppresses defenses. Curr. Biol. 17, 499–508.

Jin, P., Wood, M.D., Wu, Y., Xie, Z., and Katagiri, F. (2003). Cleavage of the *Pseudomonas syringae* type III effector AvrRpt2 requires a host factor(s) common among eukaryotes and is important for AvrRpt2 localization in the host cell. Plant Physiol. 133, 1072–1082.

Jones, J.D., and Dangl, J.L. (2006). The plant immune system. Nature 444, 323–329.

Kamoun, S. (2007). Groovy times: filamentous pathogen effectors revealed. Curr. Opin. Plant Biol. 10, 358–365.

Kang, L., Li, J., Zhao, T., Xiao, F., Tang, X., Thilmony, R., He, S., and Zhou, J.M. (2003). Interplay of the *Arabidopsis* nonhost resistance gene *NHO1* with bacterial virulence. Proc. Natl. Acad. Sci. USA 100, 3519–3524.

Kay, S., Hahn, S., Marois, E., Hause, G., and Bonas, U. (2007). A bacterial effector acts as a plant transcription factor and induces a cell size regulator. Science 318, 648–651.

Kerscher, O., Felberbaum, R., and Hochstrasser, M. (2006). Modification of proteins by ubiquitin and ubiquitin-like proteins. Annu. Rev. Cell. Dev. Biol. 22, 159–180.

Keshavarzi, M., Soylu, S., Brown, I., Bonas, U., Nicole, M., Rossiter, J., and Mansfield, J. (2004). Basal defenses induced in pepper by lipopolysaccharides are suppressed by *Xanthomonas campestris* pv. *vesicatoria*. Mol. Plant–Microbe Interact. 17, 805–815.

Kim, H.S., Desveaux, D., Singer, A.U., Patel, P., Sondek, J., and Dangl, J.L. (2005a). The *Pseudomonas syringae* effector AvrRpt2 cleaves its C-terminally acylated target, RIN4, from *Arabidopsis* membranes to block RPM1 activation. Proc. Natl. Acad. Sci. USA 102, 6496–6501.

Kim, J.F., Charkowski, A.O., Alfano, J.R., Collmer, A., and Beer, S.V. (1998). Sequences related to transposable elements and bacteriophages flank avirulence genes of *Pseudomonas syringae*. Mol. Plant–Microbe Interact. 11, 1247–1252.

Kim, M.G., da Cunha, L., McFall, A.J., Belkhadir, Y., DebRoy, S., Dangl, J.L., and Mackey, D. (2005b). Two *Pseudomonas syringae* type III effectors inhibit RIN4-regulated basal defense in *Arabidopsis*. Cell 121, 749–759.

Kim, Y.-J., Lin, N.-C., and Martin, G.B. (2002). Two distinct *Pseudomonas* effector proteins interact with the Pto kinase and activate plant immmunity. Cell 109, 589–598.

Klement, Z. (1982). Hypersensitivity, In Phytopathogenic procaryotes, M.S. Mount, and G.H. Lacy, eds. (New York: Academic Press), pp. 149–177.

Kunkel, B.N., and Brooks, D.M. (2002). Cross talk between signaling pathways in pathogen defense. Curr. Opin. Plant Biol. 5, 325–331.

Landgraf, A., Weingart, H., Tsiamis, G., and Boch, J. (2006). Different versions of *Pseudomonas syringae* pv. *tomato* DC3000 exist due to the activity of an effector transposon. Mol. Plant Pathol. 7, 355–364.

Li, H., Xu, H., Zhou, Y., Zhang, J., Long, C., Li, S., Chen, S., Zhou, J.M., and Shao, F. (2007). The phosphothreonine lyase activity of a bacterial type III effector family. Science 315, 1000–1003.

Li, X., Lin, H., Zhang, W., Zou, Y., Zhang, J., Tang, X., and Zhou, J.M. (2005). Flagellin induces innate immunity in nonhost interactions that is suppressed by *Pseudomonas syringae* effectors. Proc. Natl. Acad. Sci. USA 102, 12990–12995.

Lim, M.T., and Kunkel, B.N. (2004). The *Pseudomonas syringae* type III effector AvrRpt2 promotes virulence independently of RIN4, a predicted virulence target in *Arabidopsis thaliana*. Plant J. 40, 790–798.

Lin, N.C., Abramovitch, R.B., Kim, Y.J., and Martin, G.B. (2006). Diverse AvrPtoB homologs from several *Pseudomonas syringae* pathovars elicit Pto-dependent resistance and have similar virulence activities. Appl. Environ. Microbiol. 72, 702–712.

Lin, N.C., and Martin, G.B. (2005). An *avrPto/avrPtoB* mutant of *Pseudomonas syringae* pv. *tomato* DC3000 does not elicit Pto-mediated resistance and is less virulent on tomato. Mol. Plant–Microbe Interact. 18, 43–51.

Lindeberg, M., Cartinhour, S., Myers, C.R., Schechter, L.M., Schneider, D.J., and Collmer, A. (2006). Closing the circle on the discovery of genes encoding Hrp regulon members and type III secretion system effectors in the genomes of three model *Pseudomonas syringae* strains. Mol. Plant–Microbe Interact. 19, 1151–1158.

Lindeberg, M., Stavrinides, J., Chang, J.H., Alfano, J.R., Collmer, A., Dangl, J.L., Greenberg, J.T., Mansfield, J.W., and Guttman, D.S. (2005). Proposed guidelines for a unified nomenclature and phylogenetic analysis of type III Hop effector proteins in the plant pathogen *Pseudomonas syringae*. Mol. Plant–Microbe Interact. 18, 275–282.

Lindgren, P.B., Panopoulos, N.J., Staskawicz, B.J., and Dahlbeck, D. (1988). Genes required for pathogenicity and hypersensitivity are conserved and interchangeable among pathovers of *Pseudomonas syringae*. Mol. Gen. Genet. 211, 499–506.

Lindgren, P.B., Peet, R.C., and Panopoulos, N.J. (1986). Gene cluster of *Pseudomonas syringae* pv.'phaseolicola' controls pathogenicity of bean plants and hypersensitivity on nonhost plants. J. Bacteriol. 168, 512–522.

Lopez-Solanilla, E., Bronstein, P.A., Schneider, A.R., and Collmer, A. (2004). HopPtoN is a *Pseudomonas syringae* Hrp (type III secretion system) cysteine protease effector that suppresses pathogen-induced necrosis associated with both compatible and incompatible plant interactions. Mol. Microbiol. 54, 353–365.

Lotze, M.T., Deisseroth, A., and Rubartelli, A. (2007). Damage associated molecular pattern molecules. Clin. Immunol. 124, 1–4.

Lu, M., Tang, X., and Zhou, J.M. (2001). Arabidopsis NHO1 is required for general resistance against *Pseudomonas* bacteria. Plant Cell 13, 437–447.

Ma, W., Dong, F.F., Stavrinides, J., and Guttman, D.S. (2006). Type III effector diversification via both pathoadaptation and horizontal transfer in response to a coevolutionary arms race. PLoS Genet. 2, e209.

Mackey, D., Belkhadir, Y., Alonso, J.M., Ecker, J.R., and Dangl, J.L. (2003). *Arabidopsis* RIN4 is a target of the type III virulence effector AvrRpt2 and modulates RPS2-mediated resistance. Cell 112, 379–389.

Mackey, D., Holt III, B.F., Wiig, A., and Dangl, J.L. (2002). RIN4 interacts with *Pseudomonas syringae* type III effector molecules and is required for RPM1-mediated resistance in *Arabidopsis*. Cell 108, 743–754.

Mackey, D., and McFall, A.J. (2006). MAMPs and MIMPs: proposed classifications for inducers of innate immunity. Mol. Microbiol. 61, 1365–1371.

Marois, E., Van den Ackerveken, G., and Bonas, U. (2002). The *Xanthomonas* type III effector protein AvrBs3 modulates plant gene expression and induces cell hypertrophy in the susceptible host. Mol. Plant–Microbe Interact. 15, 637–646.

Maurer-Stroh, S., and Eisenhaber, F. (2004). Myristoylation of viral and bacterial proteins. Trends Microbiol. 12, 178–185.

Melotto, M., Underwood, W., Koczan, J., Nomura, K., and He, S.Y. (2006). Plant stomata function in innate immunity against bacterial invasion. Cell 126, 969–980.

Metz, M., Dahlbeck, D., Morales, C.Q., Al Sady, B., Clark, E.T., and Staskawicz, B.J. (2005). The conserved *Xanthomonas campestris* pv. *vesicatoria* effector protein XopX is a virulence factor and suppresses host defense in *Nicotiana benthamiana*. Plant J. *41*, 801–814.

Mittal, R., Peak-Chew, S.Y., and McMahon, H.T. (2006). Acetylation of MEK2 and I kappa B kinase (IKK) activation loop residues by YopJ inhibits signaling. Proc. Natl. Acad. Sci. USA *103*, 18574–18579.

Mucyn, T.S., Clemente, A., Andriotis, V.M., Balmuth, A.L., Oldroyd, G.E., Staskawicz, B.J., and Rathjen, J.P. (2006). The tomato NBARC-LRR protein Prf interacts with Pto kinase *in vivo* to regulate specific plant immunity. Plant Cell *18*, 2792–2806.

Mudgett, M.B., and Staskawicz, B.J. (1999). Characterization of the *Pseudomonas syringae* pv. *tomato* AvrRpt2 protein: demonstration of secretion and processing during bacterial pathogenesis. Mol. Microbiol. *32*, 927–941.

Mukherjee, S., Hao, Y.H., and Orth, K. (2007). A newly discovered post-translational modification – the acetylation of serine and threonine residues. Trends Biochem. Sci. *32*, 210–216.

Mukherjee, S., Keitany, G., Li, Y., Wang, Y., Ball, H.L., Goldsmith, E.J., and Orth, K. (2006). Yersinia YopJ acetylates and inhibits kinase activation by blocking phosphorylation. Science *312*, 1211–1214.

Nakagami, H., Pitzschke, A., and Hirt, H. (2005). Emerging MAP kinase pathways in plant stress signalling. Trends Plant Sci. *10*, 339–346.

Nambara, E., and Marion-Poll, A. (2005). Abscisic acid biosynthesis and catabolism. Annu. Rev. Plant Biol. *56*, 165–185.

Navarro, L., Dunoyer, P., Jay, F., Arnold, B., Dharmasiri, N., Estelle, M., Voinnet, O., and Jones, J.D. (2006). A plant miRNA contributes to antibacterial resistance by repressing auxin signaling. Science *312*, 436–439.

Nimchuk, Z., Marois, E., Kjemtrup, S., Leister, R.T., Katagiri, F., and Dangl, J.L. (2000). Eukaryotic fatty acylation drives plasma membrane targeting and enhances function of several type III effector proteins from *Pseudomonas syringae*. Cell *101*, 353–363.

Nimchuk, Z.L., Fisher, E.J., Desveaux, D., Chang, J.H., and Dangl, J.L. (2007). The HopX (AvrPphE) family of *Pseudomonas syringae* type III effectors require a catalytic triad and a novel N-terminal domain for function. Mol. Plant–Microbe Interact. *20*, 346–357.

Nissan, G., Manulis-Sasson, S., Weinthal, D., Mor, H., Sessa, G., and Barash, I. (2006). The type III effectors HsvG and HsvB of gall-forming *Pantoea agglomerans* determine host specificity and function as transcriptional activators. Mol. Microbiol. *61*, 1118–1131.

Nomura, K., Debroy, S., Lee, Y.H., Pumplin, N., Jones, J., and He, S.Y. (2006). A bacterial virulence protein suppresses host innate immunity to cause plant disease. Science *313*, 220–223.

Nürnberger, T., Brunner, F., Kemmerling, B., and Piater, L. (2004). Innate immunity in plants and animals: striking similarities and obvious differences. Immunol. Rev. *198*, 249–266.

Nürnberger, T., and Scheel, D. (2001). Signal transmission in the plant immune response. Trends Plant Sci. *6*, 372–379.

O'Donnell, P.J., Schmelz, E., Block, A., Miersch, O., Wasternack, C., Jones, J.B., and Klee, H.J. (2003). Multiple hormones act sequentially to mediate a susceptible tomato pathogen defense response. Plant Physiol. *133*, 1181–1189.

Oh, H.S., and Collmer, A. (2005). Basal resistance against bacteria in *Nicotiana benthamiana* leaves is accompanied by reduced vascular staining and suppressed by multiple *Pseudomonas syringae* type III secretion system effector proteins. Plant J. *44*, 348–359.

Orth, K., Palmer, L.E., Bao, Z.Q., Stewart, S., Rudolph, A.E., Bliska, J.B., and Dixon, J.E. (1999). Inhibition of the mitogen-activated protein kinase kinase superfamily by a *Yersinia* effector. Science *285*, 1920–1923.

Orth, K., Xu, Z., Mudgett, M.B., Bao, Z.Q., Palmer, L.E., Bliska, J.B., Mangel, W.F., Staskawicz, B.J., and Dixon, J.E. (2000). Disruption of signaling by *Yersinia* effector YopJ, a ubiquitin-like protein protease. Science *290*, 1594–1597.

Pallen, M.J., Beatson, S.A., and Bailey, C.M. (2005). Bioinformatics, genomics and evolution of non-flagellar type-III secretion systems: a Darwinian perspective. FEMS Microbiol. Rev. *29*, 201–229.

Park, J.E., Park, J.Y., Kim, Y.S., Staswick, P.E., Jeon, J., Yun, J., Kim, S.Y., Kim, J., Lee, Y.H., and Park, C.M. (2007). GH3-mediated auxin homeostasis links growth regulation with stress adadptation response in *Arabidopsis*. J. Biol. Chem. *282*, 10036–10046.

Pedley, K.F., and Martin, G.B. (2003). Molecular basis of *Pto*-mediated resistance to bacterial speck disease in tomato. Annu. Rev. Phytopathol. *41*, 215–243.

Pedley, K.F., and Martin, G.B. (2005). Role of mitogen-activated protein kinases in plant immunity. Curr. Opin. Plant Biol. *8*, 541–547.

Pickart, C.M., and Cohen, R.E. (2004). Proteasomes and their kin: proteases in the machine age. Nat. Rev. Mol. Cell. Biol. *5*, 177–187.

Pühler, A., Arlat, M., Becker, A., Göttfert, M., Morrissey, J.P., and O'Gara, F. (2004). What can bacterial genome research teach us about bacteria-plant interactions? Curr. Opin. Plant Biol. *7*, 137–147.

Ritter, C., and Dangl, J.L. (1996). Interference between two specific pathogen recognition events mediated by distinct plant disease resistance genes. Plant Cell *8*, 251–257.

Robert-Seilaniantz, A., Navarro, L., Bari, R., and Jones, J.D. (2007). Pathological hormone imbalances. Curr. Opin. Plant Biol. *10*, 372–379.

Robert-Seilaniantz, A., Shan, L., Zhou, J.M., and Tang, X. (2006). The *Pseudomonas syringae* pv. *tomato* DC3000 type III effector HopF2 has a putative myristoylation site required for its avirulence and virulence functions. Mol. Plant–Microbe Interact. *19*, 130–138.

Roden, J., Eardley, L., Hotson, A., Cao, Y., and Mudgett, M.B. (2004). Characterization of the *Xanthomonas* AvrXv4 effector, a SUMO protease translocated into plant cells. Mol. Plant–Microbe Interact. 17, 633–643.

Römer, P., Hahn, S., Jordan, T., Strauss, T., Bonas, U., and Lahaye, T. (2007). Plant pathogen recognition mediated by promoter activation of the pepper Bs3 resistance gene. Science 318, 645–648.

Rosebrock, T.R., Zeng, L., Brady, J.J., Abramovitch, R.B., Xiao, F., and Martin, G.B. (2007). A bacterial E3 ubiquitin ligase targets a host protein kinase to disrupt plant immunity. Nature 448, 370–374.

Schmelz, E.A., Engelberth, J., Alborn, H.T., O'Donnell, P., Sammons, M., Toshima, H., and Tumlinson, J.H., 3rd (2003). Simultaneous analysis of phytohormones, phytotoxins, and volatile organic compounds in plants. Proc. Natl. Acad. Sci. USA 100, 10552–10557.

Scofield, S.R., Tobias, C.M., Rathjen, J.P., Chang, J.H., Lavelle, D.T., Michelmore, R.W., and Staskawicz, B.J. (1996). Molecular basis of gene-for-gene specificity in bacterial speck disease of tomato. Science 274, 2063–2065.

Sessa, G., D'Ascenzo, M., and Martin, G.B. (2000). Thr38 and Ser198 are Pto autophosphorylation sites required for the AvrPto-Pto-mediated hypersensitive response. EMBO J. 19, 2257–2269.

Shan, L., Thara, V.K., Martin, G.B., Zhou, J.M., and Tang, X. (2000). The Pseudomonas AvrPto protein is differentially recognized by tomato and tobacco and is localized to the plant plasma membrane. Plant Cell 12, 2323–2338.

Shang, Y., Li, X., Cui, H., He, P., Thilmony, R., Chintamanani, S., Zwiesler-Vollick, J., Gopalan, S., Tang, X., and Zhou, J.M. (2006). RAR1, a central player in plant immunity, is targeted by Pseudomonas syringae effector AvrB. Proc. Natl. Acad. Sci. USA 103, 19200–19205.

Shao, F., Golstein, C., Ade, J., Stoutemyer, M., Dixon, J.E., and Innes, R.W. (2003). Cleavage of Arabidopsis PBS1 by a bacterial type III effector. Science 301, 1230–1233.

Shao, F., Merritt, P.M., Bao, Z., Innes, R.W., and Dixon, J.E. (2002). A Yersinia effector and a Pseudomonas avirulence protein define a family of cysteine proteases functioning in bacterial pathogenesis. Cell 109, 575–588.

Shen, Q.H., and Schulze-Lefert, P. (2007). Rumble in the nuclear jungle: compartmentalization, trafficking, and nuclear action of plant immune receptors. EMBO J. 26, 4293–4301.

Sohn, K.H., Lei, R., Nemri, A., and Jones, J.D. (2007). The downy mildew effector proteins ATR1 and ATR13 promote disease susceptibility in Arabidopsis thaliana. Plant Cell 19, 4077–4090.

Staskawicz, B., Dahlbeck, D., Keen, N., and Napoli, C. (1987). Molecular characterization of cloned avirulence genes from race 0 and race 1 of Pseudomonas syringae pv. glycinea. J. Bacteriol. 169, 5789–5794.

Staskawicz, B.J., Dahlbeck, D., and Keen, N.T. (1984). Cloned avirulence gene of Pseudomonas syringae pv. glycinea determined race-specific incompatibility on Glycine max (L.) Merr. Proc. Natl. Acad. Sci. USA 81, 6024–6028.

Staswick, P.E. (2008). JAZing up jasmonate signalling. Trends Plant Sci. 13, 66–71.

Stavrinides, J., Ma, W., and Guttman, D.S. (2006). Terminal reassortment drives the quantum evolution of type III effectors in bacterial pathogens. PLoS Pathog. 2.

Süß, C., Hempel, J., Zehner, S., Krause, A., Patschkowski, T., and Göttfert, M. (2006). Identification of genistein-inducible and type III-secreted proteins of Bradyrhizobium japonicum. J. Biotechnol. 126, 69–77.

Swarup, S., de Feyter, R., Brlansky, R.H., and Gabriel, D.W. (1991). A pathogenicity locus from Xanthomonas citri enables strains from several pathovars of X. campestris to elicit cankerlike lesions on citrus. Phytopathology 81, 802–809.

Szurek, B., Marois, E., Bonas, U., and Van den Ackerveken, G. (2001). Eukaryotic features of the Xanthomonas type III effector AvrBs3: protein domains involved in transcriptional activation and the interaction with nuclear import receptors from pepper. Plant J. 26, 523–534.

Tampakaki, A.P., Bastaki, M., Mansfield, J.W., and Panopoulos, N.J. (2002). Molecular determinants required for the avirulence function of AvrPphB in bean and other plants. Mol. Plant–Microbe Interact. 15, 292–300.

Tampakaki, A.P., Fadouloglou, V.E., Gazi, A.D., Panopoulos, N.J., and Kokkinidis, M. (2004). Conserved features of type III secretion. Cell. Microbiol. 6, 805–816.

Tang, X., Frederick, R.D., Zhou, J., Halterman, D.A., Jia, Y., and Martin, G.B. (1996). Initiation of plant disease resistance by physical interaction of AvrPto and Pto kinase. Science 274, 2060–2063.

Tao, Y., Xie, Z., Chen, W., Glazebrook, J., Chang, H.S., Han, B., Zhu, T., Zou, G., and Katagiri, F. (2003). Quantitative nature of Arabidopsis responses during compatible and incompatible interactions with the bacterial pathogen Pseudomonas syringae. Plant Cell 15, 317–330.

Thaler, J.S., and Bostock, R.M. (2004). Interactions between abscisic-acid-mediated responses and plant resistance to pathogens and insects. Ecology 85, 48–58.

Thieme, F., Szczesny, R., Urban, A., Kirchner, O., Hause, G., and Bonas, U. (2007). New type III effectors from Xanthomonas campestris pv. vesicatoria trigger plant reactions dependent on a conserved N-myristoylation motif. Mol. Plant–Microbe Interact. 20, 1250–1261.

Thordal-Christensen, H. (2003). Fresh insights into processes of nonhost resistance. Curr. Opin. Plant Biol. 6, 351–357.

Torres, M.A., Jones, J.D., and Dangl, J.L. (2006). Reactive oxygen species signaling in response to pathogens. Plant Physiol. 141, 373–378.

Troisfontaines, P., and Cornelis, G.R. (2005). Type III secretion: more systems than you think. Physiology (Bethesda) 20, 326–339.

Truman, W., Bennett, M.H., Kubigsteltig, I., Turnbull, C., and Grant, M. (2007). Arabidopsis systemic immunity uses conserved defense signaling pathways and is mediated by jasmonates. Proc. Natl. Acad. Sci. USA 104, 1075–1080.

Truman, W., de Torres-Zabala, M., and Grant, M. (2006). Type III effectors orchestrate a complex interplay between transcriptional networks to modify basal defence responses during pathogenesis and resistance. Plant J. 46, 14–33.

Tsiamis, G., Mansfield, J.W., Hockenhull, R., Jackson, R.W., Sesma, A., Athanassopoulos, E., Bennett, M.A., Stevens, C., Vivian, A., Taylor, J.D., and Murillo, J. (2000). Cultivar-specific avirulence and virulence functions assigned to avrPphF in Pseudomonas syringae pv. phaseolicola, the cause of bean halo-blight disease. EMBO J. 19, 3204–3214.

Underwood, W., Zhang, S., and He, S.Y. (2007). The Pseudomonas syringae type III effector tyrosine phosphatase HopAO1 suppresses innate immunity in Arabidopsis thaliana. Plant J. 52, 658–672.

Van den Ackerveken, G., Marois, E., and Bonas, U. (1996). Recognition of the bacterial avirulence protein AvrBs3 occurs inside the host plant cell. Cell 87, 1307–1316.

Van der Biezen, E.A., and Jones, J.D. (1998). Plant disease-resistance proteins and the gene-for-gene concept. Trends Biochem. Sci. 23, 454–456.

Vanholme, B., De Meutter, J., Tytgat, T., Van Montagu, M., Coomans, A., and Gheysen, G. (2004). Secretions of plant-parasitic nematodes: a molecular update. Gene 332, 13–27.

Verger, A., Perdomo, J., and Crossley, M. (2003). Modification with SUMO. A role in transcriptional regulation. EMBO Rep. 4, 137–142.

Viboud, G.I., and Bliska, J.B. (2005). Yersinia outer proteins: role in modulation of host cell signaling responses and pathogenesis. Annu. Rev. Microbiol. 59, 69–89.

Vivian, A., Murillo, J., and Jackson, R.W. (2001). The roles of plasmids in phytopathogenic bacteria: mobile arsenals? Microbiology 147, 763–780.

Wang, D., Pajerowska-Mukhtar, K., Hendrickson Culler, A., and Dong, X. (2007a). Salicylic acid inhibits pathogen growth in plants through repression of the auxin signaling pathway. Curr. Biol. 17, 1784–1790.

Wang, L., Tang, X., and He, C. (2007b). The bifunctional effector AvrXccC of Xanthomonas campestris pv. campestris requires plasma membrane-anchoring for host recognition. Mol. Plant Pathol. 8, 491–501.

Wei, C.F., Kvitko, B.H., Shimizu, R., Crabill, E., Alfano, J.R., Lin, N.C., Martin, G.B., Huang, H.C., and Collmer, A. (2007). A Pseudomonas syringae pv. tomato DC3000 mutant lacking the type III effector HopQ1–1 is able to cause disease in the model plant Nicotiana benthamiana. Plant J. 51, 32–46.

Wu, A.J., Andriotis, V.M., Durrant, M.C., and Rathjen, J.P. (2004). A patch of surface-exposed residues mediates negative regulation of immune signaling by tomato Pto kinase. Plant Cell 16, 2809–2821.

Xiang, T., Zong, N., Zou, Y., Wu, Y., Zhang, J., Xing, W., Li, Y., Tang, X., Zhu, L., Chai, J., and Zhou, J.M. (2008). Pseudomonas syringae effector AvrPto blocks innate immunity by targeting receptor kinases. Curr. Biol. 18, 74–80.

Xiao, F., Giavalisco, P., and Martin, G.B. (2007a). Pseudomonas syringae type III effector AvrPtoB is phosphorylated in plant cells on serine 258, promoting its virulence activity. J. Biol. Chem. 282, 30737–30744.

Xiao, F., He, P., Abramovitch, R.B., Dawson, J.E., Nicholson, L.K., Sheen, J., and Martin, G.B. (2007b). The N-terminal region of Pseudomonas type III effector AvrPtoB elicits Pto-dependent immunity and has two distinct virulence determinants. Plant J. 52, 595–614.

Xing, W., Zou, Y., Liu, Q., Liu, J., Luo, X., Huang, Q., Chen, S., Zhu, L., Bi, R., Hao, Q., et al. (2007). The structural basis for activation of plant immunity by bacterial effector protein AvrPto. Nature 449, 243–247.

Xu, L., Liu, F., Lechner, E., Genschik, P., Crosby, W.L., Ma, H., Peng, W., Huang, D., and Xie, D. (2002). The SCF(COI1) ubiquitin-ligase complexes are required for jasmonate response in Arabidopsis. Plant Cell 14, 1919–1935.

Yang, B., Sugio, A., and White, F.F. (2006). Os8N3 is a host disease-susceptibility gene for bacterial blight of rice. Proc. Natl. Acad. Sci. USA 103, 10503–10508.

Yang, S.H., and Sharrocks, A.D. (2004). SUMO promotes HDAC-mediated transcriptional repression. Mol. Cell. 13, 611–617.

Yang, Y., Yuan, Q., and Gabriel, D.W. (1996). Watersoaking function(s) of XcmH1005 are redundantly encoded by members of the Xanthomonas avr/pth gene family. Mol. Plant–Microbe Interact. 9, 105–113.

Zhang, J., Shao, F., Li, Y., Cui, H., Chen, L., Li, H., Zou, Y., Long, C., Lan, L., Chai, J., et al. (2007). A Pseudomonas syringae effector inactivates MAPKs to suppress PAMP-induced immunity in plants. Cell. Host & Microbe 1, 175–185.

Zhou, H., Monack, D.M., Kayagaki, N., Wertz, I., Yin, J., Wolf, B., and Dixit, V.M. (2005). Yersinia virulence factor YopJ acts as a deubiquitinase to inhibit NF-kappa B activation. J. Exp. Med. 202, 1327–1332.

Zhu, M., Shao, F., Innes, R.W., Dixon, J.E., and Xu, Z. (2004). The crystal structure of Pseudomonas avirulence protein AvrPphB: A papain-like fold with a distinct substrate-binding site. Proc. Natl. Acad. Sci. USA 101, 302–307.

Zhu, W., Yang, B., Chittoor, J.M., Johnson, L.B., and White, F.F. (1998). AvrXa10 contains an acidic transcriptional activation domain in the functionally conserved C terminus. Mol. Plant–Microbe Interact. 11, 824–832.

13

Cyclic Di-GMP Signalling and the Regulation of Virulence in Bacterial Plant Pathogens

J. Maxwell Dow, Yvonne McCarthy, Belén Fernandez Garcia and Robert P. Ryan

Abstract

Cyclic di-GMP is a novel second messenger that regulates a range of functions including developmental transitions, adhesion, biofilm formation and virulence in diverse bacteria including plant pathogens. Cellular levels of cyclic di-GMP are influenced by both synthesis and degradation. The GGDEF protein domain synthesizes cyclic di-GMP, whereas EAL and HD-GYP domains are involved in cyclic di-GMP hydrolysis. The majority of proteins with GGDEF, EAL and HD-GYP domains contain additional signal input domains, suggesting that their activities (and consequently cyclic di-GMP levels) are responsive to environmental cues. Cyclic di-GMP exerts its effects on certain cellular functions by binding to proteins containing a PilZ domain. This domain may occur either as a stand-alone domain, which may act as an 'adaptor' to bind other proteins, or as part of a larger protein as is found in the BcsA subunit of cellulose synthase. Some details of the organization and function of cyclic di-GMP signalling systems have emerged, where both networks of systems regulating the same functions and systems apparently dedicated to specific other tasks occur together in bacterial cells. This has lead to the controversial concept of discrete pools of cyclic di-GMP that are generated and act in a highly localized fashion.

Introduction

Cyclic di-GMP (bis-(3′–5′)-cyclic di-guanosine monophosphate) (Fig. 13.1) was originally described in 1987 as an allosteric regulator of cellulose synthesis in *Acetobacter xylinum* (now *Gluconacetobacter xylinus*) (Ross et al., 1987).

The last decade or so has seen an expanding body of work that has described a role for cyclic di-GMP signalling in regulation of many aspects of bacterial behaviour including adhesion to surfaces, aggregation and biofilm formation, developmental transitions and, importantly, the virulence of bacterial pathogens of both animals and plants. The cellular level of cyclic di-GMP results from a balance between synthesis and degradation. Three protein domains are implicated in these processes: the GGDEF domain catalyses synthesis of cyclic di-GMP from two molecules of GTP whereas EAL and HD-GYP domains catalyse hydrolysis of cyclic di-GMP, firstly to the linear nucleotide pGpG and thence at different rates to GMP (reviewed by Ryan et al., 2006b). All of these domains are named after conserved amino acid motifs. Whole genome sequencing has revealed an abundance of GGDEF and EAL domain containing proteins across the majority of bacterial species (both Gram-positive and Gram-negative) including plant pathogens (Galperin, 2005). HD-GYP domain proteins are less abundant, but still widely distributed (Galperin, 2005) and are found in plant pathogens, symbionts and plant-associated bacteria (Dow et al., 2006). Most proteins with GGDEF/EAL/HD-GYP domains contain additional signal input domains, suggesting that their activities are responsive to signals from the bacterial environment. Some of these primary environmental signals have now been defined. Transduction of these signals via the use of cyclic di-GMP as a second messenger is thus one of a number of mechanisms that serve to link environmental changes to alterations in bacterial behaviour that

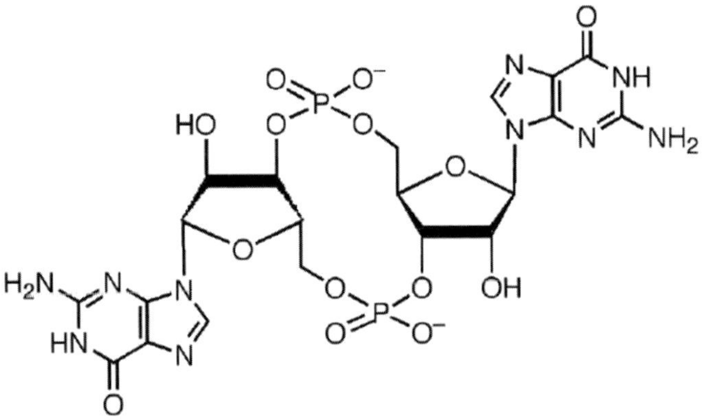

Figure 13.1 Structure of the second messenger bis-(3′–5′)-cyclic di-guanosine monophosphate (cyclic di-GMP).

can include modulation of synthesis of virulence factors. An emerging theme from a number of studies is that high levels of cyclic di-GMP promote biofilm formation and sessility, whereas low levels promote motility and the production of virulence factors (Simm et al., 2004; Römling et al., 2005).

In the following sections we review the field of cyclic di-GMP signalling and its role in the virulence of plant pathogens. We begin with a general discussion of the roles of the GGDEF, EAL and HD-GYP domains in cyclic di-GMP turnover and of the PilZ domain in cyclic di-GMP action. We then address the functional organization of cyclic di-GMP signalling systems in bacteria before going on to discuss specific examples of cyclic di-GMP regulation of virulence in plant pathogens. We also highlight cases in which a role in regulation for specific cyclic di-GMP signalling systems in other bacteria may inform work on the virulence of plant pathogens. The reader is also directed to several recent reviews of this area (Jenal, 2004; Römling et al., 2005; Jenal and Malone, 2006; Ryan et al., 2006b, Römling and Amikam, 2006; Cotter and Stibitz, 2007).

The role of GGDEF, EAL and HD-GYP domains in cyclic di-GMP turnover

The association of GGDEF and EAL domains with cyclic di-GMP turnover was first made through investigations of the regulation of cellulose synthesis in G. xylinus. Biochemical studies revealed that the level of cyclic di-GMP in G. xylinus was controlled by the opposing action of the enzymes diguanylate cyclase (DGC), which catalyses its formation and phosphodiesterase A (PDEA), which catalyses its degradation. Amino acid sequence information derived from purified DGC and PDEA allowed the cloning of three *cdg* genes encoding isoforms of DGC and three *pdeA* genes encoding isoforms of PDEA (Tal et al., 1998). All three *cdg* genes and all three *pdeA* genes encode hybrid proteins containing a GGDEF domain (formerly known as DUF1, for domain of unknown function) and a C-terminal EAL domain (formerly known as DUF2). The GGDEF and EAL nomenclature relates to conserved amino acid motifs in these domains (Galperin et al., 2001). The association of GGDEF and EAL domains with cyclic di-GMP turnover was further strengthened by the later demonstration of cyclic di-GMP phosphodiesterase activity of the recombinant PdeA1 protein (Chang et al., 2001). The GGDEF domain was first identified in PleD, a regulatory protein controlling swarmer-to-stalked-cell transition in *Caulobacter crescentus* (Hecht and Newton, 1995). The EAL domain was first described in BvgR, a repressor of virulence gene expression in *Bordetella pertussis* (Merkel and Stibitz, 1995; Merkel et al., 1998a,b). In the light of the work in G. xylinus that implicated fused GGDEF-EAL domain proteins in cyclic di-GMP turnover, these findings gave the first indication that cyclic di-GMP signalling might be more widespread in bacteria and control functions other than cellulose synthesis.

Although the work in G. xylinus associated GGDEF and EAL domains with cyclic di-GMP turnover, the precise biochemical role of these domains was not resolved. Indirect evidence for the role of the GGDEF domain in cyclic di-GMP synthesis came from *in silico* studies indicating some structural conservation with the proposed nucleotide-binding loop of eukaryotic adenylyl cyclases (Pei and Grishin, 2001). This suggestion was supported by genetic experiments in which expression of *dgc1* from G. xylinus, which encodes a GGDEF-EAL hybrid protein, as well as genes encoding proteins with GGDEF but no EAL domain from other bacteria were shown to complement a *celR2* mutant of *Rhizobium* (Ausmees *et al.*, 1999) for defects in cellulose production (Ausmees *et al.*, 2001). Subsequently it was shown that expression of genes encoding GGDEF proteins could increase the cellular levels of cyclic di-GMP in several bacteria (Paul *et al.*, 2004; Simm *et al.*, 2004, Tischler and Camilli, 2004). Direct evidence for the role of the GGDEF domain was obtained by biochemical studies of purified GGDEF domain proteins including PleD (Paul *et al.*, 2004; Ryjenkov *et al.*, 2005) and of isolated GGDEF domains from proteins from different bacterial phyla (Ryjenkov *et al.*, 2005). Each of these GGDEF domains/proteins converted two molecules of GTP to cyclic di-GMP with Mg^{2+} as a cofactor, but had no activity with other nucleotides (Fig. 13.2). The conservation of function of GGDEF domains from divergent bacteria was further demonstrated by reciprocal complementation of *hmsT*, which is involved in biofilm formation in *Yersinia pestis* and *adrA*, which is involved in cellulose synthesis in *Salmonella enterica* serovar typhimurium (Simm *et al.*, 2005). Furthermore in many of the experiments cited above, site-directed mutagenesis was used to establish that the conserved GGDEF motif residues were critical for cyclic di-GMP synthesis.

Indirect support for the role of the EAL domain in cyclic di-GMP degradation was provided by the demonstration that heterologous expression of genes encoding proteins with EAL but not GGDEF domains could reduce cellular levels of cyclic di-GMP (Simm *et al.*, 2004; Tischler and Camilli, 2004; 2005) and, conversely, that mutation of an EAL domain protein increased cellular cyclic di-GMP levels (Hisert *et al.*, 2005). Moreover the EAL domain protein HmsP of *Yersinia pestis* was shown to possess activity against the model phosphodiesterase substrate bis-(p-nitrophenol) phosphate (Bobrov *et al.*, 2005). Direct biochemical evidence for the role of the EAL domain as a cyclic di-GMP phosphodiesterase came from studies of intact proteins as well as isolated EAL domains (Christen *et al.*, 2005; Schmidt *et al.*, 2005; Tamayo *et al.*, 2005; Ryan *et al.*, 2006a). The major product of the enzymatic action of the EAL domain was the linear nucleotide pGpG, which was more slowly converted to GMP (Fig. 13.2). This activity of the EAL domain was absolutely dependent on the presence of Mg^{2+} or Mn^{2+} (Schmidt *et al.*, 2005). Again in a number of these cases mutational analysis indicated the essential role of the conserved EAL motif in enzymatic activity.

Regulatory roles for enzymatically inactive GGDEF and EAL domains

Variations in the amino acids within the GGDEF and EAL motifs occur naturally. Although domains with sequences that are divergent from the consensus may have lost enzymatic activity, they may still play a regulatory role. In a

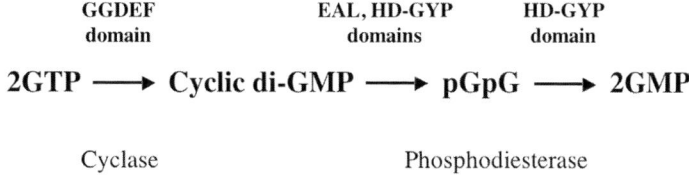

Figure 13.2 The role of GGDEF, EAL and HD-GYP domains in the synthesis and degradation of cyclic di-GMP. Synthesis of cyclic di-GMP from two molecules of GTP is catalysed by the GGDEF domain and is predicted to occur in two steps, with pppGpG as intermediate. Each step releases a molecule of inorganic pyrophosphate. The degradation of cyclic di-GMP to GMP also occurs via a two-step reaction, with the linear dinucleotide pGpG as intermediate. EAL domains characterized thus far catalyse only the first step, whereas the HD-GYP domain catalyses both steps. Other, perhaps non-specific, phosphodiesterase enzymes may also convert pGpG to 5′ GMP.

GGDEF-EAL domain protein from *Caulobacter* (CC3396), binding of GTP to an enzymatically inactive GGDEF domain (actually GEDEF) acts to regulate the activity of the protein in cyclic di-GMP hydrolysis (Christen et al., 2005). A similar activation mechanism may occur in FimX from *Pseudomonas aeruginosa*, which has the variant sequence GDSIF in the GGDEF domain (Kazmierczak et al., 2006). The same considerations may also apply to inactive EAL domains (Schmidt et al., 2005). As well as intramolecular effects, enzymatically inactive EAL or GGDEF domains could conceivably regulate cellular processes through intermolecular interactions with other proteins or other molecules, in a fashion influenced by the binding of cyclic di-GMP or GTP. CsrD from *Escherichia coli*, which has GGDEF and EAL domains both with divergent sequence motifs (HRSDF and ELM respectively), is implicated in regulation of degradation of the small regulatory RNAs CsrB and CsrC that attenuate the action of the post-transcriptional regulator CsrA (Suzuki et al., 2006). It is not known if CsrD retains any enzymatic activity against cyclic di-GMP or if the nucleotide can regulate the action of CsrD that may conceivably involve binding of the degradative enzyme RNaseE and/or the small regulatory RNAs (see also below).

The biochemical activity of GGDEF-EAL domain proteins

The definition of the biochemical functions of GGDEF and EAL domains presents a conundrum; what determines the activity of proteins such as the PDEA and DGC from G. xylinus, which contain both domains (Tal et al., 1998). One possible resolution to this paradox is that one of the two domains is non-functional but has a regulatory role (see above). A second resolution of the paradox could be that the proteins can have both activities but switch between states able to synthesize and hydrolyse cyclic di-GMP. One possible mechanism could be related to the oligomerization state. Structural analysis of the PleD regulator suggests that the GGDEF domain acts in cyclic di-GMP synthesis as a dimer (Chan et al., 2004) whereas EAL activity is apparently independent of protein oligomerization (Schmidt et al., 2005). Regulation of the oligomerization state of the GGDEF-EAL proteins, perhaps influenced by the sensory input domains, may then serve to determine which activity is expressed. A third alternative is that the two activities may be independently but dynamically set to ensure a very precise concentration emanating from a point source.

The HD-GYP domain is a second cyclic di-GMP phosphodiesterase

Bioinformatic studies have suggested that a third domain HD-GYP is also involved in cyclic di-GMP hydrolysis (Galperin et al., 1999; 2001). HD-GYP is a subgroup of the HD superfamily of metal dependent phosphohydrolases. The association of the HD-GYP domain with a CheY-like two-component receiver domain in many bacterial proteomes indicates a role in signalling (Galperin et al., 1999; 2001). A role for HD-GYP in cyclic di-GMP hydrolysis was proposed based on an examination of the distribution and numbers of GGDEF, EAL and HD-GYP domains encoded by different bacterial genomes, where several genomes encode proteins with the GGDEF and HD-GYP domains but no EAL domain (Galperin et al., 1999; 2001; Galperin, 2005).

In the plant pathogen *Xanthomonas campestris* pv. *campestris* the HD-GYP domain regulator RpfG positively regulates synthesis of extracellular enzyme virulence factors and negatively regulates biofilm formation (Slater et al., 2000; Dow et al., 2003). Expression of genes encoding EAL domain proteins in the *X. campestris* pv. *campestris rpfG* mutant restored extracellular enzymes and blocked biofilm formation. In contrast expression of genes encoding a GGDEF domain protein in wild type *X. campestris* pv. *campestris* gave a phenocopy of the *rpfG* mutant (Ryan et al., 2006a). These indirect observations were consistent with a role for the HD-GYP domain in cyclic di-GMP hydrolysis. This conclusion was supported by biochemical studies that demonstrated that the isolated domain could hydrolyse cyclic di-GMP to GMP via a linear intermediate (Ryan et al., 2006a; Fig. 13.2). This reaction depended upon Mn^{2+} for which Mg^{2+} could not substitute. Mutation of the HD residues comprising the presumed catalytic diad of the HD-GYP domain abolishes both the regulatory activity and enzymatic activity against cyclic di-GMP (Ryan et al., 2006a).

This recent finding of the occurrence of a second cyclic di-GMP phosphodiesterase unrelated to the EAL domain raises a number of questions for which there are currently no answers. Why did such an alternative activity arise? Is there any significance to the different relative activities of the EAL and HD-GYP domains against pGpG or for the different requirements for divalent metal ions? It has been suggested for example that pGpG may have a regulatory role in cyclic di-GMP signalling or may itself act as a signal molecule (Römling et al., 2005; Schmidt et al., 2005). The HD-GYP domain could serve to regulate pGpG levels, although this task may be performed by other, non-specific, nucleases. Intriguingly recent findings from yeast two-hybrid analysis have revealed a physical interaction between the HD-GYP domain of RpfG of X. axonopodis pv. citri and a subset of GGDEF domain proteins (Andrade et al., 2006). This subset of GGDEF domains is not associated with EAL domains in hybrid proteins. Although the biological relevance of such interactions has yet to be tested, these observations suggest a scenario whereby the HD-GYP domain may have evolved to modulate the activity of GGDEF domain proteins that are not associated with an EAL domain. Interestingly a small number of proteins, largely found in Clostridium and related genera, contain fused GGDEF and HD-GYP domains, whereas to our knowledge combinations of EAL and HD-GYP domains do not occur.

Mechanisms of cyclic di-GMP action-the role of PilZ domains

Relatively little is known about the mechanism(s) by which cyclic di-GMP exerts its influence on different cellular processes. The best-studied effect of cyclic di-GMP is its allosteric activation of cellulose synthesis in G. xylinus, where cyclic di-GMP was shown to bind to BcsB, the β-subunit of cellulose synthase and to an uncharacterized 200kDa protein (Mayer et al., 1991; Weinhouse et al., 1997). Recent bioinformatics studies have suggested that in contrast cyclic di-GMP binds to BcsA, the α-subunit of cellulose synthase and that PilZ (Pfam PF07238), a domain at the C-terminus of BcsA, is part of the binding site (Amikam and Galperin, 2006). PilZ was originally described as a protein involved in assembly of functional pili in P. aeruginosa (Alm et al., 1996). Multiple proteins with a PilZ domain are found in the same bacterium. For example in P. syringae pv. tomato there are seven proteins with a PilZ domain; four proteins (PSPTO_2116, PSPTO_3825, PSPTO_4335, PSPTO_4639) of approximately 100–120 amino acids constitute a 'stand-alone' domain, whereas in three proteins (PSPTO_1027 (BcsA), PSPTO_1241(Alg44), PSPTO_1923), the PilZ domain is either the C-terminal or N-terminal part of a multidomain protein.

Very recent findings have shown that PilZ domain proteins from Vibrio cholerae (Pratt et al., 2007), C. crescentus (Christen et al., 2007), P. aeruginosa (Ramelot et al., 2007) and E. coli (Ryjenkov et al., 2006) and the PilZ domain of BcsA of G. xylinus (Ryjenkov et al., 2006) do indeed bind cyclic di-GMP and that upon binding they regulate a number of bacterial functions including biofilm formation, motility, and virulence (Ryjenkov et al., 2006; Pratt et al., 2007; Christen et al., 2007). Cyclic di-GMP action through a PilZ domain protein has a role in alginate synthesis in Pseudomonas spp. (Remminghorst and Rehm, 2006; Meringhi et al., 2007). Alg44, which is required for alginate synthesis (Remminghorst and Rehm, 2006), possesses a PilZ domain that is involved in binding of cyclic di-GMP (Merighi et al., 2007). This binding is required for the action of Alg44 in alginate biosynthesis, where it may control polymerization or transport of the polysaccharide (Merighi et al., 2007). Alginate has a role in the virulence and epiphytic fitness of P. syringae pv. syringae (Yu et al., 1999) and makes a contribution to biofilm formation in P. syringae, although other extracellular polysaccharides may play a more important role (Laue et al., 2006).

By analogy with cyclic AMP, cyclic di-GMP may affect transcription by binding to transcriptional regulators, although as far as we are aware there have been no descriptions of such binding. An alternative view is that cyclic di-GMP works primarily at the post-translational level, and effects on transcription are indirect. Only further experimental work will resolve these issues.

Bacterial genomes encode multiple proteins with GGDEF, EAL and HD-GYP domains

Large scale sequencing of bacterial genomes has revealed that GGDEF and EAL domains are widely distributed and highly abundant, although they are not found in archaea (Galperin, 2005). At the time of writing there were over 5200 GGDEF domains and over 3000 EAL domains in the Pfam protein family database. The HD-GYP domain is also widely distributed although slightly less abundant, with over 250 HD-GYP domains in over 80 genomes. Most bacterial genomes encode a number of proteins with these domains, with the complexity of cyclic di-GMP signalling increasing with the size of the genome in a non-linear fashion. For example, *Xylella fastidiosa* (~ 2.7 Mb) encodes five proteins with a potential role in cyclic di-GMP signalling whereas the genome of the related *X. campestris* pv. *campestris* (~ 5 Mb) encodes 37 such proteins. The proteome of *X. campestris* pv. *campestris* includes 21 predicted proteins with a GGDEF but no EAL domain, five proteins with an EAL but no GGDEF domain, eight proteins with both GGDEF and EAL domains and three proteins with an HD-GYP domain (Ryan *et al.*, 2007).

Many GGDEF, EAL and HD-GYP domain proteins have associated regulatory/sensory input domains

As indicated above, many GGDEF, EAL and HD-GYP domain proteins have additional domains that may directly sense environmental cues (Zhulin *et al.*, 2003). These domains include PAS, which binds flavin or haem and may sense molecular oxygen or redox potential, GAF, which binds cyclic mononucleotides and other small molecular weight effectors and various membrane-associated or periplasmic domains that may be involved in sensing small molecules (Römling *et al.*, 2005). Binding of effectors to the sensory input domain is believed to affect the enzyme activity of the protein. In a number of cases, multiple sensory input domains are found, suggesting complex regulation of individual enzymes in response to a range of environmental cues. As an example, PdeA1 of *G. xylinus* and *X. campestris* pv. *campestris* has a PAS-GAF-GGDEF-EAL domain structure (Tal *et al.*, 1998; Chang *et al.*, 2001; Ryan *et al.*, 2007).

A number of GGDEF, EAL and HD-GYP domain proteins contain a CheY-like receiver (REC) domain. These proteins can be part of two-component signal transduction systems and their activity may be altered by phosphorylation. In these cases the environmental signal is presumably sensed by another element of the system, either a sensory histidine kinase or methyl-accepting chemotaxis protein. Signal transduction involves autophosphorylation of the sensory histidine kinase or of a CheA-like histidine kinase and subsequent phosphotransfer to the REC domain, which alters the activity of the enzymatic domain in cyclic di-GMP synthesis or degradation. Examples of proteins belonging to two component systems include the GGDEF domain protein PleD of *C. crescentus* (Aldridge *et al.*, 2003; Paul *et al.*, 2004), the HD-GYP domain protein RpfG of *Xanthomonas* spp. which along with the sensory histidine kinase RpfC transduces the DSF (for diffusible signal factor) cell–cell signal (Slater *et al.*, 2000; Ryan *et al.*, 2006a), and the GGDEF domain protein WspR which is found in *P. aeruginosa*, *P. fluorescens* and pathovars of *P. syringae*, where it is part of chemotaxis-like signal transduction system (Goymer, 2002; Hickman *et al.*, 2005; Malone *et al.*, 2007). These last two examples will be discussed in more detail below.

Functional organization of cyclic di-GMP signalling

As outlined above, the genomes of bacteria can encode substantial numbers of proteins with a potential role in cyclic di-GMP signalling. For example *X. campestris* pv. *campestris* at ~5 Mb encodes 37 and *P. syringae* pv. *tomato* at ~6 Mb encodes 40. These large numbers indicate that there must be considerable complexity in the organization of cyclic di-GMP signalling within a single organism, which is relatively poorly understood (Jenal, 2004; Römling *et al.* 2005). One pertinent issue is whether signalling systems are part of interactive regulatory networks or are dedicated to specific cellular tasks. A related question is whether cyclic di-GMP exists as a general pool, as a set of discrete localized pools or as both. We address these issues in the following sections.

Localization studies of the PleD regulator (CheY-CheY-GGDEF), which influences swarmer to stalk cell transitions and pole development in *C. crescentus* (Aldridge *et al.*, 2003), have shown that upon phosphorylation the protein locates to the pole of the cell where the new stalk will be formed (Paul *et al.*, 2004). Phosphorylation also activates the protein for cyclic di-GMP synthesis (Paul *et al.*, 2004). The EAL domain protein TipF of *C. crescentus* localizes to the division septum and the newborn pole after division (Huitema *et al.*, 2006). TipF is a flagellum assembly factor that relies on a second protein, TipN for proper positioning. In the absence of TipN, flagella are assembled at ectopic locations, and TipF is mislocalized to such sites. The GGDEF-EAL domain protein FimX of *P. aeruginosa* has also been shown to locate to a single pole of the cell (Huang *et al.*, 2003), an effect that depends upon both the EAL and GGDEF domains (Kazmierczak *et al.*, 2006). Previously it was shown that the DgcA and PdeA proteins of *G. xylinus* co-purified with the cellulose synthase (Ross *et al.*, 1987). These findings have led to the suggestion that the alteration of localized pools of cyclic di-GMP by specific components in cyclic di-GMP signalling may activate processes that are determined by co-localizing proteins. In other words, certain cyclic di-GMP signalling systems are dedicated to specific cellular tasks.

The existence of localized pools of cyclic di-GMP has been proposed to explain the apparently anomalous effects of mutation of some genes encoding GGDEF and EAL domain proteins on biofilm formation in *P. aeruginosa* (Hoffman *et al.*, 2005; Kulesekara *et al.*, 2006). In *P. aeruginosa* PAO1, the EAL domain protein Arr (for aminoglycoside response regulator) is required for biofilm formation in response to sub-inhibitory concentrations of the antibiotic tobramycin (Hoffman *et al.*, 2005). This suggests that cyclic di-GMP degradation is required for biofilm formation, which contradicts the consensus view from work of a number of bacterial systems that high cyclic di-GMP levels promote biofilm formation. Similarly, examination of the effects of mutation of genes encoding other GGDEF and/or EAL domain proteins in *P. aeruginosa* PAO1 and PA14 reveals a complex relationship to biofilm formation (Hoffman *et al.*, 2005; Kulesekara *et al.*, 2006). In general the results are consistent with the concept that enhanced cellular levels of cyclic di-GMP promote biofilm formation. There is however no strict correlation; for example in *P. aeruginosa* PA14, mutation of the gene encoding the GGEEF protein PA3343, which is active in cyclic di-GMP synthesis, leads to a hyperbiofilm formation whereas overexpression of PA2870 and PA3343, which both lead to increases in the cellular level of cyclic di-GMP, has no effect on biofilm formation in the wild type (Kulesekara *et al.*, 2006). These findings support the notion of localized effects of elements involved in cyclic di-GMP signalling, where synthesis or hydrolysis of the nucleotide is intimately related to its site of action. By inference, some functions contributing to biofilm formation in *P. aeruginosa* could be activated by low levels of cyclic di-GMP.

An alternative, but not mutually exclusive, view is that a number of signalling systems form a surveillance network to integrate information about various aspects of the cellular environment and to process this information by determining a cellular level of cyclic di-GMP, which may influence bacterial functions. Observations that different GGDEF, EAL or HD-GYP domain proteins have significant roles in regulation of specific bacterial processes under different environmental conditions are consistent with the notion of a an environmentally responsive network. This has been reported for the role of GGDEF and EAL domain proteins in cellulose synthesis and multicellular behaviour in *Salmonella* (Garcia *et al.*, 2004; Simm *et al.*, 2004; Römling 2005; Simm *et al.*, 2007) and for the regulation of the synthesis of extracellular enzymes by cyclic di-GMP signalling systems in *X. campestris* pv. *campestris* (Ryan *et al.*, 2007).

Recent studies in *Salmonella* have revealed a hierarchical arrangement of cyclic di-GMP signalling in the regulation of cellulose synthesis contributing to the rdar phenotype on plates (Kader *et al.*, 2006). Two GGDEF-EAL domain proteins (STM3388 and STM2123) additively contribute to the expression of the transcriptional regulator CsgD. This protein regulates transcription of *adrA*, which encodes a GGDEF domain protein that is directly implicated in regulation of cellulose synthesis, but not in the expression of *csgD* (Kader et al., 2006). These studies were extended to investigate the role of EAL domain-containing proteins (Simm et al., 2007). It was shown that mutation of four of the genes encoding EAL or GGDEF-EAL domain proteins up-regulated the expression of *csgD*. Different subsets of EAL and GGDEF-EAL domain proteins influenced the rdar phenotype on plates and multicellular behaviour in liquid media. These findings point to the co-existence of both networks of signalling systems and dedicated signalling systems in the same cell and indicate that different systems can operate independently, perhaps by influencing discrete pools of cyclic di-GMP.

Complex interplay between different cyclic di-GMP signalling systems is also evident from studies in *Xanthomonas* spp., where both specific and networking roles of different signalling systems, as well as additional complexities, have been described (Andrade et al., 2006; Ryan et al., 2007). A comprehensive mutational analysis of the role of all cyclic di-GMP signalling systems on the synthesis of virulence factors and motility on agar plates in *X. campestris* pv. *campestris* revealed that a number of signalling proteins have significant regulatory effects on the synthesis of extracellular enzymes. In each case co-ordinate effects of endoglucanase and endomannanase were seen, as has been previously reported for *rpfG* mutation, although unlike RpfG, most of the signalling proteins only had an effect under certain growth conditions (Ryan et al., 2007). In addition RpfG, which is required for full virulence to Chinese Radish, regulates the expression of genes encoding other cyclic di-GMP signalling proteins that also contribute to virulence, although it is not known whether this is a direct or indirect effect (Ryan et al., 2007). Other findings are also consistent with the concept of regulatory networks of cyclic di-GMP signalling systems involving RpfG. Using yeast two-hybrid technology, Andrade and colleagues (2006) have shown that physical interactions occur between the HD-GYP domain of RpfG of *X. axonopodis* pv. *citri* and a subset of those proteins with a GGDEF domain, but no EAL domain. Although the biological relevance of these physical interactions remains to be assessed, the findings suggest the existence of a hierarchical organization of particular cyclic di-GMP signalling systems within the *Xanthomonas* cell.

In contrast, there is evidence that other *Xanthomonas* signalling proteins are dedicated to other cellular functions. Mutation of *XC2161* and *XC2226* causes reduced and increased motility respectively, but has no detectable influence on extracellular enzyme synthesis, which remains at wild type levels. Furthermore, with the exception of RpfG, none of the signalling proteins of *X. campestris* pv. *campestris* that affect enzyme synthesis have an effect on motility. This finding is consistent with the concept that certain cyclic di-GMP signalling systems are dedicated to functions controlling motility. The effects of mutation of *rpfG* on both motility and extracellular enzyme synthesis could be due to a strong influence of the level of cyclic di-GMP, which influences several different pools of the nucleotide in the cell, or to the localization of activated RpfG to a range of cellular sites, perhaps influenced by the protein-protein interactions between HD-GYP and GGDEF domains revealed by Andrade et al. (2006). It will now be of interest to determine the cellular functions that are regulated by the interacting GGDEF domain proteins and to investigate if different HD-GYP domain proteins in *Xanthomonas* interact with the same or different subsets of GGDEF domain proteins. Although the role of the HD diad in enzymatic activity of the HD-GYP domain has been revealed, nothing is known of the other structural features that may be required for interactions with GGDEF domains.

Interplay between cyclic di-GMP signalling and other regulatory networks

It is highly likely that interplay between cyclic di-GMP signalling and other regulatory circuits within bacteria will occur. Quorum sensing regulates many of the same bacterial behaviours (multicellularity, biofilm formation and virulence) as cyclic di-GMP signalling. This has led to suggestions that cyclic di-GMP signalling and quorum sensing converge in the regulation of particular functions (Camilli and Bassler,

2006). As we will outline below, in *X. campestris* pv. *campestris* transduction of the cell–cell signal DSF involves the HD-GYP domain regulator RpfG. This is indicative of a direct link between the two types of regulatory system. However activation of a GGDEF/EAL domain protein by binding of a quorum-sensing molecule has yet to be reported and it remains likely that in many cases the regulatory interplay between cyclic di-GMP signalling and cell–cell signalling is more indirect.

Recent findings have also indicated potential interplay between cyclic di-GMP signalling and regulation by small RNAs in the Csr (carbon storage regulation) system of *E. coli*. The Csr system serves to negatively regulate glycogen synthesis, gluconeogenesis and biofilm formation and to positively regulate motility and expression of the *flhDC* flagellar master operon (Romeo, 1998; Wei *et al.*, 2001; Jackson *et al.*, 2002). The elements of the Csr system are CsrA, a 61 amino acid RNA-binding protein that acts in post-transcriptional regulation by binding at or near the ribosome binding site of particular nascent transcripts thereby promoting degradation, two non-coding small RNAs, CsrB and CsrC, which antagonize CsrA by sequestering it in a ribonucleoprotein complex (Dubey *et al.*, 2005), and CsrD, a specificity factor that targets CsrB and CsrC for degradation by RNase E (Suzuki *et al.*, 2006). CsrD, which is predicted to be membrane bound, has a GGDEF and an EAL domain both of which show divergence from the canonical sequence, not only in the signature sequences (which are HRSDF and ELM respectively), but also elsewhere in the protein. This sequence divergence may indicate that CsrD does not participate in cyclic di-GMP turnover. Furthermore expression of the GGDEF domain protein AdrA or the EAL domain protein YhjH, which both influence biofilm formation in *E coli*, do not affect Csr signalling, suggesting that CsrD does not respond to dramatic changes in cyclic di-GMP levels. This offers indirect support for the contention that CsrD does not bind cyclic di-GMP, although this has not been specifically examined. Clearly the exciting discovery of CsrD poses a series of questions about the biochemistry of the protein and molecular basis for its action. Is cyclic di-GMP involved at all? Does CsrD bind sRNA species and/or RNaseE? CsrD is predicted to have two transmembrane helices and a periplasmic domain. Does this part of the molecule recognize a specific environmental cue to modulate the activity?

The study of CsrD in *E. coli* has broader implications since proteins with similar amino acid sequence divergence in both GGDEF and EAL domains are found in a number of other bacteria, including the plant pathogen *Erwinia carotovora* subsp. *atroseptica* (Suzuki *et al.*, 2006). The role of these homologues has not yet been examined however. It is noteworthy that *Erwinia* species have a post-transcriptional regulatory system called Rsm (for repression of secondary metabolites), which is analogous to the Csr system of *E. coli*. RsmA, which is highly related to CsrA, is an RNA-binding protein that acts to suppress production of a range of extracellular plant cell wall degrading enzymes, synthesis of *N*-acyl homoserine lactone signalling molecules and virulence (Cui *et al.*, 1995). The activity of CsrA is attenuated by the regulatory RNA RsmB (Liu *et al.*, 1998), whereas the small protein RsmC negatively controls extracellular enzyme production and virulence by modulating the levels of RsmB and RsmA (Cui *et al.*, 1999). On the basis of the role of CsrD in *E. coli*, it is tempting to speculate that the CsrD homologue in *Erwinia carotovora* subsp. *atroseptica* is another player in this regulatory circuit, and hence has an influence on virulence to plants.

Cyclic di-GMP signalling and bacterial virulence

It is now established that cyclic di-GMP signalling contributes to the pathogenesis of both animal and plant pathogens. In *V. cholerae* and *P. aeruginosa*, specific GGDEF and/or EAL domain proteins are implicated in virulence in different mouse models (Tamayo *et al.*, 2005; Tischler and Camilli, 2005; Kulesekara *et al.*, 2006). In *V. cholerae*, the REC-EAL-HTH domain protein VieA positively activates expression of virulence genes *toxT*, which encodes a transcriptional regulator and *ctxAB*, which encode cholera toxin (Tischler *et al.*, 2002; Tamayo *et al.*, 2005) and negatively influences exopolysaccharide production and biofilm formation (Tischler and Camilli, 2004). A *vieA* mutant is attenuated for colonization in the infant mouse model (Tischler and Camilli, 2005). All of these effects require the cyclic di-GMP phosphodiesterase activity of VieA and are abolished by mutation of the EAL amino acid motif (Tischler and Camilli, 2005). In *Salmonella* the 'stand-alone' EAL domain protein CdgR is required for the bacterium to resist host phagocyte oxidase *in vivo* and contributes to virulence in mice (Hisert et al., 2005).

These findings extend earlier work that implicated specific EAL or GGDEF domain proteins in bacterial disease but did not address their biochemical function. In *B. pertussis*, activation of virulence factors by the BvgAS two-component system is accompanied by repression of transcription of a further set of genes, which involves the 'stand-alone' EAL domain protein BvgR (Merkel and Stibitz, 1995; Merkel et al., 1998a,b). BvgR-mediated regulation of gene expression contributes to respiratory infection of mice (Merkel et al., 1998b). In *V. anguillarum* the GGDEF domain protein VirC of contributes to the virulence to fish (Milton et al., 1995).

In the following sections we describe in some detail examples where the role of cyclic di-GMP signalling systems in regulation of virulence in plant pathogens has either been demonstrated or can be inferred from work on closely related organisms.

Cyclic di-GMP signalling and cellulose synthesis in plant pathogens and plant associated microbes

The synthesis of cellulose contributes to a number of bacterial functions such as attachment or adhesion to plant surfaces and bacterial cell–cell attachment that have a potential and in some cases demonstrable role in bacterial pathogenesis. Genome sequencing studies have revealed that many bacterial genomes carry genes encoding proteins that can catalyse the synthesis and modification of cellulose and that these are usually organized in an operon. It is highly appropriate to begin our survey of cyclic di-GMP signalling in plant pathogens with an examination of cellulose synthesis, since studies of the regulation of cellulose synthesis in *G. xylinus* lead both to the discovery of cyclic di-GMP and to the seminal findings of the role of GGDEF/EAL domain proteins in metabolism of this nucleotide (see Introduction).

The *wss* operon directs cellulose synthesis

The study of cellulose synthesis in plant-associated and plant pathogenic bacteria has been restricted to relatively few species including the pathogen *Agrobacterium tumefaciens*, symbiotic *Rhizobium* spp. and the plant growth-promoting organism *P. fluorescens*. Cellulose has been identified as a matrix component in biofilms produced by a number of plant pathogens including *Pseudomonas* spp. (Ude et al., 2007) and *Erwinia* (Yap et al., 2005), although its role in *Erwinia* is unclear. In *P. fluorescens* SBW25, synthesis of cellulose requires genes in the *wss* operon, which is composed of 10 genes (Spiers et al., 2003). WssBCE are the cellulose synthase subunits, WssD is an endoglucanase, WssFGHI are predicted to be involved in cellulose acetylation and WssA and J are predicted to play a role in the correct localization if the synthase/acetylation complex. A similar gene organization is found in *P. syringae* pv. *tomato* DC3000, although *wssI* and *J* are missing. Homologues of WssB (variously named BcsA and CelA), WssC (BcsB, CelB), WssD (CelC) and WssE (BcsC) are found in other bacteria including species of *Rhizobium* and *Agrobacterium* and some *Xanthomonas* spp. but excluding *X. campestris* pv. *campestris* and *X. oryzae* pv. *oryzae*; homologues of *wssAFGHIJ* appear to be absent in these latter bacteria. As pointed out above, cyclic di-GMP was originally described to bind the BcsB subunit of cellulose synthase of *G. xylinus* although it has now been shown that the nucleotide binds to the PilZ domain of BcsA (Ryjenkov et al., 2006). This presumably leads to the allosteric activation of the cellulose synthase complex, although the precise means by which this is achieved are unknown.

The AdrA protein, cyclic di-GMP synthesis and regulation of cellulose synthase

Bacterial genomes encode a large number of proteins with a GGDEF domain capable of the generation of cyclic di-GMP. A pertinent issue is whether specific GGDEF domain proteins have a role in controlling cellulose synthesis through modulation of cyclic di-GMP levels. In some organisms, particular systems appear to have a major role under certain conditions. In *G. xylinus*, mutation of the three *dgc* genes identified by Tal *et al*. (1998) lead to different degrees of reduction of cellulose synthesis, suggesting a major role for *dgc*1. Strains with combinatorial mutations showing progressively lower levels of cellulose than strains with single mutations. In *Salmonella*, cellulose synthesis is regulated by a hierarchical arrangement of cyclic di-GMP signalling proteins. Two GGDEF-EAL domain proteins (STM3388 and STM2123) additively contribute to the expression of the transcriptional regulator CsgD. This protein regulates transcription of *adrA*, which encodes a GGDEF domain protein that is directly implicated in regulation of cellulose synthesis, but not in the expression of *csgD* (Kader *et al*., 2006). These different systems can operate independently, perhaps by influencing discrete pools of cyclic di-GMP. The activities of the AdrA, STM3388 and STM2123 proteins may also be influenced by environmental cues. Intriguingly homologues of AdrA, which also has a MASE2 membrane-associated sensor domain in addition to the GGDEF domain, are found in *P. fluorescens* and *P. syringae* pv. *tomato*, although it is not known if they have the same function as in *Salmonella*.

The *wsp* operon and cyclic di-GMP synthesis

The wrinkly spreader phenotype of *P. fluorescens* is associated with enhanced synthesis of cellulose that is modified by acetylation (Spiers *et al*., 2003). This phenotype depends upon activation of the REC-GGDEF domain protein WspR, which is part of a chemotaxis-like sensory transduction system encoded by the *wsp* operon. The available evidence suggests that WspR is activated for cyclic di-GMP synthesis by phosphorylation of the CheY-like REC domain (Fig. 13.3). Two classes of mutation give rise to the wrinkly spreader phenotype; those in genes encoding other elements of the sensory transduction pathway that result in enhanced or constitutive activation of WspR and those introducing amino acid alterations in WspR giving a 'locked-on' conformation (Goymer, 2002; Malone *et al*., 2007). A homologous system is found in *P. aeruginosa*, where mutations in *wspF* that likely result in constitutive phosphorylation of WspR give rise to aggregation and altered colony morphology (Hickman *et al*., 2005), and in pathovars of *P. syringae*, where effects of mutations on bacterial behaviour have not yet been studied.

In *P. fluorescens*, mutations leading to the wrinkly spreader phenotype occur spontaneously and the resulting strains can show a high degree of competitive advantage over the wild type in the colonization of particular niches such as the air–liquid interface in unshaken liquid cultures and the rhizosphere. It is important to recognize that modified chemotaxis systems linked to cyclic di-GMP signalling occur in many bacteria, but are not always linked to cyclic di-GMP synthesis. For example in *X. campestris* pv. *campestris*, the operon comprising *XC1409–XC1414* encodes a modified chemotaxis system in which the likely 'output' is activation of the REC-EAL domain protein XC1411.

Cyclic di-GMP signalling and the regulation of virulence in *Xanthomonas* and related species

The study of cyclic di-GMP signalling in *Xanthomonas* spp. has been largely focused on the role of the HD-GYP domain protein RpfG, which regulates the synthesis of virulence factors, biofilm dispersal and virulence in *X. campestris* pv. *campestris*. More recently however a comprehensive functional genomic analysis of the role of all signalling proteins involved in cyclic di-GMP turnover in the virulence of *X. campestris* pv. *campestris* has been reported (Ryan *et al*., 2007). Both of these topics will be discussed in the following sections.

The *rpf* genes encode components of a cell–cell signalling system that regulates virulence, virulence factor synthesis and biofilm dynamics

The ability of *X. campestris* pv. *campestris* to cause disease in plants depends in part upon the synthesis of extracellular plant cell wall degrading enzymes and synthesis of the extracellular polysaccharide xanthan. A two-component system comprising the HD-GYP domain regulator

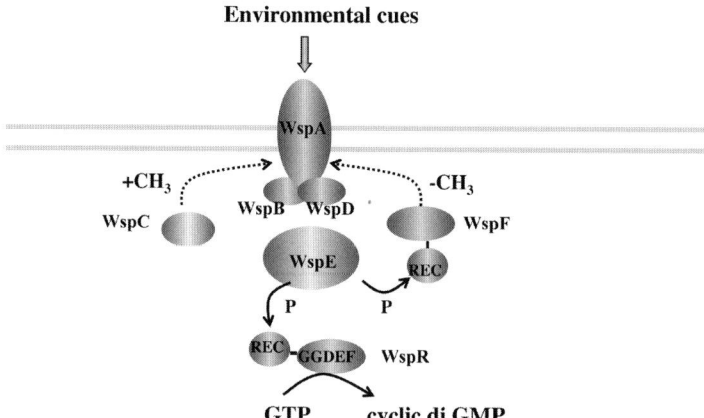

Figure 13.3 The *wsp* operon encodes components of a modified chemotaxis system involved in cyclic di-GMP generation. The figure shows proposed functions of the components of the Wsp system found in *Pseudomonas fluorescens* and *Pseudomonas aeruginosa*; homologous systems are found in pathovars of *Pseudomonas syringae*. Recognition of an environmental signal or signals by the methyl-accepting chemotaxis protein WspA located in the cytoplasmic membrane leads to autophosphorylation of the histidine kinase WspE. This effect requires interactions with the CheW-like protein WspB and the CheW fusion protein WspD. WspE transfers phosphate to the CheY-like receiver domain (REC) of the REC-GGDEF domain protein WspR, activating it for cyclic di-GMP synthesis. The activity of WspA is modulated by methylation by WspC and demethylation by WspF. Methylation increases the activity of WspA and the flux of phosphate to WspR. The demethylation activity of WspF is regulated by phosphorylation by WspE, thus providing a negative feedback loop. Mutations in *wspF* thus lead to a high level of activation of WspR and high cellular levels of cyclic di-GMP.

RpfG and the complex sensor histidine kinase RpfC acts to positively regulate virulence, the synthesis of extracellular enzymes and biofilm dispersal in *X. campestris* pv. *campestris* (Slater et al., 2000; Dow et al., 2003; Ryan et al., 2006a). The *rpfG* and *rpfC* genes are transcribed as the *rpfGHC* operon although no function has yet been ascribed to RpfH, which is related in amino acid sequence to the input domain of RpfC. Extracellular enzyme synthesis and biofilm dynamics in *X. campestris* pv. *campestris* are also regulated by cell–cell signalling, mediated by the diffusible signal molecule DSF (Barber et al., 1997; Slater et al., 2000). DSF has been characterized as the unsaturated fatty acid *cis*-11-methyl-2-dodecenoic acid (Wang et al., 2004). The synthesis of the DSF signal is partially dependent on *rpfB*, which encodes a long chain fatty acyl CoA ligase and fully dependent on *rpfF*, which encodes a protein with some similarity to enoyl CoA hydratase (Barber et al., 1997). The *rpfB* and *rpfF* genes are contiguous and adjacent to the *rpfGHC* operon, which is convergently transcribed (Barber et al., 1997; Slater et al., 2000). In addition to a positive regulatory action on virulence factor synthesis, RpfC acts to negatively regulate the synthesis of DSF.

Addition of DSF can restore extracellular enzyme synthesis to *rpfF* mutants but not to strains with mutations in *rpfG* or *rpfC*. Mutation of *rpfC*, *rpfG* or *rpfF* leads to the formation of biofilms by *X. campestris* pv. *campestris* in rich media, where the bacteria grow in matrix-enclosed aggregates. In contrast under these conditions, the wild type grows in a dispersed planktonic fashion (Dow et al., 2003). Addition of DSF causes dispersal of biofilms produced by the *rpfF* mutant but not by other *rpf* mutants. These findings are consistent with a role for RpfC/RpfG in perception and transduction of the DSF signal (reviewed in Crossman and Dow, 2004). Further support for this model has come from experiments in which the RpfC/RpfG two-component system has been re-constructed in *P. aeruginosa* and shown to confer responsiveness to exogenously added DSF as seen through effects on swarming motility (Ryan et al., 2006a).

Transduction of the DSF signal utilizes cyclic di-GMP as a second messenger

Three conserved amino acid residues of RpfC implicated in phosphorelay(H198 in the histidine kinase domain, D512 in the receiver domain and H657 in the histidine phosphotransfer domain) are essential for activation of the production of extracellular enzymes and extracellular polysaccharide virulence factors, but not for repression of DSF biosynthesis (He *et al.*, 2006a). Domain deletion and subsequent *in trans* expression analysis revealed that the receiver domain of RpfC alone was sufficient to repress DSF overproduction in an *rpfC* deletion mutant. Co-immunoprecipitation and far western blot analyses suggested an interaction between the receiver domain and RpfF, the enzyme involved in DSF biosynthesis. These data support a model in which RpfC modulates two different functions (virulence factor synthesis and DSF synthesis) by utilization of a conserved phosphorelay system and a novel domain-specific protein-protein interaction mechanism, respectively (He *et al.*, 2006a). Recognition of DSF by RpfC is proposed to lead to phosphorylation of RpfG and its activation in cyclic di-GMP hydrolysis (Ryan *et al.*, 2006a; Fig. 13.4) although this has not been directly demonstrated. The mechanism(s) by which alteration of cyclic di-GMP level influences the synthesis of virulence factors and biofilm dynamics are currently obscure.

The contribution of cell–cell signalling to biofilm formation in *Xanthomonas* depends upon growth conditions

As outlined above, experiments in rich nutrient medium suggested an effect of DSF on biofilm dispersal requiring the RpfC/RpfG two-component system, but no influence of DSF on biofilm formation. More recent experiments using GFP-labelled bacteria grown in static cultures in minimal medium present a substantially different picture (Torres *et al.*, 2007). Under these conditions, the wild type strain forms a structured biofilm, whereas *rpfF* and *rpfC* mutant strains do not. Furthermore, experiments conducted with mixed cultures indicate that the *rpfC* mutant can prevent the formation of structured biofilms by the wild type. Taken together, these findings indicate that that DSF signalling has a role in the formation of structured biofilms and that an excess of DSF prevents such biofilm formation. Although a close correlation is observed between the effects of DSF levels on structured biofilm formation in minimal medium and on virulence in the model plant *Nicotiana benthamiana* (Torres *et al.*, 2007), a direct cause and effect relationship cannot be concluded.

Work on other bacteria has established that the environment has an impact on the contribution of cell–cell signalling or quorum sensing to the development of bacterial biofilms and that quorum sensing may be integral to biofilm formation only under certain conditions (Kjelleberg and Molin, 2002; Kirisits and Parsek, 2006). The same considerations appear to apply to the role of DSF signalling in biofilm formation in *X. campestris*. One possible reason for this is that the synthesis of xanthan, which is required for biofilm formation, is considerably enhanced in rich medium in the presence of glucose, so that mutation of *rpf* genes may reduce xanthan production below a critical level for biofilm formation only in minimal medium. A second possibility is that some cyclic di-GMP signalling systems may be activated only in response to certain environmental cues, so that the quantitative contribution of DSF signalling to the overall modulation of cyclic di-GMP levels (and hence to biofilm formation) may vary with growth conditions.

Cyclic di-GMP regulates a wide range of functions in *Xanthomonas campestris*

The available evidence suggests that cyclic di-GMP regulates a wide range of functions in *X. campestris* pv. *campestris* as has been shown in other bacteria (Hickman *et al.*, 2005). Transcriptome profiling has revealed that perception of DSF has a widespread influence on *X. campestris* pv. *campestris* gene expression to include effects on genes implicated in multidrug resistance, motility, chemotaxis, tricarboxylic acid cycle and iron uptake as well as the previously established effects on genes encoding extracellular enzymes and the synthesis of EPS (He *et al.*, 2006b). A similar transcriptome analysis of the role of

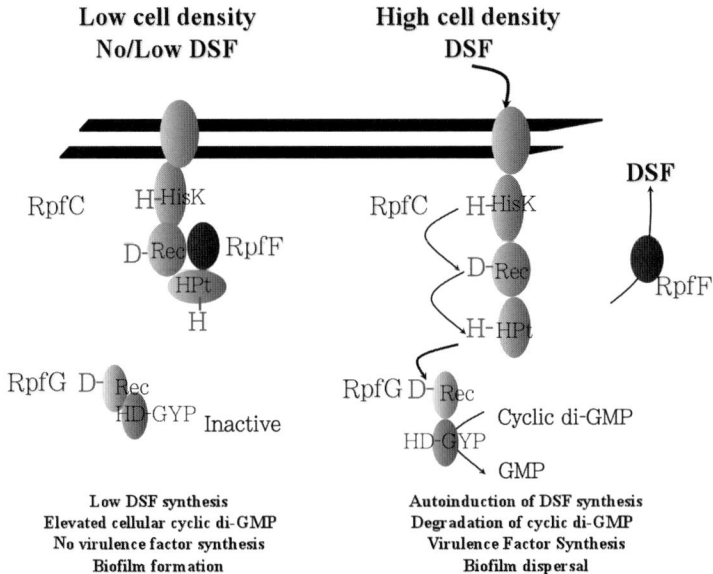

Figure 13.4 Transduction of the DSF cell–cell signal in *Xanthomonas campestris* is linked to the degradation of cyclic di-GMP by the HD-GYP domain regulator RpfG. The synthesis of the DSF signal requires RpfF and is partially dependent on RpfB. DSF perception and signal transduction involves the complex sensor RpfC and HD-GYP domain regulator RpfG. RpfC has a sensory input domain with five predicted transmembrane helices that span the cytoplasmic membrane. By analogy with a number of related sensor proteins, signal transduction may involve autophosphorylation of RpfC in response to ligand (DSF) binding, followed by phosphorelay and phosphotransfer to the cognate regulator, in this case RpfG. Phosphorylation of RpfG leads to its activation as a cyclic di-GMP phosphodiesterase, an activity associated with the HD-GYP domain. The consequent alterations in the level of cyclic di-GMP affect the synthesis of virulence factors and biofilm dispersal by as yet unknown mechanisms. In addition, RpfC acts to negatively regulate DSF synthesis by a mechanism that does not require phosphorelay, but may involve sequestration of RpfF through protein-protein interactions with the REC domain of RpfC. See text for details. Key to domains: REC, two-component receiver; HPt, histidine phosphotransfer; HisK, histidine kinase.

RpfG on gene expression has yet to be reported, although comparison of the effects of *rpfF* and *rpfG* mutation would provide a further test for the model that DSF is recognized via the RpfG/RpfC two-component system. The possibilities that RpfC may sense additional environmental cues to activate RpfG phosphorylation and that other perception system(s) for DSF occur in *X.campestris* pv. *campestris* cannot currently be excluded however.

The rpf/DSF regulatory system occurs in other Xanthomonas spp and the related pathogen Xylella fastidiosa

Genome sequencing indicates the presence of an *rpf* gene cluster with the *rpfB-rpfF-rpfC-rpfH-rpfG* gene organization of *X. campestris* pv. *campestris* strain 8004 in *X. campestris* pv. *campestris* strain 33913 and *X. campestris* pv. *vesicatoria*. In *X. oryzae* pv. *oryzae* and *X. axonopodis* pv. *citri*, the *rpfH* gene is missing. In *Xylella fastidiosa*, *rpfH* is missing and *rpfB* is located elsewhere in the chromosome. DSF activity has been detected in a number of *Xanthomonas* spp. (Barber *et al.*, 1997; Jacques *et al.*, 2005) and in *Xylella fastidiosa* (Scarpari *et al.*, 2003; Newman *et al.*, 2004). DSF from *Xylella fastidiosa* has been tentatively identified as 12-methyl-tetradecanoic acid (Colnaghi Simionato *et al.*, 2007), which lacks the cis-unsaturated bond believed to be important for the action of DSF from *X. campestris* pv. *campestris* (Wang *et al.*, 2004). A role for DSF signalling in the virulence of a number of these bacteria has now been described.

Mutation of *rpfC* or *rpfF* in different strains of *X. oryzae* pv. *oryzae* leads to a loss of virulence on rice (Tang *et al.*, 1996; Chatterjee and Sonti, 2002). The *rpfC* mutant of *X. oryzae* pv. *oryzae* strain 13751 showed a defect in EPS production, but no effects on extracellular enzyme synthesis.

In contrast the *rpfF* mutant of *Xoo* strain BXO4 showed no alteration in EPS synthesis but a defect in iron uptake-related functions. These latter findings are consonant with the recent results of transcriptome profiling from *Xcc* that show that DSF regulates a number of iron-uptake related functions (He *et al.*, 2006b). Even in the absence of strain-dependent effects, similarity in the effects of mutation of *rpfC* and *rpfF* might not be expected since RpfC may recognize additional environmental factors than DSF and it cannot be excluded that additional mechanisms of DSF signal transduction occur.

Xylella fastidiosa is a xylem-restricted bacterium that requires an insect vector to be transmitted between plants. In *Xylella fastidiosa*, mutation of *rpfF* leads to a loss of the DSF signal, which influences the interaction of *Xylella fastidiosa* with both its insect vector and host plant (Newman *et al.*, 2004). The absence of the DSF signal leads to the loss of biofilm formation in the insect vector but an increased virulence when the bacteria are mechanically inoculated into plants. These effects in plants are thus apparently opposite to those seen after mutation of *rpfF* in *X. campestris* pv. *campestris*, which leads to a reduction in virulence. Newman and colleagues (2004) have proposed that progression of disease caused by *Xylella fastidiosa* requires transitions between an aggregated state and a state able to synthesize extracellular enzymes that degrade the primary cell wall in bordered pit membranes, allowing the bacteria to colonize new xylem vessels. Newman and colleagues (2004) envisage that these transitions occur in response to the DSF signal. The hypovirulent phenotype of the *rpfF* mutant may result from the inability of mutant bacteria to release from an aggregated state and/or to degrade the pit membranes, with a consequent blockage of the vascular system and increased symptom production. Consistent with this view, transcriptome profiling showed that the wild type has elevated expression of genes involved in EPS synthesis compared to the *rpfF* mutant but reduced expression of genes encoding extracellular cell wall degrading enzymes and factors contributing to motility. As outlined above, the environment also has a considerable influence on the role of cell–cell signalling in biofilm formation in *Xylella fastidiosa*, since in contrast to the effects *in planta*, mutation of *rpfF* leads to a loss of biofilm formation in insects (Newman *et al.*, 2004).

The above examples illustrate the importance of the *rpf*/DSF regulatory system to the virulence of a number of plant pathogenic bacteria. By extension of the work in *X. campestris* pv. *campestris* DSF perception may be linked to the turnover of cyclic di-GMP in these related bacteria.

Other cyclic di-GMP signalling systems also contribute to *Xanthomonas campestris* virulence

The demonstration that the HD-GYP domain of RpfG is a cyclic di-GMP phosphodiesterase (Ryan *et al.*, 2006a) implicated cyclic di-GMP in the control of virulence, virulence factor synthesis and biofilm formation in *X. campestris* pv. *campestris*. This finding raised the question of whether other GGDEF, EAL and HD-GYP domain proteins of *X. campestris* pv. *campestris* also had a role in virulence factor synthesis and virulence to plants. The genome of *X. campestris* pv. *campestris* encodes 37 proteins with GGDEF, EAL or HD-GYP domains. Using a panel of defined mutants, Ryan and colleagues (2007) investigated the role of each of these elements in virulence to plants, in the production of virulence factors under a range of growth conditions *in vitro*, in motility and in biofilm formation.

A set of thirteen genes with an effect on virulence on Chinese radish encode signalling proteins predicted to be involved in cyclic di-GMP synthesis and degradation. Only five of these proteins significantly influenced the synthesis of extracellular enzymes virulence factors *in vitro*, although all thirteen genes are expressed in both rich and minimal medium. Within the subset of virulence genes, only *rpfG* and XC2324, which is highly related to *pdeA1* in *G. xylinus*, regulated biofilm formation *in vitro*, whereas only *rpfG* regulated motility. It remains possible that some signalling systems with a role in virulence respond to environmental cues that are only found in plants to regulate known virulence-related functions such as extracellular enzyme synthesis and biofilm formation. A second possibility is that these signalling systems regulate as yet unidentified functions that contribute to virulence.

What are the environmental cues for virulence-related signalling *in planta*? As we have seen, the HD-GYP domain regulator RpfG and cognate sensor RpfC are involved in transduction of the DSF cell–cell signal, but also perhaps other environmental signals. Other systems implicated in virulence of *X. campestris* pv. *campestris* control the synthesis of extracellular enzymes under conditions of low oxygen tension (XC2324) or in minimal medium with glutamine as a sole carbon source (XC1582, XC1841). Glutamine is the most abundant amino acid found in guttation fluid of *Arabidopsis thaliana* (Pilot *et al*., 2004). Both low oxygen tension and the presence of glutamine are therefore environmental conditions likely to be encountered by a vascular pathogen such as *X. campestris* pv. *campestris* during disease. The binding of molecular oxygen to PdeA1 of *G. xylinus*, which is highly related to XC2324, modulates its enzymatic activity in cyclic di-GMP degradation (Chang *et al*., 2001). It remains to be determined whether XC2324 is similarly directly regulated by molecular oxygen. The molecular basis of the effects of amino acid cues on the synthesis of extracellular enzymes and the signals or cues recognized by the other virulence-related systems are obscure.

Bioinformatic analyses indicate that proteins with the same domain organization as those implicated in *X. campestris* pv. *campestris* virulence are found in other plant pathogens and plant-associated microbes as well as animal pathogens. Homologues of the GGDEF-EAL domain protein XC1582 (including the N-terminal region which is as yet uncharacterized) are found in a range of agronomically important plant pathogens belonging to the genera *Pseudomonas*. Homologues of the GGDEF-EAL domain protein XC1476 with the same domain organization are found in plant beneficial *Pseudomonas* spp., plant symbiotic *Rhizobiaceae* and in the polyphyletic pathogen *P. aeruginosa*. Homologues of PdeA are found in other xanthomonads and in *Sinorhizobium meliloti*. Analysis of the GGDEF/EAL/HD-GYP domain proteins in the proteomes of sequenced xanthomonads indicates a conservation of many of those signalling proteins with a role in virulence in *X. campestris* pv. *campestris*. Comparison of the predicted proteomes of *X. oryzae* pv. *oryzae* and *X. campestris* pv. *campestris* indicates that eleven proteins present in *X. campestris* pv. *campestris* are absent in *X. oryzae* pv. *oryzae*, whereas *X. oryzae* pv. *oryzae* has an additional gene encoding a GGDEF protein that is not found in *X. campestris* pv. *campestris*. Interestingly, of the thirteen genes encoding HD-GYP, GGDEF and/or EAL domain proteins significantly contributing to virulence of *X. campestris* pv. *campestris* on Chinese radish, ten are retained in the *X. oryzae* pv. *oryzae* genome (Ryan *et al*., 2007).

Cyclic di-GMP signalling and biofilm formation

Molecular analysis of communal behaviour and biofilm formation has uncovered roles for proteins containing GGDEF and/or EAL domains in adhesion, aggregative behaviour and biofilm formation in a number of bacteria. Although it is now appreciated that biofilm formation has an important role in bacterial phytopathogenesis, there are only a few reports on the role of specific cyclic di-GMP signalling systems in the regulation of adhesion or biofilm formation in plant pathogens. Studies on a number of bacteria have shown that cyclic di-GMP can influence synthesis of polysaccharides such as alginate, cellulose, xanthan and uncharacterized polymers from *P. aeruginosa* that may be components of the extracellular matrix in biofilms of different bacteria. This regulation may be exerted at the level of transcription and/or post-translationally through allosteric effects. In addition, cyclic di-GMP can regulate other functions that contribute to biofilm dynamics such as motility, that has an established role in biofilm development, and the synthesis of enzymes implicated in biofilm dispersal.

The formation of biofilms is influenced by many aspects of the bacterial environment, including nutrient status, quorum sensing or cell–cell signals, signals from eukaryotic hosts or partners, oxygen tension and stresses such as the presence of antibiotics or host defence factors (Römling *et al*., 2005; Ryan *et al*., 2006b). In a few cases, the role of a specific cyclic di-GMP signalling system in linking perception of particular environmental cues to biofilm formation or aggregative behaviour has been described. Examples include PdeA1of *G. xylinus* which is involved in direct sensing of molecular oxygen (Chang et al., 2001), the RpfG/RpfC two component system of *X. campestris* pv. *campestris* which is involved in perception and transduction of the diffusible signal DSF (Ryan et al., 2006a), the GGDEF-EAL domain protein MbaA which mediates norspermidine effects on biofilm formation in *V. cholerae* (Karatan et al., 2005) and the EAL domain protein Arr, which has a role in biofilm formation in *P. aeruginosa* that is triggered by subinhibitory concentrations of the antibiotic tobramycin (Hoffman et al., 2005).

A consensus view is that elevated cellular levels of cyclic di-GMP promote biofilm formation and sessility, whereas reduced levels promote motility. This is consistent with many, but not all, of the reported effects of elevated cyclic di-GMP levels on extracellular polysaccharide synthesis. However several laboratories have reported effects of mutation of genes encoding particular cyclic di-GMP signalling proteins that lead to apparently anomalous effects on biofilm formation. One such example is Arr of *P. aeruginosa*, where mutation of the *arr* gene reduces biofilm formation in response to tobramycin. In the next section we discuss another example of the perception of an environmentally relevant signal that is linked to cyclic di-GMP metabolism, that of phosphate regulation of biofilm formation in *P. fluorescens*. Some mechanistic details of these effects have been uncovered and, importantly, these findings point to a novel role for cyclic di-GMP in the modulation of protein secretion linked to biofilm formation (Monds et al., 2007).

Cyclic di-GMP signalling and modulation of protein secretion

The level of inorganic phosphate in the environment regulates the formation of biofilms by *P. fluorescens* (Monds et al., 2007). The sensing of low phosphate conditions, in which attachment and biofilm formation are compromised, involves the PhoB-PhoR two-component system. Activation of this system leads to autophosphorylation of PhoR and phosphotransfer to PhoB, which then binds to pho boxes in the upstream regions of target genes to modulate their expression. Phosphorylated PhoB inhibits the expression of *lapA*, which encodes a large adhesion protein required for irreversible surface attachment and biofilm formation and the *lapEBC* operon, which encodes an ABC transporter involved in LapA secretion. The effect of PhoB on expression of these *lap* genes in *P.fluorescens* Pf-01is however insufficient to account for the loss of biofilm formation. The appreciation that other genes may be involved led to the discovery of *rapA*, which encodes an EAL/GGDEF domain protein with unusual domain organization. Expression of the *rapA* gene is positively regulated by phospho-PhoB. Genetic analysis indicates that RapA has an inhibitory effect on biofilm formation, but does not effect expression of the *lap* genes. RapA has an unusual domain organization, with a N-terminal domain EAL domain of the canonical signature sequence and a C-terminal GGDEF domain with the divergent sequence GGDDF. The effect of RapA on biofilm formation appears to be exerted though an influence of the levels of cyclic di-GMP on secretion of LapA. Since *in vivo* studies indicate that RapA acts as a cyclic di-GMP phosphodiesterase, it is inferred that cyclic di-GMP acts to positively modulate the secretion of LapA. The mechanism by which this occurs is not known.

Intriguingly a second GGDEF-EAL domain protein, LapD, also influences biofilm formation (Hinsa and O'Toole, 2006), although this effect has been investigated in another *P. fluorescens* strain, WCS365. The *lapD* gene is adjacent to the *lapA* gene but convergently transcribed. LapD is an inner membrane protein that may influence the secretion of LapA although effects of LapD on the synthesis or stability of LapA cannot be discounted. LapD has GGDEF and EAL domains with the divergent signature sequences RGGEF and KVL respectively. Whether LapD and RapA co-operate in the regulation of biofilm formation through an effect on LapA secretion remains to be determined.

These findings have a number of implications for the possible role of cyclic di-GMP signalling systems in phytopathogenesis. Homologues of RapA can be identified in a number of *Pseudomonas* species including *P. syringae* pv. *tomato* DC3000 and different strains of *P. aeruginosa*. In all of the above cases the upstream region of the gene exhibits the same organization of overlapping pho boxes as was seen for *P.fluorescens* Pf0-1. These data indicate the potential for conserved regulation of *rapA* homologues by inorganic phosphate. Whether these homologues of *rapA* are phosphate-regulated and have orthologous roles in biofilm formation in different *Pseudomonas* species remains to be examined experimentally.

Concluding remarks

It is now evident that cyclic di-GMP regulates many aspects of bacterial physiology that impinge on the ability of bacteria to cause disease. These regulated functions include the synthesis of particular virulence determinants, adhesion to surfaces and the formation of biofilms. There are as yet only a limited number of reports directly linking cyclic di-GMP signalling to bacterial pathogenesis to plants. Nevertheless, work on animal pathogens and other bacteria such as the beneficial *P. fluorescens* suggests a role of a number of particular cyclic di-GMP signalling systems in controlling factors contributing to virulence in related plant pathogens. These predictions are open to experimental verification. It is also to be hoped that in a reciprocal fashion, work on plant pathogenesis, which can utilize rapid and relatively inexpensive screening of a large numbers of mutants, can inform work on animal and human pathogens.

Despite the significant advances in our understanding of cyclic di-GMP signalling, there are still many questions that remained unanswered. Some details of the mechanisms by which cyclic di-GMP exerts its action on different cellular functions (via PilZ domains) have been defined, but presumably others remain to be discovered. Researchers are beginning to investigate the cell biology and spatial aspects of the various signalling systems, which will have consequences for our understanding of the functional organization and relationship of different signalling systems in the same cell. Detailed studies of specific signalling proteins may also allow identification of the cues or signals to which they respond by activation or inactivation.

Can the information gained from these studies underpin new strategies to control bacterial disease of plants? The ability to control cyclic di-GMP-regulated functions such as the synthesis of virulence determinants and formation and detachment from biofilms may be of considerable importance for disease control. Examples of the latter include interference with the ability of pathogens to attach and form biofilms during epiphytic growth phases on leaf surfaces, or to associate with insect vectors. Chemical agents that inhibit key virulence factors or processes are often used for therapeutic action against human pathogens. In principle, interference with the action of cyclic di-GMP signalling proteins could be used to control bacterial diseases of both animals and plants. For a variety of reasons however, bacterial diseases of plants have historically not been satisfactorily controlled by chemicals. Exceptions include those pathogens, such as *Xylella fastidiosa*, which need insect vectors and for which insecticides may disrupt the disease cycle.

Modulation of bacterial behaviour through the inappropriate or untimely activation of cyclic di-GMP signalling systems offers another possible approach to disease control. As we have seen, the synthesis of the DSF cell–cell signal in *X. campestris* pv. *campestris* is finely balanced during both structured biofilm formation *in vitro* and disease progression in plants. Disruption of this balance by co-inoculation of the wild type strain with a DSF over-producing mutant strain prevents structured biofilm formation and causes attenuation of disease. Such an outcome could be achieved through the deployment of benign phylloplane or endophytic bacteria that either produce the DSF signal or are engineered to do so. The rationale for exploring modulation of DSF levels as a route to control plant disease is based on the role of the molecule as a cell–cell signal, rather than as a trigger of a cyclic di-GMP dependent signalling system. Nevertheless, the same principles may apply to the use of other molecules or molecular mimics that inappropriately activate or inactivate cyclic di-GMP signalling pathways, with a consequent negative effect on bacterial disease. Clearly such approaches require a greater understanding of the specific environmental cues activating cyclic di-GMP signalling systems.

Acknowledgements

The work in the authors' laboratory is supported by a Principal Investigator Award from the Science Foundation of Ireland to J.M.D. Collaborative work with J.-L. Tang, J.-X. Feng and Y.-Q. He at Guangxi University, Nanning, PRC, is supported by the China-Ireland Research Collaboration Programme.

References

Aldridge, P., Paul, R., Goymer, P., Rainey, P. and Jenal, U. (2003). Role of the GGDEF regulator PleD in polar development of *Caulobacter crescentus*. Mol. Microbiol. 47, 1695–1708.

Alm, R.A., Bodero, A.J., Free, P.D. and Mattick, J.S. (1996). Identification of a novel gene, *pilZ*, essential for type 4 fimbrial biogenesis in *Pseudomonas aeruginosa*. J. Bacteriol. 178, 46–53.

Amikam, D. and Galperin, M.Y. (2006). PilZ is part of the bacterial c-di-GMP binding protein. Bioinformatics 22, 3–6.

Andrade, M.O., Alegria, M.C., Guzzo, C.R., Docena, C., Rosa, M.C., Ramos, C.H. and Farah, C.S. (2006). The HD-GYP domain of RpfG mediates a direct linkage between the Rpf quorum-sensing pathway and a subset of diguanylate cyclase proteins in the phytopathogen *Xanthomonas axonopodis* pv citri. Mol. Microbiol. 62, 537–551.

Ausmees, N., Jonsson, H., Höglund, S., Ljunggren, H. and Lindberg, M. (1999). Structural and putative regulatory genes involved in cellulose synthesis in *Rhizobium leguminosarum* bv. trifolii. Microbiology 145, 1253–1262.

Ausmees, N., Mayer, R., Weinhouse, H., Volman, G., Amikam, D., Benziman, M. and Lindberg, M. (2001). Genetic data indicate that proteins containing the GGDEF domain possess diguanylate cyclase activity. FEMS Microbiol. Lett. 204, 163–167.

Barber, C.E., Tang, J.L., Feng, J.X., Pan, M.Q., Wilson, T.J., Slater, H., Dow, J.M., Williams, P. and Daniels, M.J. (1997). A novel regulatory system required for pathogenicity of *Xanthomonas campestris* is mediated by a small diffusible signal molecule. Mol. Microbiol. 24, 555–566.

Bobrov, A.G., Kirillina, O. and Perry, R.D. (2005). The phosphodiesterase activity of the HmsP EAL domain is required for negative regulation of biofilm formation in *Yersinia pestis*. FEMS Microbiol. Lett. 247, 123–130.

Camilli, A. and Bassler, B.L. (2006). Bacterial small-molecule signaling pathways. Science 311, 1113–1116.

Chan, C., Paul, R., Samoray, D., Amiot, N.C., Giese, B., Jenal, U. and Schirmer, T. (2004). Structural basis of activity and allosteric control of diguanylate cyclase. Proc. Natl. Acad. Sci. USA 101, 17084–17089.

Chang, A. L., Tuckerman, J. R., Gonzalez, G., Mayer, R., Weinhouse, H., Volman, G. Amikam, D., Benziman, M.and Gilles-Gonzalez, M. A. (2001). Phosphodiesterase A1, a regulator of cellulose synthesis in *Acetobacter xylinum*, is a heme-based sensor. Biochemistry 40, 3420–3426.

Chatterjee, S., and Sonti, R.V. (2002). *rpfF* mutants of *Xanthomonas oryzae* pv. *oryzae* are deficient for virulence and growth under low iron conditions. Mol. Plant Microbe Interact. 15, 463–471.

Christen, M., Christen, B., Folcher, M., Schauerte, A. and Jenal, U. (2005). Identification and characterization of a cyclic di-GMP-specific phosphodiesterase and its allosteric control by GTP. J. Biol. Chem. 280, 30829–30837.

Christen, M., Christen, B., Allan, M.G., Folcher, M., Jeno, P., Grzesiek, S. and Jenal, U. (2007). DgrA is a member of a new family of cyclic diguanosine monophosphate receptors and controls flagellar motor function in *Caulobacter crescentus*. Proc. Natl. Acad. Sci. USA 104, 4112–4117.

Colnaghi Simionato, A.V., da Silva, D.S., Lambais, M.R. and Carrilho, E. (2007). Characterization of a putative *Xylella fastidiosa* diffusible signal factor by HRGC-EI-MS. J. Mass Spectrom. 42, 490–46.

Cotter, P.A. and Stibitz, S. (2007). c-di-GMP-mediated regulation of virulence and biofilm formation. Curr. Opin. Microbiol. 10, 17–23.

Crossman, L. and Dow, J.M. (2004). Biofilm formation and dispersal in *Xanthomonas campestris*. Microbes Infect. 6, 623–629.

Cui, Y., Chatterjee, A., Liu, Y., Dumenyo, C.K. and Chatterjee, A.K. (1995). Identification of a global repressor gene, *rsmA*, of *Erwinia carotovora* subsp. *carotovora* that controls extracellular enzymes, N-(3-oxohexanoyl)-L-homoserine lactone, and pathogenicity in soft-rotting *Erwinia* spp. J. Bacteriol. 177, 5108–5115.

Cui, Y., Mukherjee, A., Dumenyo, C.K., Liu, Y. and Chatterjee, A.K. (1999). *rsmC* of the soft-rotting bacterium *Erwinia carotovora* subsp. *carotovora* negatively controls extracellular enzyme and harpin(*Ecc*) production and virulence by modulating levels of regulatory RNA (*rsmB*) and RNA-binding protein (RsmA). J. Bacteriol. 181, 6042–6045.

Dow, J.M., Crossman, L., Findlay, K., He, Y.Q., Feng, J.X. and Tang, J.L. (2003). Biofilm dispersal in *Xanthomonas campestris* is controlled by cell–cell signaling and is required for full virulence to plants. Proc. Natl. Acad. Sci. USA 100, 10995–11000.

Dow, J.M., Fouhy, Y., Lucey, J.F. and Ryan, R.P. (2006). The HD-GYP domain, cyclic di-GMP signaling, and bacterial virulence to plants. Mol. Plant Microbe Interact. 19, 1378–1384.

Dubey, A.K., Baker, C.S., Romeo, T. and Babitzke, P. (2005). RNA sequence and secondary structure participate in high-affinity CsrA-RNA interaction. RNA 11, 1579–1587.

Galperin, M.Y., Natale, D.A., Aravind, L. and Koonin, E.V. (1999). A specialized version of the HD hydrolase domain implicated in signal transduction. J. Mol. Microbiol. Biotechnol. 1, 303–305.

Galperin, M.Y., Nikolskaya, A.N. and Koonin, E.V. (2001). Novel domains of the prokaryotic two-component signal transduction systems. FEMS Microbiol. Lett. 203, 11–21.

Galperin, M.Y. (2005). A census of membrane-bound and intracellular signal transduction proteins in bacteria: Bacterial IQ, extroverts and introverts. BMC Microbiology 5, 35

Garcia, B., Latasa, C., Solano, C., Portillo, F.G., Gamazo, C. and Lasa, I. (2004). Role of the GGDEF protein family in *Salmonella* cellulose biosynthesis and biofilm formation. Mol. Microbiol. 54, 264–277.

Goymer P. (2002). Characterisation of the *Pseudomonas fluorescens WspR* locus. D.Phil. thesis, University of Oxford.

He, Y.W., Wang, C., Zhou, L., Song, H., Dow, J.M. and Zhang, L.H. (2006a). Dual signaling functions of the hybrid sensor kinase RpfC of *Xanthomonas campestris* involve either phosphorelay or receiver domain-protein interaction. J. Biol. Chem. 281, 33414–33421.

He, Y.W., Xu, M., Lin, K., Ng, Y.J., Wen, C.M., Wang, L.H., Liu, Z.D., Zhang, H.B., Dong, Y.H., Dow, J.M. and Zhang, L.H. (2006b). Genome scale analysis of diffusible signal factor regulon in *Xanthomonas campestris* pv. *campestris*: identification of novel cell–cell communication-dependent genes and functions. Mol. Microbiol. 59, 610–622.

Hecht, G. B. and Newton, A. (1995). Identification of a novel response regulator required for the swarmer-to-stalked-cell transition in *Caulobacter crescentus*. J. Bacteriol. 177, 6223–6229.

Hickman, J.W., Tifrea, D.F. and Harwood, C.S. (2005). A chemosensory system that regulates biofilm formation through modulation of cyclic diguanylate levels. Proc. Natl. Acad. Sci. USA 102, 14422–14427.

Hinsa, S.M. and O'Toole, G.A. (2006). Biofilm formation by *Pseudomonas fluorescens* WCS365: a role for LapD. Microbiology 152, 1375–1383.

Hisert, K.B., MacCoss, M., Shiloh, M.U., Darwin, K.H., Singh, S., Jones, R.A., Ehrt, S., Zhang, Z., Gaffney, B.L., Gandotra, S. et al., (2005). A glutamate-alanine-leucine (EAL) domain protein of *Salmonella* controls bacterial survival in mice, antioxidant defence and killing of macrophages: role of cyclic diGMP. Mol Microbiol. 56, 1234–1245.

Hoffman, L.R., D'Argenio, D.A., MacCoss, M.J., Zhang, Z., Jones, R.A. and Miller, S.I. (2005). Aminoglycoside antibiotics induce bacterial biofilm formation. Nature 436, 1171–1175.

Huang, B., Whitchurch, C.B. and Mattick, J.S. (2003). FimX, a multidomain protein connecting environmental signals to twitching motility in *Pseudomonas aeruginosa*. J. Bacteriol, 185, 7068–7076.

Huitema, E., Pritchard, S., Matteson, D., Radhakrishnan, S.K. and Viollier, P.H. (2006). Bacterial birth scar proteins mark future flagellum assembly site. Cell 124, 1025–1037.

Jackson, D.W., Suzuki, K., Oakford, L., Simecka, J.W., Hart, M.E and Romeo, T. (2002). Biofilm formation and dispersal under the influence of the global regulator CsrA of *Escherichia coli*. J. Bacteriol. 184, 290–30.

Jacques, M.A., Josi, K., Darrasse, A. and Samson, R. (2005). *Xanthomonas axonopodis* pv. *phaseoli* var. *fuscans* is aggregated in stable biofilm population sizes in the phyllosphere of field-grown beans. Appl. Environ. Microbiol. 71, 2008–2015.

Jenal, U. (2004). Cyclic di-guanosine-monophosphate comes of age: a novel secondary messenger involved in modulating cell surface structures in bacteria? Curr. Opin. Microbiol. 7, 185–191.

Jenal, U. and Malone, J. (2006). Mechanisms of cyclic-di-GMP signaling in bacteria. Annu. Rev. Genet. 40, 385–407.

Kader, A., Simm, R., Gerstel, U., Morr, M. and Römling U. (2006). Hierarchical involvement of various GGDEF domain proteins in rdar morphotype development of *Salmonella enterica* serovar *Typhimurium*. Mol Microbiol. 60, 602–616.

Karatan, E., Duncan, T.R. and Watnick, P.I. (2005). NspS, a predicted polyamine sensor, mediates activation of *Vibrio cholerae* biofilm formation by norspermidine. J. Bacteriol. 187, 7434–7443.

Kazmierczak, B.I., Lebron, M.B. and Murray, T.S. (2006). Analysis of FimX, a phosphodiesterase that governs twitching motility in *Pseudomonas aeruginosa*. Mol. Microbiol. 60, 1026–1043.

Kirisits, M.J., and Parsek, M.R. (2006). Does *Pseudomonas aeruginosa* use intercellular signalling to build biofilm communities? Cellular Microbiol. 8, 1841–1849.

Kjelleberg, S., and Molin, S. (2002). Is there a role for quorum sensing signals in bacterial biofilms? Curr. Opin. Microbiol. 5, 254–258.

Kulesekara, H., Lee, V., Brencic, A., Liberati, N., Urbach, J., Miyata, S., Lee, D.G., Neely, A.N., Hyodo, M., Hayakawa, Y. et al., (2006). Analysis of *Pseudomonas aeruginosa* diguanylate cyclases and phosphodiesterases reveals a role for bis-(3′–5′)-cyclic-GMP in virulence. Proc. Natl. Acad. Sci. USA 103, 2839–2844.

Laue, H., Schenk, A., Li, H., Lambertsen, L., Neu, T.R., Molin, S., and Ullrich, M.S. (2006). Contribution of alginate and levan production to biofilm formation by *Pseudomonas syringae*. Microbiology 152, 2909–2918.

Liu, Y., Cui, Y., Mukherjee, A. and Chatterjee, A.K. (1998). Characterization of a novel RNA regulator of *Erwinia carotovora* ssp. *carotovora* that controls production of extracellular enzymes and secondary metabolites. Mol. Microbiol. 29, 219–234.

Malone, J.G., Williams, R., Christen, M., Jenal, U., Spiers, A.J. and Rainey, P.B. (2007). The structure-function relationship of WspR, a *Pseudomonas fluorescens* response regulator with a GGDEF output domain. Microbiology. 153, 980–994.

Mayer, R., Ross, P., Weinhouse, H., Amikam, D., Volman, G., Ohana, P., Calhoon, R.D., Wong, H.C., Emerick, A.W. and Benziman M. (1991). Polypeptide composition of bacterial cyclic diguanylic acid-dependent cellulose synthase and the occurrence of immunologically crossreacting proteins in higher plants. Proc. Natl. Acad. Sci. USA. 88, 5472–5476.

Merighi, M. Lee, V.T., Hyodo, M., Hayakawa, Y. and Lory, S. (2007). The second messenger bis-(3′-5′)-cyclic-GMP and its PilZ domain-containing receptor Alg44 are required for alginate biosynthesis in *Pseudomonas aeruginosa* Mol. Microbiol. 65, 876–895.

Merkel, T.J., and Stibitz, S. (1995). Identification of a locus required for the regulation of *bvg*-repressed genes in *Bordetella pertussis*. J. Bacteriol. 177, 2727–2736.

Merkel, T. J., Barros, C. and Stibitz, S. (1998a). Characterization of the *bvgR* locus of *Bordetella pertussis*. J. Bacteriol. 180, 1682–1690.

Merkel, T.J., Stibitz, S., Keith, J.M., Leef, M. and Shahin, R. (1998b). Contribution of regulation by the *bvg*

Milton, D.L., Norqvist, A. and Wolf-Watz, H. (1995). Sequence of a novel virulence-mediating gene, virC, from *Vibrio anguillarum*. Gene *164*, 95–100.

Monds, R.D., Newell, P.D., Gross, R.H. and O'Toole, G.A. (2007). Phosphate-dependent modulation of c-di-GMP levels regulates *Pseudomonas fluorescens* Pf0–1 biofilm formation by controlling secretion of the adhesin LapA. Mol. Microbiol. *63*, 656–679.

Morgan, R., Kohn, S., Hwang, S.H., Hassett, D.J. and Sauer, K. (2006). BdlA, a chemotaxis regulator essential for biofilm dispersion in *Pseudomonas aeruginosa*. J. Bacteriol. *188*, 7335–7343.

Newman, K.L., Almeida, R.P., Purcell, A.H., and Lindow, S.E. (2004). Cell–cell signaling controls *Xylella fastidiosa* interactions with both insects and plants. Proc. Natl. Acad. Sci. USA *101*, 1737–1742.

Paul, R., Weiser, S., Amiot, N.C., Chan, C., Schirmer, T., Giese, B. and Jenal, U. (2004). Cell cycle-dependent dynamic localization of a bacterial response regulator with a novel di-guanylate cyclase output domain. Genes Dev. *18*, 715–727.

Pei, J. and Grishin, N.V. (2001). GGDEF domain is homologous to adenylyl cyclase. Proteins *42*, 210–216.

Pilot, G., Stransky, H., Bushey, D.F., Pratelli, R., Ludewig, U., Wingate, V.P. and Frommer, W.B. (2004). Overexpression of glutamine dumper1 leads to hypersecretion of glutamine from hydathodes of Arabidopsis leaves. Plant Cell *16*, 1827–1840.

Pratt, J.T., Tamayo, R., Tischler, A.D. and Camilli, A. (2007). PilZ domain proteins bind cyclic diguanylate and regulate diverse processes in *Vibrio cholerae*. J. Biol. Chem. *282*, 12860–12870.

Ramelot, T.A., Yee, A., Cort, J.R., Semesi, A., Arrowsmith, C.H. and Kennedy, M.A. (2007). NMR structure and binding studies confirm that PA4608 from *Pseudomonas aeruginosa* is a PilZ domain and a c-di-GMP binding protein. Proteins *66*, 266–271.

Remminghorst, U. and Rehm, B.H. (2006). Alg44, a unique protein required for alginate biosynthesis in *Pseudomonas aeruginosa*. FEBS Lett. *580*, 3883–3888.

Romeo, T. (1998). Global regulation by the small RNA-binding protein CsrA and the non-coding RNA molecule CsrB. Mol. Microbiol. *29*, 1321–1330.

Römling, U. (2005). Characterization of the rdar morphotype, a multicellular behaviour in Enterobacteriaceae. Cell. Mol. Life Sci. *62*, 1–13.

Römling, U., Gomelsky, M. and Galperin, M.Y. (2005). C-di-GMP: The dawning of a novel bacterial signalling system. Mol. Microbiol. *57*, 629–639.

Römling, U. and Amikam, D. (2006). Cyclic di-GMP as a second messenger. Curr. Opin. Microbiol. *9*, 218–228.

Ross, P., Weinhouse, H., Aloni, Y., Michaeli, D., Weinberger-Ohana, P., Mayer, R., Braun, S., de Vroom, E., van der Marel, G. A., van Boom, J. H. and Benziman, M. (1987). Regulation of cellulose synthesis in *Acetobacter xylinum* by cyclic diguanylate. Nature *325*, 279–281.

Ryan, R.P., Fouhy, Y., Lucey, J.F., Crossman, L.C., Spiro, S., He, Y.W., Zhang, L.H., Heeb, S., Camara, M., Williams, P. and Dow, J.M. (2006a) Cell–cell signaling in *Xanthomonas campestris* involves an HD-GYP domain protein that functions in cyclic di-GMP turnover. Proc. Natl. Acad. Sci. USA. *103*, 6712–6717.

Ryan, R.P., Fouhy, Y., Lucey, J.F. and Dow, J.M. (2006b). Cyclic di-GMP signaling in bacteria: recent advances and new puzzles. J. Bacteriol. *188*, 8327–8334.

Ryan, R.P., Fouhy, Y., Lucey, J.F., Jiang, B.L., He, Y.Q., Feng, J.X., Tang, J.L. and Dow, J.M. (2007). Cyclic di-GMP signalling in the virulence and environmental adaptation of *Xanthomonas campestris*. Mol. Microbiol. *63*, 429–442.

Ryjenkov, D.A., Tarutina, M., Moskvin, O.M. and Gomelsky, M. (2005). Cyclic diguanylate is a ubiquitous signaling molecule in Bacteria: insights into biochemistry of the GGDEF protein domain. J. Bacteriol. *187*, 1792–1798.

Ryjenkov, D.A., Simm, R., Römling, U. and Gomelsky, M. (2006). The PilZ domain is a receptor for the second messenger c-di-GMP: the PilZ domain protein YcgR controls motility in enterobacteria. J. Biol. Chem. *281*, 30310–30314.

Scarpari, L.M., Lambais, M.R., Silva, D.S., Carraro, D.M., and Carrer, H. (2003). Expression of putative pathogenicity-related genes in *Xylella fastidiosa* grown at low and high cell density conditions in vitro. FEMS Microbiol. Lett. *222*, 83–92.

Schmidt, A.J., Ryjenkov, D.A. and Gomelsky, M. (2005). Ubiquitous protein domain EAL encodes cyclic diguanylate- specific phosphodiesterase: enzymatically active and inactive EAL domains. J. Bacteriol. *187*, 4774–4781.

Simm, R., Morr, M., Kader, A., Nimtz, M. and Römling, U. (2004). GGDEF and EAL domains inversely regulate cyclic di-GMP levels and transition from sessility to motility. Mol. Microbiol. *53*, 1123–1134.

Simm, R., Fetherston, J.D., Kader, A., Römling, U. and Perry, R.D. (2005). Phenotypic convergence mediated by GGDEF-domain-containing proteins. J. Bacteriol. *187*, 6816–23.

Simm, R., Lusch, A., Kader, A., Andersson, M. and Römling U. (2007). Role of EAL-containing proteins in multicellular behavior of *Salmonella enterica* Serovar *Typhimurium*. J. Bacteriol. *189*, 3613–3623.

Slater, H., Alvarez-Morales, A., Barber, C.E., Daniels, M.J. and Dow, J.M. (2000). A two-component system involving an HD-GYP domain protein links cell–cell signalling to pathogenicity gene expression in *Xanthomonas campestris*. Mol. Microbiol. *38*, 986–1003.

Spiers, A. J., Bohannon, J., Gehrig, S. M. and Rainey, P. B. (2003). Biofilm formation at the air-liquid interface by the *Pseudomonas fluorescens* SBW25 wrinkly spreader requires an acetylated form of cellulose. Mol. Microbiol. *50*, 15–27.

Suzuki, K., Babitzke, P., Kushner, S.R. and Romeo, T. (2006). Identification of a novel regulatory protein (CsrD) that targets the global regulatory RNAs CsrB and CsrC for degradation by RNase E. Genes Dev. *20*, 2605–2617.

Tal, R., Wong, H. C., Calhoon, R., Gelfand, D., Fear, A. L., Volman, G., Mayer, R., Ross, P., Amikam, D., Weinhouse, H., Cohen, A., Sapir, S., Ohana, P. and Benziman, M. (1998). Three *cdg* operons control cellular turnover of cyclic di-GMP in *Acetobacter xylinum*: genetic organization and occurrence of conserved domains in isoenzymes. J. Bacteriol. *180*, 4416–4425.

Tamayo, R., Tischler, A.D. and Camilli, A. (2005). The EAL domain protein VieA is a cyclic di-guanylate phosphodiesterase. J. Biol. Chem. *280*, 33324–33330.

Tang, J.L., Feng, J.X., Li, Q.Q., Wen, H.X., Zhou, D.L., Wilson, T.J., Dow, J.M., Ma, Q.S., and Daniels, M.J. (1996). Cloning and characterization of the *rpfC* gene of *Xanthomonas oryzae* pv. *oryzae*: involvement in exopolysaccharide production and virulence to rice. Mol. Plant Microbe Interact. *9*, 664–666.

Tischler, A. D., Lee, S. H. and Camilli, A. (2002). The *Vibrio cholerae vieSAB* locus encodes a pathway contributing to cholera toxin production. J. Bacteriol. *184*, 4104–4113.

Tischler, A.D. and Camilli, A. (2004). Cyclic diguanylate (c-di-GMP) regulates *Vibrio cholerae* biofilm formation. Mol. Microbiol. *53*, 857–869.

Tischler, A.D. and Camilli, A. (2005). Cyclic diguanylate regulates *Vibrio cholerae* virulence gene expression. Infect. Immun. *73*, 5873–5882.

Torres, P. S., Malamud, F., Rigano, L. A., Russo, D.M., Marano, M. R., Castagnaro, A. P., Zorreguieta, A., Bouarab, K., Dow, J. M. and Vojnov, A. A. (2007). Controlled synthesis of the DSF cell–cell signal is required for biofilm formation and virulence in *Xanthomonas campestris*. Environ. Microbiol. *9*, 2101–2109.

Ude, S., Arnold, D.L., Moon, C.D., Timms-Wilson, T. and Spiers, A.J. (2006). Biofilm formation and cellulose expression among diverse environmental *Pseudomonas* isolates. Environ. Microbiol. *8*, 1997–2011.

Wang, L.H., He, Y., Gao, Y., Wu, J.E., Dong, Y.H., He, C., Wang, S.X., Weng, L.X., Xu, J.L., Tay, L., et al. (2004). A bacterial cell–cell communication signal with cross-kingdom structural analogues. Mol. Microbiol. *51*, 903–912.

Wei, B.L., Brun-Zinkernagel, A.M., Simecka, J.W., Pruss, B.M., Babitzke, P. and Romeo T. (2001). Positive regulation of motility and *flhDC* expression by the RNA-binding protein CsrA of *Escherichia coli*. Mol. Microbiol. *40*, 245–256.

Weinhouse, H., Sapir, S., Amikam, D., Shilo, Y., Volman, G., Ohana, P. and Benziman, M. (1997). C-di-GMP-binding protein, a new factor regulating cellulose synthesis in *Acetobacter xylinum*. FEBS Lett. *416*, 207–211.

Yap, M.N., Yang, C.H., Barak, J.D., Jahn, C.E. and Charkowski, A.O. (2005). The *Erwinia chrysanthemi* type III secretion system is required for multicellular behavior. J. Bacteriol. *187*, 639–648

Yu, J., Penaloza-Vazquez, A., Chakrabarty, A.M. and Bender, C.L. (1999). Involvement of the exopolysaccharide alginate in the virulence and epiphytic fitness of *Pseudomonas syringae* pv. *syringae*. Mol. Microbiol. *33*, 712–720.

Zhulin, I.B., Nikolskaya, A.N., and Galperin, M.Y. (2003). Common extracellular sensory domains in transmembrane receptors for diverse signal transduction pathways in bacteria and archaea. J. Bacteriol. *185*, 285–294.

Gene traders: Characteristics of Native Plasmids from Plant Pathogenic Bacteria

George W. Sundin and Jesús Murillo

Abstract

The concept of bacterial plasmids as gene traders is illustrative of the role of these elements in horizontal gene transfer, and specifically in the acquisition and distribution of sequences that enable rapid evolution. Plasmids are components of the horizontal gene pool and, as such, their genetic content is potentially accessible by a wide range of organisms. Most plasmids appear to ameliorate any potential negative effect on host fitness by encoding determinants of virulence and ecological fitness that can enhance adaptation to a specific niche or can influence niche expansion. The availability of multiple complete genome sequences of bacterial phytopathogens has shown the importance of horizontally acquired gene sequences in pathogen evolution. We suspect that plasmids have played a significant role in this gene mobility and also in the delivery of acquired genes to bacterial chromosomes through plasmid integration events. The versatility of plasmids plays a critical role in the evolutionary arms race of bacterial pathogens and plants.

Introduction

Throughout history, human settlements have sprouted and flourished with commercial trade, which in turn started and increased with communications. Trade brought wealth and, perhaps more importantly, a continuous exchange of ideas and innovations, favouring the rapid evolution of societies. Likewise, 'gene trading' has been pivotal in the evolution of species and their adaptation to all kind of environments, especially in prokaryotes (Ochman et al., 2000; Doolittle, 2005). Indeed, the acquisition of several complex physiological processes by microbes, such as photosynthesis, quorum sensing or nitrogen fixation, appear to have been via horizontal gene transfer (HGT) (Boucher et al., 2003). On a genomic scale, recent studies indicate that between 1.6% and 32.6% of different prokaryotic genomes, mostly depending on their lifestyle, were acquired by HGT (Koonin et al., 2001). Although it can potentially happen through different mechanisms, this exchange of DNA has probably been greatly favoured by bacteriophages and plasmids, which can often easily cross the species barrier.

Plasmids of phytopathogenic bacteria have received much attention since the discovery that a plasmid was responsible for the tumorigenic properties of *Agrobacterium tumefaciens* and that plasmids typically encoded different virulence genes in animal pathogens (Coplin, 1989). The majority of plant pathogenic bacteria contain one to several native plasmids that can vary in size from a few kilobases to a few megabases, with the majority in the range of 30–200 kb (Coplin, 1989; Vivian et al., 2001; Sundin, 2007). The gene content of these plasmids is also highly variable, although in many instances they carry genes important for virulence or for the survival of the host bacterium. Often, these genes also show a patchy distribution in different strains of a given species and can be encountered only in some individuals or in different genomic locations, such as on another unrelated plasmid or in the chromosome. However, and given the potential ability of native plasmids to be transferred among many different, unrelated bacteria, this gene pool is potentially accessible to a wide range of bacterial species. The mobilization of this DNA among bacteria interacting with plants, and especially plant pathogenic bacteria, can have enormous implications for agriculture and is in some cases severely limiting the control of bacterial diseases, as it occurs with the emergence and distribution of genes for resistance to copper or antibiotics (Cooksey, 1990; McManus et al., 2002).

The gene content of plasmids is dynamic in nature, as these elements can usually readily acquire or lose gene sequences. We consider plasmids as 'gene traders' because plasmid-encoded sequences are favoured for horizontal dissemination, and because most plasmids are mosaic in nature, due to extensive 'gene trading'. Indeed the success of plasmids as genetic mosaics implies that gene trading is an adaptive trait. Thus, plasmids contribute to genome plasticity which, in turn, significantly affects bacterial evolution. The core genome of any bacterial species refers to genes that are ubiquitous in that species and includes housekeeping genes and genes essential for survival. These genes are almost universally encoded on chromosomes and are rarely subject to horizontal gene transfer. In contrast, the flexible genome refers to genes that are non-universal in distribution within a species and consists of genes that contribute to strain differences that are ecologically consequential and can enhance adaptation to a specific niche or colonization of a novel niche. The flexible bacterial genome comprises genes that are readily transferred among organisms. Genes of the flexible genome are usually associated with mobile genetic elements such as plasmids and bacteriophages and may be found in context with sequences encoding mobility functions such as transposons, insertion sequence (IS) elements, or enzymes such as integrases. These genes may be organized into genomic islands which are defined as mobile genetic elements that contribute to rapid changes in virulence and/or adaptive potential of bacteria (Dobrindt et al., 2004). Islands are thought to be transferred *en bloc* and are possibly evolved from integrative plasmids or bacteriophages that have lost genes required for replication and other maintenance and transfer functions (Dobrindt et al., 2004). Islands and lineage-specific regions are a prevalent feature of the *Pseudomonas syringae* flexible genome (Arnold et al., 2003; Joardar et al., 2005). Thus, the significance of DNA of the flexible genome is that acquisition of these sequences enables bacteria to evolve by 'quantum leaps' allowing rapid expansion into new ecological niches.

Gene content

In addition to the genes required for their reproduction, plasmids can potentially encode any given DNA sequence because of their facility to acquire new DNA. Plasmids can often integrate into the bacterial chromosome, or combine with other plasmids through homologous recombination of mobile elements or other repeated sequences forming a cointegrate. Imprecise excision of the integrated plasmid, for example through recombination with secondary insertions of a mobile element, then results in the incorporation of new sequences. Plasmids tend to encode genes that confer local adaptation, that is, 'adaptations to variations in environmental conditions that occur only sporadically in time or space' (Eberhard, 1989). Without trying to be exhaustive, Table 14.1 details functions coded for by native plasmids in different species of plant pathogenic bacteria.

Genomic analysis of native plasmids of *Pseudomonas syringae* pathovars (Zhao et al., 2005; Pérez-Martínez et al., 2008) showed a highly variable distribution of genes among plasmids and chromosomes, and highly dissimilar gene content even between related plasmids. However, the inherent mobility of plasmid DNA makes this gene pool potentially available to the whole bacterial species, and probably to many other species, constituting a highly dynamic and flexible metagenome that can favour adaptation to an ever changing environment.

Below we summarize representative examples of genes and functions coded by native plasmids in phytopathogenic bacteria, which were separated in sometimes arbitrary categories.

Replication, maintenance and mobilization

In general, plasmids carry at least the determinants necessary for their propagation and stable maintenance in the host cell. These determinants differ depending on the type of plasmid, but include one or more origins of replication and, generally, a gene coding for an initiation protein, often called Rep, that specifically recognizes the origin(s) of replication and initiates the replication process (Nordström, 1993). Also, plasmids with medium to low copy number tend to encode one or more systems that contribute to their stable maintenance in the cell. These systems are also of various types and can include, among others, partition systems, to distribute the plasmid molecules to all daughter cells at cell division; killer systems, to kill segregant cells that had lost the plasmid, or site-specific recombination systems that resolve multimers formed by homologous recombination among individual plasmid molecules (Nordström, 1993). An important consequence of the replication process is the prediction that two plasmids that contain homologous sequences for replication or maintenance can not be stably maintained in the same cell. In consequence, these plasmids are classified in the same *incompatibility* group. An important exception to this is the so called pPT23A family of plasmids (PFP) of *P. syringae*, which are characterized by containing a homologous replication gene sharing high levels of nucleotide identity (Murillo and Keen, 1994; Gibbon *et al.*, 1999; Ma *et al.*, 2007). In spite of this, strains of *P. syringae* very often contain two or more coexisting PFP plasmids (Sesma *et al.*, 1998).

Plasmids also frequently harbour genes for their mobilization via conjugation, a process that typically incorporates the actions of over 15 proteins performing different functions to assemble a type IV secretion system for protein or DNA delivery to other cells (see below; Llosa and de la Cruz, 2005; Backert and Meyer, 2006). The plasmid must also contain an origin of transfer, or *oriT*, where the transfer process starts and ends and that can function *in trans*; in practical terms, this means that any plasmid – or chromosomal sequence – containing an *oriT* can be mobilized by the cognate conjugation machinery. Genes required for the biosynthesis of the conjugation machinery occupy a large amount of DNA, while the *oriT* usually spreads over only a few hundreds base pairs. It is not surprising then that many native plasmids encode only partial conjugation systems (Sundin, 2007), because their deletion will not have a functional effect due to gene redundancy or because the possession of *oriT* is enough to be mobilized. Plasmids that do not encode a conjugation system or lack *oriT* can also be mobilized simply by recombining with a conjugative or mobilizable plasmid; for instance, the non-conjugative copper resistance plasmid pPT23D from *P. syringae* pv. *tomato* PT23 was mobilized by recombination through repeated sequences with the conjugative cryptic plasmid pPT23C (Cooksey, 1990). Also, the conjugative plasmid pIPO2 was isolated from the wheat rhizosphere and has the ability to mobilize a variety of IncQ plasmids to many diverse Gram-negative bacteria (van Elsas *et al.*, 1998). This form of genetic hitchhiking suggests that any DNA fragment can potentially be mobilized and illustrates the many mechanisms that facilitate gene trading among bacteria.

Pathogenicity plasmids

There are several examples of plasmids coding for one or more genes essential for disease production or, sometimes, for the production of typical symptoms on a given plant host. Although less frequently, some plasmids contain all the genetic information to transform an otherwise saprophytic bacterium into a plant pathogen; these can potentially allow the emergence of new pathogens in evolutionary quantum leaps rather than through slow adaptive evolution. Here, we will include all these plasmids into a loose group of pathogenicity plasmids and will describe some examples in detail below.

Agrobacterium tumefaciens causes crown gall on dicotyledonous plants after transferring the T-DNA portion of the tumour-inducing plasmid (pTi) to some of the plant cells (Hooykaas and Beijersbergen, 1994; see chapter by Setubal *et al.*). Plasmids pTi are essential for pathogenicity, have a size of about 200 kb and their host range is restricted to the Order *Rhizobiales*. The region required for virulence, the T-DNA, occupies about half of the pTi and can be divided into the Vir-region, that is responsible for the synthesis of a type IV secretion system for the delivery of proteins and the T-DNA into plant cells, and the T region, that encompasses the genes for hormone biosynthesis in plant cells (Hooykaas and Beijersbergen, 1994; Backert and Meyer, 2006). The T-DNA also includes genes

Table 14.1 Relevant functions coded for by native plasmids in phytopathogenic bacteria

Category or function	Products or genes	Plasmid	Bacterial species	Observations/reference
Catabolism				
Synthesis and utilization of opines	Octopine and many other opines	pRi and pTi	*Agrobacterium*	Hooykaas and Beijersbergen (1994)
Thiamine biosynthesis	*thiOGF*	pEa29	*Erwinia amylovora*	Laurent *et al.* (1989), McGhee and Jones (2000)
Pathogenicity and virulence				
Cell-to-cell adhesion	Adhesin SkARP1, sarpin family	pSKU146	*Spiroplasma kunkelii*	Davis *et al.* (2005)
Cell-cell signaling	gene *att*	pFiD188	*Rhodococcus fascians*	Linear plasmid, essential for pathogenicity, contains several other pathogenicity genes (Crespi *et al.*, 1992)
Effector genes	Many different effectors	Different plasmids	*Pantoea agglomerans* pv. *gypsophilae* and pv. *betae*, *Pseudomonas syringae* pvs., *Xanthomonas campestris*	Vivian *et al.* (2001), Grant *et al.* (2006), Barash and Manulis-Sasson (2007)
Enzymes	Endopolygalacturonase (*pehA*)	pPEC320	*Burkholderia cepacia*	Gonzalez *et al.* (1997)
	Cellulase (*celA*)	pCM1	*Clavibacter michiganensis* subsp. *michiganensis*	Gene essential for pathogenesis (Gartemann *et al.*, 2003)
	Serine protease (*pat1*)	pCM2	*C. michiganensis* subsp. *michiganensis*	Gene essential for pathogenesis (Gartemann *et al.*, 2003)
Pathogenicity island	Six to seven effector genes	pAV511 and other large plasmids	*P. syringae* pv. *phaseolicola*	This PAI is essential to induce disease on bean and soybean (Jackson *et al.*, 1999)
Phytohormones	3-Indoleacetic acid (*iaaM-iaaH-iaaL*)	pPATH	*P. agglomerans* pv. *gypsophilae* and pv. *betae*	Barash and Manulis-Sasson (2007),
		Different plasmids	*P. savastanoi* pv. *nerii* and pv. *savastanoi*	Glass and Kosuge (1988), Caponero *et al.* (1995), Pérez-Martinez *et al.* (2008)
	Cytokinins (*ipt* or *ptz*)	pRi and pTi	*Agrobacterium*	Hooykaas and Beijersbergen (1994)
		pFiD188	*R. fascians*	Contains six genes coding for an isopentenyltransferase and other genes involved in the biosynthesis of cytokinins. Linear plasmid, essential for pathogenicity, contains several other pathogenicity genes (Crespi *et al.*, 1992)

Function	Gene/product	Plasmid	Organism	Reference
		pPATH	*P. agglomerans* pv. *gypsophilae* and pv. *betae*	Barash and Manulis-Sasson (2007)
	Ethylene (efe)	Different plasmids	*P. savastanoi* pv. *nerii* and pv. *savastanoi*	Powell and Morris (1986); Caponero *et al.* (1995); Pérez-Martínez *et al.* (2008)
		pPSP1 and other	*P. syringae* pv. *cannabina*, pv. *glycinea* and pv. *phaseolicola*	Fukuda *et al.* (1992), Nagahama *et al.* (1994), Sato *et al.* (1997), Watanabe *et al.* (1998)
Phytotoxins	Coronatine	Different large plasmids	*P. syringae* pv. *atropurpurea*, pv. *glycinea*, pv. *maculicola*, pv. *morsprunorum* and pv. *tomato*	Bender *et al.* (1999)
Quorum sensing	*traI-traR*	pTi	*A. tumefaciens*	White and Winans (2007)
Type III secretion genes	*hrp/hrc* genes	pPATH	*P. agglomerans* pv. *gypsophilae* and pv. *betae*	Barash and Manulis-Sasson (2007)
Type IV secretion genes	Homologes of *vir* and *tra* genes	pSKU146	*S. kunkelii*	This plasmid is thought to be mobilizable for the transmission of pathogenicity related genes (Davis *et al.*, 2005)
	Type IVA	pXAC64	*X. axonopodis* pv. *citri*	da Silva *et al.* (2002)
	Type IVB	XF51	*Xylella fastidiosa*	Simpson *et al.* (2000)
	Types IVA and/or IVB	Different plasmids	*P. syringae* and *P. savastanoi*	Zhao *et al.* (2005), Pérez-Martínez *et al.* (2008)
Production of bacteriocins				
Agrocin 84	agrocin 84	pAgK84	*A. radiobacter* K84	Roberts *et al.* (1977)
Resistance to stresses				
Arsenic and cobalt	ND	pPSR12	*P. syringae* pv. *syringae*	The plasmid confers resistance to arsenate, cobalt and copper, and stimulates production of alginate (Kidambi *et al.*, 1995)
Cadmium	ND	pD188	*R. fascians*	Desomer *et al.* (1988)
Copper	*copABCD*, *copRS*	Many different	Different *P. syringae*, *X. arboricola*, and *X. axonopodis* pathovars	Cooksey (1990), Gardan *et al.* (1993), Cooksey (1994)
Streptomycin	*strAB*	Many different	*E. amylovora P. syringae X. campestris*	Carried by transposon Tn5393 (McManus *et al.*, 2002; Sundin, 2002)
UV light	*rulAB*	Many different	*P. syringae*	Sundin and Murillo (1999)

ND, not determined.

for the biosynthesis of specialized carbon compounds, called opines, that will be produced in huge amounts by the transformed plant cells and that can be metabolized almost exclusively by tumorigenic agrobacteria. The other half of the pTi is not required for virulence and includes determinants for replication, catabolism of opines, and conjugation, which is dependent on *tra* genes that function independently of the *vir* system (Hooykaas and Beijersbergen, 1994; White and Winans, 2007). Conjugational transfer of the pTi is the subject of complex regulation by quorum sensing and opines, indicating that the transfer of pTi to agrobacteria or other organisms can potentially only occur in the vicinity of the plant tumours. Plasmid pTi, or an artificial hyperconjugative derivative of pTi, can be transferred to many species of agrobacteria and to species of *Rhizobium* and *Phyllobacterium*, all included in the O. Rhizobiales. Acquisition of pTi converts recipient cells into tumour-inducing pathogens (Hooykaas *et al*., 1977; Teyssier-Cuvelle *et al*., 2004). Nevertheless, there are no reports of the natural dissemination of the pTi among soil bacteria.

Ralstonia solanacearum is a devastating plant pathogen that colonizes the phloem and causes bacterial wilt in a wide range of plant species (see chapter by Brown). Most of the strains of this pathogen contain a megaplasmid that is essential for pathogenicity (Boucher *et al*., 1986) and that in strain GMI1000 is a circular replicon of 2.1 megabases (Genin and Boucher, 2004). The megaplasmid has clearly originated from a plasmid, although its base composition indicates that it has coevolved with the main chromosome for a long time. During this process, the megaplasmid has captured a copy of the rDNA operon, two tRNA and a copy of elongation factor G, as well as 55 other genes involved in the biosynthesis of amino acids, nucleotides and cofactors; as a result, both the chromosome and the megaplasmid are needed for growth *in vitro* in defined minimal medium (Genin and Boucher, 2004). Although the type III secretion system (T3SS) is encoded on the megaplasmid, other pathogenicity genes are distributed among the two replicons, many of which are clustered in areas of divergent G+C content that have the characteristics of pathogenicity islands and might have been acquired through HGT. There are no reports of the transfer of the megaplasmid to other bacteria; conversely, there is a large variation in the genetic repertoire of different strains and phylotypes of *R. solanacearum*, although a large majority of the pathogenicity genes are part of the core genome (Guidot *et al*., 2007). Variable genes appear to result from differential deletion in cell lineages as well as from acquisition of genes and genomic islands that are currently dispersed on the two replicons, although the megaplasmid presents larger variability.

Pantoea agglomerans (syn. *Erwinia herbicola* and *Enterobacter cloacae*) is a saprophytic bacterium that is normally found living as an epiphyte or endophyte of many plant species. However, *P. agglomerans* pv. *betae* and pv. *gypsophilae* are plant pathogens that induce gall formation in several plant species due to the acquisition of a non-conjugative pathogenicity plasmid designated pPATH (Barash and Manulis-Sasson, 2007). The best studied pPATH, pPATH*Pag* from *P. agglomerans* pv. *gypsophilae* 824-1, has a size of around 135 kb that includes a loosely defined pathogenicity island (PAI) of about 75 kb. This PAI includes a complete T3SS that is very similar to the one described in *E. amylovora*, with six effector genes and genes for the biosynthesis of 3-indoleacetic acid and cytokinins. The PAI also includes six insertion sequences, five of which were present in pathogenic but not in non-pathogenic strains of *P. agglomerans*, as well as remnants of known gene sequences from diverse bacteria that include *Yersinia pestis* and *Xylella fastidiosa* (Guo *et al*., 2002). Since *P. agglomerans* is normally a saprophyte, it is highly likely that either pPATH or the PAI were acquired in one or few steps, transforming a tame bacterium into a plant predator.

The Gram-positive bacterium *Clavibacter michiganensis* subsp. *michiganensis* causes bacterial canker and wilt of tomato, characterized by spots on leaves, stems and fruits, wilting, and, occasionally, small cankers on leaves and leaf veins (Agrios, 1997). Strain NCPPB382 carries two plasmids, pCM1 and pCM2, that encode pathogenicity genes for a cellulase (*celA*) and a putative serine protease (*pat-1*), respectively (see Table

14.1; Gartemann *et al.*, 2003). Strain CMM100, cured of both plasmids, is still able to colonize tomato to high titers, but do not produce disease symptoms. Remarkably, strains carrying either pCM1 or pCM2 can induce wilting of tomato plants, but were reduced in virulence; in both cases, the phenotype was ascribed to either *celA* or *pat-1*. Plasmid pCM2 is somewhat unstable and appears to be lost in around 20% of the population; however, both pCM1 and pCM2 are readily transferred in infected plants, which possibly contributes to modulate virulence and maintain variability (Gartemann *et al.*, 2003).

Rhodococcus fascians is another Gram-positive bacterium, causing leafy gall on sweet pea and other plants, although it thrives as an epiphyte and endophyte (Agrios, 1997; Goethals *et al.*, 2001). Strain D188 contains a conjugative, 200 kb linear plasmid, pFiD188, that is essential for disease production and that contains three large loci with several virulence genes, including genes for the biosynthesis of cytokinins, among many others (Crespi *et al.*, 1992; Goethals *et al.*, 2001; Maes *et al.*, 2001). Strains cured of the linear plasmid have a much lower efficiency of endophytic colonization than the wild type, although they can still penetrate plants (Goethals *et al.*, 2001), perhaps suggesting that *R. fascians* could have been an endophyte before acquiring the capacity to cause disease. In this context, *R. fascians* has been proposed to use plant carbon sources that would be in high concentrations only in the leafy gall, so that the gall would represent a niche that offers the pathogen a selective advantage over other bacteria (Goethals *et al.*, 2001) and a driving force for the maintenance, and perhaps the acquisition, of the virulence genes carried by the linear plasmid.

Virulence genes

Phytopathogenic bacteria use several strategies to incite disease in their plant hosts, and there is a large variety of virulence genes that enhance their aggressiveness towards the plant (Alfano and Collmer, 1996). Virulence genes are in general dispensable for the bacterium and, perhaps for this, are often found on plasmids. Here we will comment on some general examples to illustrate their mobility by native plasmids.

Effector genes, previously known as avirulence genes, generally code for proteins that are translocated into the plant cell by a T3SS, and that in many cases act to suppress plant defence responses (Grant *et al.*, 2006; see chapter by Boch). There are many examples of effector genes carried by plasmids in species of *Pantoea*, *Pseudomonas* and *Xanthomonas*, and *R. solanacearum* (for instance, Vivian *et al.*, 1997; Vivian *et al.*, 2001; Genin and Boucher, 2004; Barash and Manulis-Sasson, 2007). Additionally, strains and plasmids of diverse *P. syringae* pathovars, and possibly of other bacterial plant pathogens, show a large variability in their effector gene load (Zhao *et al.*, 2005; Sarkar *et al.*, 2006). This, and the variable locations of these genes on plasmids and chromosome, illustrate their mobility.

Plasmid pEA29, a non-conjugative ubiquitous plasmid of *E. amylovora*, encodes genes *thiOGF* for the biosynthesis of thiamine and is required for full virulence (McGhee and Jones, 2000). Thiamine biosynthesis is required for the high level production of the exopolysaccharide amylovoran, which is essential for pathogenicity of *E. amylovora* (Sundin *et al.*, 2004; Sundin, 2007), and this is probably the reason why pEA29 is so widely distributed, although there are certain strains devoid of this plasmid that are as virulent as the strains containing it (Llop *et al.*, 2006). This example highlights the occurrence of genetic interactions that might favour the maintenance of a given plasmid.

The phytotoxin coronatine, produced by several pathovars of *P. syringae*, is a structural and functional homologue of the plant signal molecule jasmonic acid that primarily functions to suppress plant defences and favour infection (Nomura *et al.*, 2005). Genes for the biosynthesis of coronatine are included in a putative PAI that is located in large native plasmids or in the chromosome. Current evidence supports the idea that the coronatine cluster evolved as two separate clusters; the combination of these clusters was probably favoured by their enhanced contribution to virulence as a whole, and their incorporation to native plasmids fostered their mobility and possibilities for evolution (Bender *et al.*, 1999; Sesma *et al.*, 2001).

Resistance to bactericides and other stresses

Native plasmids have been instrumental in the dissemination of genes for resistance to commonly used bactericides, such as copper compounds and streptomycin (see below, Cooksey, 1990; McManus et al., 2002). Additionally, plasmids often carry determinants that help the bacterium to cope with other environmental stresses (some examples are shown in Table 14.1), such as tolerance to ultraviolet light.

Although a plant pathogen, *P. syringae* spends a good part of its life cycle as an epiphyte (Lindow and Brandl, 2003). Perhaps not surprisingly, most strains of *P. syringae* contain at least a copy of the *rulAB* genes, which confer a significant increase in bacterial survival upon exposure to UV light (Sundin and Murillo, 1999). Generally, *rulAB* are present on native plasmids of the pPT23A family, and often two or more copies are distributed among plasmids and the chromosome, which could favour recombination and gene trading. Additionally, *rulAB* was described as a hot spot for the insertion of different genes, including effector genes (Arnold et al., 2000). Although *rulAB* primarily confer resistance to an environmental stress, it is yet another example of a DNA sequence that enhances bacterial survival in different ways, favouring its acquisition, maintenance and distribution by plasmids.

Mechanisms of plasmid transfer

Bacterial conjugation is a mechanism for intercellular DNA transfer between donor and recipient cells; the 'conjugation machine' is related to other bacterial type IV secretion systems that mediate the transfer of DNA and/or proteins into prokaryotic or eukaryotic target cells (Christie et al., 2005). Conjugation is now known as a process in which the DNA is transferred as a nucleoprotein particle composed of a protein bound to the 5′ end of a ssDNA molecule (Christie, 2004). There are two main types of conjugation systems present in plant pathogenic bacteria best exemplified by the VirB/D4 system of *A. tumefaciens* and the Tra-type system found on some plasmids in *P. syringae*, including plasmids in strain DC3000. Many other plasmids, including several plasmids of the pPT23A family in *P. syringae*, comprise complete or nearly complete sets of VirB/D4 homologues (Chen et al., 2005; Zhao et al., 2005). This system is also referred to as the P-like conjugative system because a homologous VirB/D4 system is found on the IncP plasmid RP4 (Lawley et al., 2003). A detailed biochemical review of the *A. tumefaciens* conjugation machine has been published recently (Chen et al., 2005). The Tra-type system is also referred to as an I-like conjugation system based on a similar system found on the IncI1 plasmid R64 (Lawley et al., 2003). Finally, the F plasmid of *Escherichia coli* serves as the model plasmid encoding a third type of system; this system is quite different from the VirB/D4 system although many of the genes are homologues (Lawley et al., 2003).

Both Vir and Tra-type conjugation systems are found on plasmids in plant pathogenic bacteria (Fig. 14.1). The most detailed distribution studies have been done with PFPs in *P. syringae*. Two macroarray examinations have identified complete VirB/D4 systems from 15 PFPs from five *P. syringae* pathovars and *P. savastanoi* pv. savastanoi (Zhao et al., 2005; Pérez-Martínez et al., 2008). Complete Tra-type systems were detected on 11 PFPs from six *P. syringae* pathovars and *P. savastanoi* pv. savastanoi (Zhao et al., 2005; Pérez-Martínez et al., 2008). In addition, a number of plasmids contained genes from both systems. Various permutations of the VirB/D4 and Tra systems were discovered on plasmids from *E. amylovora*, *Xanthomonas campestris*, and *Xylella fastidiosa* during sequencing experiments (reviewed in Sundin, 2007). Because of the large genomic component invested in genes encoding the conjugation apparatus, it is evident that the potential to transfer to new hosts is important to the life history of plasmids from plant pathogenic bacteria. A recent phylogenetic analysis indicated that closely related PFPs (on the basis of their major replication gene *repA*) had been transferred among more distantly related *P. syringae* pathovars (Ma et al., 2007), again confirming the importance of transfer functions to plasmid biology.

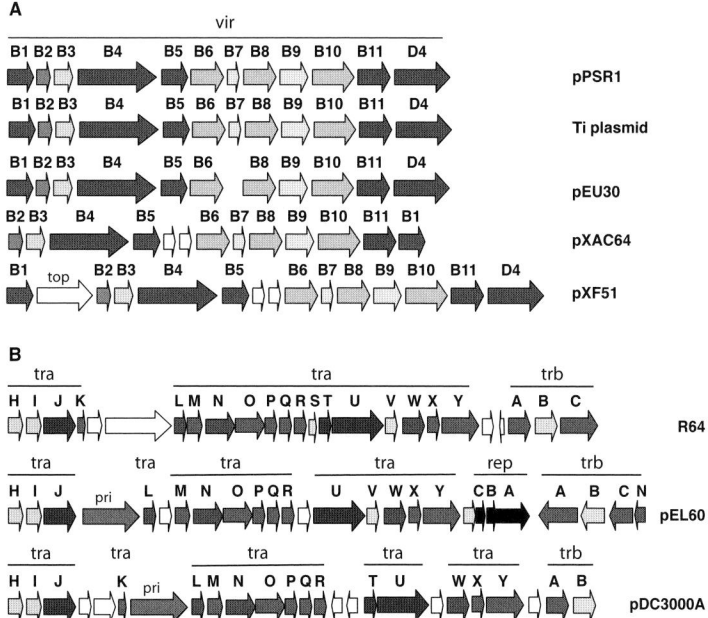

Figure 14.1 Schematic map for the Vir (A) and Tra (B) type IV secretion genes from phytopathogenic bacterial plasmids. Genes with similar functions were drawn with similar colours. Genes without colour are hypothetical with no known function. The hosts for the plasmids are as follows: pPSR1, *Pseudomonas syringae* pv. *syringae*; Ti plasmid, *Agrobacterium tumefaciens*; pEU30, *Erwinia amylovora*; pXAC64, *Xanthomonas axonopodis* pv. *citri*; pXF51, *Xylella fastidiosa*; R64, broad-host-range plasmid; pEL60, *E. amylovora*; pDC3000A, *P. syringae* pv. *tomato*.

Evidence for gene transfer mediated by plasmids

Initial examinations of gene transfer events mediated by plasmids in plant pathogenic bacteria focused on the transfer of bactericide-resistance genes or on the putative transfer of individual genetic determinants (Cooksey, 1990). With the advent of genome sequencing and the ever-increasing availability of new sequence information among a wide range of bacterial taxa, there are many examples of recent sequence transfer within species of plant pathogenic bacteria and between plant pathogens and organisms from other environmental niches. We will not attempt a comprehensive examination but instead will highlight several examples illustrating the variety of sequences and organisms involved.

Bactericide resistance

In response to repeated use of copper and/or streptomycin bactericides, plant pathogenic bacteria have evolved resistance to these compounds (reviewed in McManus *et al.*, 2002). As with known examples of antibiotic resistance in clinical bacteria, resistance evolution is typically not due to the generation of novel determinants, but occurs via the acquisition of pre-existing resistance genes. For example, the *strA-strB* genes confer streptomycin resistance in *E. amylovora*, *P. marginalis*, *P. syringae*, and *X. campestris* strains from several hosts on different continents (Sundin and Bender, 1996; Han *et al.*, 2004). The *strA-strB* genes are harboured within the transposon Tn5393; this element has transposed to important plasmids in plant pathogens including the ubiquitous plasmid pEA29 in *E. amylovora* (McGhee and Jones, 2000), and plasmids of the pPT23A family in *P. syringae* (Sundin *et al.*, 1994). Genetic and genomic analyses have now shown that Tn5393 is essentially distributed worldwide among bacterial pathogens from humans, animals, and plants, and is present in both Gram-negative and Gram-positive bacteria isolated from various environmental habitats (Table 14.2). The transfer of this element among diverse bacteria is thought to be relatively recent,

Table 14.2 Plasmids and host bacterial species harbouring intact copies of the streptomycin-resistance transposon Tn5393

Species	Source	Insertions within the Tn5393 backbone	Reference
Aeromonas salmonicida	Norway	None	L'Abee-Lund and Sorum (2000)
Alcaligenes faecalis	Italy	Tn4176, aphA6b	Mantengoli and Rossolini (2005)
Corynebacterium striatum	Japan	Tn5715, IS1250	Tauch et al. (2000)
Erwinia amylovora	USA	IS1133	Chiou and Jones (1993)
Pseudomonas sp.	Germany	none	Tauch et al. (2003)
P. aeruginosa	USA	none	Stokes et al. (2007)
P. marginalis	Korea, Japan	none	Han et al. (2004)
P. syringae pv. syringae	USA	none	Sundin and Bender (1995)
Salmonella enterica	Italy	IS1133	Pezzella et al. (2004)
Xanthomonas campestris pv. vesicatoria	Argentina	IS6100	Sundin and Bender (1995)

as the nucleotide sequence divergence of genes from Tn5393 from different bacterial species is usually < 2% (Sundin, 2002). Thus, Tn5393 is a mobile DNA element presumably accessed by plant pathogenic bacteria from an environmental gene pool in response to streptomycin selection. This transposon, and the wide range of bacteria that carry it, is an excellent example of the extent of gene trading among diverse bacteria.

Fitness-enhancing genes

Evidence for the transfer of genes that potentially influence the fitness of bacterial plant pathogens indicates that these genes are transferred among a more restricted range of bacteria, most likely because of the adaptive advantage of carriage of these determinants is more species specific. We will present several examples ranging from the transfer of individual genes to genomic islands.

Genes encoding effector proteins secreted through the T3SS collectively contribute to virulence in most genera of plant pathogenic bacteria. In *P. syringae*, it is known that plasmids of the pPT23A family typically encode one or more effectors (Zhao et al., 2005). The location of effector genes such as *virPphA* within different genetic contexts on PFPs from distinct *P. syringae* pathovars is retrospective evidence supporting the horizontal transfer of *virPphA* between plasmids and between pathovars (Jackson et al., 2002). Indeed, phylogenetic evidence strongly indicates that other PFP-encoded genes have regularly been transferred between different PFPs (Ma et al., 2007). The observation that effector genes are often located adjacent to insertion sequence elements also supports the hypothesis that these genes are mobile within and between pathogen populations (Kim et al., 1998).

The type III effector protein PthA is thought to be responsible for the elicitation of hyperplastic symptoms of citrus canker disease by *X. citri* pv. *citri* (Brunings and Gabriel, 2003), and all strains capable of causing citrus canker harbour the *pthA* gene (Cubero and Graham, 2002). El Yacoubi et al. (2007) hypothesized that the ability to cause citrus canker in phylogenetically distinct groups of pathogens including *X. axonopodis* pvs. *alii* and *citrumelo* could be due to the acquisition of *pthA* via horizontal transfer. These researchers sequenced the 37-kb indigenous plasmid pXcB from *X. citri* pv. *citri*, found that this plasmid encoded *pthA*, and demonstrated that the plasmid was highly transferable *in planta* to other *X. citri* pv. *citri* strains (El Yacoubi et al., 2007).

Pathogenicity islands (PAIs) are larger genomic regions that show evidence of acquisition by horizontal transfer and typically encode genes influencing virulence in bacterial pathogens. The 150-kb pPATH plasmid from the gall pathogen *P. agglomerans* pv. *gypsophilae* harbours a ca. 70-kb PAI that contains an intact *hrp* gene cluster encoding the structural components of a T3SS as well as genes encoding various effector proteins that are presumably secreted through the T3SS into host plant cells (Mor et al., 2001). As mentioned above, the pPATH plasmid is absolutely required for pathogenesis and is the only known example of a *hrp* gene cluster residing on a plasmid in the size range 19–200 kb. Since *P. agglomerans* is a common plant epiphyte, it has been hypothesized that *P. agglomerans* pv. *gypsophilae* is a recently evolved pathogen with plasmid-encoded pathogenicity traits superimposed on pre-existing epiphytic fitness traits of the organism (Manulis and Barash, 2003).

A 30-kb PAI encoding three type III effectors and other potential virulence genes was identified on the 150-kb plasmid pAV511 from *P. syringae* pv. *phaseolicola* following a plasmid-curing experiment that indicated the loss of pAV511 was correlated with a loss of virulence on bean (Jackson et al., 1999). The effectors either conferred virulence functions on bean or avirulence functions on some soybean cultivars (Jackson et al., 1999). A potential PAI encoding the biosynthesis of the phytotoxin coronatine is encoded on a PFP from *P. syringae* pv. *glycinea* and is located in the chromosome of *P. syringae* pv. *tomato* DC3000 (Alarcón-Chaidez et al., 1999).

Evidence of recent plasmid and gene transfer events between plant pathogens and diverse bacterial species

The carriage of closely similar gene sequences in diverse bacteria is an indication of relatively recent gene transfer events. There are several interesting examples ranging from whole plasmids to a few genes that suggest that plant pathogens can access a horizontal gene pool also accessible to organisms that do not inhabit plant surfaces. For example, the plasmid pEL60 recently sequenced from *E. amylovora* (Foster et al., 2004) is a 60-kb Inc L/M plasmid that is highly similar to the plasmid pCTX-M3 from the opportunistic human pathogen *Citrobacter freundii* (Mierzejewska et al., 2007). While pEL60 encodes mostly plasmid-specific replication and transfer functions, pCTX-M3 contains two additional insertions totalling 29.6 kb that encode several antibiotic resistance genes (Mierzejewska et al., 2007). Both *E. amylovora* and *C. freundii* are enteric bacteria, although inhabitants of distinctly different ecological niches. Nevertheless, the placement of two similar plasmids in these disparate organisms indicates the breadth of the horizontal gene pool accessible to *E. amylovora*.

Several recent examples uncovered in sequencing projects retrospectively demonstrate the movement of one to many genes between partners including a plant pathogenic bacterial species. For example, the IncP-1β plasmid pBP136 from the human pathogen *Bordetella pertussis* encodes two open reading frames (orf1 and orf2) that are not present in other IncP-1β plasmids and are highly similar with genes XF1597 and XF1596 from *Xylella fastidiosa* (Kamachi et al., 2006). Likewise, the transfer region of the *X. fastidiosa* plasmid pXF51 is highly similar to regions found on plasmids including pIPO2 and pSB102, which were exogenously isolated from bacteria inhabiting the rhizosphere of wheat and alfalfa, respectively (Schneiker et al., 2001; Tauch et al., 2002). In fact, the rhizosphere appears to be a hotspot for gene transfer and could be a source accessible to plant pathogens for genetic determinants involved in host-microbe interactions.

Experimental examples of conjugative DNA transfer among plant pathogens in the environment

The evidence presented above links plant pathogenic bacteria with a global horizontal gene pool enabling the acquisition of varied DNA sequences with the potential to exert a pronounced effect on ecological fitness. Future genomic and plasmid sequencing endeavors will undoubtedly continue to identify varied sequences in plant pathogens that have been acquired or transferred to other sources. In this section, we will review evidence from physical examinations of conjugative gene transfer in the plant environment.

Gene transfer in the phyllosphere

The aerial habitat of plants colonized by microbes is termed the phyllosphere, and bacterial populations on leaves, for example, average 10^6 to 10^7 cells cm^{-2} (Lindow and Brandl, 2003). Bacterial cells are typically aggregated on leaf surfaces at microsites of nutrient accumulation, and these cell populations in close proximity present opportunities for plasmid transfer. Plasmid transfer on leaves has been demonstrated in many microcosm systems, and some studies have shown elevated transfer rates on leaves compared to transfer in culture (for examples, see Normander et al., 1998). A nice study demonstrating horizontal gene transfer in the environment was performed by Lilley and Bailey (1997a) and involved the introduction of a genetically marked strain of P. fluorescens onto sugar beet leaves and the subsequent tracking of acquisition of mercury-resistance plasmids from the indigenous microflora. This group has studied in detail the transfer dynamics of the large mercury-resistance plasmid pQBR103 in the phyllosphere, rhizosphere and in bacterial biofilms (Lilley and Bailey, 1997b). These and other examples of horizontal transfer on leaves have led groups to suggest that leaves are important breeding grounds for microbial diversity because of their potential to serve as hot spots for gene transfer (Lindow and Brandl, 2003).

Conjugative plasmid transfer on leaves appears to occur mostly in sites such as stomates where bacterial population densities are higher (Normander et al., 1998). This observation is also true regarding conjugal transfer in the rhizosphere (Molbak et al., 2003), and Espinosa-Urgel (2004) has hypothesized the dual involvement of cell density effects and quorum sensing-regulated events on the transfer processes. It is known that conjugative transfer functions are regulated by quorum sensing in organisms such as *Agrobacterium* and *Rhizobium* (He et al., 2003; Tun-Garrido et al., 2003), however, the role of quorum sensing in gene transfer events in other bacterial genera has not been demonstrated.

Conclusions

In this chapter, we have presented a brief summary of the current knowledge regarding the significance of plasmids in phytopathogenic bacteria. Details of the gene content of these plasmids indicate the strong involvement of these elements in critical disease processes and in survival during other phases of the life history of the pathogen (such as the epiphytic phase of *P. syringae*). It is not known what drives the maintenance of plasmids in bacteria. However, the carriage of particular types of plasmids in individual pathogen species (such as the pPT23A plasmid family of *P. syringae* or pEA29 by *E. amylovora*) suggests an intimate association. It is likely that ecological fitness effects direct the ultimate plasmid profile of an organism. Negative effects on fitness would prevent the fixation of a newly acquired plasmid. What plant pathologists are observing today represents but a brief snapshot into the evolutionary history of bacterial phytopathogens and does not incorporate any long-term knowledge of plasmid dynamics in these organisms. However, retrospective evidence and evidence from genomic analyses demonstrates the amazing ability of plasmids to sample the environment for new genes that could be beneficial for the host. The adoption of newly acquired genes by bacterial pathogens will continue to influence host–pathogen interactions.

The trading of genes among plant pathogenic bacteria is facilitated by plasmids, elements that, over an evolutionary time scale, have colonized most bacterial species. Plasmids can be unobtrusive without apparent negative effect on host fitness in some environments, and can also be absolutely essential to survival or to virulence in others. Work as a gene trader favours plasmids by maintaining them as integral parts of a bacterial genome. In addition, if gene acquisition enables a quantum leap by the bacterial host, the plasmid also moves into a new ecological niche along with the host. This new move, in turn, exposes the host and its plasmid(s) to new potential mating partners. And thus the gene trade continues…

References

Agrios, G.N. (1997). Plant Pathology, 4th edn (San Diego, USA: Academic Press).

Alarcón-Chaidez, F.J., Peñaloza-Vázquez, A., Ullrich, M., and Bender, C.L. (1999). Characterization of plasmids encoding the phytotoxin coronatine in *Pseudomonas syringae*. Plasmid 42, 210–220.

Alfano, J.R., and Collmer, A. (1996). Bacterial pathogens in plants: life up against the wall. Plant Cell 8, 1683–1689.

Arnold, D.L., Jackson, R.W., and Vivian, A. (2000). Evidence for the mobility of an avirulence gene, *avrPpiA1*, between the chromosome and plasmids of races of *Pseudomonas syringae* pv. *pisi*. Mol. Plant Pathol. 1, 195–199.

Arnold, D.L., Pitman, A., and Jackson, R.W. (2003). Pathogenicity and other genomic islands in plant pathogenic bacteria. Mol. Plant Pathol. 4, 407–420.

Backert, S., and Meyer, T.F. (2006). Type IV secretion systems and their effectors in bacterial pathogenesis. Curr. Opin. Microbiol. 9, 207–217.

Barash, I., and Manulis-Sasson, S. (2007). Virulence mechanisms and host specificity of gall-forming *Pantoea agglomerans*. Trends Microbiol. 15, 538–545.

Bender, C.L., Alarcón-Chaidez, F., and Gross, D.C. (1999). *Pseudomonas syringae* phytotoxins: mode of action, regulation, and biosynthesis by peptide and polyketide synthetases. Microbiol. Mol. Biol. Rev. 63, 266–292.

Boucher, C., Martinel, A., Barberis, P., Alloing, G., and Zischek, C. (1986). Virulence genes are carried by a megaplasmid of the plant pathogen *Pseudomonas solanacearum*. Mol. Gen. Genet. 205, 270–275.

Boucher, Y., Douady, C.J., Papke, R.T., Walsh, D.A., Boudreau, M.E.R., Nesbo, C.L., Case, R.J., and Doolittle, W.F. (2003). Lateral gene transfer and the origins of prokaryotic groups. Annu. Rev. Genet. 37, 283–328.

Brunings, A.M., and Gabriel, D.W. (2003). *Xanthomonas citri*: breaking the surface. Mol. Plant Pathol. 4, 141–157.

Caponero, A., Contesini, A.M., and Iacobellis, N.S. (1995). Population diversity of *Pseudomonas syringae* subsp. *savastanoi* on olive and oleander. Plant Pathol. 44, 848–855.

Cooksey, D.A. (1990). Genetics of bactericide resistance in plant pathogenic bacteria. Annu. Rev. Phytopathol. 28, 201–219.

Cooksey, D.A. (1994). Molecular mechanisms of copper resistance and accumulation in bacteria. FEMS Microbiol. Rev. 14, 381–386.

Coplin, D.L. (1989). Plasmids and their role in the evolution of plant pathogenic bacteria. Annu. Rev. Phytopathol. 27, 187–212.

Crespi, M., Messens, E., Caplan, A.B., van Montagu, M., and Desomer, J. (1992). Fasciation induction by the phytopathogen *Rhodococcus fascians* depends upon a linear plasmid encoding a cytokinin synthase gene. EMBO J. 11, 795–804.

Cubero, J., and Graham, J.H. (2002). Genetic relationship among worldwide strains of *Xanthomonas* causing canker in citrus species and design of new primers for their identification by PCR. Appl. Environ. Microbiol. 68, 1257–1264.

Chen, I., Christie, P.J., and Dubnau, D. (2005). The ins and outs of DNA transfer in bacteria. Science 310, 1456–1460.

Chiou, C.-S., and Jones, A.L. (1993). Nucleotide sequence analysis of a transposon (Tn5393) carrying streptomycin resistance genes in *Erwinia amylovora* and other Gram-negative bacteria. J. Bacteriol. 175, 732–740.

Christie, P.J. (2004). Type IV secretion: the *Agrobacterium* VirB/D4 and related conjugation systems. Biochim. Biophys. Acta 1694, 219–234.

Christie, P.J., Atmakuri, K., Krishnamoorthy, V., Jakubowski, S., and Cascales, E. (2005). Biogenesis, architecture, and function of bacterial type IV secretion systems. Annu. Rev. Microbiol. 59, 451–485.

da Silva, A.C., Ferro, J.A., Reinach, F.C., Farah, C.S., Furlan, L.R., Quaggio, R.B., Monteiro-Vitorello, C.B., Van Sluys, M.A., Almeida, N.F., Alves, L.M., et al. (2002). Comparison of the genomes of two *Xanthomonas* pathogens with differing host specificities. Nature 417, 459–463.

Davis, R.E., Dally, E.L., Jomantiene, R., Zhao, Y., Roe, B., Lin, S., and Shao, J. (2005). Cryptic plasmid pSKU146 from the wall-less plant pathogen *Spiroplasma kunkelii* encodes an adhesin and components of a type IV translocation-related conjugation system. Plasmid 53, 179–190.

Desomer, J., Dhaese, P., and Van Montagu, M. (1988). Conjugative transfer of cadmium resistance plasmids in *Rhodococcus fascians* strains. J. Bacteriol. 170, 2401–2405.

Dobrindt, U., Hochhut, B., Hentschel, U., and Hacker, J. (2004). Genomic islands in pathogenic and environmental microorganisms. Nat. Rev. Microbiol 2, 414–424.

Doolittle, R.F. (2005). Evolutionary aspects of whole-genome biology. Curr. Opin. Struct. Biol. 15, 248–253.

Eberhard, W.G. (1989). Why do bacterial plasmids carry some genes and not others? Plasmid 21, 167–174.

El Yacoubi, B., Brunings, A.M., Yuan, Q., Shankar, S., and Gabriel, D.W. (2007). In planta horizontal transfer of a major pathogenicity effector gene. Appl. Environ. Microbiol. 73, 1612–1621.

Espinosa-Urgel, M. (2004). Plant-associated *Pseudomonas* populations: molecular biology, DNA dynamics, and gene transfer. Plasmid 52, 139–150.

Foster, G.C., McGhee, G.C., Jones, A.L., and Sundin, G.W. (2004). Nucleotide sequences, genetic organization, and distribution of pEU30 and pEL60 from *Erwinia amylovora*. Appl. Environ. Microbiol. 70, 7539–7544.

Fukuda, H., Ogawa, T., Ishihara, K., Fujii, T., Nagahama, K., Omata, T., Inoue, Y., Tanase, S., and Morino, Y. (1992). Molecular cloning in *Escherichia coli*, expression, and nucleotide sequence of the gene for the ethylene-forming enzyme of *Pseudomonas syringae* pv. *phaseolicola* PK2. Biochem. Biophys. Res. Commun. 188, 826–832.

Gardan, L., Brault, T., and Germain, E. (1993). Copper resistance of *Xanthomonas campestris* pv. *juglandis* in French walnut orchards and its association with conjugative plasmids. Acta Horticult. 311, 259–265.

Gartemann, K.H., Kirchner, O., Engemann, J., Grafen, I., Eichenlaub, R., and Burger, A. (2003). *Clavibacter michiganensis* subsp. *michiganensis*: first steps in the understanding of virulence of a Gram-positive phytopathogenic bacterium. J. Biotechnol. 106, 179–191.

Genin, S., and Boucher, C. (2004). Lessons learned from the genome analysis of *Ralstonia solanacearum*. Annu. Rev. Phytopathol. 42, 107–134.

Gibbon, M.J., Sesma, A., Canal, A., Wood, J.R., Hidalgo, E., Brown, J., Vivian, A., and Murillo, J. (1999). Replication regions from plant-pathogenic *Pseudomonas syringae* plasmids are similar to ColE2-related replicons. Microbiology 145, 325–334.

Glass, N.L., and Kosuge, T. (1988). Role of indoleacetic acid lysine synthetase in regulation of indoleacetic acid pool size and virulence of *Pseudomonas syringae* subsp. *savastanoi*. J. Bacteriol. 170, 2367–2373.

Goethals, K., Vereecke, D., Jaziri, M., Van Montagu, M., and Holsters, M. (2001). Leafy gall formation by *Rhodococcus fascians*. Annu. Rev. Phytopathol. 39, 27–52.

Gonzalez, C.F., Pettit, E.A., Valadez, V.A., and Provin, E.M. (1997). Mobilization, cloning, and sequence determination of a plasmid-encoded polygalacturonase from a phytopathogenic *Burkholderia* (*Pseudomonas*) *cepacia*. Mol. Plant–Microbe Interact. 10, 840–851.

Grant, S.R., Fisher, E.J., Chang, J.H., Mole, B.M., and Dangl, J.L. (2006). Subterfuge and manipulation: Type III effector proteins of phytopathogenic bacteria. Annu. Rev. Microbiol. 60, 425–449.

Guidot, A., Prior, P., Schoenfeld, J., Carrere, S., Genin, S., and Boucher, C. (2007). Genomic structure and phylogeny of the plant pathogen *Ralstonia solanacearum* inferred from gene distribution analysis. J. Bacteriol. 189, 377–387.

Guo, M., Manulis, S., Mor, H., and Barash, I. (2002). The presence of diverse IS elements and an *avrPphD* homologue that acts as a virulence factor on the pathogenicity plasmid of *Erwinia herbicola* pv. *gypsophilae*. Mol. Plant–Microbe Interact. 15, 709–716.

Han, H.S., Koh, Y.J., Hur, J.S., and Jung, J.S. (2004). Occurrence of the *strA-strB* streptomycin resistance genes in *Pseudomonas* species isolated from kiwifruit plants. J. Microbiol. 42, 365–368.

He, X., Chang, W., Pierce, D.L., Seib, L.O., Wagner, J., and Fuqua, C. (2003). Quorum sensing in *Rhizobium* sp. strain NGR234 regulates conjugal transfer (*tra*) gene expression and influences growth rate. J. Bacteriol. 185, 809–822.

Hooykaas, P.J.J., and Beijersbergen, A.G.M. (1994). The virulence system of *Agrobacterium tumefaciens*. Annu. Rev. Phytopathol. 32, 157–181.

Hooykaas, P.J.J., Klapwijk, P.M., Nuti, M.P., Schilperoort, R.A., and Rorsch, A. (1977). Transfer of *Agrobacterium tumefaciens* Ti plasmid to avirulent agrobacteria and to *Rhizobium ex planta*. J. Gen. Microbiol. 98, 477–484.

Jackson, R.W., Athanassopoulos, E., Tsiamis, G., Mansfield, J.W., Sesma, A., Arnold, D.L., Gibbon, M.J., Murillo, J., Taylor, J.D., and Vivian, A. (1999). Identification of a pathogenicity island, which contains genes for virulence and avirulence, on a large native plasmid in the bean pathogen *Pseudomonas syringae* pathovar phaseolicola. Proc. Natl. Acad. Sci. 96, 10875–10880.

Jackson, R.W., Mansfield, J.W., Ammouneh, H., Dutton, L.C., Wharton, B., Ortiz-Barredo, A., Arnold, D.L., Tsiamis, G., Sesma, A., Butcher, D., et al. (2002). Location and activity of members of a family of *virPphA* homologues in pathovars of *Pseudomonas syringae* and *P. savastanoi*. Mol. Plant Pathol. 3, 205–216.

Joardar, V., Lindeberg, M., Schneider, D.J., Collmer, A., and Buell, C.R. (2005). Lineage-specific regions in *Pseudomonas syringae* pv. tomato DC3000. Mol. Plant Pathol. 6, 53–64.

Kamachi, K., Sota, M., Tamai, Y., Nagata, N., Konda, T., Inoue, T., Top, E.M., and Arakawa, Y. (2006). Plasmid pBP136 from *Bordetella pertussis* represents an ancestral form of IncP-1beta plasmids without accessory mobile elements. Microbiology 152, 3477–3484.

Kidambi, S.P., Sundin, G.W., Palmer, D.A., Chakrabarty, A.M., and Bender, C.L. (1995). Copper as a signal for alginate synthesis in *Pseudomonas syringae* pv. *syringae*. Appl. Environ. Microbiol. 61, 2172–2179.

Kim, J.F., Charkowski, A.O., Alfano, J.R., Collmer, A., and Beer, S.V. (1998). Transposable elements and bacteriophage sequences flanking *Pseudomonas syringae* avirulence genes. Mol. Plant–Microbe Interact. 11, 1247–1252.

Koonin, E.V., Makarova, K.S., and Aravind, L. (2001). Horizontal gene transfer in prokaryotes: Quantification and classification. Annu. Rev. Microbiol. 55, 709–742.

L'Abee-Lund, T.M., and Sorum, H. (2000). Functional Tn5393-like transposon in the R plasmid pRAS2 from the fish pathogen *Aeromonas salmonicida* subspecies salmonicida isolated in Norway. Appl. Environ. Microbiol. 66, 5533–5535.

Laurent, J., Barny, M.-A., Kotoujansky, A., Dufriche, P., and Vanneste, J.L. (1989). Characterization of a ubiquitous plasmid in *Erwinia amylovora*. Mol. Plant–Microbe Interact. 2, 160–164.

Lawley, T.D., Klimke, W.A., Gubbins, M.J., and Frost, L.S. (2003). F factor conjugation is a true type IV secretion system. FEMS Microbiol. Lett. 224, 1–15.

Lilley, A.K., and Bailey, M.J. (1997a). The acquisition of indigenous plasmids by a genetically marked pseudomonad population colonizing the sugar beet phytosphere is related to local environmental conditions. Appl. Environ. Microbiol. 63, 1577–1583.

Lilley, A.K., and Bailey, M.J. (1997b). Impact of plasmid pQBR103 acquisition and carriage on the phytosphere fitness of *Pseudomonas fluorescens* SBW25: burden and benefit. Appl. Environ. Microbiol. 63, 1584–1587.

Lindow, S.E., and Brandl, M.T. (2003). Microbiology of the phyllosphere. Appl. Environ. Microbiol. 69, 1875–1883.

Llop, P., Donat, V., Rodríguez, M., Cabrefiga, J., Ruz, L., Palomo, J.L., Montesinos, E., and López, M.M. (2006). An indigenous virulent strain of *Erwinia amylovora* lacking the ubiquitous plasmid pEA29. Phytopathology 96, 900–907.

Llosa, M., and de la Cruz, F. (2005). Bacterial conjugation: a potential tool for genomic engineering. Res. Microbiol. 156, 1–6.

Ma, Z., Smith, J.J., Zhao, Y., Jackson, R.W., Arnold, D.L., Murillo, J., and Sundin, G.W. (2007). Phylogenetic analysis of the pPT23A plasmid family of *Pseudomonas syringae*. Appl. Environ. Microbiol. 73, 1287–1295.

Maes, T., Vereecke, D., Ritsema, T., Cornelis, K., Thu, H.N.T., Van Montagu, M., Holsters, M., and Goethals, K. (2001). The *att* locus of *Rhodococcus fascians* strain D188 is essential for full virulence on tobacco through the production of an autoregulatory compound. Mol. Microbiol. 42, 13–28.

Mantengoli, E., and Rossolini, G.M. (2005). Tn5393d, a complex Tn5393 derivative carrying the PER-1 extended-spectrum beta-lactamase gene and other resistance determinants. Antimicrob. Agents Chemother. 49, 3289–3296.

Manulis, S., and Barash, I. (2003). *Pantoea agglomerans* pvs. *gypsophilae* and *betae*, recently evolved pathogens? Mol. Plant Pathol. 4, 307–314.

McGhee, G.C., and Jones, A.L. (2000). Complete nucleotide sequence of ubiquitous plasmid pEA29 from *Erwinia amylovora* strain Ea88: Gene organization and intraspecies variation. Appl. Environ. Microbiol. 66, 4897–4907.

McManus, P.S., Stockwell, V.O., Sundin, G.W., and Jones, A.L. (2002). Antibiotic use in plant agriculture. Annu. Rev. Phytopathol. 40, 443–465.

Mierzejewska, J., Kulinska, A., and Jagura-Burdzy, G. (2007). Functional analysis of replication and stability regions of broad-host-range conjugative plasmid CTX-M3 from the IncL/M incompatibility group. Plasmid 57, 95–107.

Molbak, L., Licht, T.R., Kvist, T., Kroer, N., and Andersen, S.R. (2003). Plasmid transfer from *Pseudomonas putida* to the indigenous bacteria on alfalfa sprouts: characterization, direct quantification, and *in situ* location of transconjugant cells. Appl. Environ. Microbiol. 69, 5536–5542.

Mor, H., Manulis, S., Zuck, M., Nizan, R., Coplin, D.L., and Barash, I. (2001). Genetic organization of the *hrp* gene cluster and *dspAE/BF* operon in *Erwinia herbicola* pv. *gypsophilae*. Mol. Plant–Microbe Interact. 14, 431–436.

Murillo, J., and Keen, N.T. (1994). Two native plasmids of *Pseudomonas syringae* pathovar tomato strain PT23 share a large amount of repeated DNA, including replication sequences. Mol. Microbiol. 12, 941–950.

Nagahama, K., Yoshino, K., Matsuloa, M., Sato, M., Tanase, S., Ogawa, T., and Fukuda, H. (1994). Ethylene production by strains of the plant-pathogenic bacterium *Pseudomonas syringae* depends upon the presence of indigenous plasmids carrying homologous genes for the ethylene-forming enzyme. Microbiology 140, 2309–2313.

Nomura, K., Melotto, M., and He, S.Y. (2005). Suppression of host defense in compatible plant-*Pseudomonas syringae* interactions. Curr. Opin. Plant Biol. 8, 361–368.

Nordström, K. (1993). Plasmid replication and maintenance. In Plasmids: a Practical Approach, K.G. Hardy, ed. (New York: Oxford University Press), pp. 1–38.

Normander, B., Christensen, B.B., Molin, S., and Kroer, N. (1998). Effect of bacterial distribution and activity on conjugal gene transfer on the phylloplane of the bush bean (*Phaseolus vulgaris*). Appl. Environ. Microbiol. 64, 1902–1909.

Ochman, H., Lawrence, J.G., and Groisman, E.A. (2000). Lateral gene transfer and the nature of bacterial innovation. Nature 405, 299–304.

Pérez-Martínez, I., Zhao, Y., Murillo, J., Sundin, G.W., and Ramos, C. (2008). Global genomic analysis of *Pseudomonas savastanoi* pv. savastanoi plasmids. J. Bacteriol. 190, 625–635.

Pezzella, C., Ricci, A., DiGiannatale, E., Luzzi, I., and Carattoli, A. (2004). Tetracycline and streptomycin resistance genes, transposons, and plasmids in *Salmonella enterica* isolates from animals in Italy. Antimicrob. Agents Chemother. 48, 903–908.

Powell, G.K., and Morris, R.O. (1986). Nucleotide sequence and expression of a *Pseudomonas savastanoi* cytokinin biosynthetic gene: homology with *Agrobacterium tumefaciens tmr* and *tzs* loci. Nucleic Acids Res. 14, 2555–2565.

Roberts, W.P., Tate, M.E., and Kerr, A. (1977). Agrocin 84 is a 6-N-phosphoramidate of an adenine nucleotide analogue. Nature 265, 379–381.

Sarkar, S.F., Gordon, J.S., Martin, G.B., and Guttman, D.S. (2006). Comparative genomics of host-specific virulence in *Pseudomonas syringae*. Genetics 174, 1041–1056.

Sato, M., Watanabe, K., Yazawa, M., Takikawa, Y., and Nishiyama, K. (1997). Detection of new ethylene-producing bacteria, *Pseudomonas syringae* pvs. *cannabina* and *sesami*, by PCR amplification of genes for the ethylene-forming enzyme. Phytopathology 87, 1192–1196.

Schneiker, S., Keller, M., Droge, M., Lanka, E., Puhler, A., and Selbitschka, W. (2001). The genetic organization and evolution of the broad host range mercury resistance plasmid pSB102 isolated from a microbial population residing in the rhizosphere of alfalfa. Nucleic Acids Res. 29, 5169–5181.

Sesma, A., Sundin, G., and Murillo, J. (1998). Closely related replicons coexisting in the phytopathogen *Pseudomonas syringae* show a mosaic organization of the replication region and altered incompatibility behavior. Appl. Environ. Microbiol. 64, 3948–3953.

Sesma, A., Aizpún, M.T., Ortiz, A., Arnold, D., Vivian, A., and Murillo, J. (2001). Virulence determinants other than coronatine in *Pseudomonas syringae* pv. *tomato* PT23 are plasmid-encoded. Physiol. Mol. Plant Pathol. 58, 83–93.

Simpson, A.J., Reinach, F.C., Arruda, P., Abreu, F.A., Acencio, M., Alvarenga, R., Alves, L.M., Araya, J.E., Baia, G.S., Baptista, C.S., et al. (2000). The genome sequence of the plant pathogen *Xylella fastidiosa*. The *Xylella fastidiosa* Consortium of the Organization for Nucleotide Sequencing and Analysis. Nature 406, 151–159.

Stokes, H.W., Elbourne, L.D., and Hall, R.M. (2007). Tn*1403*, a multiple-antibiotic resistance transposon made up of three distinct transposons. Antimicrob. Agents Chemother. 51, 1827–1829.

Sundin, G.W. (2002). Distinct recent lineages of the *strA-strB* streptomycin-resistance genes in clinical and environmental bacteria. Curr. Microbiol. 45, 63–69.

Sundin, G.W. (2007). Genomic insights into the contribution of phytopathogenic bacterial plasmids to the evolutionary history of their hosts. Annu. Rev. Phytopathol. 45, 129–151.

Sundin, G.W., and Bender, C.L. (1995). Expression of the *strA-strB* streptomycin resistance genes in *Pseudomonas syringae* and *Xanthomonas campestris* and characterization of IS*6100* in *X. campestris*. Appl. Environ. Microbiol. 61, 2891–2897.

Sundin, G.W., and Bender, C.L. (1996). Dissemination of the *strA-strB* streptomycin-resistance genes among commensal and pathogenic bacteria from humans, animals, and plants. Mol. Ecol. 5, 133–143.

Sundin, G.W., and Murillo, J. (1999). Functional analysis of the *Pseudomonas syringae rulAB* determinant in tolerance to ultraviolet B (290–320 nm) radiation and distribution of *rulAB* among *P. syringae* pathovars. Environ. Microbiol. 1, 75–88.

Sundin, G.W., Demezas, D.H., and Bender, C.L. (1994). Genetic and plasmid diversity within natural populations of *Pseudomonas syringae* with various exposures to copper and streptomycin bactericides. Appl. Environ. Microbiol. *60*, 4421–4431.

Sundin, G.W., McGhee, G.C., Foster, G.C., and Jones, A.L. (2004). Genetic analysis of the ubiquitous plasmid pEA29 and two new *Erwinia amylovora* plasmids. Acta Horticult. *704*, 423–430.

Tauch, A., Krieft, S., Kalinowski, J., and Puhler, A. (2000). The 51,409-bp R-plasmid pTP10 from the multiresistant clinical isolate *Corynebacterium striatum* M82B is composed of DNA segments initially identified in soil bacteria and in plant, animal, and human pathogens. Mol. Gen. Genet. *263*, 1–11.

Tauch, A., Schneiker, S., Selbitschka, W., Puhler, A., van Overbeek, L.S., Smalla, K., Thomas, C.M., Bailey, M.J., Forney, L.J., Weightman, A., et al. (2002). The complete nucleotide sequence and environmental distribution of the cryptic, conjugative, broad-host-range plasmid pIPO2 isolated from bacteria of the wheat rhizosphere. Microbiology *148*, 1637–1653.

Tauch, A., Schluter, A., Bischoff, N., Goesmann, A., Meyer, F., and Puhler, A. (2003). The 79,370-bp conjugative plasmid pB4 consists of an IncP-1beta backbone loaded with a chromate resistance transposon, the *strA-strB* streptomycin resistance gene pair, the oxacillinase gene bla(NPS-1), and a tripartite antibiotic efflux system of the resistance-nodulation-division family. Mol. Genet. Genomics *268*, 570–584.

Teyssier-Cuvelle, S., Oger, P., Mougel, C., Groud, K., Farrand, S.K., and Nesme, X. (2004). A highly selectable and highly transferable Ti plasmid to study conjugal host range and Ti plasmid dissemination in complex ecosystems. Microb. Ecol. *48*, 10–18.

Tun-Garrido, C., Bustos, P., Gonzalez, V., and Brom, S. (2003). Conjugative transfer of p42a from *Rhizobium etli* CFN42, which is required for mobilization of the symbiotic plasmid, is regulated by quorum sensing. J. Bacteriol. *185*, 1681–1692.

van Elsas, J.D., McSpadden Gardener, B.B., Wolters, A.C., and Smit, E. (1998). Isolation, characterization, and transfer of cryptic gene-mobilizing plasmids in the wheat rhizosphere. Appl. Environ. Microbiol. *64*, 880–889.

Vivian, A., Gibbon, M.J., and Murillo, J. (1997). The molecular genetics of specificity determinants in plant pathogenic bacteria. In The gene-for-gene relationship in plant parasite interactions, I.R. Crute, E.B. Holub, and J.J. Burdon, eds. (Wallingford, U.K.: CAB International), pp. 293–328.

Vivian, A., Murillo, J., and Jackson, R.W. (2001). The role of plasmids in phytopathogenic bacteria: mobile arsenals? Microbiology *147*, 763–780.

Watanabe, K., Nagahama, K., and Sato, M. (1998). A conjugative plasmid carrying the *efe* gene for the ethylene-forming enzyme isolated from *Pseudomonas syringae* pv. *glycinea*. Phytopathology *88*, 1205–1209.

White, C.E., and Winans, S.C. (2007). Cell–cell communication in the plant pathogen *Agrobacterium tumefaciens*. Philos. Trans. R. Soc. B-Biol. Sci. *362*, 1135–1148.

Zhao, Y.F., Ma, Z.H., and Sundin, G.W. (2005). Comparative genomic analysis of the pPT23A plasmid family of *Pseudomonas syringae*. J. Bacteriol. *187*, 2113–2126.

Bioinformatics Aspects of High-throughput Sequencing Technology

Dan MacLean and David J. Studholme

Abstract

Recently developed massively high-throughput sequencing technologies look set to revolutionize the way that we practise research in microbial and plant sciences, and will necessitate greater investment in data management and analysis than in the past. In this chapter, we describe some of the informatics issues and solutions that we have come across in the early stages of using these new technologies both for resequencing and *de novo* sequencing. As well as issues directly connected with the infrastructure and technology of high-throughput sequencing, we present a brief tour of many more traditional sequence analysis tools that will continue to be relevant in the context of larger datasets.

Introduction

In the last decade, the study of microbiology has been revolutionized by the complete sequencing of hundreds of bacterial genomes. This situation was largely brought about by gradual improvements of sequencing technology and a steady decrease in costs. Now, we find ourselves on the brink of a new revolution in DNA sequencing technology; it is now possible to generate a Gigabase (1×10^9, or 1000 million bases) of sequence in a few days at a cost of around only $10 000 (USD) with new massively high-throughput sequencing technologies. These new technologies use radically different methodologies to traditional 'Sanger' sequencing and leverage modern automation and computing capabilities as well as advances in molecular biochemistry to generate huge amounts of sequence. The most prominent of the currently available platforms are 454's pyrosequencing (Margulies *et al.*, 2005), and Ilumina's Solexa platform (http://www.illumina.com/), which will be described briefly later. Such high-perfomance technology is not cheap and the initial capital costs of these machines and the accompanying support technology may be too high to make it possible for many academic research departments to purchase their own sequencing instruments. Naturally, commercial service providers are increasingly offering massively high-throughput sequencing at affordable rates making the technologies available to a wider market. Together these factors have caused a dramatic drop in the 'per base' cost of sequencing compared to more traditional methods, seemingly making the acquisition of large amounts of sequence, and by extension, genomes much cheaper and easier. However, there is a significant caveat to this assertion: that is the fundamental differences in the new technologies lead to much shorter read lengths than traditional sequencing technologies. Whereas traditional 'Sanger' sequencing routinely generates sequence reads of 500 nt or greater, 454 pyrosequencing generates reads of about 100–250 nt, whilst the Solexa method produces sequence reads of less than 50 nt. Such short read-lengths present a paradigm shift in the nature of the challenges for bioinformatic exploitation of the data.

Our laboratory has been an early adopter of massively high-throughput sequencing in the field of plant and microbial sciences. In this article, we will describe some of the practical bioinformatics issues that we have overcome in exploiting high-throughput sequencing technology to further our understanding of plant–pathogen interactions. We will focus on the applied rather than theoretical or mathematical aspects. In particular we will attempt to convey some idea of the requirements in terms of information technology (IT) infrastructure and bioinformatics resources and expertise; we hope this will be helpful to microbiology and plant science laboratories that are considering using high-throughput sequencing technologies in their own research programmes. We will also very briefly summarize some of our preliminary sequence data, though the data and their biological implications will be published in more detail elsewhere.

454 Life Sciences pyrosequencing technology

One of the first high-throughput sequencing technologies to become commercially available was that developed by 454 Life Sciences. The 454 GS20 platform was capable of generating millions of bases of nucleotide sequence in a few hours and has been successfully applied to both resequencing and *de novo* sequencing of bacterial and viral genomes, as well as SAGE and cDNA sequencing and even metagenomic sequencing of environmental samples. A comprehensive database of peer-reviewed publications based on 454 sequencing technology is maintained on the 454 Life Sciences website (http://www.454.com) and currently contains over 70 publications. The lengths of sequence reads from the 454 GS20 platform were up to about 100nt; however, the next generation of 454 sequencers, the GS FLX, are expected to generate reads of up to 250 nt.

In November 2006, staff at 454 published in *Nature* (Green *et al.* 2006) the first million base pairs of the Neanderthal Genome, and initiated the Neanderthal Genome Project to complete the sequence of the Neanderthal Genome by 2009. The 454 sequencing technology is based on a form of sequencing-by-synthesis known as pyrosequencing; the mechanism is described in detail in the manuscript of Margulies *et al.* (2005). In our laboratory we have successfully employed 454 sequencing to sequence libraries of ESTs from a eukaryotic phytopathogen (A. Rougon and J. Jones, unpublished) and numerous libraries of short non-protein-coding RNA species from plants (D. Baulcombe *et al.*, unpublished) and from the single-celled green alga, *Chlamydomonas rheinhadtii* (Molnar *et al.*, 2007).

The Solexa sequencing technology

The Solexa sequencing platform, provided commercially by Illumina (http://www.illumina.com), uses a novel reversible terminator-based sequencing chemistry to sequence millions of DNA molecules in a massively parallel fashion. The wet-lab component of this process is carried out on a machine called the Solexa 1G Genetic Analyzer, a complex microfluidics and molecular synthesis machine that outputs raw image files of measured fluorescence taken during synthesis of a fluorescently labelled DNA fragment. Prepared genomic DNA or cDNA is first fragmented randomly (e.g. by nebulization). Fragments are then attached to a Flow Cell surface, on which they undergo solid-phase amplification before sequencing using a 'four-colour DNA sequencing-by-synthesis technology'. At each of these cycles a fluorescence image is taken, representing a base read. The number of cycles used for reading sets the upper limit for the length of each eventual sequence read. In our laboratory, we routinely run 36 sequencing cycles, though it is possible to increase this number and therefore the overall read length at the expense of the accuracy of detection of the identity of the later bases in the read. The end product from a sequencing run on the Genetic Analyzer, and the actual 'raw data' in an experimental sense, is an extremely large collection of image files and accompanying data and metadata that typically comprises around 800 Gb of data. These files are functionally analogous to the chromatograms from ABI type capillary sequencers. The whole dataset must be removed from the Genetic Analyzer before beginning the next sequencing run. The IT challenge of safe, fast, reliable transfer and storage of this amount of data is a non-trivial one itself, which will be discussed in more detail later. Briefly, we transfer the raw data onto a 13Tb NetApp fibre-channel storage device (http://www.netapp.com) using File Transfer Protocol (FTP). This file storage device is mounted on the filesystem of our Blade cluster via Network File System (NFS).

The Solexa base-calling pipeline involves the execution of numerous concurrent processes, each of which requires a high level of read and write access to data held on the hard disk(s). Moderately high levels of disk activity are sustained for long periods of time, and, if applied to standard commodity disk hardware, would present a highly restrictive bottleneck. Further, it might lead to disk failures at an unacceptable frequency. Disk failures can lead to downtime at best, and unrecoverable data-loss at worst. Therefore a highly resilient, high-performance, and highly redundant disk storage system is desirable.

The process of base-calling involves distilling the large dataset (largely consisting of image files) into a set of sequence reads and accompanying quality scores; base-calling is performed by Illumina's proprietary analysis pipeline and will not be discussed in detail here. However, it is worth mentioning that the base-calling pipeline, though computationally intensive, is amenable to parallelization and we have successfully deployed it on a modest Linux-based computer cluster consisting of 15 dual processor IBM DS10 Blades, each equipped with two Opteron 64 bit processors and 4 Gb RAM, with job management being provided by Platform's LSF lsmake utilities (http://www.platform.com). Using this cluster, we are able to perform base-calling on a billion bases in a few hours. The pipeline will also run happily on fewer computers with a corresponding increase in run time, though our experience has shown that base-calling as quickly as possible is best.

Raw data from the Solexa Genetic Analyzer

Given the large size of the raw datasets, disk space is a serious issue. With one or two 800 Gb raw datasets being generated every week, we have found that it is simply not feasible to provide live storage of raw data for more than a few weeks. Either the image data has to be archived or discarded. Options for archiving range from very expensive professional archiving services, such as tape backups stored in physically secured units, to simply copying the data onto commodity hardware such as portable USB 1 Tb hard-drive, which can be bought for a few hundred dollars.

Of course, hard drives do occasionally fail, and are not immune to physical disasters such as fire, flood, and even theft. A management decision has to be made about how valuable a given dataset is and weighed against the costs and risks. However, once base-calling is completed, there is really little reason to keep the original image data, so simply archiving on USB drives will probably be sufficient in most cases. In situations where the data originate from a particularly valuable or difficult biological sample to repeat it may be that, in the case of disaster, the best course of action might be simply to repeat the sequencing run.

Sequence data

The current 454 and Solexa technologies also generate quite different amounts of sequence. A typical 454 GS FLX run is currently capable of generating around 250 000 reads of 250 nt length, a sequencing depth of about 10 x over a 6 Mbase genome. Conversely the Solexa method can generate 40 million 36 nt reads resulting in a sequencing depth of about 240x coverage.

Whatever the method by which sequence is obtained, the end product of base-calling is essentially a simple text file containing a set of short sequence reads accompanied by quality scores. The quality scores are derived from fluorescence intensities and provide an estimate of the confidence with which each base has been called.

Characterization of error distribution in Solexa reads

The company that produces the Solexa technology, Illumina, market the machine primarily as a tool for resequencing applications but the sheer amount of sequence it generates makes it tempting to use it in *de novo* genome sequencing. A major question then must be 'How accurate are the sequences produced?' The Solexa base calling software generates a PHRED equivalency score as it creates base calls that are supposed to reflect the amount of error in the sequence. Some reports and our own experience suggested that the true error rate is somewhat higher than these scores suggest, at least under some circumstances. To determine the error rates in Solexa reads generated in our laboratory empirically, we used genomic sequence reads from the *Pseudomonas syringae* pv. *syringae* B728A genome, a previously sequenced organism, and assessed the frequency and position of errors in the reads. From a collection of 11 824 049 reads, we collected the 730 845 reads (about 6%) with the highest average quality score over the whole read. Each read was aligned to the genome of B728A to find the best possible match (*i.e* the match with the fewest errors). The bases in the read were then compared with the genome at that position to identify which positions contained errors. The frequency of errors can be seen in Fig. 15.1. Error rates are lowest at the 5′ end of the read. Up to the 17th base in the read the error rate is low: under 3.5 % of sequences have an error at each given position. After this the percentage of base errors rises dramatically; at the 30th position 14% of bases have an error, and by the 36th position, the final base, 42% of all reads have this base called incorrectly. The pattern indicates that Solexa reads are extremely error prone at the 3′ end, a serious limitation when it comes to *de novo* assembly. The problem is not insurmountable though; quality scores can be used to filter reads to increase the accuracy of each read or set, either by selecting only the most accurate or trimming the lower quality ends. We have found that different strategies work well for different assembly protocols and will be discussed below. Also, it should be pointed out that the error distribution in this particular dataset might not be representative of other datasets, especially given that there have been refinements to protocols, software, and hardware since that dataset was generated.

Sequencing and assembly of bacterial genomes with Solexa reads: some case studies

With its great depth and accuracy, the most obvious application for the Solexa technology is the exploration of genetic diversity by 'resequencing' of genomes for which we already have a reference genome. However, there is a significant demand for rapid and cheap 'de novo' sequencing of previously unsequenced genomes. For example, in studying interactions between phytopathogenic bacteria and their hosts, it would be very useful to rapidly and cheaply determine the repertoire of genes encoding effectors. Effectors are genes whose products facilitate the infection of a host by a bacterium. Currently it can take many months to identify candidate effectors in the genome of a bacterial pathogen using genetic screens; however,

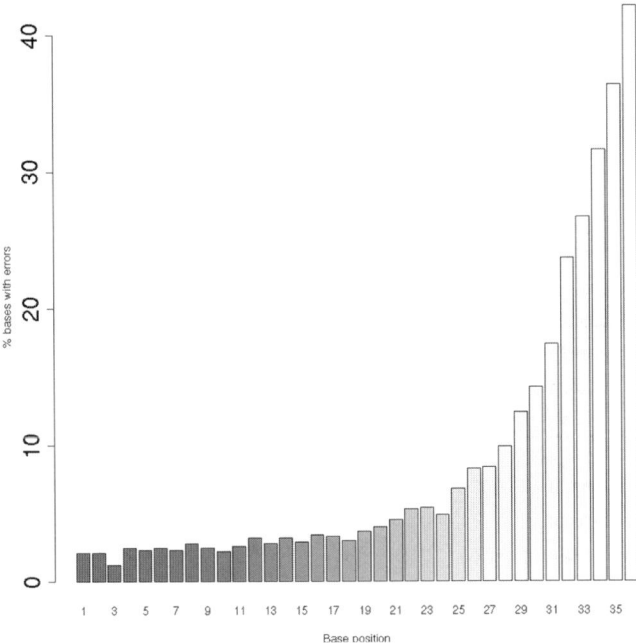

Figure 15.1 The frequency of erroneous base-calls for each cycle of a Solexa read from the genome of *Pseudomonas syringae* pv. *syringae* B728a.

sequencing the pathogen's complete genome, and performing sequence similarity searches against the catalogue of known effectors (e.g. those in the Hop database http://www.pseudomonas-syringae.org) provides a straightforward way to identify such candidates extremely quickly (and cheaply). In fact, for an application such as this, it is not even strictly necessary for the genome to be completely 'finished' (i.e. all the gaps between contigs closed to form a single complete chromosome); nor is it necessary for the sequence to be 100% accurate. Much of our work has focussed on this non-standard application of these sequencing technologies and we will discuss the challenges and methodologies that we have encountered here. The process can be divided into two phases, assembly and genomic feature hunting, or annotation. Once raw sequence has been generated, the first step is to assemble the reads into contigs.

Assembly of short sequence reads

The process of going from individual sequence reads (e.g. shotgun Sanger reads of about 500 bp) to longer contigs and complete chromosomes is served by well established sequence assembly algorithms and mature software applications e.g. the TIGR assembler (Pop and Kosack, 2004), PHRAP/PHRED (Chen and Skiena, 2000; http://bozeman.mbt.washington.edu/), and ARACHNE (Batzoglou et al., 2002), amongst others. However, these algorithms are unable to assemble contigs from very short reads, thus there has arisen an urgent need for specialized assembly algorithms capable of handling very short reads and very large quantities of reads. This need is rapidly being filled by open source software from several academic research groups: Warren et al. (2007) recently developed SSAKE. Subsequent to publication, they have developed the updated SSAKE 1.1, which they claim has enhanced handling of sequencing errors. Jeck et al. (2007) have created a SSAKE derivative called VCAKE that can utilize different read lengths and errors. Zerbino and Birney (personal communication) have developed Velvet (http://www.ebi.ac.uk/~zerbino/velvet/). All of these applications can take as input many millions of short reads (e.g. 36 mers) and assemble these into a series of contigs, though the methods by which they do this are very different. SSAKE and its derivatives, SSAKE 1.1 and VCAKE work by brute-force extension of overlapping k-mers (fragments of reads) to create contigs, relying on sheer depth of coverage whereas Velvet uses a more analytic, mathematical approach and employs a De Bruijn graph model to create contigs.

The SSAKE program (short sequence assembly by Kmer search and 3' extension) and its derivatives, (SSAKE 1.1 and VCAKE 0.5) are based on simplistic yet somewhat effective brute force algorithms. Essentially these algorithms function by creating a table of unique k-mers (read fragments of a specific length). For each read, each possible 3' subsequence is compared with the k-mers in order of abundance until the most abundant sequence that has a perfect match with the sub-sequence is found. The k-mer is then extended by the sequence. Such a method is only useful in sets of sequence with very low error-rates, meaning that Solexa datasets, with their inherent high error rate need to be filtered very stringently with strict quality controls to allow reliable contig generation. In its favour though, SSAKE is capable of handling reads of variable length, making a sliding window method of trimming reads practicable. Sliding window methods of trimming poor Solexa reads are particularly useful since sequence quality deteriorates closer to the 3' end of the read and maximum sequence can be retained if the 3' end is too poor to be assembled.

SSAKE 1.1 modifies the strict approach of its predecessor by allowing the growing contig to be trimmed at its 3' end and repeat its search, allowing for errors in sequencing to be accounted for. VCAKE 0.5 (Verified consensus assembly by K-mer extension) handles sequencing error by enforcing a consensus rule. For each possible k-mer extension it considers the next base as suggested by all reads that overlap the k-mer and extends the k-mer by the most frequently occurring base.

The Velvet assembly software implements a De Bruijn graph as its underlying model for mapping out contigs from individual sequence reads. In this algorithm, overlaps between sequence reads are identified and these overlaps are used to make an association or 'edge' between the reads. Edges are given a direction, indicating that one read is 5′ to another, thus the edges become a path down which the algorithm can walk to construct contigs. It is from these associated reads that a De Bruijn graph is formed. A De Bruijn graph is a particular type of network structure in which each read is visited only once and thus the order of reads for assembly into the contig is preserved.

In practice, these softwares generate very different assemblies when run on the same datasets. To show the efficacies of each of the softwares, we carried out a small test run using a variety of parameter settings for each program and a real Solexa dataset generated from *P. syringae* pv. *tomato* T1 genomic DNA. From a set of 71 254 789 reads (36mers) we selected the 15 723 622 reads (about 22%) with the highest mean quality scores. This bacterium is believed to have a genome size of approximately 6 Mb. We attempted to assemble this dataset using each of the available programs. The best result for SSAKE over a wide range of parameter values was an assembly of 1 413 974 contigs with average length of only 161 nt. The longest contig was only 8797 nt and the sum of all contig lengths was a massive 228 910 620 nt. Such a figure is many times the size of the genome and indicates that the assembly is riddled with errors and false extensions, rendering the contigs extremely unreliable to say the least. The later version of this software, SSAKE 1.1, has in-built error-handling features and performed much better, generating 827 993 contigs of average length nearly twice that of the earlier version at 303 nt and generating its longest contig of 73 356 nt. The sum of all contig lengths was still very large at 251 030 341, again implying that there are numerous unreliable sequence extensions. VCAKE fares somewhat better in some respects than both of these, generating 1 663 382 contigs of average length 67 nt with biggest 4958 nt and sum of contig lengths at 111 785 105. The De Bruijn graph based assembler Velvet generated 2482 contigs of average length 2425 nt with longest contig of 27 738 nt and, tellingly, a summed contig length of 6 020 014 nt, which is very close to the actual genome size of *P. syringae*. Velvet then seems to generate the most useable assemblies.

Computational requirements of assembly software

The single most important limitation on these algorithms in an IT sense is the amount of memory required; the sequences must be loaded into memory all at once and since Solexa runs can typically generate many gigabases of sequence, then many gigabytes of computer memory will be needed. In our experience the SSAKE and VCAKE algorithms will use about 6–7 Gb of memory to carry out a typical assembly of a 1 Gb read file. Memory usage by Velvet is somewhat more variable, and for a similar input will typically use 16 Gb of memory though it can use over 32 Gb of memory depending on assembly parameters and the actual sequence.

All assembly tasks are limited by processor speed, irrespective of the amount of memory. Thankfully limitations in raw compute power can often be balanced by allowing a process more time to complete, so super-processor machines can prove to be a false economy. A single, somewhat modestly specced, dual processor IBM DS10 Blade, equipped with two Opteron 64 bit processors and 32 Gb of RAM, managed to complete such jobs as described above within about 6 h when running SSAKE, in around 24 h when running VCAKE and Velvet assemblies typically take less than an hour, providing that sufficient RAM is available. We consistently found that the Velvet sequence assembly software gave superior performance to that of SSAKE on the basis of all metrics that we investigated and on all datasets. The following discussion is limited to the results of assemblies using Velvet.

Automatic annotation of the genome assembly

The last phase of *de novo* sequencing is the detection of features of the genome, such as genes and transcription factor binding sites, and annotating their position on the new contigs. The end product of assembly is a multi-sequence Fasta format file containing sequences of each of the several thousand contigs. Although this is a valuable dataset, in this form it is of limited use to our colleagues in the 'wet' laboratory. In order to exploit the data and inform their experiments they need to be able to see the data represented in a way that makes intuitive sense; often this will mean some level of interaction or browsing and interrogation of the sequence data. Interrogation of the data usually begins with sequence similarity searches, such as BLAST, that compare the assembly sequence with known sequences, such as those of genes of interest. It is quite straightforward for a competent bioinformatician to convert a multi-sequence Fasta file into a searchable BLAST database and to provide a simple web-based user interface for the BLAST tool using, for example, Perl-CGI scripts, assuming that a CGI-enabled webserver is available. Using such a tool, the biologist should be able to identify contigs showing sequence similarity to their genes of interest, enabling them to design PCR primers for cloning and/or gene knock-outs.

Visualizing and presenting genome assemblies

In our experience, bench biologists find it very helpful to visualize the sequence data in the form of a genome browser. At a basic level these are systems for creating complex renderings of the relationships between sequence entities such as BLAST hits and contigs and presenting them in an interactive format. Fortunately, this process is facilitated by a combination of several excellent open source informatics tools: (i) the Generic Genome Browser (Stein *et al.*, 2002), (ii) the Apache webserver, (iii) the Perl scripting language and the BioPerl extensions to the Perl language and (iv) the MySQL Relational Database Management System (RDBMS). There are other alternative solutions, for example the Ensembl browser (Stalker *et al.*, 2004) could replace the Generic Genome Browser, and there are numerous alternative webservers and RDBMS; but this is probably the most commonly encountered combination of tools for the task in actual use in academic research environments. The task for the practical bioinformatician is to generate some preliminary annotation of the sequence in an automated fashion using a range of sequence analysis tools, and build this annotation into a relational database that is accessible to the Generic Genome Browser software.

The Gbrowse application consists of a bundle of Perl CGI scripts and modules that are deployed on a webserver and interact with a database of DNA sequence and sequence features. The bioinformatician's job is to build such a database of genomic annotation. Central to the process of incorporating sequence feature data into the annotation database is the GFF file format specification (http://www.sanger.ac.uk/Software/formats/GFF). This file format provides a simple yet powerful way of describing sequence features including their start and end positions, plus or minus strand, reading frame, score. Once annotations are expressed in GFF format they can be uploaded into the genome browser database using scripts that are supplied with BioPerl. The steps for creation of a Gbrowse genome browser are well documented elsewhere (Stein *et al.*, 2002) so instead we will deal with strategies for creating the annotations for loading into the browser itself.

Sequence assembly in practice

In both methods there are several parameters, whose optimal values cannot be readily predicted *ab initio*. Therefore, our approach is to perform a parameter scan: we run the assembly multiple times, each time using a different combination of parameter values, and then select the 'best' resulting assembly empirically. How do we choose the 'best' assembly? There are many metrics that could be used to measure the quality of an assembly. These include the number of contigs, mean contig-length, median contig-length, length of the longest contig, sum of lengths of all contigs, and the percentage identity between the contigs and the 'true' sequence. Of course, the last of these metrics can only be measured if we have a previously sequenced reference genome; for *de novo* sequencing we do not have such a reference sequence, by definition. All of the other metrics can be readily calculated from a truly *de novo* sequence assembly. In our laboratory, we have to date generated genome sequence assemblies for several previously unsequenced phytopathogenic bacteria: *Xanthomonas campestris* pv. *musearum*, *X. campestris* pv. *vasculorum* and *P. syringae* pv. *syringae* T1, from Solexa sequences alone. We also generated genome sequence assemblies from Solexa sequence reads for the previously sequenced *P. syringae* pv. *syringae* B728a (see Fig. 15.2). These measures will mean little if the actual gene content of the assembled contigs is poor. To assess the efficiency of reconstruction and retrieval of known genes in a known sequence we carried out BLAST searches of our 'best' B728A assembly to identify known gene sequences. The BLAST searches identified that 99 % of B728A genes (5072 / 5090 cDNA sequences) were present in the contigs and that over 99 % of these genes had 100% identity to the original sequence (Fig. 15.2), showing that gene retrieval can in fact be very good. To assess the integrity of the assembly and determine which areas of the genome may have been missed, the contigs were compared with the whole genome sequence, again with BLAST. The majority of the genome was covered by contig fragments that matched the genome, although some gaps are apparent. The small gaps that did occur in the sequence were spread evenly around the genome (Fig. 15.2) and do not appear to have been a result of failure of read coverage over specific parts of the genome, such as repeat sequences. Seeded assembly of the entire genome with sequences from BLAST alignment of contigs with the single whole genome sequence of B728a resulted in generating a sequence that aligned over 100% of the genome with small gaps at seemingly random intervals (Fig. 15.2) resulting in continuous coverage of over 84% of the B728A genome.

Identification of protein-coding genes

The most frequently asked question of any new genome sequence will, of course, be something

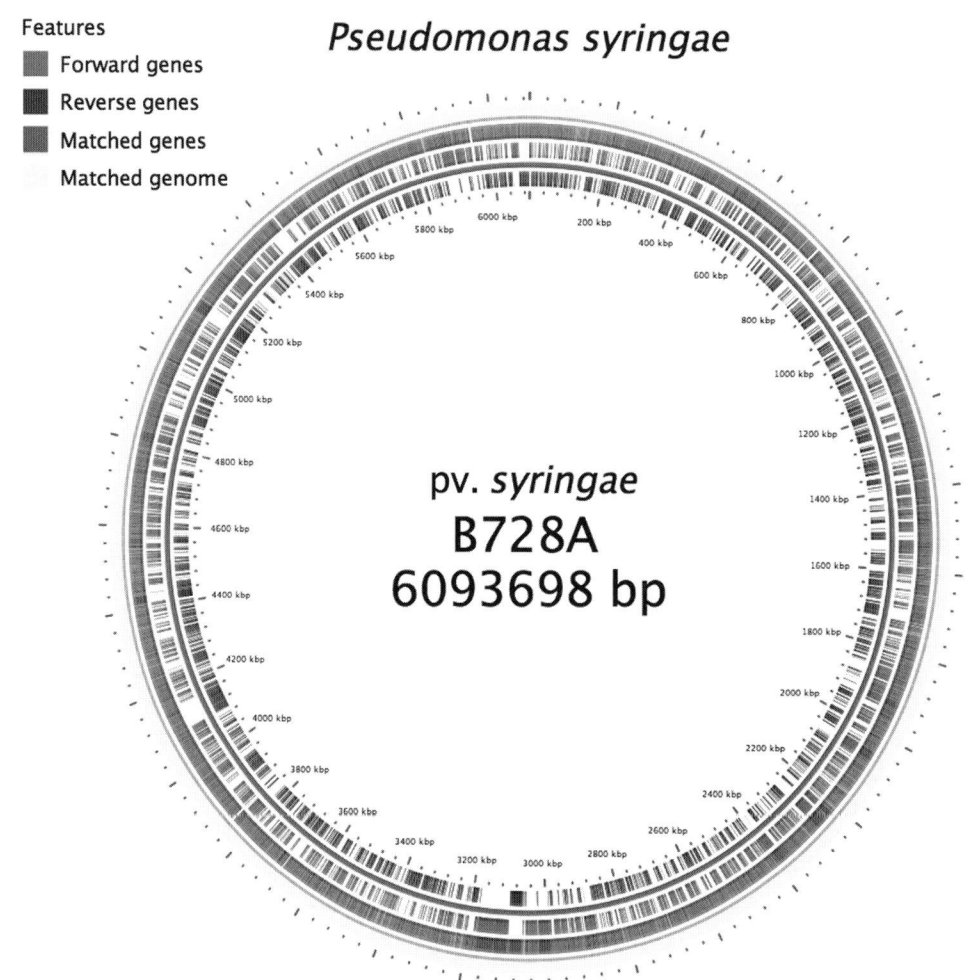

Figure 15.2 The genome of *Pseudomonas syringae* pv. *syringae* B728a assembled from 36 nt Solexa reads, compared with the previously published complete genome sequence. Track legend, from inside to outside: Forward orientation genes; Reverse orientation genes; Matched genes; Matched genome.

along the lines of, 'What genes has it got in it?'. Answering this question allows us to take advantage of the many technologies and resources developed over recent years for sequences derived from traditional Sanger sequencing projects. Current approaches tend to rely on some sort of knowledge of what a gene or protein is, and rely either on identifying areas within the contigs that have significant sequence identity to existing genes or proteins or in determining the positions of likely open-reading frames by identifying patterns of markers such as TATA boxes and ribosome binding sites. Many thousands of gene sequences have been identified and the exponential rate of growth of accessions to public sequence databases is widely acknowledged. As a result, the utility of sequence comparison protocols for identifying genes within a sequence should not be underestimated. Although more full discussions of sequence databases and sequence search programs and parameters can be read elsewhere it is useful for us to discuss briefly some of the examples that we have found to be of greatest utility and some potential pitfalls for the beginner.

Identifying genes with identity to known proteins

As a first pass method of protein gene identification and general sequence annotation the best reference is arguably the SwissProt database (UniProt, 2007), a curated protein sequence database with a high level of annotation and minimal redundancy. Using this resource with appropriate search software, such as BLAST (Altschul, 1990; Altschul et al., 1997), it should be possible to identify the sequences in contigs from a *de novo* project. The main advantage of SwissProt is its robustness; since it is human-curated, then the sequences it contains are less likely to be falsely called 'hypothetical proteins' originating from an error in gene prediction programs. Also the deep level of annotation can make adding human readable and interpretable sequence information to *de novo* assemblies much easier. Another strength of this database is its wide scope; sequences from all organisms are entered into it, so matches in a *de novo* assembly are not restricted to proteins from closely related species. The related TrEMBL database

also provides as much annotation information as the SwissProt database but the sequences in it have not (yet) been checked manually. For many organisms now, particularly bacteria, there will be annotated sequence from at least one closely related species, which one would expect to share a large proportion of genes with the organism of interest.

At first glance it may seem that the most effective way to find protein coding genes would be simply to search against the largest possible database and annotate based on the results. We would advise against this wide-scale approach as an initial strategy since many databases contain hypothetical proteins and fragments of genes that may spuriously match the *de novo* contigs. Similarly, the sequences may have arisen from similarity searches themselves and annotations attached to them may not be the most reliable available.

The most widely used sequence similarity search tool is by far BLAST (Altschul, 1990; Altschul et al., 1997), with which almost all molecular biologists will be familiar at some level. There are four main 'flavours' of BLAST, though the most useful in the task of finding protein coding sequence in an initially nucleotide sequence will be the *blastx* (translated nucleotide query) or *tblastn* (translated nucleotide database) variants, which compare protein sequence against protein sequence, rather than nucleotide versus nucleotide. The advantage of this is clear, protein sequences are much less divergent than their associated nucleotide sequences and matches are therefore easier to identify. However, all variants of BLAST have an almost bewildering array of parameters to set. We have found it useful to modify default settings for number of matches returned, e-score, low-complexity filtering and masking. For further reading on BLAST and the appropriate settings for different parameters see Korf, Yandell and Bedell (2003).

Identification of genes *de novo*

Many methods exist to predict the position of genes in sequence without comparing to known sequence. Pre-eminent among these is Glimmer (Gene Locator and Interpolated Markov Model ER) (Delcher et al., 1999). Glimmer uses interpolated Markov models (IMMs) to identify the coding regions in sequence and distinguish them from non-coding DNA. Briefly, the Glimmer algorithm creates a model known as an ICM (interpolated context model) that defines a probability for each base in all possible 9-mers of a training set (i.e. a set of coding genes from a known sequence) from a probability distribution for the previous 8 bases. The ICM is then used to infer regions of the *de novo* sequence-containing genes by identifying all open reading frames in all 6 translation frames that are longer than a specified minimum. The method works best when related genome sequence is available or when at least some of the genes in a genome are identified. Glimmer can be used as a next step after known genes have been identified using homology based searches.

Detection of non-coding RNAs in genomic sequence is also often necessary. Thankfully, computational methods for doing this are mature and excellent methods have been available for many years. tRNA-scan is one such system that claims identification of greater than 99% tRNAs in genomic DNA, with fewer than one false detection in 15 gigabases of sequence. It achieves these excellent results by virtue of the highly conserved sequence of tRNAs on one hand and their very predictable secondary structure on the other. These attributes make tRNA much easier to detect than other, more variable, genomic features. Other non-coding but highly conserved genomic features such as rRNAs can be identified by using sequence similarity programs. These methods can often be augmented by utilizing approaches such as trained profile HMMs to identifty the promoters of different Polymerases that transcribe different classes of RNAs. Such techniques are more generally applied to motif finding and are discussed more fully below.

Sometimes, in specific cases read coverage or retrieval of genes will be insufficient to generate complete fragments of a genomic region of interest. In such scenarios there are techniques that can rescue a difficult situation. One approach that we find useful is a technique that we call Blast-Seeded Assembly (BSA). After contigs have been prepared as well as possible, a reference sequence, such as a gene of interest from a related organism, is used in sequence similarity search against the contigs. The fragments from the contigs that match the reference sequence are then aligned across the reference and a consensus for the full sequence of the gene as it is in the organism of interest is generated, resulting in the best match to the gene of interest that can be assembled from the contigs. The approach can also be applied using a whole genome sequence as a reference, allowing the best matches to a whole genome sequence to be generated from contigs that may be too small to be useful for other searches. The BSA approach is limited in that the sequence of interest may in reality contain features that are absent from the reference sequence so cannot possibly be included in the BSA assembled sequence.

The BSA approach can be usefully applied to the smaller contigs generated from the SSAKE class of assemblers. These assemblers do not work well with errors or generate such large contigs as Velvet in many cases, but can be used in lower memory capable computing environments. To demonstrate the efficacy of the method we will discuss a case of assembly of reads from *P. syringae* pv. *syringae* B728A using SSAKE and a BSA combination. The Solexa run produced 10.03 million raw reads from B728a (~ 60× coverage of the genome). A subset of the raw B728a reads was selected based on the PHRED equivalent quality scores: a sliding window approach was used, moving across the read and terminating the read at the first point the average quality score within a 5 nt window dropped below a threshold of 1 under the maximum quality score. A minimum allowed sequence length of 22 nt was also used. Termination of reads at these points reduced the coverage substantially, reducing genome coverage to 5.4 million reads (~ 20× coverage), with average read length dropping from 36 nt to 26 nt. We carried out BSA with these contigs. Seeding assemblies with BLAST in this way allowed us to recreate 99.5% of B728A genes (5068/5090 cDNA sequences) with over 96% of these genes having 100% identity to the B728A gene: this result is very similar to that achieved with contigs created by Velvet. Despite this success in a difficult situation, we still feel that BSA is not an approach that should be relied upon for frequent primary sequence assembly and should be used sparingly.

Identification of transcription factor binding sites (TFBS) in genome sequence

A living cell is much more than just an assemblage of gene products; it is the timely and selective expression of genes that result in the characteristic behaviours of living cells and distinguishes them from mere collections of protein. The expression of genes at the most basic level is controlled at a transcriptional level and it is the regulatory networks of interactions between transcription factors and their cognate binding sites that facilitate this. To describe, let alone understand, the complex network of interactions, we must first identify the system's components, including the transcription factor binding sites (TFBS).

There are basically two computational approaches for predicting and identifying TFBS in genomic DNA sequence. The most favourable scenario is that we have a catalogue of experimentally confirmed examples of binding sites for a given regulatory protein. Then we can use this catalogue as a 'training set' to build some model (e.g. a regular expression, matrix, or profile HMM). This model can then be used to search DNA sequence for similar sequences, and thus other genes that might be under the control of the same regulatory protein. We refer to this process as 'pattern matching'. The alternative situation is that we have a collection of genes, of which at least a subset are believed to be co-regulated. Identification of co-regulation may come from transcriptome experiments, or some other form of evidence or prediction. In this case, a computational model would be trained with a collection of upstream regions from these genes with the goal of identifying probable shared TFBS, e.g. by identifying sequence strings that are over-represented in these upstream regions. We refer to this process as 'pattern discovery'. Once such novel patterns have been discovered, we can then apply the process of pattern matching against these newly discovered TFBS sequence motifs.

Genome sequences are generated much faster than TFBS can be confirmed by experiment; fortunately there are well-established computational tools for modelling and detecting TFBS motifs and these fall into two main categories, sequence-identity based and statistical sequence model approaches. Given a 'training set' of confirmed sequences of a given TFBS, it is fairly straightforward to identify close matches in a genome sequence. The methods for identifying matches include (i) exact matching to a consensus sequence, (ii) similarity searching using BLAST (iii) pattern matching using regular expressions, (iv) similarity searching against a Position-Specific Weight Matrix (PSWM), and (v) similarity searching against a Profile Hidden Markov Model (HMM). For example, Fig. 15.3A illustrates an alignment of Hrp boxes found upstream of effector genes in *P. syringae* pv. *tomato* DC3000 (see also chapters by Arnold et al. and Boch). The Hrp box represents the binding site for the HrpL sigma factor, which is responsbible for intitiating transcription of many genes associated with type III secretion in *P. syringae*. Clearly, the Hrp box sequences in Fig. 15.3A share some features in common. It is possible to summarize the mutiple alignment in a single line by the consensus sequence gtGGAACcgantcangnnnnaggnCcACtcAgna. It is very easy, using scripting language such as Perl, Python or Ruby, to search a nucleotide sequence for exact matches to a consensus sequence using regular expressions. However, in distilling the 26 sequences in the multiple alignment down into a single consensus sequence, much of the information content has been discarded. For example, at position 8 in the alignment the residue T occurs in five of the sequences, but this information is not represented in consensus sequence. An intuitively visual way of representing the sequence information is the sequence logo, as illustrated in Fig. 15.3B. Sequence logos (Crookes et al., 2004) are a useful way of describing a sequence motif or a sequence alignment. A PSWM is a simple statistical model that encapsulates most of the information from the alignment and can be used to scan a nucleotide sequence for sites that more or less closely match the motif. The two main limitations of the PSWM model are (i) that it assumes the identity of each position in the motif is independent of all the other positions, and (ii) it does not cope well with motifs of variable length (i.e. alignments that contain gaps). A more flexible statistical model, the profile HMM, adequately models insertions and deletions, but, like the PSWM, assumes independence between each position.

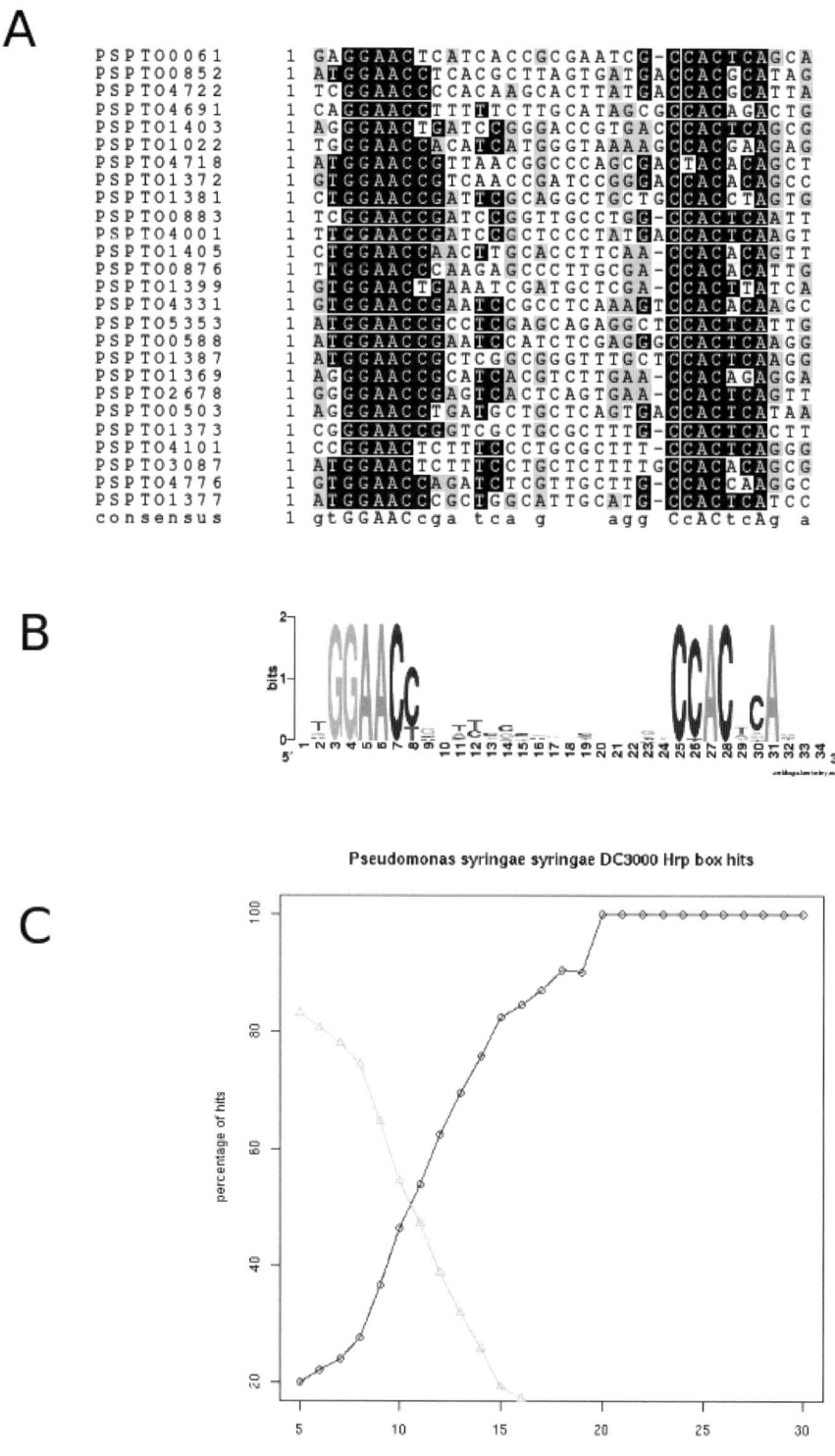

Figure 15.3 The Hrp box. Panel A shows an alignment of confirmed and likely Hrp box sequences from *Pseudomonas syringae* tomato DC3000. Panel B illustrates the aligned sequences as a seqence logo (Crooks et al., 2004). Panel C is a plot for estimating false-positive rates against HMMER scores for the Hrp box HMM against the DC3000 genome. For various threshold scores, we have plotted the percentage of hits that are intergenic (triangles) and percentage of hits that are in the incorrect orientation with respect to their adjacent genes (circles).

PSWMs have proved to be effective and convenient tools for the prediction of TFBS in bacteria. For example the TRACTOR_DB database (Perez *et al.*, 2007) uses a PSWM as the basis for a comprehensive resource for predicted regulons in a wide range of bacterial species.

Hertz and Stormo (1999) developed the Patser software package for the analysis of nucleotide sequence using PSWMs. This freely available software is quite widely used. There are modules in BioPerl for performing Patser searches and parsing the results, making it quite covenient to automate large numbers of such searches. Patser was used to perform the PSWM searches that make up the content of AthaMap, an online resource for TFBS in *Arabidopsis thaliana* (Steffens *et al.*, 2004). The Patser software is equally applicable for analysing prokaryotic DNA sequence.

The theory and statistical properties of profile HMMs have been described in great detail elsewhere (e.g. Eddy, 1998, 2004). For the present purposes, we can treat a profile HMM as a 'black box' that encapsulates an information-rich model of a multiple sequence alignment. In practical terms, the use of profile HMMs in sequence analysis is greatly facilitated by Sean Eddy's HMMER package (http://hmmer.janelia.org). This suite of tools is widely used for analysis of protein sequences and nucleotide sequences and is a central component of such resources as Pfam and Interpro. For analysis of nucleotide sequences, HMMER1.8.5 is preferred over the HMMER2 lineage of versions, which is more optimized towards analysis of peptide sequences. Using the *hmmb* tool in HMMER1.8.5, it is easy to generate a profile HMM from a multiple sequence alignment. It is then straightforward to scan both strands of a sequence database (e.g. a genome sequence, or a set of contigs) against the profile HMM using the *hmmls* tool from HMMER1.8.5. The *hmmls* tool reports all matches to the profile HMM scoring greater than the user-specified threshold score. The selection of an appropriate threshold score is crucial. Selection of too low a threshold will result in false positive matches to the motif, that is reduced selectivity; selection of too stringent a threshold score will result in false negatives, that is reduced sensitivity. When selecting a threshold score, it is useful to have an estimate of the false-negative rates and false positive rates. An estimate of the false-negative rate for a given threshold value can be obtained by counting how many of the sequences from the original 'training set' alignment score pass the threshold. One approach for estimating the rate of false positives is to assume that most TFBS fall within the intergenic regions of the prokaryotic genome rather than within protein-coding genes. In this case, most of the intragenic matches will be false positives. If we plot (Fig. 15.3C) the percentage of hits that fall in intergenic regions against threshold score, we find that for HMMs that have discriminatory power, the percentage of hits that are intergenic increases with score. In the case of the Hrp box, for hits scoring > 16, at least 80% of the hits fall in the intergenic region, even though intergenic regions account for less than 10% of the genome sequence as a whole.

Comparative genomics and genetic diversity

It is often said that plants and their microbial pathogens are locked into an evolutionary 'arms race' (e.g. Dodds *et al.*, 2006; Guttman *et al.*, 2006; Ma *et al.*, 2006). This metaphor describes the situation where the host evolves defence mechanisms, only to have them overcome by the pathogen's 'effectors', that is secreted molecules that subvert the host's defences. Subsequently, the host evolves the ability to recognize these effectors and so the iterative process continues. This 'arms race' scenario implies that pathogen genes encoding effector proteins are under strong diversifying selection, more so than, for example, the pathogen's house-keeping metabolic genes. Therefore, if we want to identify functional effectors and to understand the dynamic evolutionary interactions and adaptations of the pathogen, then we must first explore and describe its genetic diversity. High-throughput sequencing is ideally suited for the purpose of quickly identifying polymorphisms between individuals and races of pathogen, especially when a fully sequenced reference genome is available. Resequencing of different pathogen strains should reveal regions showing diversifying selection due to an 'arms race'. Identification of candidate effectors should therefore be possible, even without sequence homology to previously described effectors. The process of detecting diversifying selection involves measuring the relative frequencies of non-synonymous and synonymous mutations (dN/dS). Probably the most widely used software for this purpose is the PAML package (Yang, 2007).

Sequencing of cDNA

An important aspect of understanding plant–pathogen interactions is the identification of pathogen and host genes that are expressed during the various stages of the infection cycle. In this respect, sequencing of cDNAs (or ESTs) can be instructive. Our colleagues (A. Rougon and J. Jones, unpublished) generated a library of approximately 260 000 ESTs from *Hyaloperonospora parasitica*-infected plant tissue using 454 sequencing. We were then able to align 23% of these to a publicly available draft genome sequence for this eukaryotic phytopathogen (a further 61% aligned to the host genome). Subsequently, our colleagues (E. Kemen and J. Jones, unpublished) generated a further library of 3 million ESTs from this material from a single lane of Solexa sequencing. This library consisted of reads of just 36 nt, but we were able to unambiguously align about 20% of the reads to the *H. parasitica* genome using the SSAHA (Ning *et al.*, 2001) sequence similarity searching tool. These alignments were uploaded into a Generic Genome Browser database, along with predicted gene calls encoding candidate effectors. It was then possible to identify, in an automated high-throughput manner, candidate effector genes that were supported by evidence of expression *in planta*.

Sequencing of ESTs provides qualitative evidence for gene expression. However, it is often desirable to have quantitative estimates of gene expression levels. Microarrays can provide such a tool for quantitatively assaying relative expression levels. However, microarrays are not available for all interesting and important phytopathogens. An alternative approach is the Serial Analysis of Gene Expression (SAGE)-based mRNA profiling method (Velculescu *et al.*, 1995). This method involves sequencing thousands of short sequence tags, generated by restriction digestion of template DNA. In principle, the high-throughput sequencing technologies should be ideally suited to rapidly sequencing such tags in a quantitative manner.

Conclusions

It is almost unquestionable that the new sequencing technologies will have a dramatic impact on microbiology and plant sciences. These technologies present new challenges and opportunities for applied bioinformaticians. Some of these challenges are qualitatively novel and require innovation of new algorithms, e.g. assembly from short reads. However, the proper exploitation of large datasets also depends on judicious use of the tried and tested contents of the bioinformatican's toolbox.

Acknowledgements

DM and DJS are supported by the Gatsby Foundation. The authors are grateful to numerous colleagues in The Sainsbury Laboratory. The informatics work described here would not be possible without the preparation of DNA samples and operation of the sequencing instrument by colleagues including Alejandra Rougon, Jodie Pike and Eric Kemen. We also acknowledge the excellent IT technical support provided by Michael Burrell and strategic and scientific leadership by Jonathan Jones and David Baulcombe. We are grateful to Jeff Chang for help with building the Hrp box alignment and to Daniel Zerbino for making available his Velvet software prior to publication.

References

UniProt. (2007). The Universal Protein Resource (UniProt). Nucleic Acids Res. 35, D193-D197.

Altschul, S.F., Gish, W., Miller, W., Myers, E.W., Lipman, D.J. (1990). Basic local alignment search tool. J. Mol. Biol. *215*, 403–410.

Altschul, S.F., Madden, T.L., Schaffer, A.A., Zhang, J., Zhang, Z. *et al.* (1997). Gapped BLAST and PSI-BLAST: a new generation of protein database search programs. Nucleic Acids Res. *25*, 3389–3402.

Batzoglou, S., Jaffe, D.B., Stanley, K., Butler, J., Gnerre, S. *et al.* (2002). ARACHNE: a whole-genome shotgun assembler. Genome Res. *12*, 177–189.

Chen, T., and Skiena, S.S. (2000). A case study in genome-level fragment assembly. Bioinformatics *16*, 494–500.

Crooks, G.E., Hon, G., Chandonia, J.M., and Brenner, S.E. (2004). WebLogo: a sequence logo generator. Genome Res. *14*, 1188–1190.

Dodds, P.N., Lawrence, G.J., Catanzariti, A.M., Teh, T., Wang, C.I., Ayliffe, M.A., Kobe, B., and Ellis, J.G.

(2006). Direct protein interaction underlies gene-for-gene specificity and coevolution of the flax resistance genes and flax rust avirulence genes. Proc. Natl. Acad. Sci. USA. *103*, 8888–8893.

Delcher, A.L., Harmon, D., Kasif, S., White, O., and Salzberg, S.L. (1999). Improved microbial gene identification with GLIMMER. Nucleic Acids Res. *27*, 4636–4641.

Eddy, S.R. (1998). Profile hidden Markov models. Bioinformatics *14*, 755–763.

Eddy, S.R. (2004). What is a hidden Markov model? Nat. Biotechnol. *22*, 1315–1316.

Green, R.E., Krause, J., Ptak, S.E., Briggs, A.W., Ronan, M.T., Simons, J.F., Du, L., Egholm, M., Rothberg, J.M., Paunovic, M., and Pääbo, S. (2006). Analysis of one million base pairs of Neanderthal DNA. Nature *444*, 330–336.

Guttman, D.S., Gropp, S.J., Morgan, R.L., and Wang, P.W. (2006). Diversifying selection drives the evolution of the type III secretion system pilus of *Pseudomonas syringae*. Mol. Biol. Evol. *23*, 2342–2354.

Hertz, G.Z. and Stormo, G.D. (1999). Identifying DNA and protein patterns with statistically significant alignments of multiple sequences. Bioinformatics. *15*, 563–577.

Jeck, W.R., Reinhardt, J.A., Baltrus, D.A., Hickenbotham, M.T., Magrini, V., Mardis, E.R., Dangl, J.L. and Jones, C.D. (2007). Extending assembly of short DNA sequences to handle error. Bioinformatics *23*, 2942–2944.

Korf, I., Yandell, M., and Bedell, J. (2003). BLAST: An Essential Guide to the Basic Local Alignment Search Tool. (Sebastopol, Ca, USA: O'Reilly).

Ma, W., Dong, F.F., Stavrinides, J., Guttman, D.S. (2006). Type III effector diversification via both pathoadaptation and horizontal transfer in response to a coevolutionary arms race. PLoS Genet. *2*, e209.

Margulies, M., Egholm, M., Altman, W.E., Attiya, S., Bader, J.S., Bemben, L.A., Berka, J., Braverman, M.S., Chen, Y.J., Chen, Z., Dewell, S.B., Du, L., Fierro, J.M., Gomes, X.V., Godwin, B.C., He, W., Helgesen, S., Ho, C.H., Irzyk, G.P., Jando, S.C., Alenquer, M.L., Jarvie, T.P., Jirage, K.B., Kim, J.B., Knight, J.R., Lanza, J.R., Leamon, J.H., Lefkowitz, S.M., Lei, M., Li, J., Lohman, K.L., Lu, H., Makhijani, V.B., McDade, K.E., McKenna, M.P., Myers, E.W., Nickerson, E., Nobile, J.R., Plant, R., Puc, B.P., Ronan, M.T., Roth, G.T., Sarkis, G.J., Simons, J.F., Simpson, J.W., Srinivasan, M., Tartaro, K.R., Tomasz, A., Vogt, K.A., Volkmer, G.A., Wang, S.H., Wang, Y., Weiner, M.P., Yu, P., Begley, R.F., and Rothberg, J.M. (2005). Genome sequencing in microfabricated high-density picolitre reactors. Nature *437*, 376–380.

Molnar, A., Schwach, F., Studholme, D.J., Thuenemann, E.C., and Baulcombe, D.C. (2007). miRNAs control gene expression in the single-cell alga *Chlamydomonas reinhardtii*. Nature *447*, 1126–1129.

Ning, Z., Cox, A.J., and Mullikin, J.C. (2001). SSAHA: a fast search method for large DNA databases. Genome Res. *11*, 1725–1729.

Perez, A.G., Angarica, V.E., Vasconcelos, A.T. and Collado-Vides, J. (2007) Tractor_DB (version 2.0): a database of regulatory interactions in gamma-proteobacterial genomes. Nucleic Acids Res. *35*, D132–136.

Pop, M. and Kosack, D. (2004). Using the TIGR assembler in shotgun sequencing projects. Methods Mol. Biol. *255*, 279–294.

Stalker J., Gibbins, B., Meidl, P., Smith, J., Spooner, W. et al. (2004). The Ensembl Web site: mechanics of a genome browser. Genome Res. *14*, 951–955.

Stein, L.D., Mungall, C., Shu, S., Caudy, M., Mangone, M., Day, A.,

Nickerson, E., Stajich, J.E., Harris, T.W., Arva, A., and Lewis, S. (2002). The generic genome browser: a building block for a model organism system database. Genome Res. *12*, 1599–1610.

Velculescu, V.E., Zhang, L., Vogelstein, B., and Kinzler, K.W. (1995). Serial analysis of gene expression. Science *270*, 484–487.

Warren, R.L., Sutton, G.G., Jones, S.J., and Holt, R.A. (2007). Assembling millions of short DNA sequences using SSAKE. Bioinformatics *23*, 500–501.

Yang, Z. PAML 4: phylogenetic analysis by maximum likelihood. (2007). Mol. Biol. Evol. *24*, 1586–1591.

Index

16S rDNA 22, 41, 50, 53, 100, 116, 176

A
454 sequencing 164, 311–313, 324
Abscisic acid 262
Actinobacteria 2, 3, 12, 135
Adaptation 10, 15, 19, 51, 63–64, 66, 70, 82, 113, 118–119, 140, 153, 156, 171, 295–296, 323
    pathoadaptation 17–19
    nutritional adaptation 71
    to apoplast 70, 73–75
    to stress 123, 234, 261–262
    to host 142, 152, 155, 166–167, 191
AdrA 282
Aerotaxis (Rs) 194
Agrobacterium 2, 11, 22, 67, 91
    cellulose synthesis 282
    chromosome evolution 98, 104
    genomics 91
    genome structure 97
    plasmid 107, 295, 297, 300, 303
    peptidoglycan elicitor 232
    radiobacter 96
    tumefaciens 93
    vitis 95, 97
    wss genes 282
ANI 47
Apoplast 63, 71, 77, 210, 235, 248, 253, 261, 263
    entry by bacteria 68
    features 69
    physiology 79
Arms race 1, 16, 182, 295, 323
Auxin 92, 243, 249, 260, 261
Average nucleotide identity *see* ANI
Avirulence 71, 153, 168, 170, 196, 211, 215, 218, 219, 229, 244, 254, 301, 305

B
Bacterial aggregation 67
Bacterial colonization of apoplast 64
Bacterial origin 1
Bacterial taxa 38
Bacterial wilt 53, 175–176, 178, 180, 186, 188, 195, 300
Bactericide resistance 303

Basal defence 16, 69, 194, 227, 235, 241, 261–264
    ABA 80, 262
    suppression 217–218, 229, 242–244, 248–257
Biocontrol 91
Biofilm formation 16, 67, 118–120, 195, 273–277, 279–290, 306
Bioinformatics 156, 207, 211, 217, 277, 311–312
Biological control 96

C
Candidatus 46
    Liberibacter 2, 21–22
    Phlomobacter 2
Carbon metabolism 80
cDNA sequencing 324
Cellulose synthesis 93, 273–275, 277, 279–280, 282–283, 288
CGH 48–49, 51, 53–54, 56, 183
Chemotaxis 194
Chloroplast 245
Chromosome evolution 103, 107
Citrus canker 22–24, 65, 77, 149–150, 304
Classification 38
Clavibacter 2, 12, 135–136, 141, 143, 298, 300
Codification 43
Cold shock protein 16, 227, 229, 234
Comparative genome hybridization 15 *see also* CGH
Comparative genomics 199, 323
Enterobacteria 163–165, 167
    Leifsonia 143
    Pseudomonas 220
    Ralstonia 175, 183
    Xanthomonadales 113, 147, 152–157
Conjugative DNA transfer 305
Cyclic di-GMP 273–290
    and biofilm 288
    and protein secretion 289
    signalling 280
    turnover 274
    virulence 281

D
Diversity 37
DNA shuffling 220

DNA–DNA hybridization 38, 41, 46, 148
DSF signal 119, 154–155, 278, 281, 290
    and Rpf 284–288
    signal transduction 285–286
    Xylella 286

E

EAL domain 273
Effector 210, 241
    activities 243, 249
    enzymatic activity 216
    evolution 262
    function 215
    localization 245
    manipulation of gene expression 259
    manipulation of post-translational modification 257
    mimicry 217
    motif 246
    Pseudomonas syringae 203
    screens 211, 213, 244
    targets 248
Elicitor 68–69, 227, 229–235
Elongation factor Tu 16, 227–229, 233, 241
Enteric pathogens 10–11, 305
Enterobacteria 10, 41, 163–165, 167–168, 171, 300
EPS 4, 141, 235, 285–287 *see also* Extracellular polysaccharide
apoplast 67, 69, 75–77
    Ralstonia 186–188, 191–193
    Xylella 118
Erwinia 2, 10, 41, 67, 282
amylovora 11, 74, 163–164, 298, 303–304
    chrysanthemi 7, 64
    carotovora 7–8, 64, 121, 281
    herbicola 142, 167, 300
Ethylene 262
Evolution 5, 91, 113, 218, 242
    genome 97
Pseudomonas syringae 203
Exoenzymes 76
Extracellular polysaccharides (exopolysaccharide) 63, 81, 95, 277, 285 *see also* EPS
    amylovoran 301
    biofilm 281, 283, 288–289,
    Xanthomonas 148, 153, 155–156,

F

Firmicutes 2–3
Flagellin 16, 141, 194–195, 227–229, 231–234, 241–242

G

Gene expression 45, 79, 164, 231, 235, 259, 263, 274, 282, 324
    apoplast 72–75, 79, 81–82
    enterobacteria 167–170
    Pseudomonas 213, 217
    Ralstonia 175, 191, 193–194,
    type III secretion 20
    Xanthomonas 153, 156, 285–286
    Xylella 119
Gene identification from sequences 319
Generalists 1, 12–14, 16

Genome assembly 316
Genome decay 135, 137–138
Genome evolution 70, 91, 95–97, 120, 127, 150, 155–156
Genome sequencing 311, 314
Genome visualization 317
GenomeDiagram 166
Genomic island 11, 156, 296, 300, 304
    Agrobacterium 97
    Leifsonia 137, 39–141, 143
    Pectobacterium 166–167
    PPHGI-1 18, 220
    Ralstonia 183–185, 198
    Xylella 122–123, 126
GGDEF domain 273–284, 287–289

H

HD-GYP domain 273–281, 283, 286–288
HGT 4, 8, 11–12, 17, 19, 120–122, 126, 155–156, 184, 295, 300
High-throughput sequencing 311
Horizontal gene transfer 47, 55, 104, 163, 166–168, 171, 263, 296, 306 *see also* HGT
Host immunity 16
Hrp box 210, 212–214, 321–323
Huanglongbing 21, 23

I

Identification of protein coding genes 318
Innate immunity 16, 216, 227–228, 230, 232–233, 235, 253,
IVET 74, 168–169, 186, 191, 193, 195–197, 212–213

J

Jasmonic acid 76, 80, 243, 260–261, 301

L

Lateral transfer 139
Leifsonia xyli 2, 12, 135–141, 143
    adaptation 142
    energy metabolism 142
    genome 137
    secretion systems 142
Lipopolysaccharide 4, 16, 69, 123, 153–154, 194, 227–229, 241

M

MAMP 16, 20, 64, 68–69, 77, 80, 227–235, 241–243, 248, 252–254, 256–257, 261
    immune responses 234
Metabolic profiling 73
Minimal genome set 120, 122–123, 126, 135, 183
MLSA 45, 47–51, 54–55
MLST 38, 44, 47–48, 51–53, 156, 184
Mobile genetic elements 4, 113, 120, 219–220, 263, 296
Motility 92, 118, 122, 155, 167, 171, 180, 206, 232, 274, 277, 285, 287–289
    cyclic di-GMP 277, 280–281
    flagellum 9
    swarming 67, 284
    swimming 194, 195
    twitching 154, 193, 195
Multi locus sequence typing *see* MLST

Multidrug efflux pumps 188
Mutualists 10–11, 107

N

Niche conversion 137
Nitrogen metabolism 78
Nutrient utilization 73
Nutritional adaptation 71

O

Operational species definition 37, 45–47, 50, 54

P

PAMP 16, 194, 227, 228, 231, 241
Pantoea 2, 10, 67, 77, 142, 163–164, 245, 247, 298, 200–201
Pathoadaptation 17–19
Pathogenicity islands 4, 45, 64, 165, 210, 298, 300, 304
Pathogenicity plasmids 297, 300
Pattern recognition receptors *See* PRR
Pectobacterium 6–7, 10–11, 64–65, 76, 163–164, 166
Peptidoglycan 16, 69, 209, 231
Phage 4, 105, 169, 263, 295–296
    Leifsonia 139–140
    Pectobacterium 166–167
    Ralstonia 175, 178–179, 181, 184–185
    Xanthomonas 150
    Xylella 117, 123, 126–127, 129
Phosphodiesterase 251, 274–277, 282, 286–289
Phyllosphere gene transfer 305
Phylogeny 1, 22, 24, 37, 40, 51, 95, 168, 236
    ANI 48
    MLST 47
    Plasmid 302, 304
    Ralstonia 175–176, 178–179, 189
    Xanthomonas 148, 150, 152, 155–156
Pierce's disease 15, 113
PilZ domain 273–274, 277, 282, 290
Plant
    cytoplasm 245
    defence 16, 243
    hormone 260
    innate immunity 227
    nucleus 245
    plasma membrane 245
    surface colonization 66
Plasmid 107, 295
    genes 296, 298
    maintenance 296
    mobilization 296
    replication 296
    resistance genes 302
    transfer 302
    transmission 107
    virulence genes 301
Plasmid-mediated gene transfer 303
Polyphasic
    classification 42
    analysis 44, 50, 148
Proteobacteria 2–3, 14, 24, 50, 64–65, 91, 97–98, 105, 113, 121, 147, 156, 163, 175, 192
Proteomics 64, 74, 154, 156, 164, 170, 199
PRR 227–229, 231–232, 234–235

Pseudomonas 2, 3, 40, 64–65, 67, 121, 140, 277, 282, 288–290
    aeruginosa 94, 121, 138, 171, 236, 276, 284
    diversity 50
    fluorescens 71, 253, 284
    genome 204
    LOS/LPS 230–231
    nutritional adaptation 72–74
    plasmid 296, 298, 301, 303–304
    putida 178
    secretion 6
    syringae 203–221, 244, 247, 284, 296, 314, 318, 322
    taxonomy 50

Q

Quorum sensing 7, 186, 206, 295, 299–300, 306
Agrobacterium 95, 107
    apoplast 76–77
    cyclic di-GMP 280–281, 285, 288
    Pectobacterium 168, 170
    Ralstonia 190, 192–193
    Xanthomonas 5, 7

R

Ralstonia solanacearum 2, 13, 65–66, 121, 142, 175–199, 217, 221, 230
    ANI 48 *see also* ANI
    chromosome evolution 106
    core genome 182
    diversity 50
    effector 244–245, 247
    evolution 184
    genetic variation 196
    genome sequence 178
    genomic islands 185
    genomic plasticity 182
    genomic variation 196–198
    host specificity 195
    life cycle 180, 187
    phenotype conversion 191
    phylotypes 177
    plasmid 300
    secreted proteins 186
    species complex 53–54
    taxonomy 53, 104–105, 176
    virulence strategies 186
Ratoon stunting disease 12 *see also* RSD
Recombination 48, 53, 55, 70, 107, 123, 126, 302
    homologous 44, 51–52, 296–297
    illegitimate 129
    rates 49–50, 52
    site-specific 129, 297
Resistance gene defence suppression 254
RIVET 74
RSD 135–136, 143

S

Salicylic acid 68, 76, 94, 170, 217, 231, 243, 252, 260–261
Sequence assembly 315
    software 316
    practicalities 317

Signalling
- bacterial 77, 92, 94, 154, 155, 186, 187, 194, 203, 273–274, 276–290
- plant defence 8, 20, 69, 80, 81, 94, 140, 217, 228, 231, 233–235, 242–245, 248–252, 254–255, 257, 260–262

Soil nitrogen 79
Solexa sequencing 311–318, 320, 324
- error distribution 313
- genome sequencing 314
- raw data 313

Specialists 1, 12, 14, 16
Species concepts 37, 43–47, 49–56, 148
Streptomycin resistance 299, 302–304
Superoxide dismutase 94, 118, 227, 229, 234
Systematics 37–38, 41, 43, 46

T

T3SS 8–10, 12–13, 20, 24, 80, 242–245, 248, 256, 263–264 *see also* type III secretion
- effectors 244, 259, 261–262
- molecular signatures 16–17
- *Pectobacterium* 166–167, 171
- plasmid 300–301, 304
- *Pseudomonas* 15, 252–254

Taxonomy 37–38, 40, 42–46, 50, 52, 54–56, 65, 148, 150, 155, 176, 178, 184–185
Terminal reassortment 18–19, 220, 263
Toll-like receptors 228
Toxin 4, 7, 11, 51, 78, 81, 118, 153, 164, 170, 206–207, 215, 253, 281
- ADPRT 218
- albicidin 142
- apoplast 63–64, 66, 68, 73, 75–76
- carotovoricin 168
- coronafacic acid 166
- coronatine 20, 75, 80, 261, 299, 301, 305
- phaseolotoxin 76, 205
- syringomycin 5, 76, 205
- syringopeptin 5
- tabtoxin 76
- thaxtomin 12

Transcription factor binding site 321
Transcriptomics 64, 74, 82, 168
tRNA 5, 8, 97, 123, 126–129, 137, 140, 178, 184–185, 197, 210, 300, 320
Type I secretion 5–6, 77, 93, 204–206
Type II secretion 6, 20, 77, 119, 155, 168–170, 186–188, 205–207
Type III secretion 5–6, 8, 81, 96, 117, 188, 228–229, 235
- ABA 80
- *Agrobacterium* 96–97
- apoplast 63–64, 74–75
- *Dickeya* 169
- *Erwinia* 171
- effectors 241–264, 304–305
- gene array 49
- evolution 17–18
- molecular signatures 16

Pectobacterium 166
plasmid 299–300
Pseudomonas 203–209, 211–212, 214, 216–220, 321
Ralstonia 178, 182–183, 186, 188–190, 193
Xanthomonas 153–155
Type IV secretion 5–7, 9, 24, 164, 207, 264, 297, 299, 302–303
pilus 117–118, 156
- plasmid-borne 11, 95, 107–108, 123
- *Xanthomonas* 154
Type V secretion 207
Type VI secretion 5, 104, 171, 203, 207–208

W

Wsp chemosensory system 284
wsp genes 283–284
wss genes 282

X

Xanthomonad genome 114
Xanthomonas 2, 54, 113, 117, 129, 138–139, 141, 168, 170, 179, 217, 228, 244
- *albilineans* 142
- ANI 48
- apoplast 64–65
- *axonopodis* 22–23
- biofilm 285
- cellulose 282
- comparative genomics 155
- cyclic di-GMP 280, 283, 287
- diversity 50
- DSF cell–cell signalling 286
- functional genomics 154
- gene classification 157
- genome sequence 114–115, 119–123, 127, 147–158, 317
- HD-GYP 276, 278
- heterologous expression 153
- LOS 230
- mobilomics 126
- *ohr* gene 94
- *oryzae* 8, 234
- pathoadaptation 19
- plasmid 298, 301–304
- random mutagenesis 153
- secretion 209
- virulence 244–248, 251, 253–254, 258–260
- taxonomy 52–53, 116

Xylella fastidiosa 2, 113–129, 141–143, 179, 290
- ABC transporter 138
- apoplast 64–65, 73, 79
- chromosome 126
- cyclic di-GMP 278
- DSF 286–287
- genome 138–140, 152–153, 155–156
- phylogeny 152
- plasmid 299–300, 302–303, 305
- sharpshooter 15